PROFESSIONAL

ENGINEER'S

EXAMINATION QUESTIONS

AND ANSWERS

PROFESSIONAL ENGINEER'S EXAMINATION QUESTIONS AND ANSWERS

WILLIAM S. LA LONDE, JR.

Licensed Professional Engineer, New Jersey, New York
Chairman Emeritus, Department of Civil and
Environmental Engineering
New Jersey Institute of Technology, Newark, N.J.

WILLIAM J. STACK-STAIKIDIS

Licensed Professional Engineer, New Jersey, New York
Head, Structures Division, Department of Civil
and Environmental Engineering
New Jersey Institute of Technology, Newark, N.J.

THIRD EDITION

McGRAW-HILL BOOK COMPANY

New York   St. Louis   San Francisco   Auckland   Bogotá   Düsseldorf
Johannesburg   London   Madrid   Mexico   Montreal
New Delhi   Panama   Paris   São Paulo   Singapore
Sydney   Tokyo   Toronto

Library of Congress Cataloging in Publication Data

La Londe, William S.
 Professional engineer's examination questions and
answers.

 1. Engineering—Examinations, questions, etc.
2. Engineering economy—Problems, exercises, etc.
I. Stack-Staikidis, William J., date. joint author.
II. Title.
TA159.L3   1976   620'.0076   76-10231
ISBN 0-07-036093-6

1234567890   MUBP   785432109876

*The editors for this book were Tyler G. Hicks and Betty Gatewood,
and the production supervisor was Frank P. Bellantoni.
It was set in Modern by Bi-Comp.*

Printed by The Murray Printing Company and bound by The Book Press.

CONTENTS

PREFACE

This book has been prepared to help both the newly graduated engineer and the older practicing engineer in their review of engineering fundamentals in preparation for the professional engineer examinations as part requirement for the engineer-in-training certificate or the professional engineer license.

About 80 per cent of the problems appearing in this book were taken directly from sample question sheets or past professional engineer examinations as given by the many state boards of professional engineer examiners throughout the United States. These problems may be considered a composite of the many examinations which have been given in the past, but they purposely do not represent the examination questions of any state. The problems that have been added were included so as to give better coverage of some of the subject matter for those preparing to take future professional engineer examinations.

Recent professional engineer license examinations indicated a need to extend the scope of this book to provide added coverage in some parts of the several professional fields. The author also received letters from users of this book suggesting more practice questions and answers in certain sections. Further, some state boards of professional engineers have adopted a multiple-choice format of questions for the general engineering examination area. This second edition, through the addition of one hundred questions and answers, attempts to meet these needs.

Grateful acknowledgment is given to the many state boards of professional engineers or engineering examiners and to the National Council of Engineering Examiners for contributing specimen or sample questions or information on the scope of examinations and of registration; and to professional engineers R. E. Anderson, S. Fishman, J. Joffe, W. Jordan, G. C. Keeffe, F. G. Lehman, R. D. Mangasarian, C. L. Mantell, E. Miller, A. A. Nims, J. L. Polaner, J. M. Robbins, F. A. Russell, and L. Shapiro, all of the faculty of the Newark College of Engineering, for the services which they contributed toward the solutions of the problems in their fields of specialization, and to the past users of this book who spotted and took the time to report discrepancies in the earlier printings.

William S. La Londe, Jr.

ACKNOWLEDGMENTS FOR THE THIRD EDITION

Since the First Edition was published in 1956, engineering knowledge has expanded and with it the codes of practice. The questions asked in the professional engineering license examinations have accordingly kept pace with these changes. This Third Edition brings this book in step with these changes. About twenty percent of the questions and answers in the professional engineering areas were either modified or new material substituted to reflect new engineering knowledge and codes of practice.

Of particular importance to me was my success in enlisting professional engineer William J. Stack-Staikidis to serve as co-author. He showed his ability in the work he did for this new edition. We both are grateful to professional engineers J. Carluccio, R. D. Mangasarian, E. Miller, and D. P. Tassios, all of the faculty of the New Jersey Institute of Technology for their contributions in the fields of their specialization to ensure that the questions and answers here presented represent the latest to be found in the professional engineering license examinations.

HOW TO USE THIS BOOK

It is suggested that the General Information in Part One be reviewed to estimate your eligibility and qualifications for making application for the engineer-in-training certificate or professional engineer's license.

The definitions set forth in the Model Law for the Registration of Professional Engineers and Land Surveyors should be helpful. Your professional experience should be carefully scrutinized and compared with the Experience Qualification Guide prepared by the Committee on Qualifying Experience of the National Council of Engineering Examiners. After you have done this and you feel that you have had the prerequisite training and years of professional experience to qualify you to apply for the professional engineer's license, then write to the secretary of your state board of registration for professional engineers and land surveyors and request an application blank and information with respect to the laws covering the requirements. After you have filled out and submitted your application blank, you will be informed by the license board whether there will be any written examinations or not. Some license boards will give bona fide applicants sample or specimen questions, and all will at least inform the applicant as to the scope of the examinations.

Having at least learned the scope of your examination, you will find many problems worked out in this book that will be helpful to you in reviewing the engineering fundamentals with which you should be familiar. It is the intent here only to help you review what you know. Space limitations prevent the derivation of formulas or explanations concerning the subject material which may be found in standard engineering texts. Formulas and symbols will be identified as well as source material taken from special tables. The solutions to these problems should be studied and used as a guide for minimum clarity and explanation when later working out the problems of your particular examination. So often engineers, when taking these examinations, present their work so nonunderstandingly as to make it difficult for the examiner to follow the specific solutions.

In studying for the examinations, you should have a sufficient duplication of problems to do a certain number with the solutions as practice and a certain number without the solutions to test yourself for competency and speed. There is no equal in the preparation for these license examinations to the working of a sufficiently large number of these problems.

In practically all the state board examinations, the applicant will be permitted some choice. It is desirable at the beginning of each examination to calmly and thoughtfully read over all the questions and to pick out those you wish to do; and it might be a good idea to do first those problems which you can do readily. If eight problems are to be done in four hours, try first to budget your time with not over 15 or 20 minutes per problem. This will allow an opportunity to go back and review the problems afterwards. It is also better to answer six problems correctly than to try to get all eight problems answered with only part scores. Sometimes if one has difficulty with a problem, it is well not to work on it too long as a suggestion in another problem or the feeling of confidence gained from the solution of a number of problems will stimulate the thinking or relax the pressure, and the solution may be obtained when one returns to the problem. Neatness and carefulness in the arithmetical work and in making tabulations will insure the best results. Never omit a check if it is possible to make one.

Sometimes needed information is purposely omitted from the examination question to test the examinee's ability to size up the situation and to supply reasonable data. In answering such a question, merely indicate your assumptions and proceed from that point. It is well to be careful, however, not to make assumptions where none are needed.

A satisfactory speed for the solution of these problems, indicative of reasonable familiarity with the fundamentals of any given section, is to be able to work two or three problems per hour. As much work as possible should be done without the use of texts or notes, as in many cases their use or the use of books such as this one are not permitted at the examinations. It is best not to bank on the use of more than a handbook of tables and formulas, and where the necessary information is supplied to the questions, perhaps not more than a slide rule will be permitted. Where books are permitted, the questions may be more searching, and unless one has a knowledge of his subject, much time can be lost browsing in books looking for the answer. The solutions to the problems in this book are purposely separated from the questions in order to encourage the reviewer as far as it may be possible to do his own thinking and work.

Those who are preparing for the engineer-in-training examinations would do well to review all questions and solutions 1.101 *to* 1.825, 5.01 *to* 5.40, *and* 7.01 *to* 7.75.

PROFESSIONAL

ENGINEER'S

EXAMINATION QUESTIONS

AND ANSWERS

PART ONE

General Information

LEGAL REGISTRATION

The office of the National Council of Engineering Examiners will furnish upon request any specific information available regarding the legal registration or licensing of professional engineers that may be required in addition to the following.

Engineers who wish to secure legal registration or professional license in any particular state must communicate directly with the state board or department in that state and request its application form and regulations. All the states, Alaska, Hawaii, Puerto Rico, and the provinces of Canada have registration laws regulating the practice of engineering.

The engineering registration laws are not uniform due to local conditions, date of enactment, etc. Most state laws, however, require either graduation in an accredited engineering curriculum, a written engineering examination, adequate and satisfactory engineering experience, or a combination of these.

INTERSTATE REGISTRATION

Legal registration or professional license in any one state or certification by the National Council of Engineering Examiners does not permit or authorize the practice of engineering in other states. Most states, however, grant registration to registrants of other states or temporary permits to practice, provided they have the required qualifications and comply with the regulations of those states.

NATIONAL ENGINEERING CERTIFICATION

The National Engineering Certification is a service of the National Council of Engineering Examiners. It was established primarily to minimize the effort and expense of registered engineers desiring to secure registration or license in more than one state. The registration laws or regulations of most states specifically provide for acceptance of the Certificates of Qualification as competent evidence. However, such certification is not binding upon the State Boards and is not a form of registration or license. The "Information Booklet" regarding NCEE Services, Council Records, Certification, and the Procedures, Requirements, and Schedule of Fees may be secured from the Executive Director, NCEE, P.O. Box 5000, Seneca, South Carolina 29678.

A MODEL LAW FOR THE REGISTRATION OF PROFESSIONAL ENGINEERS AND LAND SURVEYORS AND PROVIDING FOR THE CERTIFICATION OF ENGINEERS-IN-TRAINING

DEFINITIONS

Engineer   The term Engineer as used in this Act should mean a Professional Engineer as hereinafter defined.

Professional Engineer   The term Professional Engineer within the meaning and intent of this Act shall mean a person who, by reason of his special knowledge of the mathematical and physical sciences and the principles and methods of engineering analysis and design, acquired by professional education and practical experience, is qualified to practice engineering as hereinafter defined, as attested by his legal registration as a Professional Engineer.

Engineer-in-Training   The term Engineer-in-Training as used in this Act shall mean a candidate for registration as Professional Engineer who is a graduate in an approved engineering curriculum of four years or more from a school or college approved by the Board as of satisfactory standing, or who has had four years or more of experience in engineering work of a character satisfactory to the Board; and who, in addition, has successfully passed the examination in the fundamental engineering subjects prior to completion of the requisite years of experience in engineering work, as provided in Section 14 of this Act, and who shall have received from the Board, as hereinafter defined, a certificate stating that he has successfully passed this portion of the professional examinations.

Practice of Engineering   The term Practice of Engineering within the meaning and intent of this Act shall mean any professional service or creative work requiring engineering education, training, and experience and the application of special knowledge of the mathematical, physical, and engineering sciences to such professional services or creative work as consultation, investigation, evaluation, planning, design, and supervision of construction for the purpose of assuring compliance with specifications and design, in connection with any public or private utilities, structures, buildings, machines, equipment, processes, works, or projects.

A person shall be construed to practice or offer to practice engineering, within the meaning and intent of this Act, who practices any branch of the profession of engineering; or who, by verbal claim, sign, advertisement, letterhead, card, or in any other way represents himself to be a professional engineer, or through the use of some other title implies that he is a professional engineer; or who holds himself out as able to perform, or who does perform any engineering service or work or any other professional service designated by the practitioner or recognized by educational authorities as engineering.

Land Surveyor   The term Land Surveyor as used in this Act shall mean a person who engages in the practice of land surveying as hereinafter defined.

Land Surveying   The practice of land surveying within the meaning and intent of this Act includes surveying of areas for their correct determination and description and for conveyancing, or for the establishment or reestablishment of land boundaries and the plotting of lands and subdivisions thereof.

SUGGESTED EXPERIENCE QUALIFICATION GUIDE

CHEMICAL ENGINEERING

Professional Experience

Design, Specifications and/or Construction: Equipment for chemical enterprises; processing plants; plant layouts; equipment; economic balances; production planning; pilot plants; heat-transmission apparatus.

Development: Processes; pilot plants; refrigeration systems for food processing.

Research: Laboratory; pilot plants; director; markets; unit operations; kinetics; processes; calculation and correlation of physical properties.

Responsible charge of broader fields of chemical engineering; executive.

Consultation; appraisals; evaluations; reports; economics; chemical patent laws.

Operation of pilot plants; product testing; technical service; technical sales.

Teaching full time, at college level; editing and writing.

Subprofessional Experience

Construction: Process equipment; pilot plants; piping systems.

Operation: Shift operator; special chemicals manufacture (solutions); pilot plants; trouble shooting; glass blower.

Drafting: Flow-sheet layout.

Instrument making and servicing; routine analyses; routine sampling and tests; latheman.

Analyst: Routine, under direction; laboratory assistant (commercial, college); computations, under direction; data taking.

Sales: Routine, of standard equipment.

CIVIL ENGINEERING

Professional Experience

Design and/or Construction: Highways; streets; subways; tunnels; drainage; drainage structures; sewerage; sewage-disposal structures; railroad spurs and turnouts; railroads; rivers, harbors, and waterfront; waterworks; water supply; water power; airports and/or airways; bridges; dams; irrigation; irrigation structures; water purification systems; incinerators; storage elevators and structures; flood control.

Development: Structures for regions; ports; wharves and docks; irrigation works; flood control; drainage.

Investigations; appraisals and evaluations; cost analysis; consultations; testing; research and economic studies; reports.

Teaching at college level, full time in an accredited college of engineering.

Editing and writing.

Surveys: Topographic; hydrographic; earthwork; triangulation; railroad; irrigation; drainage; water power.

Independent responsibility and supervision of the work of others.

Subprofessional Experience

Fieldwork: Instrumentman; chainman; rodman; chief of party; recorder; surveyman; levelman; staking out foundations; staking out dredging ranges; railroad lines and turnouts; sounding; instrumental observations.

Drafting: Plotting field notes; map drafting and tracing; tracing designs; structural detailing; plotting soundings.

Construction: Inspection; timekeeping; supervision, without authority to change designs.

Computations: Routine, under direction; taking off quantities; monthly estimates.

Cost Data: Routine assembling and computing; estimating, under direction.

Testing: Routine testing of materials of construction; water.

Teaching: As an associate without full responsibility in an engineering college.

ELECTRICAL ENGINEERING

Professional Experience

Design, Specifications and/or Construction: Power plants; electronic and transmission equipment; application of electrical equipment to industry; electric service and supply; communication systems; control systems and servomechanisms; manufacturing plants; operating procedures; distribution networks; transportation; radio circuits; major electrical installations.

Development and Production: New techniques and devices; electrical and electronic devices.

Supervision and/or Management: Power plants; manufacturing plants.

Research and investigations in electrical fields; water, steam, and electric power; electronics; new techniques and devices; high-voltage cables.

Consultation; appraisals; valuations; estimates; cost analysis; reports; determination of rates; economics and cost figures; calculations of system and device performance.

Sales and application of equipment to industry; service engineering; testing, when not of a routine character.

Editing and writing on electrical-engineering subjects.

Teaching full time, at college level, in an accredited school.

Subprofessional Experience

Construction and Installation: Construction of electrical installations, under competent direction; electronic and radio equipment; wiring, lineman; switchboard wiring; electrician; meterman; armature winding.

Inspection: Routine, under direction.

Installation: Heating, ventilating, and air conditioning; radio equipment.

Drafting: Elementary layout; circuit drafting; detailing.

Testing: Routine; performance of electrical machinery.

Operation: Radio; power stations; substations; switchboards.

Repairs: Motors; generators; electrical devices; electronic and radio devices.

Maintenance: Electronic and radio devices; radio and television technician.

Apprentice: Work as junior in fields of illumination, communication, and transportation.

Sales: Routine; almost all customer service; inventory.

Teaching: Manual skills; instructor in accredited school.

MECHANICAL ENGINEERING

Professional Experience

Design and/or Construction: Machines; machinery and mill layouts; heating, ventilating, and air conditioning; power plants; power-plant equipment; industrial layouts; refrigeration; tools and processes; designs involving application of existing equipment and principles; internal-combustion engines.

  Development of industrial plants and processes; consultation.

  Appraisals; investigations; research and economic studies; cost analyses; reports.

Operation: Major mechanical installations; manufacturing plants; power plants; testing; application of instruments and control to manufacturing processes.

  Editing and writing on mechanical-engineering subjects.

  Teaching full time, at college level, in accredited schools.

Subprofessional Experience

Construction and Installation: Machinery; heating; ventilating and air conditioning; mechanical structures.

Operation: Heating, ventilating, and air-conditioning equipment; power plants; mechanical manufacturing plant; stationary machinery (small operations); foundry and machine shops.

Drafting: Detailing; shop drawings; checking shop drawings; tracing; layout work; design of tools, jigs, and fixtures.

Recording; routine computing under competent direction; laboratory assistant; routine testing under competent direction; inspection of materials, etc.

Maintenance; repairs; welding.

Teaching as an assistant without full responsibility, in an accredited college.

SANITARY ENGINEERING

Professional Experience

Design, Specifications, and/or Construction: Water-treatment plants; sewage-disposal plants; waterworks; sewers; incinerators; distribution systems; sewage- and industrial-waste-treatment plants.

Consultations; reports for proposed works; original research for processes and equipment; review of plans for works for state and Federal authorities; planning and analysis of surveys and pollution data; advising on water-purification plants; counseling municipal officials concerning needed sanitary facilities; planning and supervising environmental sanitation work in sanitation of food, milk, etc.; investigations; studies; reports and evaluation of proposed or existing water systems or sewerage; industrial-hygiene surveys and reports; research in sanitary engineering or public health.

Abatement procedures on stream pollution; supervision of works for large cities; supervision of pollution surveys; control of rodents and insects.

Teaching full time, at college level, in an accredited school.

Editing and writing.

Subprofessional Experience

Routine laboratory tests; installation of machinery; gathering records for reports; stream gaging; surveying; plant operation and maintenance; record keeping; sanitarian's inspection; mapping; inspection; routine chemical and bacteriological tests; pest control; collection of data; take off; sampling; establishing of lines and grades; inspection of eating and drinking establishments; routine vessel-sanitation inspections; routine watering-point inspections; orientation periods in public-health agencies; inspection of construction.

Maintenance and operation of filter and sewage plant; routine laboratory examinations.

Same as Civil Engineering.

STRUCTURAL ENGINEERING

Professional Experience

Design, specifications, and/or construction on structures; hydraulic design; stress analysis; soils analysis; mechanical analysis.

Research on structures; testing; layout studies and comparisons; new design procedures and methods of analysis; reports; consultations; investigation evaluations.

Project supervision; field execution of projects; senior structural draftsman; senior stress analyst; senior engineer on materials testing; senior concrete laboratory technician; resident (field) engineer 'on bridges, dams, and buildings; research engineer.

Same as Civil Engineering.

Teaching full time, at college level, in an accredited school.

Subprofessional Experience

Structural drafting, detailing; drafting layout and details; junior stress analyst; computer; checking detail drawings; checking design computations; collection of data for reports; inspector; junior laboratory engineer; research assistant; junior concrete technician.

TRAFFIC ENGINEERING

Professional Experience

Design of traffic islands; design of new and augmented control systems; design of relief systems; design of highway lighting; traffic-engineering planning and roadway design; design of mechanical traffic control.

Planning and analyzing surveys and traffic counts; interpretation of field data; computing earning power of proposed routes; establishing traffic-count stations; transportation studies (major or complete); traffic-education programs.

Traffic-engineering organization and administration. Teaching, at college level.

Subprofessional Experience

Traffic counts; compiling of data; timing of automatic signals; plotting traffic-flow charts; plotting traffic-dispersion charts; origin and destination data; weighting stations; accident study and tabulation; sign and signal checking.

ADDRESSES OF REGISTRATION BOARDS

Alabama State Board of Registration for Professional Engineers
and Land Surveyors
Room 606 Administration Building
64 North Union Street, Montgomery, Alabama 36104

Alaska State Board of Registration for Architects, Engineers, and
Land Surveyors
Pouch D, Juneau, Alaska 99801

Arizona State Board of Technical Registration
1645 West Jefferson Street, Phoenix, Arizona 85007

Arkansas State Board of Registration for Professional Engineers
and Land Surveyors
P.O. Box 2541, Little Rock, Arkansas 72203

California State Board of Registration for Professional Engineers
1006 Fourth Street, Sixth Floor, Sacramento, California 95814

Canal Zone Board of Registration for Architects and Professional
Engineers
P.O. Box 223, Balboa Heights, Canal Zone

Colorado State Board of Registration for Professional Engineers
and Land Surveyors
Room 127, State Service Building
1525 Sherman Street, Denver, Colorado 80203

Connecticut State Board of Registration for Professional Engi-
neers and Land Surveyors
Room 533, State Office Building, 165 Capitol Avenue, Hartford,
Connecticut 06115

Delaware Association of Professional Engineers
1508 Pennsylvania Avenue, Wilmington, Delaware 19806

Delaware State Board of Registration for Professional Engineers
and Land Surveyors
1228 North Scott Street, Wilmington, Delaware 19806

District of Columbia Board of Registration for Professional
Engineers
614 H Street, N.W., Room 109, Washington, D.C. 20001

Florida State Board of Professional Engineers and Land
Surveyors
Suite 100, 6990 Lake Ellenor Drive, Orlando, Florida 32809

Georgia State Board of Registration for Professional Engineers
and Land Surveyors
166 Pryor Street, S.W., Atlanta, Georgia 30303

Guam Board of Registration for Professional Engineers, Architects, and Land Surveyors
Department of Public Works, Govt. of Guam, Agana, Guam
96910

Hawaii State Board of Registration for Professional Engineers,
Architects, Land Surveyors, and Landscape Architects
P.O. Box 3469, Honolulu, Hawaii 96801

Idaho State Board of Engineering Examiners
842 La Cassia Drive, Boise, Idaho 83705

Illinois Department of Registration & Education, Professional
Engineers' Examining Committee
628 East Adams Street, Springfield, Illinois 62786

Indiana State Board of Registration for Professional Engineers
and Land Surveyors
1021 State Office Building, Indianapolis, Indiana 46204

Iowa State Board of Engineering Examiners
State Capitol Complex, 821 Des Moines Street, Des Moines,
Iowa 50319

Kansas State Board of Engineering Examiners
535 Kansas Avenue, Topeka, Kansas 66603

Kentucky State Board of Registration for Professional Engineers
and Land Surveyors
University Station Box 5075, Lexington, Kentucky 40506

Louisiana State Board of Registration for Professional Engineers
and Land Surveyors
1055 St. Charles Avenue, Suite 415, New Orleans, Louisiana 70130

Maine State Board of Registration for Professional Engineers
State House, Augusta, Maine 04330

Maryland State Board of Registration for Professional Engineers
and Professional Land Surveyors
1 South Calvert Street, 8th floor, Baltimore, Maryland 21202

Massachusetts State Board of Registration of Professional Engineers and Land Surveyors
Room 1512 Leverett Saltonstall Building
100 Cambridge Street, Boston, Massachusetts 02202

Michigan Department of Licensing and Regulation, State of
Michigan Board of Registration for Professional Engineers and
Board of Registration for Land Surveyors
1116 South Washington Avenue, Lansing, Michigan 48926

Minnesota State Board of Registration for Architects, Engineers,
Land Surveyors, and Landscape Architecture
5th Floor, Metro Square, 7th and Roberts Streets, St. Paul,
Minnesota 55101

Mississippi State Board of Registration for Professional Engineers
and Land Surveyors
P.O. Box 3, Jackson, Mississippi 39205

Missouri Board for Architects, Professional Engineers, and Land
Surveyors
P.O. Box 184, Jefferson City, Missouri 65101

Montana State Board of Professional Engineers and Land Surveyors
La Londe Building, Helena, Montana 59601

Nebraska State Board of Examiners for Professional Engineers and Architects
512 Terminal Building, 941 "O" Street, Lincoln, Nebraska 68508

Nevada State Board of Registered Professional Engineers
P.O. Box 5208, Reno, Nevada 89503

New Hampshire State Board of Registration for Professional Engineers
c/o Secretary of State, State House, Concord, New Hampshire 03301

New Jersey State Board of Professional Engineers and Land Surveyors
1100 Raymond Boulevard, Newark, New Jersey 07102

New Mexico State Board of Registration for Professional Engineers and Land Surveyors
P.O. Box 4847, Santa Fe, New Mexico 87501

New York State Board for Engineering and Land Surveying
The State Education Department
99 Washington Avenue, Albany, New York 12230

North Carolina State Board of Registration for Professional Engineers and Land Surveyors
1307 Glenwood Avenue, Suite 152, Raleigh, North Carolina 27605

North Dakota State Board of Registration for Professional Engineers and Land Surveyors
P.O. Box 1264, Minot, North Dakota 58701

Ohio State Board of Registration for Professional Engineers and Surveyors
180 East Broad Street, Suite 1014, Columbus, Ohio 43215

Oklahoma State Board of Registration for Professional Engineers
and Land Surveyors
401 United Founders Tower
5900 Mosteller Drive, Oklahoma City, Oklahoma 73112

Oregon State Board of Engineering Examiners
Department of Commerce
325 13th Street, N.E., Suite 605, Salem, Oregon 97310

Pennsylvania State Registration Board for Professional Engineers
279 Boas Street, Room 404, Box 2649, Harrisburg, Pennsylvania
17120

Puerto Rico Board of Examiners of Engineers, Architects, and
Surveyors
P.O. Box 3271, San Juan, Puerto Rico 00904

Rhode Island State Board of Registration for Professional Engineers and Land Surveyors
20 State Office Building, Providence, Rhode Island 02903

South Carolina State Board of Engineering Examiners
710 Palmetto State Life Building, Columbia, South Carolina
29201

South Dakota State Board of Engineering and Architectural
Examiners
2040 West Main Street, Suite 212, Rapid City, South Dakota
57701

Tennessee State Board of Architectural and Engineering
Examiners
550 Capitol Hill Building
301 7th Avenue, North, Nashville, Tennessee 37219

Texas State Board of Registration for Professional Engineers
Room 200, 1400 Congress, Austin, Texas 78701

Utah Representative Committee of Professional Engineers and
Land Surveyors
330 East 4th South, Room 210, Salt Lake City, Utah 84111

Vermont State Board of Registration for Professional Engineers
Norwich University, Northfield, Vermont 05663

Virginia State Board of Architects, Professional Engineers, and
Land Surveyors
P.O. Box 1-X, Richmond, Virginia 23202

Virgin Islands Board for Architects, Engineers, and Land
Surveyors
Public Works Department
P.O. Box 476, St. Thomas, Virgin Islands 00801

Washington State Board of Registration for Professional Engineers and Land Surveyors
P.O. Box 649, Olympia, Washington 98504

West Virginia State Board of Registration for Professional
Engineers
Room 411, 1800 East Washington Street, Charleston, West
Virginia 25305

Wisconsin State Board of Architects, Professional Engineers,
Designers, and Land Surveyors
201 East Washington Avenue, Madison, Wisconsin 53702

Wyoming State Board of Examining Engineers State Office
Building, East, Cheyenne, Wyoming 82002

CANADA

Canadian Council of Professional Engineers
401—116 Albert Street, Ottawa, Ontario KIP 5G3

Association of Professional Engineers, Geologists, and Geo-
physicists of Alberta
215 One Thornton Court, Edmonton, Alberta

Association of Professional Engineers of the Province of British
Columbia
2210 West 12th Avenue, Vancouver 9, British Columbia

Association of Professional Engineers of the Province of Manitoba
710—177 Lombard Avenue, Winnipeg, Manitoba R3B 2W9

Association of Professional Engineers of the Province of New
Brunswick
123 York Street, Fredericton, New Brunswick

Association of Professional Engineers of the Province of
Newfoundland
P.O. Box 31, St. John's Newfoundland

Association of Professional Engineers of the Province of Nova
Scotia
P.O. Box 129, Halifax, Nova Scotia

Association of Professional Engineers of the Province of Ontario
1027 Yonge Street, Toronto, Ontario, M4W 3E5

Association of Professional Engineers of the Province of Prince
Edward Island
242 North River Road, Charlottetown, Prince Edward Island

Order of Engineers of Quebec
1100—2075 University Street, Montreal, Quebec H3A 1K8

Association of Professional Engineers of the Province of
Saskatchewan
2220 Twelfth Avenue, Regina, Saskatchewan

Association of Professional Engineers of the Yukon Territory,
P.O. Box 4125, Whitehorse, Yukon Territory

PART TWO

Questions from Professional
Engineer's Examinations

BASIC FUNDAMENTALS

CHEMISTRY, MATERIALS SCIENCE, AND NUCLEONICS

1.101 The weight of sulfuric acid and of sodium chloride that would be required to prepare 1 liter of hydrochloric acid, density 1.201, is: (*a*) 595.3, (*b*) 646.8, (*c*) 673.2, or (*d*) 692.4 grams? HCl of density 1.201 contains 40.09 per cent by weight HCl. Chemical equation and atomic weights are as indicated.

$$2NaCl + H_2SO_4 \rightarrow Na_2SO_4 + 2HCl$$

Atomic weights: Na, 23; Cl, 35.46; H, 1.008; O, 16; S, 32.

1.102 Explain:
   a. Oxidation and reduction
   b. The periodic law of the chemical elements
   c. Radioactivity
   d. Acidity

1.103 Cupric sulfide (CuS) ore may be roasted in air, the sulfur driven off as sulfur dioxide and later converted to sulfuric acid. The pounds of acid (93 per cent H_2SO_4) that can be made from 1 ton of ore that shows 60 per cent copper, allowing for 10 per cent loss of sulfur during the process is: (*a*) 1,528.5, (*b*) 1,645.8, (*c*) 1,775.0, or (*d*) 1,893.8 lb?

Atomic weights: Cu, 63.6; H, 1.00; S, 32; O, 16.

1.104 Describe the commercial method for the production of sodium hydroxide (caustic soda) from soda ash and indicate the

23

type of equipment used for each step in the process. **What engineering problems are involved?**

1.105 A solution of 4.00 grams of an unknown substance in 200 grams benzene is found to freeze at 4.82°C; the benzene freezes at 5.49°C. The molecular weight of the substance is: (*a*) 135.7, (*b*) 141.4, (*c*) 147.1, or (*d*) 152.8?

1.106 An aqueous solution of sodium sulfate is saturated at 32°C. The temperature to which the solution must be cooled to crystallize 60 per cent of the solution as $Na_2SO_4 \cdot 10H_2O$ is: (*a*) 18.7, (*b*) 16.6, (*c*) 15.4, or (*d*) 14.7°C?

1.107 It has been determined by trial mixing that a $1:1\frac{1}{2}:3$ mixture (by volume) with 5 gal water per sack of cement produces a satisfactory mixture of concrete with a 2-in. slump. What quantity of mixed concrete in cubic feet will a barrel of cement produce if the following is known about the cement, sand, and stone aggregate?

	Voids	Unit weight, lb
Cement	0.513	94
Sand	0.360	120
Stone	0.330	105

1.108 The number of cubic feet of a gas composed of 28 per cent CO and 72 per cent N_2 (by volume) and measured at 85°F and 100 psig required to reduce 1 ton of ore which is 84 per cent Fe_3O_4 to metallic iron is: (*a*) 4975.3, (*b*) 5060.3, (*c*) 5145.1, or (*d*) 5230.0 cu ft? The reactions are:

$$Fe_3O_4 + CO \rightarrow 3FeO + CO_2$$
$$FeO + CO \rightarrow Fe + CO_2$$

Atomic weights: C, 12.00; Fe, 55.84; O, 16.00; N, 14.008

1.109 Balance the chemical equation given below. Atomic weights: Al, 26.97; Fe, 55.84; S, 32.06; O, 16.00.

$$?Al + ?Fe_2(SO_4)_3 \rightarrow Al_2(SO_4)_3 + ?Fe SO_4$$

In each of the following questions, 1.110 to 1.119, select one, and only one, correct answer:

1.110   The chemical most commonly used to speed sedimentation of sewage is: (1) sulfuric acid; (2) copper sulfate; (3) **lime**; (4) methylene blue; (5) sodium permanganate.

1.111   The gas from sludge digestion tanks is mainly composed of: (1) nitrogen; (2) hydrogen sulfide; (3) carbon dioxide; (4) carbon monoxide; (5) methane.

1.112   The quantity of chlorine in parts per million required to satisfactorily chlorinate sewage is usually: (1) 0–25; (2) 30–60; (3) 65–90; (4) 95–120; (5) 125–150.

1.113   The ratio of oxygen available to the oxygen required for stabilization of sewage is called: (1) biochemical oxygen demand; (2) oxygen-ion concentration; (3) relative stability; (4) bacterial-stability factor; (5) concentration factor.

1.114   Most of the bacteria in sewage are: (1) parasitic; (2) saprophytic; (3) pathogenic; (4) anaerobic; (5) dangerous.

1.115   In the design of grit chambers: (1) temperature is an important factor; (2) baffles are essential; (3) there should be a 5 to 1 ratio of length to depth; (4) the detention period should be at least 30 min; (5) the maximum velocity of flow is 1 ft per sec.

1.116   In an Imhoff tank: (1) the effluent contains very little dissolved oxygen; (2) there are no settling compartments; (3) the sludge and fresh sewage are well mixed to give complete digestion; (4) the sludge and raw sewage are not mixed; (5) much larger settling compartments are required than for an equivalent septic tank.

1.117   The per cent of total solids in most domestic sewage is approximately: (1) 0.01; (2) 0.1; (3) 1.0; (4) 10; (5) 100.

1.118   Intermittent sand filters are primarily used to: (1) remove offensive odors; (2) supply fertilizer to farmers; (3) oxidize putrescible matter; (4) neutralize sludge; (5) remove solids from sewage.

1.119 To divert excessive flow from combined sewers designers often use: (1) a Cipoletti weir; (2) a V-notch weir; (3) a broad-crested weir; (4) a submerged weir; (5) a leaping weir.

1.120 A batch of concrete consisted of 200 lb fine aggregate, 350 lb coarse aggregate, 94 lb cement, and 5 gal water. The specific gravity of the sand and gravel may be taken as 2.65 and that of the cement as 3.10.

  a. What was the total volume of concrete produced?

  b. What was the weight of concrete in place per cubic foot?

  c. How much by weight of each ingredient is required to produce 1 cu yd concrete?

1.121 The most important factor in determining high-temperature behavior of an alloy is: (*a*) dispersion, (*b*) ionization, (*c*) crystallization, (*d*) composition, or (*e*) endurance?

1.122 At relatively high temperatures and low rates of strains, structures will perform better if their material is: (*a*) fine-grained, (*b*) course-grained, or (*c*) their behavior is independent of the grain?

1.123 With regard to corrosion of metals, passivation is the process that: (*a*) intensifies deterioration, (*b*) intensifies deterioration temporarily, (*c*) changes the composition of the metal, (*d*) inhibits further deterioration, or (*e*) alters the grain size of the metal?

1.124 In the process of pair formation, a pair cannot be formed unless the quantum has an energy greater than: (*a*) 0.5 MeV, (*b*) $2m_0C^2$, (*c*) $\frac{1}{2}mV^2$, (*d*) $\frac{h\nu}{C}$, or (*e*) 3.5 MeV?

1.125 One of the two types of non-material nuclear radiation is: (*a*) transmutation radiation, (*b*) Walton radiation, (*c*) gamma radiation, (*d*) betatron radiation, or (*e*) Rutherford radiation?

ELECTRICITY

1.201 An electric truck is propelled by a d-c motor driven by a 110-volt storage battery. The truck is required to exert a tractive effort of 200 lb at a speed of 5 mph. The over-all efficiency of the motor and drive is 70 per cent. The current taken from the battery is: (*a*) 25.8, (*b*) 28.0, (*c*) 30.2, or (*d*) 32.4 amp?

1.202 In the electric circuit shown, if the total resistance from A to B is 16 ohms, find the resistance R_2.

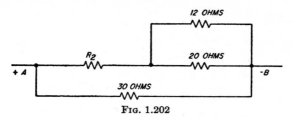

Fig. 1.202

1.203 Determine the current supplied by the battery in the network shown. Resistance in ohms are as indicated.

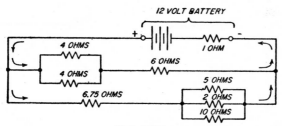

Fig. 1.203

1.204 Two coils of wire have resistance of 5 and 17 ohms respectively. Find the voltage used by each and the current flowing through each if the two are connected in parallel to the terminals of a 6-volt battery.

1.205 In the electrolytic refining of copper, the original weight of the cathode is 15.0 lb. After 15 days in a bath of $CuSO_4$ and free H_2SO_4, the weight of the electrode is 300 lb. Assuming 100 per cent efficiency, what total ampere-hours are required and what is the current flow? If the cathode is 40 by 36 in., what is the current density?

1.206 A 50-kva 2,300/230-volt 60-cycle single-phase transformer has a core loss of 400 watts and a full-load copper loss of 600 watts. Calculate its efficiency at 100 per cent load and 90 per cent power factor. At what per cent load will maximum efficiency occur?

1.207 Two groups of electric lights are connected in series across a 220-volt line. The first group has 25 lights in parallel, and the second group has 15 lights in parallel. Each light will carry 0.9 amp at 110 volts. Assuming the resistance of the lights to be constant at varying voltages, what is the potential difference across each group? What is the current in each light?

1.208 A d-c load of 10 kw at 120 volts is to be supplied from a copper line which has a diameter of 0.204 in. The distance from generator to load is 500 ft.

1. Is the required voltage of the generator: (1*a*) 141.1, (1*b*) 147.3, (1*c*) 153.5, or (1*d*) 159.7 volts?

2. Is the power loss in the line: (2*a*) 1665, (2*b*) 1710, (2*c*) 1755, or (2*d*) 1800 watts?

1.209 Two condensers of 2 and 10 μf capacitance are connected in parallel, and this combination is connected in series with a third condenser of 8 μf capacitance across a 120-volt d-c supply circuit. Find the charge of each condenser and the voltage across each. What is the total energy stored in the three condensers?

1.210 *a.* Two storage cells connected in parallel supply jointly a current of 20 amp to an external circuit. One cell has an emf of 2.1 volts and the other an emf of 2.0 volts, and each has an internal resistance of 0.05 ohm. Find the current flowing through each cell.

b. Explain why the specific gravity of the electrolyte of a lead storage cell decreases as it discharges.

1.211 A d-c generator has four poles with faces 6 in. square and has a flux density in the air gap of 40,000 lines per sq in. The machine has 360 conductors arranged in four parallel paths through the armature. The emf of the generator when driven at 1,200 rpm is: (*a*) 97.1, (*b*) 99.3, (*c*) 101.5, or (*d*) 103.7 volts?

1.212 Two storage batteries of 10 and 15 volts and an internal resistance of 1 and 2 ohms respectively have their positive terminals attached to one end of a variable resistance and their negative terminals to the other end. What must be the value of this resistance in order that the same current will flow from each battery?

1.213 Could two 30-watt and one 60-watt 120-volt lamps be connected across a 240-volt circuit in such a way as to give normal illumination? If so, show circuit diagram and explain.

1.214 Electrical resistances of 7 ohms and of 11 ohms are connected in parallel and the combination is then placed in series with a resistance of 15 ohms. The whole combination is connected across 110-volt d-c mains. What current flows through each resistance?

1.215 A d-c motor having an efficiency of 85 per cent is rated 5 hp and 220 volts. It is located 150 ft from the supply lines. If a maximum voltage drop of 5 per cent based on rated voltage is permitted, recommend the wire size for overhead wiring.

1.216 How many 100-watt lamps can be placed in a 115-volt circuit, as shown, which is protected by a 15-amp fuse? (The fuse "burns out" and opens the circuit when the current through it rises to a value of 15 amp.)

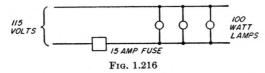

FIG. 1.216

1.217 If two voltmeters are connected in series to a power line and the first voltmeter, which has a resistance of 16,000 ohms, reads 60 volts, what is the voltage between the wires of the power lines provided the second voltmeter's resistance is 20,000 ohms?

1.218 A circuit consists of three devices and a 30-volt battery having an internal resistance of 1 ohm. One of the devices, which has 7 ohms resistance, is connected in series with the other two, and these, one of 3 ohms and the other of 6 ohms resistance, are connected in parallel. Find:

 a. The voltage drop on the 7-ohm device
 b. The power expended in the 3-ohm device
 c. The current in the 6-ohm device

1.219 A 10-hp 220-volt d-c shunt motor has armature and field resistances of 0.25 and 100 ohms respectively. The full-load efficiency is 83 per cent. The value of the starting resistance, in order that the starting current will not exceed 200 per cent of the full-load value, is: (*a*) 2.2, (*b*) 2.5, (*c*) 2.8, or (*d*) 3.1 ohms?

1.220 A circuit consisting of a 63-ohm resistance, a 0.15-henry inductance, and a 15-μf capacitor connected in series, is placed on a 110-volt 60-cycle supply line. Find the value of the current in the circuit.

1.221 A three-phase Y-connected induction motor operating on 220 volts has a power factor of 70 per cent and an efficiency of 82 per cent when delivering 8 hp.

1. The voltage across each phase is: (1*a*) 127.0, (1*b*) 133.5, (1*c*) 140.0, or (1*d*) 146.5 volts?

2. The current in each phase is: (2*a*) 18.9, (2*b*) 21.7, (2*c*) 24.5, or (2*d*) 27.3 amp?

1.222 Explain how the plate resistance, the amplification factor, and the mutual conductance may be calculated for a three-element electron tube from a family of plate-characteristic curves for the tube.

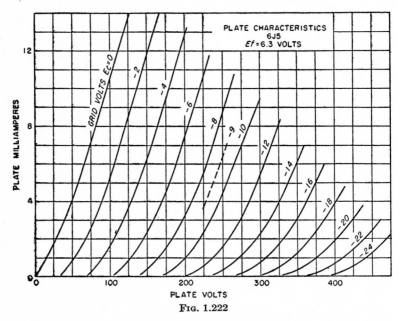

FIG. 1.222

1.223 If the current from a short-circuited 1.5-volt dry cell is 25 amp, what is the internal resistance of the cell? The resistance of the short circuit is negligible. If this cell is connected through a 1-ohm external circuit, find the current and the reading of a voltmeter connected to its terminals.

1.224 A series L-C-R circuit with inductance of 100 henrys has a transient resonant frequency of 5 cps.

   a. What is the period?

   b. The capacitance?

Neglect effect of R on frequency.

1.225 Same circuit as in 1.224. The condenser is given an initial charge Q, and at time $t = 0$, is switched to discharge through L and R in series.

   a. Write the differential equation.

   b. What is the physical meaning of each term?

FLUID MECHANICS

1.301 Compute the total normal pressure on the interior sides and bottom of a cylindrical water tower 40 ft in diameter containing water to a depth of 60 ft.

1.302 A vertical gate 4 ft wide and 6 ft high hinged at the upper edge is kept closed by the pressure of water standing 8 ft deep over its top edge. What force applied normally at the bottom of the gate would be required to open it?

1.303 How far below the water surface is it necessary to immerse a vertical plane surface 3 ft square, two edges of square being horizontal, in order that center of pressure shall be but 1 in. from center of gravity?

1.304 Water at a temperature of 80°F flows through two separate pipes 10 and 12 in. in diameter. If the mean velocity of flow in the 12-in. pipe is 6 fps,

   1. Will the velocity of flow in the 10-in. pipe be: (1*a*) 6.6, (1*b*) 6.8, (1*c*) 7.0, or (1*d*) 7.2 fps?

   2. Will the Reynolds number for the 12-in. pipe be: (2*a*) 624,000, (2*b*) 636,000, (2*c*) 648,000, or (2*d*) 660,000?

1.305 If the water in the 10-in. pipe of question 1.304 is replaced by oil having a specific gravity of 0.80 and a μ value of 0.000042, will the oil's velocity for similarity in the two flows be: (*a*) 21.0, (*b*) 21.5, (*c*) 22.0, or (*d*) 22.5 fps?

1.306 A flat-bottomed scow is built with vertical sides and straight sloping ends. Its length on deck is 80 ft, and on the bottom, 65 ft; its width is 20 ft, and its vertical depth is 12 ft How much water will it draw if it weighs 250 tons?

1.307 A tank with vertical sides is 4 ft square, 10 ft deep, and is filled to a depth of 9 ft with water. By how much, if at all, will the pressure on one side of the tank be changed if a cube of wood, specific gravity 0.5, measuring 2 ft on an edge is placed in the water so as to float with one face horizontal?

1.308 Water from a reservoir is pumped over a hill through a pipe 3 ft in diameter, and a pressure of 30 psi is maintained at the summit, where the pipe is 300 ft above the reservoir. The quantity pumped is 49.5 cfs, and by reason of friction in the pump and pipe there is 10 ft of head lost between reservoir and summit. What amount of energy must be furnished the water each second by the pump?

1.309 A centrifugal pump draws water from a pit through a vertical 12-in. pipe which extends below the water surface. It discharges into a 6-in. horizontal pipe 15 ft above the water surface. While pumping 2 cfs, a pressure gage on the discharge pipe reads 24 psi, and a gage on the suction pipe registers 5 psi below atmosphere. Both gages are close to the pump and are separated by a vertical distance of 5 ft. What is the horsepower output of the pump?

1.310 A steel box, rectangular in plan, floats with a draft of 4 ft. If the box is 20 ft long, 10 ft wide, and 6 ft deep, compute the time necessary to sink it to its top edge by opening a standard orifice, 6 in. in diameter, in its bottom. Neglect the thickness of the vertical sides, and assume $c = 0.60$.

1.311 A rectangular weir with end contractions has a crest 10 ft long and 1.00 ft above the bottom of the channel. If the channel width be 16 ft, what amount of water will be discharged under a head of 0.875 ft? Neglect the velocity of approach.

1.312 A suppressed weir having a crest length of 7.00 ft, a height of 1.00 ft, discharges under a head of 0.875 ft. Compute the rate of discharge. Neglect the velocity of approach.

1.313 Two reservoirs with a difference in level of 100 ft are connected by 12,000 ft of 12-in. cast-iron pipe. Compute the rate of discharge.

1.314   An 18-in. cast-iron pipe is discharging 7,500 gpm. At a point 4,000 ft from the supplying reservoir (measured on pipe) the center of the pipe is 180 ft below the reservoir surface. What pressure, in pounds per square inch, is to be expected there?

1.315   A pump at elevation 1,000 is pumping 2.00 cfs through 8,000 ft of 6-in. pipe to a reservoir whose level is at elevation 1,250. What pressure will be found in the pipe at a point where the elevation is 1,100 ft above datum and the distance (measured along the pipe) from the pump is 3,200 ft? Assume $f = 0.0225$.

1.316   From a reservoir A whose surface is at elevation 800 water is pumped through 4,000 ft of 12-in. pipe across a valley to a second reservoir B whose level is at elevation 900. If, during pumping, the pressure is 100 psi at a point M on the pipe, midway of its length and at elevation 700, compute the rate of discharge and the horsepower exerted by the pumps ($f = 0.02$).

1.317   Reservoir A is at elevation 1,000 ft above datum. From reservoir A an 8-in. pipeline leads 3,000 ft to point Y at elevation 800, at which point it branches into two lines: a 6-in. line running 2,000 ft to reservoir B, elevation 850, and a 6-in. line running 1,000 ft to reservoir C, elevation 875. At what rate will water be delivered to each reservoir? Assume $f = 0.02$ in all cases.

1.318   Reservoir R is at grade 400. From reservoir R an 18-in. pipe, which is to carry 7 cfs of water, leads 2,000 ft to grade 300 at point Y. It there divides and branch A, 12 in. in diameter, leads 13,000 ft to reservoir A, which is at grade 250. Branch B leads 4,000 ft to reservoir B, which is at grade 50.

    Neglect all losses except from friction, assume f as 0.02 in each case, and find the diameter of branch B.

1.319   A 48-in. main, carrying 75.4 cfs, branches at a point A into two pipes, one 2,000 ft long, 3 ft in diameter, and one 6,000 ft long, 2 ft in diameter. Both pipes come together at a point B and continue as a single 48-in. pipe. The following value of f may be assumed: $f = 0.0210$, 0.022, and 0.023 respectively for the 48-in., 36-in., and 24-in. pipes. Compute the rate of flow in the branch pipes.

1.320   A pipeline 25,000 ft long and 5 ft in diameter supplies eight nozzles with water from a reservoir whose level is 600 ft above the nozzles. Each nozzle has an opening 3 in. in diameter and a coefficient of discharge and velocity of 0.95. Assuming $f = 0.017$, find the aggregate horsepower available in the jets.

1.321 A venturi meter has an area ratio of 9 to 1, the larger diameter being 24 in. During flow the recorded pressure head in the large section is 36 ft, and that at the throat 9 ft. If c is 0.99, what rate of discharge through the meter is indicated?

1.322 A canal has a bottom width of 30 ft and side slopes of 3 horizontal to 1 vertical. If the water depth is 4 ft and the slope 1 in 1,000, compute the probable velocity and discharge, using Kutter's C and $n = 0.02$.

1.323 A cylindrical vessel open at the top is 4 ft high and 4 ft in diameter. It is revolved about its center vertical axis at the rate of 56 rpm. A piezometer tube is attached to the side of the vessel and stands vertically 3 ft from the cylinder axis. If the vessel was previously filled with water to the top edge,

    1. Will the water rise: ($1a$) 2.14, ($1b$) 2.68, ($1c$) 3.75, or ($1d$) 4.82 ft above the brim of the vessel?

    2. Will the loss of water be: ($2a$) 6.72, ($2b$) 10.09, ($2c$) 11.77, or ($2d$) 13.45 cu ft?

1.324 A flat sign board is circular and 20 ft in diameter. Weight of air is 0.0807 lb/cu ft. There is a drag coefficient of 1.12 for a wind force normal to the sign surface. For a wind velocity of 40 mph, will the normal force on the sign be: (a) 1,480, (b) 1,520, (c) 1,600, or (d) 1,720 lb?

1.325 Water leaves the toe of a spillway with a horizontal velocity of 30 fps, and a depth of 0.80 ft, flowing directly onto a level concrete apron.

    1. If a jump is to occur on the apron, must the depth attained on the apron be: ($1a$) 6.3, ($1b$) 6.8, ($1c$) 7.2, or ($1d$) 7.5 ft?

    2. Will the horizontal velocity of the water after the jump be: ($2a$) 2.65, ($2b$) 3.81, ($2c$) 4.97, or ($2d$) 6.13 fps?

    3. If the stream is 200 ft wide, will the horsepower absorbed by the jump be: ($3a$) 4,300, ($3b$) 4,380, ($3c$) 4,460, or ($3d$) 4,540 hp.?

    4. If the length (L in feet) in which the jump will occur is approximately 4.8 times the depth after the jump occurs, will this length be: ($4a$) 36.00, ($4b$) 34.56, ($4c$) 32.64, or ($4d$) 30.24 ft?

MATHEMATICS AND MEASUREMENTS

1.401 An observer wishes to determine the height of a tower. He takes sights at the top of the tower from points A and B, which are 50 ft apart, at the same elevation, and on a direct line with the tower. The vertical angle at point A is 30° and at point B is 40°. The height of the tower is: (a) 89.51, (b) 92.54, (c) 95.38, or (d) 97.33 ft?

1.402 Express the equation of the straight line passing through points $x_1 = 2, y_1 = 2$ and $x_2 = 4, y_2 = 3$ in the form $y = mx + b$.

1.403 Solve

$$x^2 + y^2 = 5z$$
$$x^2 - y^2 = 3z$$

How many and what numerical values for x, y, and z will satisfy these simultaneous equations?

1.404 The area of a circle is 89.42 sq in. What is
  a. The diameter of the circle?
  b. The circumference?
  c. The length of a side of a regular hexagon inscribed in this circle?

1.405 A vein of ore has a strike of N45°E., i.e., it intersects level ground along a line having this bearing. The vein dips westerly at an angle of 15° with the horizontal. A drift (slightly sloping mine opening following the vein) has been opened in the vein on a bearing of N30°E. Determine the per cent of grade of this drift.

1.406 The distance between two points measured with a steel tape was recorded as 916.58 ft. Later, the tape was checked and found to be only 99.9 ft long. What is the true distance between the points?

1.407 Solve for x (show all work):

$$4 + \frac{x+3}{x-3} - \frac{4x^2}{x^2-9} = \frac{x+9}{x+3}$$

1.408 A circular piece of tin 24 in. in diameter has a triangular hole 9 by 12 by 15 in. The vertex of the triangle at the intersection of the 9- and 12-in. sides is at the center of the circle. Find the position of the center of gravity of the piece of tin.

1.409 Compute the moment of inertia of the area shown with respect to its centroidal x and y axes.

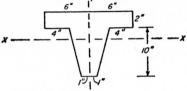

FIG. 1.409

1.410 Locate the centroids of the accompanying figure with respect to the axes shown and determine the moments of inertia with respect to the axes shown.

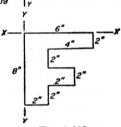

FIG. 1.410

1.411 Two planes leave Cleveland for Jacksonville, a distance of 900 miles. The four-motored plane (A) travels at a ground speed of 90 mph faster than the two-motored plane (B). Plane A arrives in Jacksonville 2 hr 15 min ahead of plane B. What were the respective ground speeds of the two planes?

1.412 Twice the sum of two numbers is 28. The sum of the squares of the two numbers is 100. The product of the two numbers is: (a) 42, (b) 48, (c) 54, or (d) 56?

1.413 An engineer was told that a survey had been made of a certain rectangular field but the dimensions had been lost. An assistant remembered that if the field had been 100 ft longer and 25 ft narrower, the area would have been increased 2,500 sq ft, and that if it had been 100 ft shorter and 50 ft wider, the area would have been decreased 5,000 sq ft. What was the area of the field and what were its dimensions?

1.414 If the outside diameter of a locomotive driving wheel minus its steel tire is 62.378 in. and the inside diameter of the steel tire at 65°F is 62.263 in., the temperature (Fahrenheit) to which the tire must be heated to just fit the wheel is: (*a*) 323, (*b*) 335, (*c*) 349, or (*d*) 362°F?

1.415   A certain steel tape is known to be 100.000 ft long at a temperature of 70°F. When the tape is at a temperature of 10°F, what tape reading corresponds to a distance of 90.000 ft?

1.416   Eight men can dig 150 ft of trench in 7 hr. Three men can backfill 100 ft of the trench in 4 hr. The time that it will take 10 men to dig and fill 200 ft of trench is: (*a*) 9 hr 52 min, (*b*) 10 hr 1 min, (*c*) 10 hr 24 min, or (*d*) 10 hr 46 min?

1.417   A man owns two square lots of unequal size, together containing 15,025 sq ft. If the lots were contiguous, it would require 530 ft of fence to embrace them in a single enclosure of six sides. Find the area of each lot.

1.418   Given the semicubical parabola $y^3 = x^3$.

   a. Find the area of the part enclosed by the line $x = 4$ and the x axis.

   b. Find the area under the curve $y = x^3 + 3x^2$ and the x axis between $x = 1$ and $x = 3$.

   c. Differentiate the following:

$$4x^2 + 17x$$

and $$ax^2 + b^{1/2}$$

   d. Integrate the following:

$$(7x^3 + 4x^2)\,dx$$

and $$x \cos (2x^2 + 7)\,dx$$

1.419   A line was measured with a steel tape when the temperature was 30°C. The measured length of the line was found to be 1,256.271 ft. The tape was afterward tested when the temperature was 10°C and it was found to be 100.042 ft long. The true length of the line if the coefficient of linear expansion of tape was 0.000011 per °C was: (*a*) 1,257.075, (*b*) 1,259,038, (*c*) 1,263.045, or (*d*) 1,275.042 ft?

1.420   The sum of three numbers in arithmetical progression is 45. If 2 is added to the first number, 3 to the second, and 7 to the third, the new numbers are in geometrical progression. Find the numbers.

1.421   Solve algebraically:

a.
$$4x^2 + 7y^2 = 32$$
$$11y^2 - 3x^2 = 41$$

b.
$$x + 2y - z = 6$$
$$2x - y + 3z = -13$$
$$3x - 2y + 3z = -16$$

1.422   a. Given $\cos 2A = 2\cos^2 A - 1$, find: $\cos 75°$

b. In a circle of unit radius, sketch a central angle B in the second quadrant and show lines that are equal in length to $\sin A$; $\cos A$; $\tan A$; in a.

1.423   A box is to be constructed from a piece of zinc 20 in. square by cutting equal squares from each corner and turning up the zinc to form the sides. What is the volume of the largest box that can be so constructed? Solve by use of calculus.

1.424   Determine the volume of a right-truncated triangular prism with the following dimensions: let the corners of the triangular base be defined by a, b, and c; the length of $ab = 10$ ft, $bc = 9$ ft, and $ca = 12$ ft. The sides at a, b, and c are perpendicular to the triangular base and have heights of 8.6 ft, 7.1 ft, and 5.5 ft, respectively.

1.425   A 6 per cent upgrade meets a 3 per cent downgrade at elevation 100.00 at station 10 + 00. If the parabolic vertical curve is 600 ft long, find the station of the high point of the curve. Plot the profile of the curve, giving all pertinent information thereon.

MECHANICS (KINETICS)

1.501   An elevator weighing 2,000 lb attains an upward velocity of 16 fps in 4 sec with uniform acceleration. The tension in the supporting cables is: (a) 2,165, (b) 2,250, (c) 2,345, or (d) 2,478 lb?

1.502   Solve for the tension in the cables of the elevator described in problem 1.501 if the tension is reduced so that the elevator comes to rest in a distance of 5 ft.

1.503   A body weighing 40 lb starts from rest and slides down a plane at an angle of 30° with the horizontal for which the coefficient of friction $f = 0.3$. How far will it move during the third second? How long will it require for it to move 60 ft?

1.504   A car and its load weigh 6,000 lb and the center of gravity is 2 ft from the ground and midway between the front and rear wheels, which are 120 in. apart. The car is brought to rest from a speed of 30 mph in 5 sec by means of the brakes. Compute the normal pressures on each of the front wheels and on each of the rear wheels.

1.505   A body weighing 1,000 lb falls 6 in. and strikes a 2,000-lb (per in.) spring. The deformation of the spring is: (*a*) 3.0, (*b*) 3.5, (*c*) 3.8, or (*d*) 4.2 in.?

1.506   A freight car weighing 100,000 lb is moving with a velocity of 2 fps when it strikes a bumping post. Assuming the drawbar spring on the car to take all of the compression, what must be the scale of the spring in order that the compression shall not exceed 2.5 in.?

1.507   If the coefficient of friction under the blocks *M* and *N* is 0.20, what is the acceleration of the blocks, the tension in the cord, and the time to move 15 ft from rest?

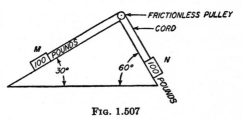

Fɪɢ. 1.507

1.508   The three bodies, *X*, *Y*, and *Z*, weigh 10, 20, and 30 lb, respectively. If they are supported in the position shown and then released simultaneously, what will be the velocity and position of each body 1 sec later? (Neglect mass of cords and pulleys.)

Fɪɢ. 1.508

1.509   A 60-ton freight car starts from rest and runs 100 ft down a 1 per cent grade, then strikes a bumping post. If train resistance is 10 lb per ton, what is the velocity of striking? If the drawbar spring has a scale of 100,000 lb per in. and is assumed to take all of the impact, how much will it be shortened?

1.510   A car weighing 40 tons is switched to a 2 per cent upgrade with a velocity of 30 mph. If train resistance is 10 lb per ton, how far up the grade will it go? If the car is then allowed to run back, what velocity will it have at the foot of the grade? If the track is then level, how far from the foot of the grade will it run?

1.511   Determine the superelevation of the outer rail of a railroad track on a 10° curve to give equal rail pressures on a car when moving with a speed of 45 mph. If a car weighing 100,000 lb with its center of gravity 5 ft above the track has a speed of 60 mph on this curve, and if the track now has a superelevation of only 8 in., what is the pressure of the outer rail on the car normal to the ties?   (A 10° curve is one on which a chord 100 ft long subtends an angle of 10° at the center.)

1.512   A concrete highway curve with a radius of 500 ft is banked to give lateral pressure equivalent to $f = 0.15$. For what coefficient of friction will skidding impend for a speed of 60 mph?

1.513   A girder weighing 20,000 lb is suspended by a cable 100 ft long. What horizontal pull is necessary to hold it 5 ft from the vertical position? What is the tension in the cable as the girder is allowed to swing back through its vertical position?

1.514   A shot is fired at an angle of 45° with the horizontal and a velocity of 300 ft per sec. Find the height and the range of the projectile.

1.515   Compute the theoretical muzzle velocity required to give a projectile a maximum range of 40 miles. Compute the maximum height to which the projectile will rise.

1.516   A cast-iron governor ball 3 in. in diameter has its center 18 in. from the point of support. Neglecting the weight of the arm itself, find the tension in the arm and the speed of rotation if the angle with the vertical axis is 60°.

1.517   A car is moving on a horizontal track around a curve of 1,000 ft radius with a speed of 60 mph. A weight of 25 lb is suspended from the ceiling by a 6 ft cord. What is the tension in the cord, the angle the cord makes with the vertical, and the horizontal displacement of the weight?

1.518   The muzzle velocity of a projectile is 1,500 fps and the distance of the target is 10 miles. The angle of elevation of the gun must be: (a)21°59′, (b) 22°41′, (c) 24°33′, or (d) 25°18′?

1.519 A cast-iron cylinder 24 in. in outside diameter and 16 in. high has a wall thickness of 1 in. Cast iron weighs 450 lb per cu ft. The moment of inertia of the cylinder with respect to its geometric axis is: (a) 10.2, (b) 9.7, (c) 9.2, or (d) 8.7?

1.520 A steel disk 40 in. in diameter and 4 in. thick has a cylindrical hole 10 in. in diameter at the center and another 12 in. in diameter, 12 in. from the center. What is the moment of inertia with respect to the geometric axis of the 40-in. disk? Steel weighs 490 lb per cu ft.

1.521 A flywheel is brought from rest up to a speed of 1,500 rpm in 1 min. What is the average angular acceleration α, and the number of revolutions until the wheel stops? What is the velocity at the end of 40 sec?

1.522 The rim of a 36-in. wheel on a brake-shoe testing machine has a speed of 60 mph when the brake is dropped. It comes to rest when the rim has traveled a tangential distance of 500 ft. What is the angular acceleration and the number of revolutions?

1.523 A steel hemisphere 6 in. in diameter is to rotate about an axis parallel to its axis of symmetry and 12 in. distant from it. The axis of the rod which connects it to the axis of rotation is 0.5 in. from the diametral plane. Compute the induced bending moment in the rod when the hemisphere is rotating at a speed of 300 rpm. How far from the diametral plane should the rod be attached so that there will be no bending moment in it?

1.524 A steel rod 24 in. long is rotating about a vertical axis through one end. If it stands at an angle of 60° with the axis, the speed at which it is rotating is: (a) 72, (b) 69, (c) 66, or (d) 64 rpm?

1.525 A cast-iron flywheel 12 ft in outside diameter has a rim 4 in. thick and 18 in. wide. If the flywheel is rotating at 240 rpm, what is the unit of centrifugal tensile stress in the rim if the tension in the arms is neglected? Cast iron weighs 450 lb per cu ft.

MECHANICS (STATICS)

1.601 Find the magnitude and direction of the resultant of a set of four vertical forces in the same plane. The first force is

10 lb upward; the next force, 5 ft away, is 10 lb downward; the third, 4 ft beyond the second, is 10 lb downward; the fourth, 3 ft beyond the third, is 10 lb upward.

1.602 A long wall stands on three rows of piles with the rows spaced 3 ft center to center. The spacings of piles parallel to the wall are 2 ft 6 in., 4 ft 0 in., and 6 ft 0 in. center to center in the first, second, and third rows, respectively. The resultant of the vertical loads is located midway between the first and second rows and amounts to 20,000 lb per lin ft of wall. What are the loads per pile in each row?

1.603 A beam 12 ft long and simply supported at each end, has a uniformly distributed load of 1,000 lb per ft extending from the left end to a point 4 ft away. There is also a clockwise couple of 10,000 ft-lb applied at the center of the beam. Draw the shear and moment diagrams for the beam and give all the necessary values. Neglect the weight of the beam.

1.604 Draw shear and moment diagrams for a 20-ft beam, simply supported at the ends. The beam carries a uniform load of 1,000 lb per ft (including its own weight) and three concentrated loads of 2,000, 4,000, and 6,000 lb acting respectively at the left quarter point, the center point, and the right quarter point.

1.605 A beam 20 ft long carries a uniform load of 2,000 lb per lin ft, including its own weight, on two supports 14 ft apart. The right end of the beam cantilevers 2 ft and the left end cantilevers 4 ft. Draw the shear and moment diagrams and compute the positions and amounts of maximum bending moments.

1.606 A beam 24 ft long rests on two supports, one at its right end and the other 6 ft from its left end. The beam carries a uniform load of 1,000 lb per lin ft over its entire length and a concentrated load of 15,000 lb at the middle of the 18-ft span. Calculate the reactions, draw the shear and moment diagrams, and calculate the position and amount of maximum bending moment.

1.607 The left-hand half of a beam on a 30-ft span has a uniform load of 4,000 lb per lin ft. A concentrated load of 30,000 lb is located 10 ft from the right end. Draw the shear and moment diagrams.

1.608   A beam carries a concentrated load P applied 6 ft from the right end. The beam is partially restrained at the left end and is simply supported at the right end. The bending moment at the partially restrained end is $PL/16$ in magnitude. The beam is 16 ft long and has a 4 by 16 in. cross section. The beam is made of wood which has a modulus of elasticity $E = 16 \times 10^5$ psi. If the maximum induced bending stress is 1,350 psi,

1. Will the magnitude of P be: (1a) 5,060, (1b) 5,269, (1c) 5,478, or (1d) 5,687 lb?

2. Will the deflection at the point of load application be: (2a) 0.276, (2b) 0.300, (2c) 0.324, or (2d) 0.348 in.?

1.609 I is uniform throughout.

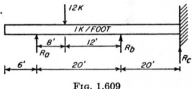

Fig. 1.609

1. The maximum positive moment is: (1a) 70.48, (1b) 73.45, (1c) 75.43, or (1d) 78.41 ft-kips?

2. The maximum negative moment is: (2a) 53.85, (2b) 58.65, (2c) 60.75, or (2d) 63.45 ft-kips?

1.610   Draw the shear and moment diagrams for the beam ABC.

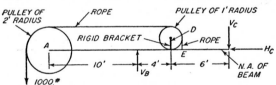

Fig. 1.610

1.611   A 1,000-lb weight A is to be supported by two wires. The first wire BA runs to a ring at point B, 3 ft above A and 4 ft to the left. The second wire CA runs to a ring at C, 4 ft above A and 3 ft to the right. If points A, B, and C are in the same vertical plane, find the horizontal and vertical components of the stress in the wires at points B and C and the stress in each wire.

1.612   A painter's scaffold 20 ft long and weighing 150 lb is supported in a horizontal position by vertical ropes attached at equal distances from the ends of the scaffold. Find the greatest distance from the ends that the ropes may be attached so as to permit a 200-lb man to stand safely at one end of the scaffold.

1.613 A 20-ft ladder stands on a rough horizontal floor and leans against a vertical, smooth wall, the foot of the ladder being 8 ft out from the wall. Halfway up the ladder is a weight of 150 lb. If the weight of the ladder is neglected, the horizontal component of the reaction against the bottom end of the ladder is: (*a*) 31.5, (*b*) 32.75, (*c*) 35.83, or (*d*) 39.85 lb?

1.614 A rectangular masonry wall is 6 ft thick and 15 ft high. The concrete weighs 150 lb per cu ft. Assuming that there is no seepage causing uplift against the bottom of the wall and that water weighs 62.5 lb per cu ft, the height the water could rise behind the wall without causing an intensity of pressure at the toe greater than twice the average pressure on the base is: (*a*) 10.9, (*b*) 11.4, (*c*) 11.8, or (*d*) 12.3 ft?

1.615 Is this wall safe against over-turning about point *A* if the masonry weighs 150 lb per cu ft and the earth weighs 100 lb per cu ft? The horizontal pressure of the earth against the masonry wall is equal to *cwh*, where *cw* is equal to 30 psf and *h* is the distance down from the top of the wall in feet.

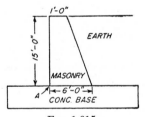

Fig. 1.615

1.616 A truck is placed on a 30-ft span to give the greatest bending moment. The loads are 4 tons on the front axle and 16 tons on the rear axle. The distance between axles is 14 ft. As the truck moves on the span,

1. The maximum bending moment induced by the loads is: (1*a*) 133.8, (1*b*) 127.3, (1*c*) 125.8, or (1*d*) 123.3 ft-tons?

2. The maximum shear induced by the loads is: (2*a*) 18.13, (2*b*) 19.83, (2*c*) 21.73, or (2*d*) 23.5 tons?

1.617 A tractor and semitrailer with axle loads as shown is a live load on a 100-ft simple span bridge girder.

a. How far is the 8-kip load from the end of the span when this load gives maximum moment?

b. Calculate the maximum moment due to this loading.

c. Determine the maximum shear due to this load.

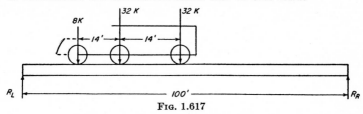

FIG. 1.617

1.618 Draw shear and moment diagrams for beams *AB*, *CD*, and *EF*. Assume all beams as "simple beams."

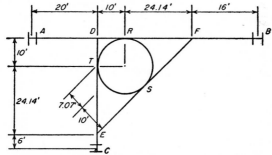

FIG. 1.618. Tank weighing 100 kips is supported at *R*, *S*, and *T* only.

1.619 Find the tension and compression in each member.

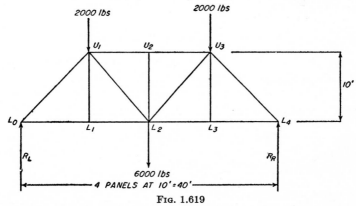

FIG. 1.619

1.620 In the roof truss shown, determine the stresses in members L_0L_1, U_1U_2, and U_1L_2. State which members are in tension or compression.

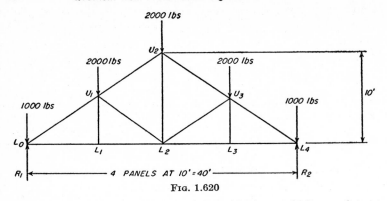

FIG. 1.620

1.621 Determine the stress in the members of the roof truss shown.

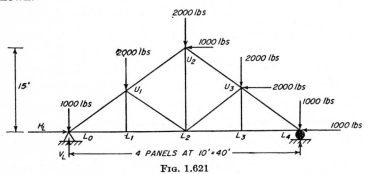

FIG. 1.621

1.622 For the deck truss shown compute the magnitude and direction (tension or compression) of the stress in the member U_2U_3, and in the diagonals U_0L_1 and U_1L_2.

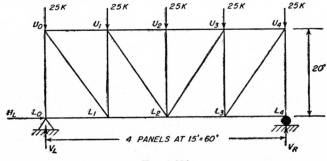

FIG. 1.622

1.623 Determine the amount and character of the stress in each member of the truss shown. Place answers on your sketch of truss.

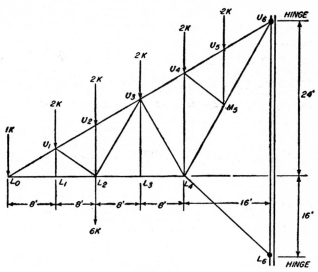

FIG. 1.623

1.624 Determine the kind and amount of stress in each member of this truss.

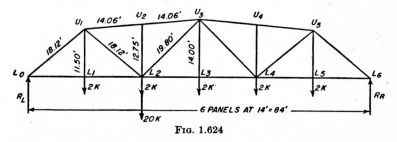

FIG. 1.624

1.625 The space structure shown in Fig. 1.625 supports a 2,400-lb horizontal load. Determine:

    a. The vertical component of stresses in member AB
    b. The X component of stress in member AD
    c. The Y component of stress in member AC
    d. The total stress in member AD

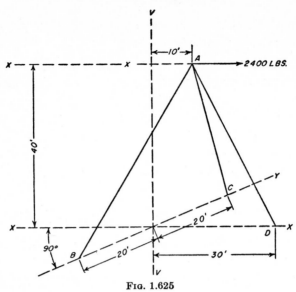

FIG. 1.625

THERMODYNAMICS

1.701   Define: (*a*) latent heat (*b*) specific heat (*c*) heat of fusion (*d*) Btu (*e*) absolute temperature (*f*) calorie (*g*) watt.

1.702   Thirty pounds of ice at 32°F is placed in 100 lb of water at 100°F. (The latent heat of ice may be taken as 144 Btu per lb.) If no heat is lost or added to the mixture, the temperature when equilibrium is reached is: (*a*) 48, (*b*) 49, (*c*) 50, or (*d*) 51°F?

1.703   A small swimming pool 25 by 75 ft is to be filled with water at a temperature of 70°F to a depth of 5 ft. Hot water at 160°F and cold water at 40°F are available. How many cubic feet of each (neglecting the thermal capacity of the tank itself) should be used in filling the tank?

1.704   A partly filled barrel contains 300 lb of water and 100 lb of ice at 32°F. How many pounds of steam at 212°F must be run into the barrel to bring its contents up to 80°F?

1.705   Air is compressed in a diesel engine from an initial pressure of 13 psia and a temperature of 120°F to one-twelfth of its initial volume. Calculate the final temperature and pressure assuming the compression to be adiabatic.

1.706 An automobile tire is inflated to 32 psig pressure at 50°F. After being driven, the temperature rises to 75°F. Assuming that the volume remains constant, the final gage pressure is: (*a*) 33.0, (*b*) 34.4, (*c*) 37.3, or (*d*) 39.2 psig?

1.707 If 100 cu ft of atmospheric air (pressure 14.7 psi) at zero Fahrenheit temperature are compressed to a volume of 1 cu ft at temperature 200°F, what will be the pressure of the compressed air in pounds per square inch?

1.708 *a.* Distinguish between higher heating value and lower heating value of fuel.

b. Why does the lower heating value at constant volume differ from that at constant pressure?

1.709 Air at 90°F dry bulb and 65°F wet bulb travels across the surface of a pond of water which is at 70°F. Will any of the 70°F water be evaporated into the air stream or will any of the water vapor in the air be condensed? Explain fully the physical laws upon which your answer is based.

1.710 *a.* If you wished the temperature distribution within a room to be as even as possible, would you blow hot air into the room near the floor or the ceilings? Reasons.

b. To heat the room most effectively, should the steam radiators be painted black or with aluminum paint? Reasons.

1.711 A steam turbine carrying a full load of 50,000 kw uses 569,000 lb steam per hour. The engine efficiency is 75 per cent and its exhaust steam is at 1 in. Hg abs and it has an enthalpy of 950 Btu per lb. What is the temperature and pressure of the steam at the throttle?

1.712 The use of electricity for melting snow in a driveway 10 ft wide by 50 ft long is being considered. At 2 cents a kilowatt-hour, what would be the cost of melting 6 in. of snow? (Assume the following data: weight of snow, 10 lb per cu ft; temperature, 32°; efficiency of operation, 50 per cent.)

1.713 A steam boiler on test generates 885,000 lb of steam in a 4-hr period. The average steam pressure is 400 psia, the average steam temperature is 700°F, and the average temperature of the feedwater supplied to the boiler is 280°F. If the boiler efficiency for the period is 82.5 per cent, and if the coal has a heating value of 13,850 Btu per lb as fired, find the average amount of coal burned in short tons per hour.

1.714 A hot-water heater consists of a 20-ft length of copper pipe of ½-in. average diameter and $\frac{1}{16}$-in. thickness. The outer surface of the pipe is maintained at 212°. What is the capacity of the coil in gallons of water per minute, if water is fed into the coil at 40° and is expected to emerge heated to 150°? [Assume the conductivity of copper to be 2100 Btu/(sq ft)(deg)(hr)(in.) thickness.]

1.715 If coal having a heat of combustion of 14,000 Btu per lb is used in a heating plant of 50 per cent efficiency, how many pounds of steam of 50 per cent quality and 212°F temperature can be made per pound of this coal from water whose initial temperature is 70°F?

1.716 A single-cylinder, double-acting, reciprocating steam engine has a 6-in. bore, and 8-in. stroke, and a piston-rod diameter of $1\frac{1}{4}$ in. The average mean effective pressure found from the indicator cards is 62 psi for each end of the cylinder. The engine operates at 300 rpm and with a mechanical efficiency of 83 per cent. If the engine is directly coupled to a generator having an efficiency of 92 per cent, find the generator output in kilowatts.

1.717 Steam is admitted to the cylinder of an engine in such a manner that the average pressure is 120 psi. The diameter of the piston is 10 in. and the length of stroke is 12 in. How much work can be done during one revolution, assuming that steam is admitted to each side of the piston in succession? What is the horsepower of the engine when it is making 300 rpm?

1.718 A volume of 400 cc of air is measured at a pressure of 740 mm Hg abs and a temperature of 18°C. The volume at 760 mm abs and 0°C is: (a) 352, (b) 358, (c) 366, or (d) 369 cc?

1.719 What is the temperature of 2 liters of water at 30°C after 500 cal of heat have been added to it?

1.720 A heat engine (Carnot cycle) has its intake and exhaust temperatures 157°C and 100°C, respectively. Its efficiency is: (a) 12.75, (b) 13.25, (c) 14.38, or (d) 16.33 per cent?

1.721 A 300-watt water heater is attached to a water faucet. If the water runs at a rate that permits it to be heated from 60 to 120°F, how long will it take to obtain a gallon and what will be the cost at 5 cents a kilowatt-hour? Assume that 75 per cent of the electrical energy will be utilized in heating the water.

1.722 By means of insulation, the loss in heat through a roof per square foot is reduced from 0.40 to 0.18 Btu per hr for each degree of difference between inside and outside temperatures. The area of the roof is 10,000 sq ft and the average difference between inside and outside temperatures is 35° during the heating season of 5,000 hr. If the heating value of coal is 13,000 Btu per lb and the efficiency of the heating plant is 60 per cent, find the value of the coal saved per season at $15 per ton.

1.723 The quality of steam that gives up 475 Btu per lb while condensing to water at a constant pressure of 20 psig is: (a) 50.7, (b) 52.3, (c) 53.6, or (d) 56.2 per cent?

1.724 A single-cylinder, double-acting, reciprocating steam engine has a 12-in. diameter piston with an 18-in. stroke. The piston-rod diameter is 2 in. Indicator cards show a mean effective pressure of 70 psi for both the head end and the crank end. The engine operates at 350 rpm with an efficiency of 92 per cent. Determine the horsepower output.

1.725 A 300-hp engine is given a brake test. The brakes are water cooled. At what rate must water at 80°F flow through the brakes if the water must not rise above 180°F?

GENERAL, MULTIPLE CHOICE

In each of the following questions, 1.801 to 1.825, select one, and only one, correct answer.

1.801 The phase-shifting network shown in Fig. 1.801 is used as a portion of a control circuit.

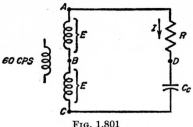

FIG. 1.801

1. Is the value of R which will yield a voltage E_{AD} that leads the voltage E_{CA} by 45° (1a) 2,630, (1b) 2,640, (1c) 2,650, or (1d) 2,660 ohms?

2. Is the magnitude of $|E_{AD}|$ as compared with $|E|$ (2a) 1.414, (2b) 1.500, (2c) 1.732, or (2d) 2.00?

1.802 A series R-L-C circuit resonates at a frequency of 10^6 cps. Its band width between half-power points is 0.1 times the resonant frequency. If it draws 10 watts from a 100-volt rms source at resonant frequency:

1. Is the value of R (1a) 1,000, (1b) 1,100, (1c) 1,200, or (1d) 1,250 ohms?

2. Is the value of L (2a) 1.29, (2b) 1.59, (2c) 1.95, or (2d) 2.59 $\times 10^{-3}$ henrys?

3. Is the value of C 15.9 $\times$ (3a) 10^{-3}, (3b) 10^{-6}, (3c) 10^{-9}, or (3d) 10^{-12} farad?

1.803 Two transformers receive power from a three-phase three-wire 2,200-volt line. The load on the first transformer consists of induction motors which draw 80 kw at a power factor of 0.8. The second transformer supplies a balanced lighting load of 50 kw at 120 volts and unity pf. Neglecting exciting currents of the transformers and their losses:

1. Is the total load (1a) $80\underline{/36.8°}$, (1b) $100\underline{/36.8°}$, (1c) $130\underline{/27.5°}$, or (1d) $143\underline{/24.7°}$ kva?

2. Is the power factor (2a) 0.866, (2b) 0.886, (2c) 0.900, or (2d) 0.910?

3. Is the primary current per phase (3a) 35.0, (3b) 36.7, (3c) 37.6, or (3d) 38.5 amp?

1.804 Figure 1.804 shows the equivalent circuit of a series-connected coil with resistance and capacitor with leakage conductance. The values are as follows: $L = 0.001$ henry, $Q_{coil} = 2$.

The product $LC = 10^{-12}$ sec² and the time constant of the
capacitor alone is 2×10^{-6} sec.

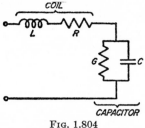

Fig. 1.804

1. Are the values of the circuit parameters R (ohms), C (farad),
G (mho), respectively, (1a) 500, 1×10^{-9}, 5×10^{-4}; (1b) 600,
1×10^{-9}, 5×10^{-4}; (1c) 700, 1×10^{-10}, 5×10^{-3}; or (1d) 500,
1×10^{-10}, 5×10^{-3}?

2. Is the series-resonant frequency (2a) 0.856, (2b) 0.860,
(2c) 0.865, or (2d) 0.885 × 10⁶ radians per sec?

3. Is the effective resistance of the circuit at resonance (3a)
433, (3b) 500, (3c) 865, or (3d) 933 ohms?

1.805 A three-wire 60-cps three-phase power-transmission line
has a balanced Y-connected load taking 1,000 kw at 56.4 amp,
0.85 pf lagging. The impedance of each conductor of the line
is $1.0 + j10$ ohms.

1. Is the voltage required at the sending end of the line (1a)
8,800, (1b) 11,000, (1c) 12,700, or (1d) 13,200 volts line to line?

2. Is the regulation of the line (2a) 5.0 per cent, (2b) 5.2 per
cent, (2c) 5.5 per cent, or (2d) 5.8 per cent?

1.806 The two intersecting streets shown in Fig. 1.806 are to
be connected by a simple curve. The center line of the curve is

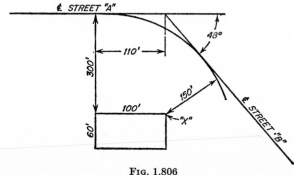

Fig. 1.806

to be so located that the distance from the curve to the corner x of the building is 150 ft. Is the radius (a) 1,040, (b) 1,125, (c) 1,210, or (d) 1,297 ft?

1.807   Is the deflection angle "alpha" to the intersection of the center line of a road (N85°E tangent) and of a property line bearing N9°E through coordinates 280N and 990E (Fig. 1.807) (a) 20°00', (b) 20°29', (c) 20°47', or (d) 21°00'?

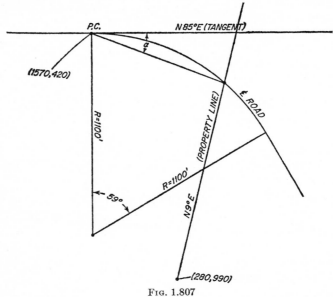

FIG. 1.807

1.808   If a string weighs 0.005 lb per ft and is under a tension of 25 lb, the velocity of a transverse pulse will be (a) 157, (b) 282, (c) 365, or (d) 400 fps?

1.809   If air is of mean molecular weight, γ is taken as 1.40, R as 8.3×10^7 ergs per mole C°, and T at 27°C, is the velocity of longitudinal waves in air (a) 1,100, (b) 1,130, (c) 1,150, or (d) 1,160 fps?

1.810   A railroad train is traveling at 60 mph.

1 and 2. If the frequency of the note emitted by the locomotive whistle is 500 cps, will the wavelength of the sound waves in front of the locomotive be (1a) 2.00, (1b) 2.05, (1c) 2.09, or (1d)

2.12 ft? Will the wavelength in back of the locomotive be (2a) 2.05, (2b) 2.12, (2c) 2.26, or (2d) 2.44 ft?

3 and 4. If a listener is standing at a crossing will the frequency of the sound from the approaching train be (3a) 5.38, (3b) 5.42, (3c) 5.46, or (3d) 5.50 cps? Will the frequency of the sound from the receding train be (4a) 4.60, (4b) 4.62, (4c) 4.64, or (4d) 4.66 cps?

5 and 6. If a listener is traveling on a train with a speed of 30 mph toward the approaching first train, will the frequency of the sound from the approaching train be (5a) 5.20, (5b) 5.28, (5c) 5.42, or (5d) 5.62 cps? Will the frequency of the sound from the receding train be (6a) 4.46, (6b) 4.64, (6c) 4.80, or (6d) 4.82 cps?

1.811 1. Is the ratio of the intensity of two sound waves, whose intensity levels differ by 10 db, (1a) 10, (1b) 20, (1c) 30, or (1d) 40?

2. If the difference of the intensity levels is doubled, is the ratio of the intensity of the two sound waves (2a) 20, (2b) 40, (2c) 70, or (2d) 100?

3. Is the difference between the intensity levels of two sound waves whose ratio of intensity is 4 to 1, (3a) 3, (3b) 6, (3c) 9, or (3d) 12 db?

4. Is the difference between the pressure levels of two sound waves whose ratio of pressure amplitudes is 6 to 1, (4a) 11.80, (4b) 14.81, (4c) 15.56, or (4d) 16.26 db?

1.812 A 5-ft diameter welded steel penstock pipe 1,000 ft long carries 300 cu ft of water per sec from a reservoir to a turbine. The difference of elevation between the reservoir surface and the tailrace level of the turbine draft tube is 125 ft.

1. If the turbine develops 3,500 bhp, will the over-all efficiency for the plant be (1a) 79.0, (1b) 80.6, (1c) 82.2, or (1d) 83.3 per cent?

2. Will the efficiency of the turbine be (2a) 88.0, (2b) 90.0, (2c) 91.0, or (2d) 91.5 per cent?

1.813 A 12-in. diameter steel pipeline traversing approximately level country carries oil having a viscosity of 400 Saybolt sec and an API gravity of 40 deg. Pumps develop a head of 1,000 ft and their inlet pressure must not fall below 6.5 psi. If the pumping rate is 2,000 gal per min, will the distance between pumping stations be (a) 43,700, (b) 50,500, (c) 57,900, or (d) 61,200 ft?

1.814 Air flows through a smooth steel pipe, 4 in. in diameter, at a temperature of 100°F. At two points 1,000 ft apart the gage pressures are 100 and 60 psi. Is the weight of air flowing per second (a) 1.82, (b) 2.94, (c) 4.06, or (d) 5.36 lb?

1.815 Capstans are extensively used on ships to create a pull on a line, and they are essentially a power-driven cylinder about which n turns of a line are wrapped, with the last wrapping being held taut by a man pulling on it. If the coefficient of friction between the capstan and the rope may be taken as 0.12, the pull to be created is 4,000 lb, and the pull exerted by a man is 50 lb, will the required number of turns of the rope be (a) 6, (b) 8, (c) 9, or (d) 9.5?

1.816 An axle for a given machine is to transmit 80 hp at 300 rpm and at the same time support a load of 1,000 lb canti-levered 30 in. beyond the support. If axial forces are neglected, and the maximum shear stress due to torque and moment loading is not to exceed 10,000 psi and the maximum tension or compression due to torque and moment loading is not to exceed 20,000 psi, will the diameter of the shaft be (a) 2.250, (b) 2.375, (c) 2.500, or (d) 2.625 in.?

1.817 A W36×280 steel beam is cut as shown in Fig. 1.817a and welded to form the beam shown in Fig. 1.817b. Based only on the section modulus, is the new welded beam (a) 1.41, (b) 1.52,

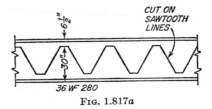

Fɪɢ. 1.817a

(c) 1.63, or (d) 1.74 times as strong as the original beam? The properties of the original beam are: area = 82.32 sq in.; depth out to out of flanges = 36.50 in.; flange width = 16.595 in.; flange thickness = 1.570 in.; web thickness = 0.885 in.; moment of inertia = 18,819 in.⁴; and section modulus = 1,031 in³.

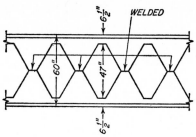

1.818 The hoisting cable for a vertical mine shaft is 600 ft long from the drum in the head frame to the load at the bottom of the shaft.

1. If the cable has a cross section of 1 sq in. and weighs 3.4 lb per ft, will the cable have a modulus of elasticity of (1a) 27, (1b) 28, (1c) 30.2, or (1d) 33.5 × 10⁶ psi if it stretches 0.222 ft in picking up a 5-ton load?

2. Will the stretch due to its own weight be (2a) 0.0113, (2b) 0.0226, (2c) 0.0452, or (2d) 0.0904 ft?

1.819 An endless steel ring 4 in. wide and ¼ in. thick is 3 ft in mean diameter. Its inside diameter is $\frac{1}{32}$ in. smaller, when unheated, than the diameter of the iron casting it is to encircle.

1. If the ring is to be heated above the shop temperature so that it will clear the casting by $\frac{1}{32}$ in., will the rise in temperature be (1a) 250, (1b) 300, (1c) 350, or (1d) 400°F?

2. Will the tension in the ring, when it cools, be (2a) 12,000, (2b) 20,000, (2c) 26,000, or (2d) 30,000 lb?

3. Will the pressure of the band against the casting be (3a) 361, (3b) 722, (3c) 903, or (3d) 994 psi?

1.820 A cable consists of an equal number of steel and copper wire strands, with the cross section of the steel and the copper each being equal to 1 sq in. The cable suspends a total load of 30,000 lb and all wires were taut, even, and unstressed (except for their own weight at the time of loading).

1. Do the steel strands carry (1a) 17,720, (1b) 18,270, (1c) 19,170, or (1d) 20,970 lb of load?

2. If the temperature rises 60°F after loading, do the copper strands carry (2a) 8,130, (2b) 9,030, (2c) 9,930, or (2d) 10,830 lb of load?

1.821 Four pounds of water at 200°F is mixed with 2 lb of water at 60°F at 14.7 psia. Is the increase of entropy of the total mass of water due to the mixing process (a) 0.041, (b) 0.062, (c) 0.093, or (d) 0.104 units?

1.822 An air turbine operates between a pressure of 60 psia and 15 psia, and receives 1 lb per sec of air at a temperature of 1200°F.

1. For an ideal turbine, is the developed horsepower (1a) 138.8, (1b) 188.5, (1c) 318.3, or (1d) 381.5?

2. When operating under the conditions above the turbine develops 150 hp and has a discharge temperature of 300°F. The turbine blades are water-cooled with water entering at 50°F and leaving at 100°F. With the constants for air assumed as c_p = 0.240 Btu per °F per lb, R = 53.3 ft per °R, and K = 1.4, is the rate of water flow in gallons per minute under these conditions (2a) 14.85, (2b) 15.40, (2c) 15.85, or (2d) 16.20?

1.823 Seven pounds of steam at atmospheric pressure, super-heated to 242°F, is introduced simultaneously with 8 lb of ice at 25°F into a copper calorimeter which weighs 5 lb and which contains 50 lb of water at 60°F. The heats of fusion and of vaporization for water are 144 and 970 Btu per lb, respectively. The thermal capacities in Btu per pound per °F may be taken as follows: steam 0.48; ice 0.50; and copper 0.093. Neglecting heat losses to all bodies other than the calorimeter itself, is the resulting temperature of the mixture (a) 135°F, (b) 148°F, (c) 157°F, or (d) 160°F?

1.824 By means of insulation, the loss in heat through a roof per square foot is reduced from 0.40 to 0.18 Btu per hr for each degree difference between inside and outside temperatures. The area of the roof is 10,000 sq ft and the average difference between inside and outside temperature is 35°F during the heating season of 5,000 hr. If the heating value of coal is 13,000 Btu per lb and the efficiency of the heating plant is 60 per cent, is the value of the coal saved per season at $15 per ton (a) $360, (b) $370, (c) $395, or (d) $410?

1.825 Ammonia enters the cooler of an ammonia refrigerating machine at 0°F and leaves it at 15°F. When operating at 10 tons ice-melting capacity and 15 input hp at the compressor:

1. For a temperature rise from 50°F at entrance to 70°F at exit in the condenser, will the condensing water supplied be (*1a*) 15.8, (*1b*) 16.1, (*1c*) 16.5, or (*1d*) 16.9 gal per min?

2. Assuming the liquid ammonia to leave the condenser at 70°F, will the ideal coefficient of performance be (*2a*) 6.57, (*2b*) 9.20, (*2c*) 25.5, or (*2d*) 30.6?

3. Will the actual coefficient of performance be (*3a*) 0.571, (*3b*) 1.57, (*3c*) 2.30, or (*3d*) 3.14?

CHEMICAL ENGINEERING

2.01 A boiler is fired with a coal containing 75 per cent carbon and 8 per cent ash, burned under such conditions that the elimination of combustible matter from the refuse is complete. The air enters the furnace at 90°F with a relative humidity of 80 per cent. The vapor pressure of water at 90°F is 36 mm Hg. The flue gases go to the stack at 680°F. The average flue-gas analysis shows 12.6 per cent CO_2, 6.2 per cent O_2, and 1 per cent CO.

Calculate:

a. The per cent of excess air
b. The complete analysis of the fuel
c. The cubic feet of stack gas per pound of coal
d. The cubic feet of air per pound of coal

2.02 A mixture of ammonia and air at a pressure of 745 mm Hg and a temperature of 40°C contains 4.9 per cent NH_3 by volume. The gas is passed at a rate of 100 cfm through an absorption tower in which only ammonia is removed. The gases leave the tower at a pressure of 740 mm Hg, a temperature of 20°C, and contain 0.13 per cent NH_3 by volume.

Using the simple gas law, calculate:

a. The rate of flow of gas leaving the tower in cubic feet per minute.
b. The weight of ammonia absorbed in the tower per minute.

2.03 A fractionating column is operating at 1 atm to produce a product of ethanol and water which leaves the top plate at

78.41°C. The bottom product is to contain 1 mole per cent ethanol. The feed contains 17 mole per cent ethanol and is introduced at its boiling point. The feed rate is 44 moles per hr. Reflux is returned to the top plate at a rate of 31 moles per hr and at the top-plate temperature.

a. If plate efficiency is 60 per cent, calculate the number of actual plates.

b. Calculate the rates of heat transfer in reboiler and condenser.

c. Find the feed-plate location.

2.04 An absorption tower packed with ½-in. Berl saddles is to be used for scrubbing a very small amount of ammonia out of air at 70°F and about 1 atm pressure, with water at 70°F as the scrubbing liquid. The tower is 12 in. in internal diameter and has a packed height of 7 ft. The water rate is 5 gal per min. Experiment has shown that, below the loading point, the resistance offered by the wetted packing at this water rate is 1.5 times that of dry packing of similar size and shape and at the same rate of flow of gas. The tower is to be operated with a gas velocity of 1.5 ft per sec, based on the empty tower; this is below the loading point.

What would be the expected drop in pressure through the tower in inches of water at 70°F?

2.05 It is desired to strip ammonia from a water solution which is 20 per cent by weight of ammonia. Eight thousand cu ft per hr of air-ammonia mixture which is 10 per cent ammonia by volume (dry basis) is to be produced. The entering air contains no ammonia and the exhausted solution contains 2 per cent ammonia by weight. The gas and liquor rates are such that $K_g a$ is 7 lb-moles/(hr) (cu ft) (atm). What volume of packing is required if the operation is isothermal at 20°C? If $K_g a$ varies as $G^{0.8}$, what relative height of packing would be required if the gas rate and the liquor rate were decreased by 25 per cent?

2.06 A multipass cooler for a continuous ethyl alcohol still consists of a copper tube $1\frac{3}{16}$ in. inside diameter with walls $\frac{1}{16}$ in. thick and 24 ft long, surrounded by a standard 2-in. steel pipe. The hot alcohol flows through the inner tube, and the cooling water flows through the outer pipe counter-current to the alcohol. The cooler consists of a number of these double pipes in series. If the alcohol is to be cooled from 172°F to 70°F, using cooling water entering at 50°F and leaving at 80°F, how many pipes in series will be needed to handle 200 gal alcohol per hour?

2.07 A 50-mole per cent solution of methanol in water is to be rectified to give a distillate containing 95 mole per cent methanol and a residue containing 95 mole per cent water. The feed enters the column at 20°C. A reflux ratio 40 per cent greater than the minimum is to be used.

    a. What is the minimum reflux ratio?

    b. How many perfect plates are needed for the desired separation under the above conditions?

    c. If the feed is 2,000 lb per hr, calculate the pounds of product and residue per hour.

    d. How much steam, at 5 psig, is required per hour?

2.08 The heats of combustion $(-\Delta H)$ of crystalline succinic acid and succinic anhydride are 356.2 and 369.4 kg-cal per gram-mole respectively, and the entropy of the acid is 42.0 entropy units at 25°C. The entropy of the anhydride is estimated to be 35.0 units at 25°C. Calculate the partial pressure of water in the equilibrium system

$$C_4H_6O_4 \rightarrow C_4H_4O_3 + H_2O$$
$$\text{solid} \qquad\quad \text{solid} \qquad \text{gas}$$

at 500 and 700°K. On the basis of these calculations, does the straight thermal dehydration of the acid to the anhydride appear to be commercially feasible or would it be necessary to use powerful desiccating agents? Assume the molal heat capacity of the acid is equal to that of the anhydride.

2.09 An experimental run made in a pilot-plant flow reactor, having an internal volume of 5 cu ft, converted 50 per cent of material A according to the homogeneous gas reaction $2A \rightarrow R$. The conditions were atmospheric pressure, 500°C, and a feed rate of 1 lb-mole per hr, gas at standard conditions.

For design purposes it is desired to find the volume of reactor required to treat 10,000 cu ft per hr of feed gases at 5 atm and 300°C with 25 per cent conversion.

Assume the gases behave as perfect gases, isothermal conditions for the reactor, and an activation energy of 30,000 cal per gram-mole for the homogeneous reaction.

2.10 A storage tank at 100°F is part full of a liquid mixture which contains 50 mole per cent *n*-butane and 50 mole per cent *n*-pentane. The vapor space contains only butane and pentane vapors. What is the pressure in the vapor space, and what is the composition of the vapors? At 100°F, pure *n*-butane has a vapor pressure of 3.0 atm; at 100°F pure *n*-pentane has a vapor pressure of 0.9 atm.

2.11 What effect does chemical equilibrium between SO_2 and SO_3 have on operating conditions used and yields obtained in sulfuric acid manufacture?

2.12 The autoclave process for hydrolyzing monochlorbenzene to phenol at 1,500 psia and 350°C in the presence of NaOH and CuCl is apparently a unimolecular reaction with a velocity constant of 0.0098 min^{-1}. What reaction time is required to obtain a conversion of 99 per cent using a batch of 1,000 lb monochlorbenzene?

2.13 Give examples of the following classes of organic chemicals which are in large-scale production and use: (1) hydrocarbon, (2) alcohol, (3) ketone, (4) ester, and (5) metallo-organic compounds. For each example, indicate (*a*) commercial method of production, (*b*) raw materials, and (*c*) use.

2.14 Outline the safety rules and practices for the protection of life, health, and property that should be observed in a plant producing toxic and highly flammable chemicals.

2.15 *a.* Describe how glycerin is usually obtained from fats.

b. Account for the presence of sodium chloride in the raw glycerin.

2.16 What weight of steel containing 0.020 per cent manganese should be taken so that the final solution will contain 10 ppm manganese?

2.17 What are the products of the following reactions?

a. Oxidation of formaldehyde   *d.* Saponification of a fat
b. Hydrolysis of an ether   *e.* Inversion of sucrose
c. Hydrolysis of the CN group

2.18 A plate-and-frame filter press with frames 1 in. thick is filtering a noncompressible slurry under constant pressure. The cake is washed with three successive portions of wash water of about the same viscosity as the filtrate; the pressure drop through the press during washing is the same as that during filtration. The time required for filtration only, to form a cake that fills the frame completely, is 2 hr. The time required for washing is 30 min. The time required for dumping, cleaning, and reassembling the press is 30 min. The resistance offered by the press and the cloth may be considered negligible in comparison with the resistance offered by the cake. By what percentage and in which direction would the substitution of frames 1½ in. thick change the average output from the press, in gallons of filtrate per hour of total operating time? Assume that the composition of the sludge, the filtration pressure, and the cleaning time are the same in both cases, and that the volume of wash water used is proportional to the volume of filtrate.

2.19 A mixture of the following composition is to be distilled in a continuous still:

Benzene	60 per cent
Toluene	25 per cent
Xylene	15 per cent

The overhead product is to be substantially pure benzene. The bottoms product is to be substantially pure xylene. From some plate in the column a side stream is to be taken off to a smaller stripper still that delivers substantially pure toluene as a bottoms product.

a. Draw a neat diagram showing, at least qualitatively, the concentrations of the several components on the plate of the main still as ordinates, and as abscissas the plate numbers in the main still. Indicate the position of the feed plate and the plate from which the side stream is taken.

b. Draw a neat diagram showing the main still with feed line, side-stream drawoff, and overhead and waste lines; and the stripper column with feed line, overhead lines, and drawoff line. Where would the overhead from the stripper still be sent?

2.20 Discuss the use and treatment of water in industrial chemical operations including particularly the compounds causing hardness, advantages and disadvantages of various methods of softening (including equations) and treating, and the reasons for these equations.

2.21 Consider the following startup problem: A vessel 10 ft high and 4 ft in diameter is to be filled with liquid to a level of 6 ft. It is initially empty. The filling is to be accomplished automatically, by using a control loop as shown in Fig. 2.21.

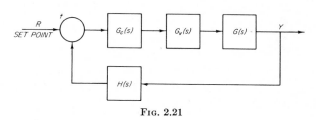

FIG. 2.21

The vessel transfer function: $G(s) = \dfrac{1}{4\pi s}$ ft/(cu ft)(min)

Linear valve: $G_v(s) = 3$ (cu ft)(min)/psi
Proportional controller: $G_c(s) = 6$ psi/psi
Measuring element: $H(s) = 0.5$ psi/ft

   a. Write the equation that expresses the liquid level Y as a function of the controller set point (the reference input) R.

   b. It is suggested that a simple way to fill the vessel to the desired level is to turn up the controller set point in a stepwise manner and allow the system to respond. If the size of this step is γ_0 psi, find the time domain response $y(t)$. What value of γ_0 would you suggest?

2.22 Two different strengths of *n*-butanol and water are accumulated continuously in intermediate tankage, and then processed for 10 days in a butanol recovery plant generating *n*-butanol having a purity of 99.3 wt per cent. The intermediate tankage facilities contain two tanks, one containing a rich butanol solution (77.9 wt per cent *n*-butanol) and the other a water-rich solution (92.3 wt per cent water).

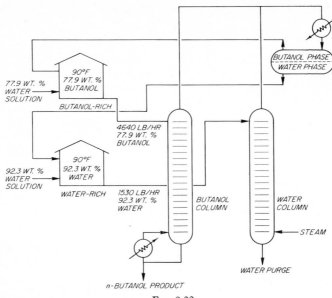

Fig. 2.22

The proposed process configuration to recover the butanol is shown in Fig. 2.22 giving the flow arrangement. It should be noted that since the butanol solutions to be processed are of radically different strengths, there are two feed streams to the columns of the recovery system, rather than a single feed to the decanter. The water purge to the dirty water sewer has been set at 500 ppm by weight (exclusive of further dilution by stripping steam).

For the proposed process configuation operating at essentially a constant pressure of 760 mm Hg, what are:

    a. The recycle stream rates and compositions from the decanter

    b. The condenser and reboiler heat loads

    c. The water column stripping steam rate predicted on a butanol-rich-solution feed rate of 4,640 lb per hr and a water-rich-solution feed rate of 1,530 lb per hr exclusive of the recycle requirements?

2.23 What are the raw materials used in the Fischer-Tropsch process for motor-fuel manufacture? What main reaction is involved? What by-products result?

2.24 Name or describe the specific types of equipment you would select to do the following jobs:

   a. Concentrate a solution of material whose solubility decreases with increase in temperature.

   b. Crystallize a material whose solubility increases greatly with temperature.

   c. Crystallize a material whose solubility is independent of temperature.

   d. Pump a liquid near its boiling point.

   e. Dry a heat-sensitive solid product.

   f. Transfer heat to and/or cool a highly viscous fluid.

   g. Distill a liquid subject to heat polymerization.

   h. Dry a material that "case hardens" easily.

2.25 The sulfur-burning process in a sulfuric acid plant comprises the following operations:

   a. Sulfur is conveyed from storage pile to melting tank.

   b. Sulfur is melted by means of steam coils.

   c. Sulfur is burned in a spray-type burner with dry air.

Indicate the equipment needed for adequate automatic control of the process.

2.26 Fully discuss the following problems relating to phosphate utilization:

   a. The use of CO_2 as an oxidizing agent to oxidize $P_{4(gas)}$ to the oxide for preparation of phosphoric acid.

   b. The use of steam and silica for defluorinating phosphate for animal nutritional purposes.

   c. The use of $Na_3PO_4 \cdot 12H_2O$ in boiler waters.

2.27 A paper plant requires $CaCO_3$ essentially free of all soluble impurities. Powdered $CaCO_3$ is available containing 1 per cent NaOH. A two-step, continuous-countercurrent washing system with 24,000 gal per day of pure water is to be used for removing the NaOH when feeding 10 tons per day of powdered $CaCO_3$ with the slurry discharge containing 0.091 tons of water per ton of $CaCO_3$ is removed in underflow. What is the caustic content of the washed and dried $CaCO_3$?

2.28 A shell and tube heat exchanger has four tube passes and one shell pass. Each tube pass has 13 copper tubes, 1 in. OD,

16 BWG, each tube being 6 ft long. The exchanger is to be used to heat 60,000 lb per hr of an aqueous solution entering at 60°F. Saturated steam at psig will be used as the heating medium; the steam condenses outside the tubes. Estimate the temperature of the solution leaving the exchanger.

Assume negligible heat losses and no scale deposits. Properties of the solution are the same as for water. The steam side coefficient of heat transfer is 2000 Btu/(hr)(sq ft)(°F)

2.29   Water at 68°F is supplied from a mountain lake to a pipe line consisting of 4-in. schedule 80 steel pipe. The pipe line consists of 1,000 equivalent feet of pipe from the lake surface to a point B, which is 400 ft lower in elevation. At this point the pipe branches into two lines both 3,000 ft long (equivalent length). One of these branch lines discharges into an open water tower at a point which is 500 ft below the lake surface. The other line feeds into the nozzles of a turbine in a powerhouse at a point 700 ft below the lake surface. The pressure at the inlet of the turbine nozzles must be maintained at 80.6 psig. All expansion and contraction losses and all changes in kinetic energy may be neglected. In the single line of 1,000 ft feeding the two branches, will the total flow of water be: (a) 625, (b) 635, (c) 645, or (d) 655 gal per min?

2.30   It is required to heat 64,000 lb per hr of air from 60 to 160°F at atmospheric pressure. It is decided to use a heater made of ¾ in. ASA Schedule 40 steel pipes with saturated steam at 235°F condensing inside the vertical pipes which are 10 ft long. The air will be blown across these tubes arranged on an equilateral triangular pitch of 3 in. The side walls of the casing will be spaced so that there will be a clearance of 1½ in. from the nearest tube (no clearance at the top and bottom).

Power delivered to the air-blower system will cost 2 cents per kwhr. Fixed charges will be .009 cents hr per sq ft of outside surface. A safety factor of 1.25 will be used. The specific heat and viscosity of the air may be considered constant at 0.25 and 0.0455 lb/(hr)(ft), respectively. The average between the mean air density and the density at the blower outlet may be taken as 0.072 lb per cu ft.

1. Would the optimum number of tubes per row be: (1a) 12, (1b) 10, (1c) 8, or (1d) 6?

2. Would the optimum number of rows of tubes be: (2a) 48, (2b) 54, (2c) 60, or (2d) 66?

2.31 A dilute mixture of ethyl acetate in air is the product of a certain drying operation. Outline possible methods of recovery of the ethyl acetate for reuse as a solvent.

2.32 In a hydroforming process, toluene, benzene, and other aromatic materials are produced from naptha feed. After the toluene is separated from other components it is condensed and cooled as shown in the flow sheet of Fig. 2.32. 48,000 lb per day of

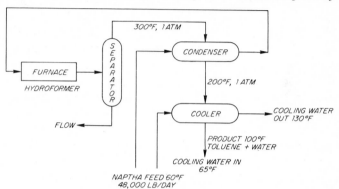

Fig. 2.32   Flow sheet.

naptha feed is fed into the system. For every 100 lb of feed, 27.5 lb of a toluene and water mixture (9.1 per cent by weight water) are produced as overhead vapor and condensed in the condenser using the feed stream as a cooling medium.

1. Will the temperature of the feed stream after it leaves the condenser be: (1a) 170, (1b) 180, (1c) 190, or (1d) 200°F?

2. Will the number of pounds of cooling water required per hour in the cooler be: (2a) 340, (2b) 360, (2c) 375, or (2d) 385 lb per hr?

Additional data

Component	C_p, Btu/ (lb)(°F)	N.B.P., °F	Latent heat of vaporization, Btu per lb
Water (liquid)	1.0	212	970
Steam	0.5		
Toluene (liquid)	0.4	260	100
Toluene (vapor)	0.3		
Naptha feed liquid	0.5		

2.33 A tower 4 ft in diameter is packed to a depth of 30 ft with 1-in. Berl saddles. The bed porosity (volume fraction of voids) is 0.69. For this packing the ratio of particle surface to particle volume is 76 ft^{-1}. Air at 14.7 psia and 70°F is to be forced through the bed at a rate of 1,000 cfm. Will the pressure drop through the bed expressed in inches of water be: (a) 5.21, (b) 5.31, (c) 5.46, or (d) 5.66 in. H_2O?

2.34 An absorption tower containing wooden grids is to be used for absorbing SO_2 in a sodium sulfite solution. A mixture of air and SO_2 will enter the tower at a rate of 70,000 cfm at a temperature of 250°F and a pressure of 1.1 atm. The concentration of SO_2 in the entering gas is specified, and a given fraction of the entering SO_2 must be removed in the absorption tower. The molecular weight of the entering gas mixture may be assumed to be 29.1. Under the specified design conditions, the number of transfer units necessary varies with the superficial gas velocity as follows:

Number of transfer units = $0.32G_s^{0.18}$ where G_s is the entering gas velocity, lb/(hr)(sq ft), based on the cross-sectional area of the empty tower. The height of a tranfer unit is constant at 15 ft. The cost for the installed tower is $1 per cu ft of inside volume, and annual fixed charges amount to 20 per cent of the initial cost. Variable operating charges for the absorbent, blower power, and pumping power are represented by the following equation:

Total variable operating costs (dollars per hour) =
$$1.8G_s^2 \times 10^{-8} + \frac{81}{G_s} + \frac{4.8}{G_s^{0.8}}.$$

The unit is to operate 8,000 hr per year. Under conditions of minimum annual cost,

1. Would the diameter of the absorption tower be: (1a) 12.6, (1b) 13.1, (1c) 13.5, or (1d) 13.8 ft?

2. Would the height of the absorption tower be: (2a) 18.5, (2b) 19.0, (2c) 20.0, or (2d) 21.5 ft?

2.35 A gas system has two parallel lines 50 miles long, one 20 in. and the other 22 in. in diameter. Gas at 555 psig and 40°F is fed to the system amounting to 6 MM per hr measured at 30 in. and 60°F. Its specific gravity is 0.65. The viscosity at 500 psi and 40°F is 2.5×10^{-7} lb-sec per sq ft. The deviation from Boyle's law is $+0.00022P$ for the weight of a cubic foot at P psi. The 20-in. line has an efficiency of 92 per cent and the other, 88 per cent. What is the outlet pressure?

2.36 Two similar vertical-tube evaporators have been arranged as a double effect in the evaporation of a caustic soda liquor from 10 to 16 per cent, the first effect being operated as the high pressure one and the second at 28 in. Hg vacuum. The steam to the first effect is at 5 psi and the feed liquor is at 70°F. Operation is continuous. Given an adequate supply of steam (5 psi) and cooling water, and having available some old heat exchangers, an old water-tube condenser of small capacity, pumps, pipes, fittings, etc., how would you adapt the equipment for a 10 per cent increase in production? If you had to adapt the equipment to handle a temporary 50 per cent overload, how would you do it? Substantiate your proposed setup.

2.37 An evaporator (single effect) is removing 500 lb per hr of water from a colloidal suspension which deposits scale on the steam chest. During one 8-hr shift each week it is cleaned, which increases the over-all heat-transfer coefficient from 50 to 225. Assuming that the scale is deposited uniformly throughout the week and none flakes off, calculate:

  a. The mean heat-transfer coefficient for the week.
  b. How often the evaporator should be cleaned for maximum capacity.

2.38 Specify the type of operation and type of slurry for which each of the following filters would be adapted:
  a. Plate and frame   *c.* Sweetland
  b. Rotary continuous   *d.* Moore

2.39 Give a brief discussion of the processes of age hardening and precipitation hardening.

2.40 Describe briefly the main types of instruments used for checking the efficiency of combustion of the flue gases. State what you consider to be their relative advantages or disadvantages.

2.41 Give a discussion of "temperature measurement" indicating the range of application of instruments of different types, the accuracy obtainable, and the precautions necessary in practice to ensure accurate results.

2.42   Give at least four different methods of removing dust particles from a gas and compare their engineering requirements. Given a gas of 50,000-cfm flow with 50 lb per min of a hygroscopic dust, averaging 10 microns size, at a temperature of 95°C and a dew point of 70°C, what method would you recommend and why?

2.43   It is desired to recover ethanol from an air-ethanol mixture, that is 5 per cent by volume of ethanol and 95 per cent by volume of air at 90°F, by compressing the mixture to 100 psig and then cooling the gas at constant pressure to 78.8°F. Estimate the per cent recovery of the ethanol.

2.44   6,000 lb of a material goes through a crusher and grinder per hour in succession (on the same power drive). Screen analysis from the crusher shows a surface area of product of 500 sq ft per lb. Screen analysis of the grinder product indicates a surface area of 4,200 sq ft per lb. The Rittinger number of the material processed is 163 sq in. per ft-lb. The efficiency of the crusher is estimated to be 25 per cent while that of the grinder is 30 per cent. Estimate the total power to be delivered to the equipment.

2.45   Tank storage is to be provided for the following chemicals and solutions: (*a*) concentrated nitric acid, (*b*) 15 per cent sodium chloride solution, (*c*) benzol, (*d*) 50 per cent caustic soda solution, and (*e*) 90 per cent mixed acid. Specify the material of construction for each tank.

2.46   A leaf filter press is operating at constant pressure, building up a 1-in. noncompressible cake in 5 hr and is delivering 3.8 cu ft of filtrate per sq ft of filtering surface. Cleaning requires 30 min. If it is required to wash the cake with a volume of water equal to one-half the volume of the filtrate, what will be the maximum capacity of the press obtainable by varying the cake thickness? The nature of the sludge and the pressure remain unchanged.

2.47   A solid granular product is to be dried in recirculating air. The moist air bled off from the recirculating stream is at 120°F and has a dew point of 110°F. The air admitted to the dryer has a dry-bulb temperature of 100°F, and a wet-bulb temperature of 75°F.
Assuming that there is no loss of heat from the dryer, how

much heat must be supplied for each pound of water removed from the cake? (Neglect heat of wetting of the solid.)

2.48 The thermal system shown below is controlled by a proportional controller. The following data apply:

1. W = flow rate of liquid through the tanks, 400 lb per min.
2. Density of liquid = 50 lb per cu ft.
3. V = holdup volume of each tank, 4 cu ft.
4. Transducer: A change of 1°F causes the controller pen to move 0.25 in.
5. Final control element: A change of 1 psi from the controller changes the heat input q by 400 Btu/min.
6. Heat capacity of liquid = 1 Btu/(lb)(°F).
The temperature of the inlet stream may vary.

a. Draw a block diagram of the control system with appropriate transfer function in each block.

b. From the block diagram, determine the overall transfer function relating the temperature T in the tank 2 to a change in the set point.

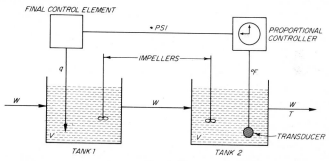

FIG. 2.48

2.49 In the case of absorption-tower design give or explain:

a. The data required to determine the packed volume.

b. The procedure to be followed to determine the packed volume.

c. The way in which the design of a stripping tower differs from that of an absorption tower.

d. The advantages and disadvantages of using the height-of-transfer-unit method instead of the method of operating lines.

e. For the case of a stripping tower followed by a rectifying tower, the economic factors that must be considered and the optimum value that must be determined.

2.50   As an engineer, you are asked to design the evaporating system to concentrate 50,000 lb per day of caustic solutions from 8 to 50 per cent. Briefly outline the procedure to be followed, including the data required, the method of determining the best setup, and materials of construction.

2.51   In a mercury-steam binary cycle, the saturated mercury vapor leaves the boiler at 60 psia ($t = 836.1°F$), and it is sent through the mercury turbine where it leaves at a pressure of 2 in. Hg abs. The mercury turbine is adiabatic and reversible. From the mercury turbine the mercury goes through the mercury condenser–steam boiler where it leaves as a saturated liquid at 2 in. Hg abs.

In the condenser-boiler, saturated steam is generated at 360 psia, and the steam turbine exhausts at 1.513 in. Hg abs. The steam turbine is adiabatic and reversible. Saturated liquid mercury at 2 in. Hg abs. is pumped into the mercury boiler and saturated liquid water at 1.513 in. Hg abs. is pumped into the condenser-boiler. Pump work may be considered negligible.

The properties of mercury are as follows:

At 60 psia:
$$h_f = 30 \text{ Btu/lb} \qquad h_{fg} = 118.6 \text{ Btu/lb} \qquad S_g = 0.1277$$

At 2 in. Hg abs.:
$$h_f = 15.85 \text{ Btu/lb} \qquad h_{fg} = 126.95 \text{ Btu/lb}$$
$$S_f = 0.02323 \qquad S_{fg} = 0.1385$$

All entropy values are in Btu/(lb)(°F).

1. Would the work of the mercury cycle be: (1a) 36.90, (1b) 37.00, (1c) 37.15, or (1d) 37.35 Btu per lb of mercury?

2. Would the work of the steam cycle be: (2a) 382.0, (2b) 383.5, (2c) 385.5, or (2d) 388.0 Btu per lb of steam?

3. Would the pounds of mercury per pound of steam be: (3a) 11.98, (3b) 12.48, (3c) 12.88, or (3d) 13.18?

4. Would the overall thermal efficiency of the binary cycle be: (4a) 46.7, (4b) 48.4, (4c) 50.2, or (4d) 52.1 per cent?

2.52   Product S is formed according to the liquid phase reaction $A \rightarrow S$ with $-r_A = kC_A$, $k = 2$ min^{-1}. A stream containing 5 moles per liter of A is available at cost of $10 per liter. The product S is sold at $2.50 per mole. Operating cost is $10 per day and no recycling of unreacted A can be used. Given a backmix reactor, what conversion of A should we use to have the maximum profit per day?

2.53 *a.* Describe the use of exchange resins in demineralizing water.

b. Compare the operation of two-bed and mixed-bed units.

2.54 Corrosion of steel piping which carries mixed process wastes has been found to be severe. What procedure should be followed to determine the type of material that should be specified for a replacement line?

2.55 *a.* What would one specify for an aluminum alloy showing good mechanical strength and superior corrosion resistance?

b. Describe the relative effectiveness of (1) iron, (2) copper, (3) zinc-coated, and (4) aluminum fasteners for securing aluminum sheet (2S).

c. Define "creep" in reference to metals and alloys and give an illustration of its importance in design considerations.

2.56 A mixture of hydrocarbons contains 20 mole per cent methane, 20 mole per cent ethane, 30 mole per cent propane, 10 mole per cent *n*-butane, 10 mole per cent isobutane, and 10 mole per cent *n*-pentane. If the material is flashed at 100°F and 115 psia, what fraction leaves the separator as a liquid? What is the composition of the liquid and vapor?

2.57 A solution containing 35 per cent by weight of ethanol (and 65 per cent water) is supplied to a fractionating column. The feed is a saturated liquid. The top product contains 80 per cent by weight of ethanol and the bottom product contains 5 per cent by weight of ethanol. The condenser operates as a total condenser and the reflux enters the top plate at 120°F.

a. If 2 moles of liquid at 120°F are returned for each mole of product removed, determine the slope of the operating line to be used in a McCabe-Thiele diagram.

b. Compare the amount of heat that must be supplied in the reboiler per pound of feed when the reflux is returned at 120°F with that required when the reflux is returned as a saturated liquid.

2.58 A reactor, which is 10 ft high and whose internal diameter is 1 ft, is to be used for contacting a spherical catalyst with a gas mixture whose density is 0.500 lb per cu ft and whose viscosity is 0.03 centipoise. The catalyst density is 152 lb per cu ft, its diameter is 0.174 in., and the static-bed porosity is 0.40. What gas velocity will be necessary to give a porosity of 0.6, and how many feet of static bed can be accommodated if a factor of 100 per cent "free space" is allowed for disengagement of solids?

2.59 A single-stage single-acting compressor has an 8-in. bore and a 10-in. stroke and turns at 200 rpm. The compressor takes in dry saturated ammonia vapor at 0°F and compresses the ammonia adiabatically but irreversibly to 140 psia. The actual shaft work of the compressor is 20 per cent more than if the compression was reversible. The volumetric efficiency of the compressor is 88 per cent. The ammonia leaving the compressor enters a condenser where it is cooled and condensed, leaving the condenser as a liquid at 70°F. The potential and kinetic energy changes are negligible. Calculate with the aid of ammonia tables or charts:

a. The pounds per minute of ammonia handled by the compressor

b. The horsepower input to the compressor if the mechanical efficiency is 85 per cent

c. The Btu per minute of heat removed in the condenser

d. The quality of the resulting vapor-liquid mixture if the ammonia leaving the condenser is throttled to a pressure of 30.42 psia

2.60 An atmospheric rotary dryer handles 10 tons of wet crystalline salt per day, reducing the moisture content from 10 to 1 per cent by means of countercurrent flow of hot air entering at 225°F dry bulb and 110°F wet bulb and leaving at 150°F dry bulb. Calculate:

a. The number of tons of dry product per 24 hr

b. The number of pounds of water removed from the salt per hour

c. The humidity of the air entering and leaving the dryer

2.61 What is the power cost for the production of electrolytic copper sheet weighing 8 oz per sq ft if a sheet 36 in. wide and 100 ft long is produced at 2.5 volts for the deposition and the current efficiency of the operation is 85 per cent? Power is 1 cent per kwhr on the transformer, with transformer losses of 3 per cent, and with 90 per cent conversion efficiency at the germanium rectifier.

2.62   How many pounds of chlorine would be produced per 24-hr day from a 20,000-amp Hooker cell operating at 6 volts and 190°F? The anodes are 3 by 3 ft by 1.5 in. and are fabricated from graphite. The current is furnished by a germanium rectifier which is supplied with 1,320-volt three-phase alternating current, through aluminum bus bars. The chlorine is produced at 700 mm of Hg, contains 98 per cent chlorine, 1 per cent oxygen, and is saturated with water vapor. The current efficiency is 92 per cent and the energy efficiency is 46 per cent. The feed to the cell is saturated sodium chloride at 60°F.

2.63   In a batch process, 100 lb of carbon monoxide gas are compressed adiabatically from 80°F and 15 psia to a final temperature of 600°F. The heat capacity in Btu per lb-mole per °F is given as $c_p = 9.46 - \dfrac{3.29 \times 10^3}{T} + \dfrac{1.07 \times 10^6}{T^2}$ where T is in degrees Rankine. If the gas is considered to be ideal, calculate:
   a. The work of compression in Btu
   b. The final pressure if the pressure is reversible

2.64   One thousand pounds of a solution, containing 50 per cent by weight of component A and 50 per cent by weight of component C, is to be extracted with a solvent B in order to remove at least 80 per cent of the component C initially present in the 1,000 lb of solution. The extraction is to be carried out isothermally by using 200 lb of solvent B in each stage of a multistage cocurrent extraction. Determine the number of stages required based upon the following data:

Equation

Point	Weight, per cent A	Weight, per cent B	Weight, per cent C
1	85	2	13
2	65	5	30
3	46	8	46
4	35	10	55
5	20	20	60
6	8	35	57
7	3	47	50
8	2	55	43
9	1	79	20

Tie Line

	Weight per cent	Point ①	Point ②	Point ③
Raffinate	A	59	74.1	88.4
	B	6	3.9	1.6
	C	35	22.0	10.0
Extract	A	5	1.8	1.3
	B	40	58.0	73.2
	C	55	40.2	25.5

2.65 Define, discuss, or describe: (*a*) two-position control; (*b*) proportional control; and (*c*) derivative or preact control.

CIVIL ENGINEERING

3.01 *a.* When fresh concrete is exposed to rapid-drying conditions, what precautions shall be taken to keep the concrete moist and for how long a period?

b. In freezing weather, what precautions shall be taken to prepare the materials for making concrete and with what care and for what length of time shall the concrete be protected after pouring?

c. Give an approximate time limit in days before you would remove forms and supports from

    1. Arch centers   4. Walls

    2. Centering underbeams 5. Columns

    3. Floor slabs   6. Sides of beams and other parts

d. What limit of time would you recommend from the time water is first added to the concrete mix until it reaches its final position in the forms?

e. Certain concrete columns are 40 ft high and 40 in. in diameter. Explain how you would go about filling the forms with concrete to insure an "A" grade of concrete.

3.02 It is planned to construct a portland-cement reinforced-concrete pavement in two lanes each 10 ft wide. The pavement is to be laid on a subgrade which has a large amount of clay in it.

a. Explain in detail the work which should be done to prevent frost heave and to provide adequate drainage of the subgrade.

b. What factors will have an effect on the strength of the concrete? What steps should be taken to insure adequate strength. What mix would you recommend?

c. What is the purpose of placing reinforcing steel in concrete pavements? Where should it be placed within the slab? Why? Explain how you would design the reinforcing steel for a concrete pavement.

d. What are the advantages and disadvantages of using vibration in placing concrete for a pavement. What is the essential reason for vibrating concrete?

e. Describe four methods of curing concrete pavements. How is each method supposed to insure adequate curing? Which method do you consider to be the best? Why?

3.03 Using the information given below, calculate the cost of completing the job in the least possible time. The durations shown are in days.

Activity	Normal cost	Crash cost	Immediately follows activity	Normal duration, days	Crash duration, days
A	$1,000	$1,100	None	3	2
B	2,000	2,000	A	4	4
C	1,600	1,600	B	4	4
D	250	300	C	3	2
E	500	600	A	2	1
F	600	800	B, E	3	1
G	800	1,100	F, I	3	1
H	3,000	3,800	A	6	2
I	400	700	H	2	1
J	1,000	1,000	I	4	4
K	500	500	C, G, J	2	2
L	500	1,000	D	2	1
M	600	600	K, L	2	2

3.04 The following statements are concerned with engineering construction. Briefly discuss the validity of each statement. Simply indicating a true or false for each one is not a fully adequate answer.

a. In anticipating the volume of rock that will be loosened by a blast, the effective depth of the drilled hole will be about 2 ft greater than the actual depth of the drilled hole.

b. The rate of drilling in rock will vary directly with the air pressure supplied.

c. The usual code specifications identify when to stop driving by using the same criteria and limitations whether the pile is driven with a single-acting, double-acting, or vibratory hammer or with a sonic pile driver.

d. Tandem bowl scrapers with all wheels driven by individual electric motors placed on each wheel are truly self-loading and can be used to excavate soft earth without requiring a prime mover.

e. Elevating scrapers will carry a fuller, denser load in the bowl than conventional scrapers of the same rated volume.

f. It is correct to expect that a crawler-mounted lifting crane rated at 20 tons can operate safely as a clamshell in a circle of 60-ft diameter if the combined weight of the bucket and earth does not exceed 10 tons.

g. The horsepower of excavating equipment varies inversely as the altitude above sea level.

h. The horsepower of excavating equipment varies inversely with air temperature.

3.05 The following statements are concerned with construction estimating and management.

a. Many contactors submit figure bids (say $497,988.45) rather than rounding off ($498,000). Why is this done?

b. Where bids are asked for on a unit price basis, many contractors prepare a lump sum total first and then divide this into unit price bids. What are the advantages of this to the contractor?

c. Give four reasons why bids may vary considerably for a job. Do not include incompetence of estimators, lack of knowledge, political collusion, or attempts at fraud.

d. In preparing monthly estimates for progress payments, would you include materials delivered and stored but not yet installed?

e. What are the advantages and disadvantages of leasing construction equipment rather than purchasing it?

3.06 A badly cracked portland-cement concrete pavement is to be surfaced with bituminous concrete. The existing pavement is to be used as a base for the new surface.

Describe in detail:

a. The steps you would take to insure no further failure of the base.

b. The preparation of the base to receive the bituminous concrete surface.

c. The preparation (including mix recommended) and laying of the wearing course.

Sketch a cross section of the finished road.

3.07 It is planned to construct a bituminous macadam road.

a. Outline the procedure which should be followed in constructing a bituminous macadam road.

b. Upon what does the strength and durability of a bituminous macadam road depend?

c. What are the principal maintenance problems of this type of road? What provisions should be made for handling these problems?

3.08 It is proposed to construct a limited-access bypass around a town of 4,200 persons. The route involved is a state route carrying 27,000 vehicles per day. Objections have come from three citizens' groups: (1) the local merchants object because they fear a loss in business; (2) residents of the suburbs object because 41 parcels of land must be taken and the state highway "will ruin the neighborhood and pose a hazard to their children"; and (3) a taxpayers' group on the statewide level objects because they feel the bypass to be unnecessary and a heavy drain on tax funds. You are the representative of the state highway department who has been chosen to justify the construction of the bypass and placate these three groups. Prepare a brief written statement outlining the advantages of the bypass and providing answers to the objections raised.

3.09 For each lettered item select the numbered description which most nearly identifies or describes it. Each numbered description may be used only once.

    a. Freeway   *f.* Stream friction
    b. Outer connection   *g.* Over-all speed
    c. Induced traffic   *h.* Internal study
    d. Practical capacity   *i.* Live parking
    e. Critical density   *j.* Desire line

1. Traffic increase due to population growth
2. Retarding effect caused by intersecting streams of traffic
3. Right-turn ramp from one through roadway to a second at a grade separation
4. Retarding effect by units traveling in the same direction
5. Maximum hourly vehicle volume without unreasonable delay or hazard or feeling of constriction
6. Total distance traversed divided by travel time
7. Traffic density just before complete stagnation
8. Traffic increase due to presence of a new facility
9. Total distance traversed divided by running time
10. O and D study by home interview
11. Expressway with fully controlled access
12. O and D study inside cordon line
13. Traffic density at maximum flow
14. Parking other than all day
15. A connecting highway which skirts a congested area
16. Plot of vehicles per capita vs. average vehicle cost
17. Highway on which no toll is charged
18. Maximum hourly vehicle volume under prevailing conditions
19. Straight line between origin and destination
20. Parking with operator in attendance

3.10 An aircraft-runway pavement is to be designed for a maximum single-wheel load of 60,000 lb with a tire pressure of 200 psi. The pavement is to be a hot-mix bituminous concrete. The following materials are available to be used in the base course: (*a*) well-graded sandy gravel, compacted CBR 80; (*b*) silty sandy gravel, compacted CBR 55; (*c*) silty sand, compacted CBR 35; (*d*) sandy clay, compacted CBR 15. The subgrade is a clayey sand with a compacted CBR of 10. The relative cost of these materials is given below. The cost is for a compacted layer 4 in. thick or thinner. Thicker layers must be constructed in two or more lifts. The relative pavement cost is for a compacted layer 3 in. thick or thinner. Thicker pavements must be constructed in two or more lifts.

Material	Relative cost in place
Compacted subgrade	1.0
Sandy clay	2.0
Silty sand	4.5
Silty sandy gravel	5.5
Well-graded sandy gravel	7.0
Hot-mix bituminous concrete	20.0

1. Design a suitable cross section for the pavement and base courses. Maximum economy is desirable within the limits of good design practice.

2. Describe the pavement course or courses.

3.11 A road is to be built that will run above an existing cable conduit from station 97 + 00 to about station 103 + 50. The pipe is horizontal in grade and its top is at elevation 94.42 ft. The P.V.I. of a symmetrical parabolic vertical curve will be at station 100 + 00 and at elev. 95.00. The highway grade descends at a rate of −2 per cent to the P.V.I. and then rises at a rate of +1.5 per cent leaving the P.V.I. Will the length of a symmetrical parabolic vertical curve, such that the minimum cover above the conduit will be 4.00 ft, be: (a) 700, (b) 750, (c) 800, or (d) 850 ft?

3.12 Answer the following true or false.

1. Correctly determined superelevation designed to take care of centrifugal force on curves varies directly as the radius and inversely as the square of the speed.

2. All state plane coordinate systems in the United States use Lambert or Transverse Mercator projections.

3. If a transit is set up at the mid-point of a highway curve, and a backsight is taken to the quarter point, the deflection angle to the P.T. is one-half the angle formed at the center of the curve by radii drawn to the quarter point and to the P.T.

4. A manhole at the quarter point of a 60-ft-wide street with an 8-in. parabolic crown is 2 in. below the center line of the street.

5. In a polyconic projection map all meridians are shown curved.

6. In a survey using the New Jersey state coordinate system, all the bearings shown would refer to the same meridian.

7. A rising curve on a highway mass diagram represents fill and a descending part represents cut.

8. Building lines are usually staked out to show the center of walls.

9. Zero per cent grades on streets are undesirable in most road layouts.

10. The average velocity of the water in a vertical section taken in a stream is usually found between the center and the surface.

3.13 A car skidded going into an intersection, struck a pedestrian, and continued until it hit a tree. Based on the damage to the front of the car, it is estimated that the car was doing 5 mph at impact with the tree. The length of the skid marks was measured at 130 ft. The road is on a downhill grade of -5 per cent. A test car skidded 45 ft on the same section of road when braked from a speed of 25 mph to a halt. Was the probable speed of the car involved in the accident when the brakes were applied: (a) 42.2, (b) 42.7, (c) 43.1, or (d) 43.4 mph?

3.14 On a certain section of a busy four-lane highway it has been observed that 1 out of every 12 motorists who make illegal left turns are caught and ticketed by the police.

1. Will the probability of being apprehended once a work week (5 days), if you commit the violation twice each day, be: (1a) 36.9, (1b) 37.4, (1c) 37.8, or (1d) 38.1 per cent?

2. Will the probability of never being caught in any given week be: (2a) 41.9, (2b) 42.4, (2c) 42.8, or (2d) 43.1 per cent?

3.15 Select from Group B the item which is described in Group A. Select carefully since Group B items may be used only once.

Group A

1. A method of trip distribution involving the use of existing volumes of interzonal traffic to measure friction prior to expanding the volume between two points in proportion to an interactance factor.

2. The allocation of traffic flows among routes available.

3. A diamond highway interchange in which the intersecting conflicts are changed to weaving conflicts.

4. A street which serves internal traffic movements within an area and connects this area with major arterials.

5. A method of signal timing whose purpose is to first clear the vehicles desiring to turn left at an intersection.

6. A signal system in which the signal faces which control a given street will be green according to a time schedule which will permit continuous operation of vehicles as is possible along the street.

7. The maximum number of passenger cars that can pass a given point per hour under the most nearly ideal roadway and traffic conditions.

8. The difference between the observed speed and the standard speed for that particular type of street.

9. A name applied to the value obtained by dividing the hourly volume of traffic by the average speed.

10. A value which may be used as a good guide in establishing upper speed limits.

Group B

a. Collector street
b. Simple progression
c. Local street
d. Growth Factor Method
e. Two-quadrant cloverleaf
f. Traffic assignment
g. Parkway speed
h. 85th percentile value
i. Simultaneous progression
j. Basic capacity
k. Scramble system

l. Bridge rotary
m. Possible capacity
n. Desire line method
o. Travel speed
p. Volume
q. Gravity method
r. 15th percentile value
s. Delay rate
t. Density
u. Advance green

3.16 A town having a present population of 10,000 is planning to construct a rapid-sand-filtration plant. From the following data calculate (1) the size of the filter units, (2) the pounds of alum required per day, and (3) the size of the distribution reservoir required for fire protection.

a. Estimated future population is 14,000.

b. Consumption per capita is 100 gal per day.

c. Maximum rate of consumption is 225 per cent of average rate.

d. Filters to operate at a rate of 125,000,000 gal per day per acre.

e. Wash water equals 4 per cent of total water filtered.

f. Average hours of filter operation between washings equals 7½.

g. Time required to wash filter and restore it to satisfactory operating condition equals ½ hr.

h. Average alum dosage is 4½ grains per gal.

i. Fire protection to meet requirements of National Board of Fire Underwriters.

3.17 A grit chamber installation is to be provided for a city with an average dry-weather flow of 1,000,000 gal per day and a storm flow of 3,000,000 gal per day. What values would you choose for the following: (*a*) minimum detention period; (*b*) maximum velocity of flow; (*c*) length of units; (*d*) top width of units; (*e*) maximum depth of units; (*f*) number of units; (*g*) grit storage capacity; (*h*) method of cleaning; (*i*) average interval between cleanings. Give the reasons for your choice.

3.18 A sewage has a suspended solids content of 250 ppm. A sedimentation tank with a retention period of 1.5 hr is planned.

a. What will be the reduction in suspended solids?

b. If the average sewage flow is 500,000 gal per day, what must be the capacity of the tank in gallons?

c. What dimensions would you use if the tank is to be rectangular in plan?

d. If a hopper-bottom tank is to be used and the sludge is to be removed once daily, what should be the capacity of the sludge hoppers, assuming the water content of the sludge to be 95 per cent?

3.19 In designing sewerage systems:

a. What maximum spacing may be used between manholes? What factor controls this spacing?

b. What period of design should be used for sewers? For treatment works? Why should there be a difference?

    c. What per capita flow of sanitary sewage should be provided for?

    d. What minimum velocity of flow should be used for sanitary sewers? For storm drains?

    e. What minimum size of pipe should be used for sanitary sewers? For storm drains? Why?

    f. What method is best adapted for the determination of the size of a storm drain? Explain briefly.

    g. Under what conditions should catch basins be used instead of inlets?

    h. An 8-in. line and a 15-in. sewer come into a manhole. A 24-in. pipe leaves it. What should be the relative elevations of the three inverts?

3.20 You are called upon to design the storm-water drains for a subdivision. The rational method of design is to be employed. State clearly and in logical order the information which you would gather, the use to which you would put this data, and each successive step in design from the upper inlet to the lower end of the line where the drain discharges into a stream.

3.21 Prepare an outline showing each successive step in gathering the data for and preparing a design of the sanitary sewers for a small subdivision.

3.22   *a.* Describe the separate sludge-digestion process used for sewage treatment.

    b. Name two methods of dewatering sludge. Give advantages of each.

3.23 In each of the following select one, and only one, correct answer.

    a. In the active-sludge process of sewage treatment, it is essential to have: (1) a dosing tank; (2) an adequate supply of air; (3) an Imhoff tank; (4) flocculation; (5) an acid sludge.

    b. For protection of aquatic life in a fresh-water stream, sewage effluent should never lower the dissolved-oxygen content lower than: (1) 1 ppm; (2) 5 ppm; (3) 10 ppm; (4) 15 ppm; (5) 20 ppm.

c. The biochemical treatment of sewage effluents is essentially a process of (1) reduction; (2) dehydration; (3) polymerization; (4) oxidation; (5) alkalinization.

d. The process of lagooning is primarily a means of: (1) increasing the capacity of storage reservoirs; (2) increasing flow of sewage through Imhoff tanks; (3) reducing the excessive flow in sewers; (4) disposing of sludge; (5) rendering sludge suitable for fertilizing purposes.

e. In treating turbid waters a popular coagulant is: (1) calcium sulfate; (2) chlorine; (3) iodine; (4) ferric sulfate; (5) pulverized coke.

f. In most well-designed sewer systems: (1) manholes are generally equipped with regulators; (2) sewers never flow full; (3) sewage ejectors are used to accelerate flow and prevent deposition at low-flow periods; (4) catch basins are not usually essential; (5) manholes are always placed over the center lines of the sewer below.

g. Chlorine is used in the treatment of sewage to: (1) cause bulking of activated sludge; (2) help grease separation; (3) aid flocculation; (4) increase the biochemical oxygen demand; (5) reduce the production of ecologies.

h. The quantity of grit in sewage depends largely on: (1) the extent to which the sewered area is built up; (2) the number of industrial plants in the area; (3) the chemicals used in the precipitation process; (4) the strength of the sewage; (5) the period of detention.

i. Septic tanks are primarily used for: (1) the aerobic decomposition of deposited sewage solids; (2) separation of deposited solids; (3) separation of oil and grease scums; (4) anaerobic decomposition of deposited solids; (5) the nitrification of raw sewage.

j. The gas from Imhoff tanks is mainly composed of: (1) carbon dioxide; (2) methane; (3) nitrogen; (4) ethane; (5) hydrogen sulfide.

3.24 Show by sketch the plan and sections for a septic tank and subsurface irrigation system (nitrification field) for a country home housing 10 people—soil is fairly porous. Give dimensions and data for design.

3.25 Describe how malaria is transmitted. If appointed sanitary engineer to a county health unit, describe just what steps you would take to reduce the number of cases of malaria in your county.

3.26 *a.* A state has a population of 2,000,000. For the year 1968 the physicians of the state reported 2,891 cases of tuberculosis with 1,304 deaths from this disease. Compute: (1) case rate; (2) death rate; (3) fatality rate.

  b. Define morbidity as used in vital statistics.

3.27 *a.* What diseases are prevented by protecting the dairy herd and what measures are applied in such protection?

  b. What are the protective measures applied to the "milk line"?

  c. Name a disease thus prevented.

  d. What is the principal danger of dirty milk and what specific measures can be taken to eliminate this danger?

  e. Name two methods of milk pasteurization that require the use of recording thermometers. Why is one method used more commonly than the other?

3.28 Give the method of transmission of

  a. Bubonic plague   *f.* Undulant fever
  b. Murine typhus fever   *g.* Tularemia
  c. Malaria   *h.* Yellow fever
  d. Dengue fever   *i.* Rocky Mountain spotted fever
  e. Smallpox

3.29 In investigating a typhoid fever outbreak in a county of 10,000 population, what steps would you take and what data would you collect in determining the sources of the infection?

3.30 Name two methods of municipal sanitary garbage disposal applicable to a city or town. Compare the two methods giving the application of each with advantages and disadvantages. *Note:* Open refuse dumps and feeding of garbage to hogs not to be considered.

3.31 Explain how to calculate (*a*) the yield of a watershed for power or water-supply purposes, and (*b*) the necessary storage capacity to provide a given rate of delivery.

3.32   Discuss (*a*) the importance of leakage and waste in a water-supply distribution system, (*b*) the most satisfactory methods to reduce such water loss, and (*c*) the effectiveness of these methods or the percentage of waste reduction that may be realized from them.

3.33   Describe the general features of a rapid sand water filtration plant.

3.34   *a.* Explain the chemical reactions that take place in a zeolite softener.

    b. What is the effect of hydrogen sulfide in water that is to be chlorinated?

    c. Explain the lime-soda method of softening water and write the chemical equations to show just what chemical reactions take place.

3.35   For what purpose is each of the following chemicals used in the process of water treatment?

    a. Chloramine   *f.* Aluminum sulfate

    b. Sodium hexametaphosphate   *g.* Hydrated lime

    c. Sodium bisulfite   *h.* Sodium chloride

    d. Sulfur dioxide   *i.* Sodium carbonate

    e. Activated carbon

3.36   *a.* What do the following terms mean when applied to pumps: (1) single stage, (2) centrifugal, (3) multistage, (4) positive displacement, (5) reciprocating, (6) double acting, (7) rotary.

    b. For what service conditions would you recommend the use of a single-stage centrifugal pump?

    c. State the service conditions for which you would recommend a two-stage pump instead of a single stage.

3.37   *a.* A report on a proposed municipal water supply shows the presence of *B. coli, coli-aerogenes,* or gives a colon index. What is the significance of these terms?

    b. Under what conditions do anaerobic bacteria thrive? Under what conditions do aerobic bacteria thrive?

    c. The report on a prospective water supply shows carbonate hardness—100 ppm. What is the significance of this information? Will softening be necessary for municipal use?

3.38 In carrying out a triangulation network for the horizontal control of an extended survey there are several factors which will influence the accuracy with which the distances between stations can be computed in addition to the precision with which the angles in the scheme are measured. State what these factors are and the manner in which each affects the determination of the lengths.

3.39 A water-purification plant is situated on the bank of a large river. It is a run-of-the-river plant without reservoir storage. The raw water has considerable color and is quite turbid but is reasonably soft. Upstream, the river passes through several shallow lakes which have a heavy algal growth during warm weather and which are much used for swimming in season. Below the lakes, the effluent from two sewage-treatment plants, using the activated-sludge process and chlorination, flows into the river. The water to be treated is diverted through a short intake canal to low-lift pumps which lift it sufficiently to provide for gravity flow through the plant. Following treatment, high-lift pumps force the water to a large distribution reservoir. Outline your suggestions for a plant to render this water potable, listing all necessary units, processes, and facilities.

3.40 A dam for a water storage is to be constructed across a river. The available length of spillway is controlled by geologic considerations. In designing the spillway section, it is necessary to determine the maximum head of water over it during the design flood. Studies of past floods have provided data for a distribution graph at the site. From these data and from analysis of the expected runoff from the peak storm, determined from long-term precipitation records, the inflow hydrograph for design has been plotted. Assuming the reservoir to be full at the beginning of the flood period, show each step in determining the crest height on the spillway during the passage of this flood, using the mass-curve method.

3.41 The following statements refer to water purification and treatment. Indicate whether the statements are true or false.

 a. Color is fully removed by slow sand filters.

 b. Activated carbon may be used for taste and odor control without subsequent filtration.

c. Turbidity is normally removed by adding a coagulant prior to sedimentation.

d. Copper sulfate is used in the shallow portions of reservoirs for algae control.

e. Activated carbon is applied to water supplies to reduce hardness.

f. The usual rate of flow through rapid sand filters is 2 gal per sq ft of surface area per min.

g. Bacteria counts of coliform organisms are usually greater in well supplies than in surface supplies.

h. The zeolite process reduces the hardness of a water supply by no more than 50 per cent.

i. Aeration of water is effective in CO_2 removal.

j. Recent installations have tended to utilize slow sand rather than rapid sand filters.

k. Sodium carbonate or sodium sulfate do not cause hardness in water.

l. Freezing does not necessarily kill spore-forming pathogens.

m. Boiling for at least an hour kills all pathogenic bacteria.

n. The presence of *E. coli* in a water supply is proof of the presence of typhoid bacilli.

o. The confirmed test for *E. coli*, when positive, is indicative of intestinal pollution.

p. Hard surface waters which require filtration for removal of pollutants are normally treated by zeolite process for the removal of hardness.

q. Newly laid water mains should be disinfected before being placed in service.

r. A raw water which is treated with alum requires the addition of soda ash when the natural alkalinity is high.

s. Fluoridation of water in the amount of 1 ppm is effective in reducing the incidence of dental caries in children.

t. A water with a pH below 7 is alkaline.

u. The lime–soda ash process is well adapted to the softening of hard surface supplies.

v. The most commonly used coagulant in water treatment is ferric chloride.

w. Highly turbid water is usually highly colored and normally requires treatment for both conditions.

x. The major water-borne diseases are typhoid, malaria, encephalitis, and scarlet fever.

y. Typhus is not a water-borne disease.

3.42 *a.* What are the major causes of precipitation in northern New Jersey and why is there no marked seasonal variation in precipitation in this area?

b. In determining the precipitation on a watershed from the records of Weather Bureau stations in or near the watershed, what methods may be used and what are the conditions which control the method selected?

c. What is the cause of the marked seasonal variation in precipitation in the San Francisco area?

d. Why does the Upper Mississippi Valley receive most of its precipitation during the crop-growing season?

3.43 *a.* Discuss in detail the effects of:
1. Discharging high-temperature wastes into a sewer
2. Discharging sawdust from a mill into a stream
3. Discharging precipitates from lime-soda water-softening plants into a stream
4. Discharging extremely fine suspended solids from an industrial plant into a river
5. Discharging oily wastes into a sewerage system

b. Discuss possible measures to prevent undesirable results in each of the above situations.

3.44 *a.* When a relatively small volume of raw sewage is disposed of by dilution into a large body of water, what are the physical, chemical, and biological processes which result in the eventual clarification and purification of the polluted water?

b. What considerations limit the disposal of sewage by dilution?

3.45 Classify sewage-treatment processes under the following headings:

a. Methods or devices which separate floating and settleable solids from the liquid

b. Processes for the aerobic oxidation of the finer suspended, colloidal, and dissolved organic compounds

  c. Disinfection of sewage effluents

  d. Digestion of sewage sludges

  e. Methods for dewatering, drying, and disposing of either raw or digested sludge

  f. Final disposal of sewage effluents

3.46 With reference to sewerage and sewage treatment:

  a. You find that at a given disposal plant the Department of Health requires complete removal of all floating solids, at least 60 per cent removal of suspended solids, and an E. coli count of less than 100 per cc. Would you consider that the river below the point of discharge was used for (1) a potable water supply, (2) bathing, (3) irrigation, (4) industrial purposes, or (5) none of the above?

  b. On entering the storage area at a plant you find the floor partially covered with an oily brown liquid which has leaked out of a crack in the lining of a rubber-lined tank. What is the liquid and for what use was it intended?

  c. What conclusions would you draw from observing a series of light hollow metal balls, varying in size from small to large, being dropped, one after the other, into a sewer at a manhole?

  d. For what purpose would copper sulfate be dumped into a sewer?

  e. What would you infer if you saw a lighted candle being lowered into a sewer?

  f. For about what minimum velocity, sewer flowing half full, would you design a sanitary sewer?

  g. Would you lay sewer pipe upgrade or downgrade?

  h. In designing sewers, what governs the location and spacing of manholes?

  i. Why does a circular gravity sewer not flow full under peak flows?

  j. Why are septic tanks not commonly used in treating a municipal sewage? Where are they used?

3.47 A community has decided to use an activated sludge treatment plant to treat the sewage. The BOD is 300 mg per liter (milligram per liter = ppm or part per million) and the design flow is 5 mgd (million gallons per day). (Also one part per million gallons is one gallon or 8.33 lb per million gallons.) The aeration tanks are required to have a minimum detention period of 6 hr at 125 per cent of the design flow. The maximum BOD loading of each aeration tank is not to exceed 38 lb per day per 1,000 cu ft of aeration tank capacity. Use four aeration tanks each having a design flow of 1.25 mgd. The primary settling tank is designed to remove 30 per cent of the raw sewage BOD. Aeration capacity at standard pressure and temperature shall be at least 1.5 cu ft per gallon of incoming raw sewage. The air-diffuser system shall be able to deliver 150 per cent of normal requirements. Consider such requirements to be 1,000 cu ft per lb BOD to be removed from the sewage entering the aeration tanks.

1. Will the volume of each aeration tank be: (1a) 51,250, (1b) 52,500, (1c) 54,000, or (1d) 55,750 cu ft?

2. Will the aeration capacity in cfm be: (2a) 4,700, (2b) 5,000, (2c) 5,200, or (2d) 5,300 cfm?

3. Will the air diffuser capacity be: (3a) 2,280, (3b) 2,320, (3c) 2,390, or (3d) 2,480 cfm?

3.48   400 mgd of sewage is discharged into a stream which has a minimum rate of flow of 2,000 cfs. The temperature of the sewage and of the diluting water is 15°C. The 5-day BOD at 20°C of the sewage is 150 ppm and 0.5 ppm for the river diluting water. The sewage contains no dissolved oxygen and the stream diluting water is 100 per cent saturated with oxygen. Determine the dissolved oxygen deficit 80 miles downstream from the point of loading if the velocity of the stream is 20 miles per day. $K_1 = 0.08$ and $K_2 = 0.28$ at 15°C; $K_1 = 0.10$ at 20°C. Will the dissolved oxygen deficit be: (a) 7.50, (b) 7.80, (c) 8.20, or (d) 8.70 ppm?

3.49   Select the one choice that will best complete statements a through j.

a. Liquid waste is considered to be
   1. A Newtonian fluid   3. An ideal plastic
   2. A Non-Newtonian fluid   4. A real fluid

b. In a situation where the topography is flat and wastes must be pumped to the treatment plant, the type of system to use is: (1) separate, (2) combined, (3) storm, (4) lateral.

c. The estimate of demand for the design basis of a treatment plant should be about: (1) 5 years, (2) 30 years, (3) 50 years, (4) 70 years.

   d. Storm and sanitary sewers
       1. May be placed in the same trench
       2. Should not be in the same trench
       3. Should never be needed in separate systems
       4. Should be located in easements through backyards
   e. Good practice in the design of storm sewers for residential areas would indicate the use of
       1. 10-year storm figures   3. 100-year storm figures
       2. 50-year storm figures   4. All storm figures
   f. When sewers are placed in rock cuts
       1. Steel pipe must be used.
       2. Concrete foundations are used.
       3. Trenches may remain open for weeks.
       4. Neither steel pipe, concrete foundations, nor leaving the trench open for weeks is practical.
   g. Water pollution control on navigable waters is a responsibility of
       1. A civilian agency of the federal government
       2. The U.S. Army
       3. The U.S. Navy
       4. The individual states
   h. The strength of a sewage is related to
       1. Its grit content
       2. The nuisance-producing items possessed
       3. The BOD
       4. The turbidity
   i. The organic portion of sewage may best be estimated by measurement of
       1. The total solids   3. The fixed solids
       2. The volatile solids   4. The pH
   j. Extra width in a sewer trench over and above that necessary for placing the pipe will
       1. Benefit the alignment
       2. Increase the loads on the pipe
       3. Help support the trench
       4. Reduce the loads on the pipe

3.50 Calculate the carbonate and non-carbonate hardness (in terms of mg per liter of $CaCo_3$) of a water that has been analyzed as follows:

Na^{++} = 20 mg per liter	Cl^- = 40 mg per liter
Ca^{++} = 20 mg per liter	NO_3^- = 1 mg per liter
Mg^{++} = 15 mg per liter	HCO_3^- = 5 mg per liter
Sr^{++} = 5 mg per liter	SO_4^{--} = 16 mg per liter
	CO_3^{--} = 10 mg per liter

3.51 Name and describe four soil tests and tell what each is a measure of.

3.52 Answer briefly:

a. How would you take care of a minor leak in the sides of a single-wall interlocking steel-sheet cofferdam?

b. What precautions would you take prior to unwatering the above cofferdam?

c. If the lunch whistle blew at a time when a pile-driving crew had only partly driven a pile would you permit the crew to stop for lunch? Why?

d. If you were ordering stone to riprap a stream bank near a bridge, what size stone would you try to get?

e. What is the advantage of prefabricating timber that is to be pressure creosoted? How would you treat the exposed timber in holes to be bored in the creosoted timbers?

3.53 Answer the following briefly:

a. In digging a trench for a sewer in hardpan and in the dry would it be necessary to use sheeting and bracing? Would this apply to any depth trench?

b. How would you protect steel from rusting if welding is to be done upon it at a later date?

c. Is "oiling the forms" harmful or beneficial to the concrete?

d. What is meant by "gunite" and how is it applied?

e. In structural welding, what advantage is there to the use of an alternating-current machine over a direct-current machine?

3.54 A consolidation test has been run in a consolidation apparatus whose internal dimensions are 1.258 in. in height and 4.23 in. in diameter. At the end of the test a determination of moisture content gives the following data:

Weight of containers plus wet soil	663.2 grams
Weight of container plus dry soil	582.0 grams
Weight of container	194.4 grams

A specific-gravity determination on part of the test specimen gives the following:

Weight of bottle plus water plus soil	723.22 grams
Weight of bottle plus water	687.39 grams
Weight of dish plus dry soil	175.50 grams
Weight of dish	118.77 grams

The test specimen has a wet weight of 468.8 grams. The consolidation-test results are summarized as follows:

Applied pressure, kg per cm²	Dial reading, in.	Specimen height, in.
0	0.4200	
¼	0.4131	
½	0.4087	
1	0.4010	1.235
2	0.3828	
4	0.3427	1.178
8	0.2914	

Draw the voids ratio vs. pressure diagram for this test.

3.55 A concentrated wheel load of 7,000 lb on the surface of an embankment has cartesian coordinates (0,0). The horizontal top of a culvert has corner coordinates (+2, +3), (+2, +7), (+5, +3), and (+5, +7), with all coordinates being in feet. If the top of the culvert is 9 ft beneath the surface of the embankment, what is the total load imposed on the culvert by the wheel, assuming the soil to be horizontally layered?

3.56 It is necessary to make foundation designs for two reinforced-concrete columns which support a 12-story apartment house. The columns are spaced 22-ft 0-in. center to center. One column carries a load of 350 kips and one a load of 450 kips. Depth to firm bedrock beneath the 350-kip column is 9 ft; beneath the 450-kip column 13 ft. Discuss the possible alternate foundation designs, explaining how you would select trial dimensions, what each foundation consists of, the stresses which must be investigated, permissible values for these stresses, where critical sections occur, how these stresses may be computed, where reinforcing steel is necessary, how it should be placed, how the necessary quantity may be computed, etc.

3.57 A sample of soil weighing 587.4 lb is removed from a test pit. 298.1 lb of water will just fill the pit. A sample of the soil weighing 112.4 grams is dried in an oven, and its weight after drying is 103.6 grams. Assume the specific gravity of solids is 2.67. Maximum attainable unit weight of the soil is 117.4 lb per cu ft and minimum attainable is 103.9 lb per cu ft.

Find: *a.* Wet unit weight
      b. Dry unit weight
      c. Moisture content
      d. Voids ratio
      e. Porosity
      f. Degree of saturation
      g. Relative density

3.58 The following excerpts are from a geologist's report on the route of a proposed Interstate highway. Discuss the engineering significance and the effect, if any, on the costs of construction, possible slowdowns, and the availability of construction materials.

a. "Section 4 will cross an ancient meander flood plain from 19 + 60 to about 26 + 00 . . . ."

b. ". . . the section in the area near the town of Carlton is being built on a glacial lake bed. Station 97 + 00 is in the center of a beach ridge extending to station 102 + 50."

c. "Most of the section northeast of Carlton is glacial till . . . obviously young and bedrock controlled. There is every indication that limestone is beneath the till."

d. "This section passes through the barchanes (sand dunes) which are quite extensive."

e. "The mile long section of proposed highway outlined in red is traversing along an esker about 60 ft deep."

3.59 Determine the required depth of penetration and the maximum bending moment in the cantilever sheet piling shown in Fig. 3.59.

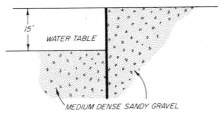

15'

WATER TABLE

MEDIUM DENSE SANDY GRAVEL

Fig. 3.59

3.60 In reinforced concrete:

a. What is meant by "balanced design"?

b. What is meant by "the transformed section"?

c. What locates the neutral axis
   1. In a new design?
   2. In an investigation of an existing beam?

d. What assumption is made as to the variation of intensity of fiber stress in a beam?

e. In column design how much additional concrete coverage of outermost steel is usually specified for fireproofing?

f. In a T beam, what part of the cross-sectional area carries the shear?

g. What function are the stirrups called upon to play?

h. Why are stirrups not needed in slabs?

i. Distinguish between beam shear and punching shear.

j. How far above the ground should you place the reinforcement in a footing?

3.61 In the two lists given below, are various kinds of tests and short statements about the value of each test. For each test in List 1, select a statement from List 2 that is most appropriate.

List 1

   1. Tension tests
   2. Compression tests
   3. Transverse tests
   4. Impact tests

5. Shear tests
6. Torsion tests

List 2

a. With brittle materials a criterion of strength, called the modulus of rupture, together with the flexibility and toughness may be determined.

b. With ductile materials the strength, ductility, toughness, and modulus of elasticity may be determined. These tests are most valuable of all mechanical tests for ductile materials.

c. A valuable measure of shock resistance for brittle and ductile materials.

d. These tests afford an imperfect measure due to the existence of bending stresses.

e. This test is of great value in determining the strength of such materials as wood, concrete, cast iron, and brick.

f. The only test wherein the shearing modulus of elasticity may be determined for ductile materials.

3.62   Answer or define briefly:

In plate girders:

a. What is meant by a sole plate and what should be its minimum thickness?

b. State three limiting conditions for the spacing of stiffener angles.

c. Where must stiffener angles have milled ends?

d. How much of the web area may be used for flange area?

e. Anchor bolts are usually of what minimum diameter and embedded to what minimum length?

In truss design:

f. What is meant by "combined stresses"?

g. What is meant by "alternate stresses"?

h. What is a gusset plate?

i. What is a filler plate?

j. What are stay plates?

3.63   At both approaches to a bridge it is necessary to construct fills of a maximum height of 20 ft and a base width of 49 ft with side slopes of 2:1.

a. Describe at least three methods of compacting fills and discuss the advantages and disadvantages of each.

b. What is meant by the term shrinkage, when used in connection with earthwork? What determines the amount of shrinkage? Why is it generally necessary to use a shrinkage factor in balancing cuts and fills? How is this factor used?

c. A temporary pavement is to be laid on the approaches. Describe in detail the construction of the type of pavement you would use. Upon what does the strength and durability of your pavement depend? Why would you recommend constructing this type of pavement in preference to some others which might have been used?

3.64 Show by means of suitable ruled sketches and full explanation the procedure you would use in laying out a control survey system for a bridge of four spans across a wide river. Both banks are unobstructed. Piers are to be constructed within rectangular cofferdams. Suitable range flags must be located from which the piles for the cofferdams can be driven on proper lines in both directions. The control must be adequate to permit the location of finished work within the cofferdams with the necessary precision for steel arch construction.

3.65 Describe the method which you would use to carry an important traverse along a concrete highway. The traverse is to connect a survey of valuable property with a pair of Geodetic Control Survey stations about one-half mile away so that the coordinates of the property corners may be determined. It is desired to have the linear measurements correct to 1 in 20,000. List the equipment and personnel which you would use to carry out this work with the required accuracy and with safety. List the corrections which it would be necessary to apply to the tape measurements as actually made.

ELECTRICAL ENGINEERING

4.01   Design a lighting layout for a parking lot in a downtown area. This lot measures 200 by 300 ft. The projectors may be mounted on 25-ft poles located around the outer edge of the lot.

4.02   Two 500-watt, bare incandescent lamps, each having symmetrical downward intensity of 800 candlepower, are mounted outdoors 10 ft above a rectangular table. The table measures 6 by 16 ft and the lamps are located directly above the long centerline and 4 ft in from each end. Determine the horizontal illumination at (*a*) the center of the table, (*b*) the center of each end, and (*c*) each corner.

4.03   A retail store maintains a 20 ft-c illumination at counter height. The luminaires use 750-watt P. S. Mazda lamps. What per cent of power cost could be saved if the existing luminaires were replaced with fluorescent units and the level raised to 30 ft-c? (Take fluorescent lumens per watt as 42.)

4.04   Specify the lighting installation, including the foot-candle intensity, wattage of lamps, type reflectors, and the wiring, for a drafting room 20 by 30 by 12 ft? The drafting tables are 4 by 3 ft and are raised 3 ft from the floor.

4.05   A manufacturer of engines is erecting a machining and assembly plant which has a work floor 40 by 250 ft with a ceiling height of 18 ft. There are columns spaced 20 ft apart which support beams 18 in. deep and 10 in. wide extending from side to

side. The floor is occupied by lathes, grinders, drills, etc., to a
height of 5 ft 6 in. Three-phase power is available at 120/208
volts 60 cycles.

   a. What illumination in foot-candles is required?

   b. What general type of lighting and kind of fixtures would be
suitable?

   c. Specify lamp size and arrangement of fixtures.

   d. What circuit arrangement would be suitable?

   e. What conditioning of walls and ceilings should be recommended?

4.06 Would you permit two type T #6 wires and one bare #6
wire to be run in a ¾-in. conduit as service conductors? Length
of run is 90 ft and the equivalent of three right-angle bends are
included.

4.07 A three-phase transmission line 50 miles long has an impedance of $5 + j40$ ohms per phase. Its "nominal" voltage rating
is 345 kv (line-to-line voltage). The terminal voltages at both
ends of the line are held at the following levels:

$$|\bar{V}_1| = 345 \text{ kv} \qquad |\bar{V}_2| = 360 \text{ kv}$$

Assume that $\bar{V}_1$ leads $\bar{V}_2$ by 10°.

   a. Compute the real and reactive line powers in each end of the
line.

   b. Compute the total line losses, $P_{\text{loss}} + jQ_{\text{loss}}$. Express all
powers in total three-phase values.

4.08 A 250-volt 10-kw d-c generator is separately excited.
It has an effective armature circuit resistance of 0.5 ohm and
inductance of 0.1 henry when it is supplying rated current.
Suddenly the terminals beyond its protective circuit breaker
are short-circuited with a short-circuit resistance of 0.2 ohm.
The breaker operates 0.02 sec after the fault occurs. Neglecting
saturation of the magnetic circuit, determine the maximum
current to which the generator is subjected.

4.09 The field of a 120-volt d-c generator draws 8.4 amp.
What resistance is required of a field discharge resistor such that
the induced voltage, upon disconnecting the field from the 120-

volt line, will be limited to one-third of the standard high-potential test voltage?

4.10 A symmetrical three-conductor cable is enclosed in a grounded metal sheath. The capacitance between the three conductors connected together and the sheath is 0.5 μf. The capacitance between one conductor and the other two connected together and also connected to the sheath is 0.6 μf. Calculate the charging current per conductor when a 60-cycle three-phase 25-kv voltage between conductors is applied to the cable.

4.11 The U.S. Navy possesses a self-cooled transformer, intended for use in arctic surroundings and having the hypothetical ratings 500 kva, 60 Hz, 1,000/500 volts. What will be the new ratings of this transformer if used in a tropical climate with the same voltage as before but at 50 Hz? Work with the following assumptions:

  a. In the new installation only 80 per cent of the total losses accepted in the arctic installation are tolerated.

  b. In the original installations the total losses amounted to 0.5 per cent of rated power and were distributed as follows:

Copper loss	1.25 kw measured at 100 per cent current
Eddy-current loss	0.625 kw measured at 100 per cent voltage
Hysteresis loss	0.625 kw measured at 100 per cent voltage

4.12 A 60-cycle 200-kva three-winding transformer is rated at 2,400 volts primary voltage, and there are two secondary windings, one rated at 600 volts and the other at 240 volts. There are 200 primary turns. The rating of each secondary winding is 100 kva, one-half that of the transformer. Determine (*a*) turns in each secondary winding; (*b*) rated primary current at unity power factor; (*c*) rated primary current at 0.8 pf, lagging current; (*d*) rated current of the 600- and 240-volt secondary windings; (*e*) primary current when the rated current, pf = 1, flows in the 240-volt winding and the rated current, pf = 0.7 lagging, flows in the 600-volt winding.

4.13 A 15-kva 2,300/230-volt transformer is given a short-circuit test by impressing 65 volts on the high-tension side with the 230-volt side short-circuited. The power input for rated current is 350 watts. If the core loss is 245 watts,

Find:

a. Per cent voltage regulation at full load and unity power factor, and

b. The maximum efficiency that could be expected from this transformer at full load.

4.14 Two transformers, each rated at 100 kva, 2,200/220 volts, and 60 cycles, operate in parallel to supply a 200-kva load at a lagging power factor of 0.80. The short-circuit tests on these transformers show that, with power supplied to the 2,200-volt windings, there is rated current in each transformer under the following conditions:

> Transformer A—1,000 watts at 72 volts
> Transformer B—1,100 watts at 75 volts

Calculate the current in the 220-volt coils of each transformer under the above conditions, in percentage of the total. Neglect the exciting currents.

4.15 Two transformers are in parallel sharing a common load. Unit 1 is rated at 45 Mva, 120 to 24 kv, $\Delta - Y$ with an equivalent impedance of $j0.08$ ohms per unit on its own base. Unit 2 is rated at 30 Mva, 120 to 24 kv, $\Delta - Y$, with $Z_e = j0.07$ per unit on its own base. To what total Mva load would you limit the transformers to prevent either unit from being overloaded? Give the unit split in each unit.

4.16 A turbine-driven a-c generator is to deliver power to an electric furnace located 10 miles from the turbine. The turbine generator is 60-cycle three-phase star-connected, with 11,000 volts between terminals. The electric furnace is three-phase, operates at 200 volts between electrodes, and takes 3,000 kva at 80 per cent lagging power factor. Specify the number, size, type, and location of the transformers required; and the voltage, size, and spacing of conductors of the high-tension line, the loss of which is not to exceed 5 per cent of the delivered power. (Draw diagram and show all essential calculations.)

4.17 A load of 25,000 kw at 85 per cent pf is located 25 miles from a large substation at which 60-cps power is available at

12.5, 66, and 115 kv. Choose voltage of transmission line and size of conductor you would recommend and give reasons for your choice. Voltage regulation and power loss are not to exceed 8 and 6 per cent respectively.

4.18 Draw a one-line diagram of essential electrical circuits and apparatus for a generating station consisting of two 50,000-kw generators, with direct-connected exciters, two step-up transformers, one high-voltage bus, no low-voltage bus, and two outgoing transmission lines which connect to a system having other generating stations. Indicate location of current and potential transformers and name instruments and relays connected to their secondaries. Instrument and relay circuits are not to be shown.

4.19 An industrial plant is supplied from a three-phase transmission line having a capacity of 10,000 kva. The present plant load is 5,000 kw, balanced, three phase, and at a lagging power factor such that the transmission line is loaded to capacity. The plant must increase its load and this will be done with 30 induction motors which run at 80 per cent pf. In order to make use of the line to its full capacity in active power a synchronous condenser having losses of 300 kw (at the load at which it will operate) will be installed in the plant.

a. Determine the maximum kva in new induction motors that can be added without overloading the line.

b. What is the kva rating of the synchronous condenser required for this service?

c. Draw a power vector diagram.

4.20 An industrial load consists of:

One 50-hp induction motor; load, 30 hp; efficiency, 0.86; and pf, 0.70.

One 100-hp induction motor; load, 75 hp; efficiency, 0.89; and pf, 0.80.

Two 15-hp induction motors; load, 15 hp; efficiency, 0.92; and pf, 0.85.

One 300-hp induction motor; load, 310 hp; efficiency, 0.92; pf, 0.85.

Lighting load, 33 kw.

Synchronous motor (to be added): 500 hp, 0.80 pf, leading current; load, 300 hp; and efficiency, 0.925, exclusive of field loss.

Determine for load, over-all, (a) kilowatts; (b) kva; (c) kilovars; (d) power factor.

4.21 A 15-Mva 8.5 kv three-phase generator has a subtransient reactance of 20 per cent. It is connected through a $\Delta - Y$ transformer to a high-voltage transmission line having a total series reactance of 70 ohms. At the load end of the line is a $Y - Y$ step-down transformer. Both transformer banks are composed of single-phase transformers connected for three-phase operation. Each of the three transformers composing each bank is rated 6,667 kva, 10 to 100 kv, with a reactance of 10 per cent. The load, represented as impedance, is drawing 10,000 kva at 12.5 kv and 80 per cent pf lagging. Choose a base of 10 Mva, 12.5 kv in the load circuit and draw the impedance diagram showing all impedances in the unit. Will the voltage at the terminals of the generator be: (a) 7.37, (b) 7.43, (c) 7.55, or (d) 7.79 kv?

4.22 A telephone line consisting of two #12 standard copper wires spaced 1 ft apart has the following parameters: $r = 10.44$ ohms per loop mile; $l = 0.00366$ henry per loop mile; $c = 0.00838 \times 10^{-6}$ f per loop mile and $g = 0.300 \times 10^{-6}$ mho per loop mile.
Determine: (a) the wavelength in miles; (b) the phase velocity as a percentage of the speed of light; (c) the loss in decibels per loop mile; and (d) the inductive loading per loop mile required for producing distortionless transmission.

4.23 An ideal transmission line 15.25 wavelengths long connects a generator, with an internal impedance of $100 + j0$ ohms, to a load of $200 + j200$ ohms. The load is matched to the line by means of a stub line bridged across the load. The frequency of operation is 600 mc. The line and the stub are to have the same characteristic impedance. Calculate the characteristic impedance of the line and the location and the minimum length of the stub for a proper termination of the line.

4.24 It is desired to transmit over a single-phase line 10 miles long, a load of 1,200 kw 0.85 pf lagging current, the loss not exceeding 7.5 per cent of the power delivered. The conductors are spaced 18 in. on center. The voltage at the load is 11,000 volts 50 cycles. (a) Draw a vector diagram. Determine: (b) the smallest size AWG solid-copper conductor, (c) the resistance per wire,

(d) the reactance per wire, (e) the sending-end voltage, (f) the line regulation, and (g) the efficiency.

4.25 A 60-cycle three-phase 100-mile transmission line with a resistance and reactance of 0.35 and 0.8 ohm respectively per mile of wire, draws a charging current of 0.5 amp per mile when the line is tested at 100,000 volts to neutral. Calculate the efficiency when this line delivers 50,000 kw at 132,000 volts and when the power factor is 0.80 lagging.

4.26 A Y-connected generator rated at 220 volts has 0.2 ohm resistance and 2.0 ohms reactance per phase. The generator is connected by lines each having an impedance of $2.06/29.05°$ ohms to a Y-Y transformer bank. Each transformer has a total equivalent impedance referred to the high side of $100/60°$ ohms, and the transformer bank is connected through lines each of which has a resistance of 50 ohms and an inductive reactance of 100 ohms. If the ratio of transformation is 6 and the low-voltage side is connected to the generator lines, calculate the actual fault current for a three-phase symmetrical short circuit at the load.

4.27 The data for a 335-hp 2,000-volt three-phase six-pole 50-cycle Y-connected induction motor are as follows: ohmic resistance per phase of stator, 0.165 ohm; rotor, 0.0127 ohm; and ratio of transformation, 4:1. No-load test: line voltage, 2,000 volts; line current, 15.3 amp; and power, 10,100 watts. The friction and windage of the motor are 2,000 watts. Blocked test: line voltage, 440 volts; line current, 170 amp; power, 40,500 watts. When the slip is 0.015, determine (a) the current in stator and rotor, (b) motor output, (c) speed, (d) torque developed by rotor and torque at pulley, (e) power factor, and (f) efficiency.

4.28 A shunt-wound generator rated at 35 kw 125 volts is to be converted into a series generator to produce the same terminal voltage at rated load. The machine has six poles and each field spool has 480 turns and carries 11 amp. Calculate the number of turns on each spool of the series winding, neglecting drop in series field.

4.29 A 2,500-kva three-phase 60-cycle 6,600-volt alternator has a field resistance of 0.43 ohm and an armature resistance of 0.072 between each terminal and the neutral. The windings are

Y-connected. The field current at full-load unity power factor is 200 amp, and at full-load 0.80 pf lagging it is 240 amp. The friction loss is 35 kw and the core loss, 47.5 kw. Assume friction and core loss constant at either unity power factor or 0.80 pf lagging.

     a. Calculate the full-load efficiency at unity power factor.

     b. Calculate the full-load efficiency at 0.80 pf.

4.30   For the periodic wave given, what are the indications on the following types of voltmeters? Neglect meter losses.

     a. Dynamometer

     b. Iron vane

     c. Peak above average, calibrated in rms of sine wave

     d. Peak above average, calibrated in rms of a sine wave with leads reversed from that of part *c*.

Fig. 4.30

4.31   A 10-hp 550-volt 60-cps three-phase induction motor has a starting torque of 160 per cent of full-load torque and a starting current of 425 per cent full-load current.

     a. What voltage is required to limit the starting current to full-load value?

     b. If the motor is used on a 440-volt 60-cps system, what is the starting torque and starting current expressed in per cent of full-load values?

4.32   A 120-volt d-c motor rated at 5 hp has a field resistance of 50 ohms and an armature circuit resistance of 0.8 ohm. Approximately what starter resistance will be required to limit the armature current at starting to 200 per cent of rated armature current?

4.33   A 208-volt 60-cycle four-wire three-phase source of power supplies a 208-volt balanced motor load of 28.8 kw with a lagging power factor of 80 per cent. In addition, it supplies three 120-volt 60-amp resistive loads to neutral. If the circuit breaker should disconnect the phase load from one line to neutral, what would be the total currents in the four lines?

4.34 An amplifier with a voltage gain of 100 has a maximum variation in gain of 10 per cent. In order to limit the gain variation, two of these amplifiers were cascaded, and the over-all gain brought back down to 100 by placing a negative feedback loop around the two similar amplifiers. Will the maximum variation in gain for the cascaded amplifier be: (*a*) 0.1, (*b*) 0.2, (*c*) 0.4, or (*d*) 0.7 per cent?

4.35 The input to a 600-volt 100-hp synchronous motor is measured by the two-wattmeter method. One wattmeter reads plus 62.5 kw, and the other reads plus 30.5 kw. The motor is known to be taking a leading current. Determine the line current and the output if the efficiency of the motor at this load is 0.90 exclusive of the d-c field loss.

4.36 Two three-phase alternators are operating in parallel to supply a 4,000-kw unity power-factor load at 6,600 volts. The current delivered by alternator No. 1 is 200 amp at 0.85 pf leading. Find the power, power factor, and current for alternator No. 2.

4.37 Two identical three-phase 13.8-kv 100-Mva 60-Hz turbine generators operate in parallel to supply a load of 150 Mva, 0.8 pf, current lagging, at rated voltage and rated frequency. The synchronous reactance of each machine is 1.1 per unit. Both machines share the real and reactive power equally.

    1. Will the generated phase voltage be: (1*a*) 13.0, (1*b*) 13.5, (1*c*) 13.8, or (1*d*) 13.9 kv per phase?

    2. Will the torque angle be: (2*a*) 23.00, (2*b*) 23.50, (2*c*) 23.85, or (2*d*) 24.10°?

4.38 *a.* Name and describe in some detail three commercial methods of starting single-phase induction motors.

    b. To how large a rating are single-phase induction motors available?

    c. How would you reverse the direction of rotation of a repulsion-starting single-phase induction motor?

    d. How does the universal motor differ from a d-c series motor?

4.39 A permanent-magnet loudspeaker is equipped with a 10-ohm 10-turn voice coil wound on a tube 1 in. in diameter. The flux density in the air gap is 10,000 lines per sq in. If the impedance of the voice coil is a pure resistance, what will be the maximum thrust delivered to the cone if the impressed signal is 20 db above 0.005 watt?

4.40   Given: $\beta = 40$, $V_{BE} = 0.7$ volt, $I_{CO} = 0$, $R_a = 90$ kΩ, $R_b = 10$ kΩ, $R_c = 5$ kΩ, and $R_e = 1$ kΩ.

   1. Will the quiescent value of I_c be: (1a) 0.86, (1b) 0.94, (1c) 1.00, or (1d) 1.04 ma?

   2. Will the quiescent value of V_{CE} be: (2a) 13.76, (2b) 14.00, (2c) 14.12, or (2d) 14.18 volts?

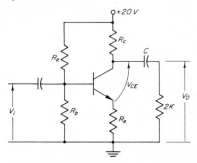

FIG. 4.40

4.41   Given: $V_{BE} = 0.7$ volt, $I_{CO} = 0$, and $\beta = 10$.
To have an operating point of $V_{CE} = 10$ volts and $I_C = 2$ ma,

   1. Will R_c be: (1a) 2.3, (1b) 2.6, (1c) 2.8, or (1d) 2.9 kΩ?

   2. Will R_a be: (2a) 20.0, (2b) 21.0, (2c) 21.5, or (2d) 21.75 kΩ?

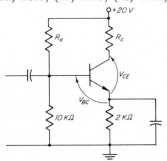

FIG. 4.41

4.42   A stack of selenium rectifier disks is rated at 2 amp average and 20 volts inverse peak per disk. A rectifier to operate from a 208 volt three-phase four-wire without transformers is desired.

Specify the number of disks per stack, and the number of single stacks required, together with the voltage and current capacity assuming 15 per cent voltage regulation from no load to full load, for the following cases:

a. The rectifier that will give the largest output without using stacks in parallel.

b. The rectifier that gives the smallest voltage ripples.

c. The rectifier that requires the smallest number of disks and still draws a balanced load. How does the ripple in this case compare with that of b?

4.43 A single-phase condenser motor, operating at rated load, takes 2.50 amp from a 220-volt line; the current in the 6.75-μf condenser is 1.30 amp, and the current in the main winding of the motor is 1.45 amp. The total power input is 550 watts.

a. What is the apparent impedance of the two windings of the motor?

b. What is the power delivered to each winding of the motor?

4.44 Given: $\beta = 40$ and $r_i = 1$ kΩ (transistor input impedence). r_0 can be neglected. Capacitors have negligible impedence.

a. Draw the small-signal model.

b. Find the current gain i_0/i_s at mid-frequency.

c. Find the voltage gain v_0/v_{be} at mid-frequency.

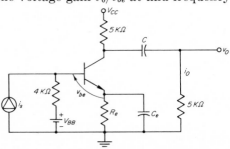

FIG. 4.44

4.45 The capacitor shown in Fig. 4.45 is initially charged in the direction shown. The capacitor and resistor are then connected to a d-c voltage of 200 volts by closing the switch at $t = 0$. Derive the equation for the current $i(t)$, sketch its graph, and find the current 2.25 msec after switching occurs.

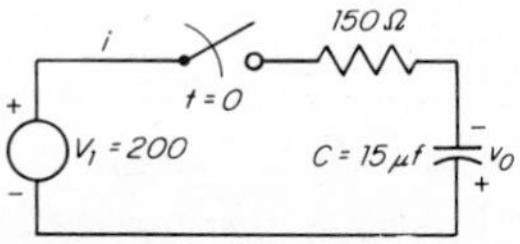

FIG. 4.45

4.46 A wave trap for use in blocking carrier current frequencies as furnished by the manufacturer consists of an air core coil and a variable capacitor pack connected in parallel. The inductance of the coil is 260 microhenrys.

Calculate the capacitance in microfarads required to tune the trap for 60 kc.

4.47 Given: $g_m = 10^{-2}$ mho, $R_d = 20$ kΩ, $r_d = 30$ kΩ, and $C = 0.01$ μf. For the FET shown:

    a. Draw the small-signal model neglecting the transistor capacitances.

    b. Find the voltage gain v_0/v_{gs} in terms of angular frequency w.

    c. Sketch A_{db} versus w with values shown.

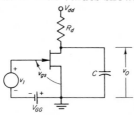

Fig. 4.47

4.48 The coil in the circuit shown has a series resistance of 12.6 ohms and the initial current of 9.2 amp is present when it has been connected to the d-c source for a long time. A make-before-break switch is thrown to the lower position at $t = 0$.

    a. At 2 sec after the switch is thrown, the voltage at terminals *a-b* is 20 volts. Find the inductance of the coil.

    b. What is the induced voltage at this time?

    c. What is the greatest voltage across the terminals *a-b* resulting from switching?

    d. How long before the transient has decreased to 1 per cent?

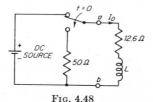

Fig. 4.48

4.49 A two-element 115-volt 5-amp polyphase watt-hour meter, having a basic watt-hour constant of $K_H = \frac{2}{3}$, is connected to a three-phase three-wire circuit through 100/5-amp current transformers and 2,300/115-volt potential transformers. A stop-watch check shows that the meter disk is making 15 revolutions in 50 sec. What is the kilowatt load on the circuit?

4.50 Explain what means are generally employed to protect a transmission line against lightning and switching disturbance.

4.51 The four corners of a bridge circuit are marked respectively A, B, C, and D. Between A and B there is a resistor of 1,000 ohms in parallel with a capacitor of 0.053 μf; between B and C there is a resistor of 1,500 ohms in series with a capacitor of 0.53 μf; and between D and A there is a capacitor of 0.265 μf. An impedance, for insertion between C and D, is to be designed so that the bridge will balance when it is supplied from a voltage source which has a frequency of 796 cps.

   a. Determine the coefficients of the impedance which must be inserted between C and D for balance.

   b. After the bridge is balanced by inserting the proper impedance between C and D, the value of the capacitance between D and A is changed to 0.270 μf. An infinite impedance detector is connected across B and D. Determine the voltage across the detector as a per cent of the voltage supplied to the bridge across A and C.

4.52 The following data are for the two machines of a properly matched motor-generator set:

   Motor: 10 hp, 1,480 rpm full load, 220 volts, 42 amp line full load, shunt field resistance = 110 ohms, armature resistance = 0.50 ohms, brush voltage = 1.0 volt total, stray power = 500 watts.

   Generator: 6 kw, 48 volts, 1,480 rpm, shunt field resistance = 8.0 ohms, armature resistance = 0.01528 ohm, brush voltage = 2.0 volts total, stray power = 648 watts.

   a. Determine the over-all efficiency of the complete unit when the set is operating at full load.

   b. Compute the motor power, line current, and speed when the set is operating at no load.

   c. If the feeder which supplies the motor is 150 ft from the main, what size wire should be used?

4.53 A multistage amplifier is shown in Fig. 4.53. All capacitors have negligible impedence. The effect of R_a and R_b can be neglected. Draw the small-signal model and find the voltage gain v_2/v_1 at mid-frequencies. Transistors are identical with a current gain of β and an input resistance of r_i.

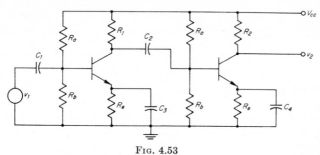

FIG. 4.53

4.54. Figure 4.54 shows a single-line plan of a three-phase electric-power system consisting of two generating stations and

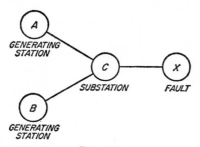

FIG. 4.54

three transmission lines. The problem is to determine the short-circuit currents which will flow in the system if a three-phase fault occurs at the location indicated. The ratings, voltages, and reactances of the generators and transmission lines are as follows:

Item	Kva rating	Kv line to line	Ohms reactance
Generator A	30,000	13.2	1.10
Generator B	35,000	13.2	0.95
Line AC	30,000	13.2	0.50
Line BC	35,000	13.2	0.30
Line C to fault	65,000	13.2	0.35

4.55 On the local 208-volt three-phase 60-cycle network distribution system with neutral, a manufacturer connects a three-phase 208-volt motor and a single-phase 120-volt motor. The three-phase motor is rated as follows: 15 hp, 208 volts, 1,740 rpm, 87 per cent efficiency, and 86.6 per cent pf. The single-phase motor is rated as follows: 3.5 hp, 115 volts, 1,750 rpm, 85 per cent efficiency, and 80 per cent pf. When the machines are operating with full loads, how much current is in each line and in the neutral?

4.56 A three-phase Y-connected power system, with neutral grounded, supplies the load shown in Fig. 4.56. The line voltages are as follows: $E_{AB} = 1,000\underline{/0°}$, $E_{BC} = 1,000\underline{/120°}$, $E_{CA} = 1,000\underline{/240°}$. The generator and line impedances are negligible.

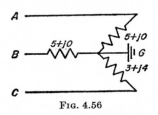

FIG. 4.56

If the neutral connection at the load were accidentally open-circuited, what would be the voltage E_{OG}?

4.57 The class A power amplifier shown in Fig. 4.57 has $I_C = 40$ ma and is designed to give greatest power to the load. Operation is under maximum signal conditions.

a. Sketch the a-c and d-c load lines showing the Q point and the values at each end of the a-c load line.

b. Find the value of R_L.

c. Find the power to the load.

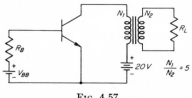

FIG. 4.57

4.58 When 2,200 volts, 60 cps, is applied to a transformer, the core loss is 240 watts. When the frequency is changed to 25 cps and the flux density is maintained constant, the iron loss falls to 75 watts. Find:

   a. The magnitude of the applied voltage at 25 cps

   b. The eddy-current and the hysteresis loss at 60 cps

4.59   For the tuned amplifier: $R_L = 100$ kΩ, $L = 1$ millihenry, r_{DS} is large, $g_m = 5 \times 10^{-3}$ mho, $w_0 = 10^6$ radians per sec center frequency, $r = 10$ ohms (coil resistance). Neglect transistor capacitances. Coupling and bypass capacitors are large.

   a. What is the value of C?

   b. Find the bandwidth.

   c. Find the voltage gain at the resonant frequency.

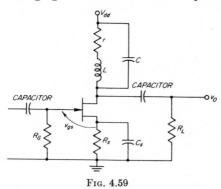

Fig. 4.59

4.60   Design a low-pass audio-frequency filter of "constant-K" type with T configuration which will match a 600-ohm resistance load and have a cutoff frequency of 1,000 cps.

ENGINEERING ECONOMICS AND BUSINESS RELATIONS

5.01   In production control, the following items are considered:
  a. Standardization of equipment
  b. Various raw materials used
  c. Speed at which operations are performed
  d. Sequence of operations
  e. Preparation of work orders
Explain why each of these items is necessary.

5.02   Modern factory production in all its variety and complexity has been developed and brought to its present mechanical perfection principally through the application of the laws of specialization. State the four laws of specialization as applied to the job, the individual, the machine, and the product.

5.03   Give a definition of the following phrases in connection with company-union labor agreements:
  a. Craft union   *f.* Industry-wide bargaining
  b. Industrial union   *g.* Down-time pay
  c. Open shop   *h.* Maintenance of membership
  d. Closed shop   *i.* Check-off
  e. Union shop   *j.* Seniority

5.04   What is meant by the following phrases in connection with company-union labor agreements?

a. Plant-wide seniority
b. Departmental seniority
c. Superseniority and synthetic seniority
d. Probationary employees
e. Grievance
f. Arbitration of grievances
g. Call-in pay
h. Portal-to-portal pay
i. Featherbedding
j. Boycotts

5.05 What are standard costs? How are they established and used?

5.06 Explain the law of diminishing returns using the variation in the height of a proposed office building as an example.

5.07 In what ways is safety related to plant layout?

5.08 What is meant by "preventive maintenance"? Outline briefly a procedure of operation for such a maintenance program in a particular industrial plant of your choice.

5.09 A small manufacturer with a maximum factory output of 1,200 machines per year prepared the following estimate of manufacturing expenses at various outputs:

Output	Direct labor and material	Indirect factory costs	Selling and administrative costs	Total cost	Unit cost
0	0	$28,000	$26,000	$ 54,000	
200	$16,000	30,000	27,000	73,000	$365.00
400	32,000	32,000	28,000	92,000	230.00
600	48,000	34,000	29,000	111,000	185.00
800	64,000	36,000	30,000	130,000	162.50
1,000	80,000	38,000	31,000	149,000	149.00
1,200	96,000	40,000	32,000	168,000	140.00

At the start of the year, after obtaining contrasts with reliable buyers for 600 machines at $160 each, it becomes evident that due to business conditions it will not be possible to sell any more machines in the domestic market for the remainder of the year regardless of price.

A reliable foreign purchaser offers to buy 600 machines at $100 each.

Disregarding all questions of long-run policy with respect to foreign sales, should the offer be accepted? Why?

5.10   A shoe manufacturer produces a pair of shoes at a labor cost of 90 cents a pair and a material cost of 80 cents a pair. The fixed charges on the business are $90,000 a month and the variable costs are 40 cents a pair. If the shoes sell for $3 a pair, how many pairs must be produced each month for the manufacturer to break even?

5.11   Define depreciation rate and give the general formula for determining it by (*a*) the straight-line method, and (*b*) the sinking-fund method. The Interstate Commerce Commission prescribes the former in maintaining the accounts of public utilities, while engineers customarily use the latter in determining the economic propriety of major undertakings. Explain why, although there is a radical difference in the rates developed by the two formulas, both of these choices are appropriate and adequate for their respective purposes.

5.12   *a.* To what extent are general partners liable for debts of the firm?

b. In what way would the formation of a limited partnership change the liability status of those who enter the firm as limited partners?

c. What is the purpose of providing for cumulative voting at stockholders' meetings instead of following the more usual method of giving one vote for each share?

5.13   Explain or discuss the differences of the following terms:

a. Partnership and corporation	*f.* Bill of lading
b. Stock watering	*g.* Promissory note
c. Certified check	*h.* Liquidated damages and unliquidated damages
d. Corporation bond	*i.* Money and barter
e. Letter of credit	

5.14   What are the essential elements of a contract?

5.15   What conditions may render a contract illegal?

5.16   *a.* When may contracts be terminated before execution?

b. Under what conditions may a contract be taken out of the hands of a contractor?

5.17   State five rules of ethical behavior for a professional engineer.

5.18 Is it proper or improper for a professional engineer (*a*) to participate in competitive bidding against his colleagues to secure a professional engagement that is to go to the lowest bidder, or (*b*) to accept an engagement to review the work of another professional engineer without his knowledge? Give reasons for your answers.

5.19 An old, light-capacity highway bridge may be strengthened at a cost of $60,000 or it may be replaced by a modern bridge of sufficient capacity at a cost of $250,000. The present net salvage value of the old bridge is $25,000. It is estimated that the reinforced bridge will last for five years after which replacement will be necessary. At the end of the five years, the net salvage value of the reinforced bridge will be $10,000 and the net salvage value of the new bridge will be $100,000 after 30 years. The additional cost of maintenance and inspection of the old bridge will be $1,500 per year. Assuming interest at 8 per cent and depreciation on a straight line basis, state whether it is more economical to reinforce the old bridge or to replace it. Show the calculations on which you base your reply.

5.20 Which is the more economical: (*a*) a building costing $200,000 which will have to have an addition built 10 years from now at an additional cost of $200,000, or (*b*) a complete building constructed now at a cost of $300,000? Interest at 10 per cent. Assume life of the building is so long that depreciation may be neglected.

5.21 Two proposed types of furnishings are to be compared as to capitalized cost at 8 per cent interest in perpetuity. Type *A* has a life of 10 years; initial cost and cost of renewal is $10,000; annual maintenance is $100; repairs every five years are $500; no salvage value. Type *B* has a life of 15 years; initial cost and cost of renewal is $15,000; no annual maintenance cost; repairs every five years are $200; salvage value is $3,000. Determine the capitalized cost of each.

5.22 A debt of $1,000 is to be paid off in five equal yearly payments, each payment combining an amortization installment and interest at 8 per cent on the previously unpaid balance of the debt. What should be the amount of each payment?

5.23 A syndicate wishes to purchase an oil well which, estimates indicate, will produce a net income of $500,000 per year for 20 years. What should the syndicate pay for the well if, out of this net income, a return of 10 per cent on the investment is desired and a sinking fund will be established at 6 per cent interest to recover the investment?

5.24 Two methods, A and B, of conveying water are being studied. Method A requires a tunnel, first cost $1,000,000; life perpetual, annual operation, and upkeep are $25,000. Method B requires a ditch plus a flume. First cost of ditch is $350,000; life perpetual, annual operation, and upkeep are $20,000. First cost of the flume is $200,000, life, 10 years; salvage value is $30,000; annual operation and upkeep are $40,000. Compare the two methods for perpetual service, assuming an interest rate of 8 per cent.

5.25 A $1,000,000 issue of 6 per cent 15-year bonds was sold at 90. If miscellaneous initial expenses of the financing were $20,000 and a yearly expense of $2,000 is incurred, what is the true cost to the nearest 0.1 per cent that the company is paying for the money it borrowed?

5.26 A corporation has total outstanding stock consisting of 10,000 shares of common and 1,000 shares of simply participating, $100 par, 5 per cent cumulative preferred. The corporation distributed $2,500 in dividends in 1965 but suffered a slight loss in 1966 and just broke even in 1967. During 1966 and 1967, it paid no dividends. In 1968, it had a net income of $78,500 applicable to dividends and in 1969, $77,000.

  a. What was the payment on each share of common stock for the year 1968?

  b. What was the payment on each share of preferred stock for the year 1969?

  If the incorporation and stock certificates carry no specifications beyond that given in the above statement,

  c. How many voting shares are there?

  d. How many shares are redeemable?

5.27 A manufacturer is analyzing his inventory control. 15,000 units of an item are purchased every year. The average daily requirement is 50 units, and the normal time for delivery after placement of a standard order of 2,000 units is 30 days. No more than 1,800 units or less than 200 units are used in any 30-day period.

1. Should the normal ordering point in units be: (1a) 1,800, (1b) 1,850, (1c) 1,900, or (1d) 2,000 units?

2. Should the maximum number of units of storage space be: (2a) 1,800, (2b) 2,400, (2c) 3,600, or (2d) 4,200 units?

3. Should the normal maximum inventory limit in units be: (3a) 2,000, (3b) 2,300, (3c) 2,600, or (3d) 3,000 units?

4. Is the average inventory in number of units: (4a) 2,250, (4b) 1,800, (4c) 1,550, or (4d) 1,300 units?

5. Is the annual inventory rate of turnover: (5a) 11.28, (5b) 11.40, (5c) 11.54, or (5d) 11.70?

5.28   In addition to the data in question 5.27, assume the cost of handling a purchase is $7.50, the storage cost for average inventory per unit, per annum, is $0.30, and the carrying charges (percentage of average annual inventory valuation) are 30 per cent. If the item in question may be purchased in lots of 1,500, 2,000, 2,500, or 3,000 units at unit prices of $2.00, $1.85, $1.75, or $1.70, respectively:

1. Would the value of the average inventory if a lot of 2,000 units is purchased be: (1a) $2,405, (1b) $2,475, (1c) $2,545, or (1d) $2,615?

2. Would the unit buying expense if a lot of 3,000 units is purchased be: (2a) $0.00500, (2b) $0.00375, (2c) $0.00300, or (2d) $0.00250?

3. Would the cost to store 2,000 units for 1 year at the average inventory per unit be: (3a) $400, (3b) $500, (3c) $600, or (3d) $700?

4. Would the inventory carrying charge for an order of 2,500 units be: (4a) $813.75, (4b) $836.50, (4c) $859.25, or (4d) $882.00?

5.29   In keeping with the data in questions 5.27 and 5.28, would the economic order size be: (a) 2,500, (b) 3,000, (c) 3,500, or (d) 4,000 units?

5.30   A manufacturer has the following commodity inventory account:

June 1	Balance on hand	500 units at \$100 per unit
June 12	Purchased	500 units at \$125 per unit
June 25	Delivered to shop	600 units
July 6	Purchased	500 units at \$110 per unit
July 15	Delivered to shop	600 units
Aug. 3	Purchased	500 units at \$105 per unit
Aug. 19	Delivered to shop	600 units

1. If the last-in first-out method of inventory valuation is used, would the value of the inventory on August 20th be: (1a) \$17,000, (1b) \$18,000, (1c) \$19,000, or (1d) \$20,000?

2. If the first-in first-out method of inventory valuation is used, would the value of the inventory on August 20th be: (2a) \$19,000, (2b) \$21,000, (2c) \$24,000, or (2d) \$28,000?

3. If the weighted-average method of inventory valuation is used, would the value of the inventory on July 31st be: (3a) \$33,333.33, (3b) \$37,500, (3c) \$41,000, or (3d) \$45,000?

5.31 A company is planning to buy a special fixture for \$1,000. Their data indicated that: the estimated saving in direct labor cost would be 5 cents; the interest rate would be 6 per cent; the rate for fixed charges would be 6 per cent; the rate for upkeep would be 13 per cent; the life of the equipment could be assumed to be 4 years; and the overhead saving due to direct labor saved would be 40 per cent. The estimated cost of each setup should average \$50.

1. In order to return the cost out of earnings in 4 years, must the number of pieces to be put through in one lot per year be: (1a) 7,675, (1b) 7,857, (1c) 8,040, or (1d) 8,224 pieces?

2. If the company plans to produce 9,200 pieces per year, will the number of years to amortize the cost of the equipment be: (2a) 1, (2b) 2, (2c) 3, or (2d) 4 years?

3. If the fixture produces 10,000 pieces per year, will the resulting gross profit per year be: (3a) \$110, (3b) \$120, (3c) \$135, or (3d) \$150?

4. If 12,000 pieces are to be manufactured in four lots per year, could the money invested in this fixture be: (4a) \$1,155, (4b) \$1,215, (4c) \$1,280, or (4d) \$1,350?

5.32 The balance sheet of a company is as follows:

Assets			*Liabilities*	
Current:			Current:	
Cash	$ 2,000		Accounts payable	$ 8,000
Accounts receivable	3,000		Notes payable	1,000
Inventory	8,000		Accrued taxes	1,000
Total current assets	$13,000		Total current liabilities	$10,000
Fixed:			Capital:	
Land	$ 1,000		Common stock, $10	
Buildings less depreciation	15,000		par, 2,500 shares outstanding	$25,000
Machinery less depreciation	9,000		Surplus	4,000
Total fixed assets	$25,000		Tangible net worth	$29,000
Deferred charges	1,000			
Total assets	$39,000		Total liabilities	$39,000

1. Would the value of the ratio of the total current liabilities to the tangible net worth be: (1*a*) 0.345, (1*b*) 0.350, (1*c*) 0.355, or (1*d*) 0.360?

2. Would the fixed assets to tangible net worth be: (2*a*) 0.856, (2*b*) 0.861, (2*c*) 0.866, or (2*d*) 0.871?

3. If the net sales for the balance period were $30,000, would the ratio of net sales to net working capital be: (3*a*) 8.5, (3*b*) 9.0, (3*c*) 10.0, or (3*d*) 11.5?

5.33 A chemical is purchased for use as a raw material in a manufacturing plant. The costs involved in making a purchase are $21 per purchase order regardless of the size of the order. During the year 3,000 gal of this chemical is consumed at a fairly uniform rate. The chemical is purchased and stored in 50-gal drums, and the purchase price per gallon, including freight, is $3.30. Annual storage costs are estimated as 50 cents per drum of maximum inventory and the annual carrying charges are estimated as 12 per cent on average inventory. To assure continuous operations, at least 200 gal should be maintained on hand at all times as an emergency stock. What is the most economical size of lot to purchase in terms of drums?

5.34 A general contractor is required to install and operate a temporary well-point system during a 6-month phase of construction of a riverside powerhouse, from Apr. 1 through Sept. 30, 1973. The necessary equipment will cost $1,000 per month to rent. A pump operator will have to be in attendance continuously and must be paid an hourly wage of $9.00 for each 8-hr weekday shift, $13.50 for each 8-hr Saturday shift, and $18.00 for each 8-hr Sunday and legal holiday shift. Payroll taxes and insurance are 13 per cent of wages. Fuel is estimated at $40 per day. Overhead and maintenance charges are 15 per cent of wages, fuel, and rental charges. Payment for successful completion of the well-pointing operation will be one lump sum at the end of the 6-month period. If financing costs the contractor 8 per cent per annum and he desires a profit and contingency of 10 per cent of his costs, what would be his lump-sum bid for the well-pointing operation?

5.35 A father wishes to develop a fund for his newborn son's college education. The fund is to pay $5,000 on the eighteenth, nineteenth, twentieth, and twenty-first birthdays of his son. The fund will be built up by the deposit of a fixed sum on the son's first to seventeenth birthdays. If the fund earns 4 per cent, what should the yearly deposit into the fund be?

5.36 A lot was purchased in January, 1960 for $5,000. Taxes and assessments were charged at the end of each year as follows: $100 for 1960, $100 for 1961, $200 for 1962, $100 for 1963, $100 for 1964, and $100 for 1965. The owner paid the charges for 1960 and 1961, but not for subsequent years. At the end of 1965 the lot was sold for $10,000, the seller paying back charges at 7 per cent interest compounded annually and also paying a commission of 5 per cent to an agent for handling the sale. What rate of return was realized on the investment?

5.37 A man owns a building on which there is a $100,000 mortgage which earns 6 per cent per annum. The mortgage is being paid for in 20 equal year-end payments. After making 3 payments, the man desires to reduce his payments by refinancing the balance of the debt with a 30-year mortgage at 8 per cent, and to be retired by equal annual payments. What would be the reduction in the yearly payment?

5.38 How much would the owner of a building be justified in paying for a sprinkler system that will save $2,000 a year in insurance premiums if the system has to be renewed every 20 years and has a salvage value equal to 10 per cent of its cost, if money is worth 5 per cent?

5.39 A new boiler has just been installed. It is expected that there will be no maintenance charges until the end of the eleventh year, when $500 will be spent on the boiler and $500 will be spent at the end of each successive year until the boiler is scrapped at the end of its nineteenth year of service. What sum of money set aside at the time of installation of the boiler at 4 per cent will take care of all maintenance expenses for the boiler?

5.40 An automobile costs $3,000. It is run approximately the same distance each year, with transportation costs at a minimum. If annual expenses for maintenance are $100 at the end of the first year and increase $100 per year each year thereafter, and if the trade-in value is $1,800 at the end of the first year and this decreases uniformly $200 each year thereafter, when should the car be traded (in full number of years)?

MECHANICAL ENGINEERING

6.01 At the beginning of compression in a cold-air Otto cycle, the temperature is 100°F, the pressure is 13.75 psi abs, and the volume is 1 cu ft. At the end of compression, the pressure is 155 psi abs. The heat supplied to the cycle is 50 Btu. Find:

   a. The compression ratio
   b. The per cent of clearance
   c. The pressure and temperature just after combustion
   d. The net work per cycle in foot-pounds
   e. The heat rejected from the cycle
   f. The m.e.p.

Show sketches of *pv* and *Ts* diagrams noting pertinent points in cycle.

6.02 A 6-in. steel main has an inside diameter of 4.897 in. and an outside diameter of 6.625 in. It is insulated at the outside with asbestos. The steam temperature is 300°F and the air temperature outside is 70°F.

$$h_s(\text{steam}) = 20 \text{ Btu/(sq ft)(hr)(°F)}$$
$$h_a(\text{air}) = 6 \text{ Btu/(sq ft)(hr)(°F)}$$
$$k(\text{asbestos}) = 0.06 \text{ Btu/(ft)(hr)(°F)}$$
$$k(\text{steel}) = 30 \text{ Btu/(ft)(hr)(°F)}$$

In order to limit the heat loss to 10 Btu/(sq ft)(h), should the thickness of the asbestos be: (*a*) 8.25, (*b*) 8.50, (*c*) 8.75, or (*d*) 9.00 in.?

6.03 A turbine generator operates on the reheat-regenerative cycle with one reheat and two stages of feedwater heating, with pressures and temperatures as indicated below. Heater no. 1 (nearest the condenser) is of the "open" or direct-contact type. Heater no. 2 is of the "closed" or surface type and drains from this heater are "cascaded," i.e., flow back to heater no. 1. Operation may be assumed to be ideal except for departures indicated below.

Steam is generated at 700 psia and some superheat, and reaches the turbine without heat loss at $p = 655$ psia and $t = 740°F$. Resuperheating takes place at $p = 109$ psia to a temperature of $700°F$. Extractions for feedwater heating occur at $p = 109$ psia and 15 psia. Condenser pressure is 1 psia. In the closed (no. 2) heater the temperature difference between the condensed steam leaving and the feedwater leaving is $10°F$.

To reduce the number of values to be looked up in the steam tables, the following partial list of values may be assumed to be correct:

At entrance to boiler	$h = 296.8$ Btu per lb
At first extraction point	$h = 1188.7$ Btu per lb
After resuperheating	$h = 1378.4$ Btu per lb
At second extraction point	$h = 1177.6$ Btu per lb
At entrance to condenser	$h = 973.0$ Btu per lb
Condensate leaving condenser	$h = 69.7$ Btu per lb
Feedwater leaving no. 1 heater	$h = 181.1$ Btu per lb
Drains leaving no. 2 heater	$h = 305.0$ Btu per lb

To be done: Assuming a flow of 1 lb of steam to the turbine:

a. Sketch a line diagram of apparatus, naming each piece. Use a consistent system of numbers or letters to indicate state entering or leaving each piece of apparatus.

b. Sketch an *h-s* diagram for the cycle, identifying pertinent points in terms of the symbols used in (*a*).

c. Determine the initial temperature of the steam leaving the boiler.

d. Determine the temperature of the feedwater leaving the last heater.

e. Determine the amount of steam that must be extracted from the turbine for heater no. 1, assuming that extraction for heater no. 2 = 0.1262 lb per lb of steam supplied to the turbine.

f. If the internal efficiency of the turbine before and after resuperheating is 80 per cent, estimate the quality of the steam at exhaust.

g. For the Rankine cycle of operation with no resuperheating and no extraction, what would be the *h* of the exhaust steam?

6.04 Calculate the heat loss in Btu per sq ft per hr through a wall construction of 8-in. cinder block with 2 in. of cork and ½ in. of plaster on one side, and ½ in. of plaster on the other side. Assume still air on both sides having a temperature difference of 40°F.

6.05 The cross section of a wall of a cold storage room consists of 16 in. of brick ($K = 3.6$), and air space equivalent to the resistance of about 8 in. of brick, 8 more in. of brick, 4 in. of corkboard ($K = 0.3$) and 1.5 in. of cement plaster ($K = 2.3$).

Find the quantity of heat transmitted through the wall by conduction per day per square foot of wall surface when the temperatures of the inner and outer wall surfaces are 30 and 70°F, respectively.

6.06 Air passes through a duct at the rate of 1,800 cu ft per min, the dry-bulb temperature being 70°F and the relative humidity 40 per cent. It is proposed to raise the humidity of the air to 75 per cent by blowing into it (*a*) water spray at 50°F or (*b*) saturated steam at atmospheric pressure. How many pounds of water or steam will be required and what will be the final dry- and wet-bulb temperatures in each case? Assume adiabatic conditions and thorough mixing so as to produce uniform conditions.

6.07 Water enters a cooling tower at 126°F and leaves at 80°F. Air enters at 85°F and relative humidity 47 per cent and leaves in a saturated condition at 115°F. Determine:

a. Weight and volume of air needed per pound of water cooled.

b. Amount of water that can be cooled with 2,000 cfm of free air.

6.08 An auditorium seating 1,800 people is to be maintained at 78°F dry-bulb and 67° wet-bulb temperature when outdoor air is at 90°F dry-bulb and 75° wet-bulb. Solar heat load is 120,000 Btu per hr. Determine:

    a. Cubic feet per minute of outdoor air required for ventilation.

    b. Volume (in cfm) of conditioned air at 65°F dry-bulb that must be circulated to carry the total sensible heat load.

    c. Wet-bulb temperature of the conditioned air to absorb the moisture load.

6.09 A small store has a sensible heat load of 53,100 Btu per hr and a latent heat load of 14,400 Btu per hr. Indoor conditions of 75°F dry bulb and 50 per cent relative humidity are to be maintained when ventilating air amounting to 1,200 cu ft per min at 91°F dry bulb and 75°F wet bulb is used. Duct work and apparatus are arranged as shown in Fig. 6.09. The air conditioner has an apparatus dew point of 50.5°F and the conditioned air leaves the fan at 61°F dry bulb. The fan energy may be neglected.

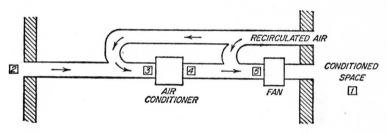

F<small>IG</small>. 6.09

Find:

    a. The tons of refrigeration required

    b. The volume of bypass air in cubic feet per minute

6.10 In an auditorium maintained at a temperature not to exceed 75°F and a relative humidity not to exceed 60 per cent, a sensible heat load of 450,000 Btu per hour, and 1,200,000 grains of moisture per hour must be removed. Air is supplied to the auditorium at 65°F.

    a. How many pounds of air per hour must be supplied?

    b. What is the dew-point temperature and the relative humidity of the entering air?

    c. How much latent heat load is picked up in the auditorium?

6.11 A two-pass surface condenser is to be designed using an over-all heat-transfer coefficient of 480 Btu per hr per deg F per sq ft of outside tube surface. The tubes are to be 1 in. outside diameter with $\frac{1}{16}$-in. walls. Entering circulating-water velocity is to be 6 fps. Steam enters the condenser at a rate of 1,000,000 lb per hr at a pressure of 1 psia and an enthalpy of 1090 Btu per lb. Condensate leaves as saturated liquid at 1 psia. Circulating water enters the condenser at 85°F and leaves at 95°F. Calculate the required number and length of condenser tubes.

6.12 Diesel fuel, $C_{12}H_{26}$, when injected into compressed air burns at a 30 to 1 air fuel ratio by weight. If the engine exhausts at 600°F into a heating system where the room temperature is 75°F what per cent of the fuel supplied is available to the heating system? From the combustion equation, the gas leaving the engine is made up of: $12CO_2$, $13H_2O$, $69.56N_2$, and 88.06 air.

6.13 Exhaust steam at 1 psig pressure and containing 12 per cent moisture is fed into an open-type feed-water heater. How many pounds of this steam are required to raise the temperature of 1,000 lb of water from 60 to 200°F? The barometric pressure is 30 in. Hg.

6.14 A tubular-type air heater has 31,154 sq ft of heating surface and heats 357,000 lb per hr of air from 80 to 479°F. Flue gas enters at 725°F. and leaves at 405°F. What is the heat-transfer rate, in Btu/(sq ft)(hr)(deg) mean temperature, difference? What is the weight of flue gas passing through?

6.15 A single-stage impulse turbine receives steam having an available energy of 100 Btu per lb. Five per cent of the maximum theoretical nozzle exit velocity is lost through nozzle friction; 10 per cent of the entering relative velocity is lost by the steam in passing through the blading. The nozzle makes an angle of 20° with the plane of the wheel. The blading is symmetrical, i.e., equal entrance and exit angles. Blade speed is 900 ft per sec. Assume nozzle inlet velocity negligible. Calculate:

   a. Steam velocity leaving nozzle.
   b. Relative velocity entering blade.
   c. Relative velocity leaving blade.
   d. Absolute exit velocity.

e. Blade entrance angle.

f. Force exerted on blade by each pound of steam flowing per second.

g. Blade horsepower if 1,800 lb of steam flow per hr.

h. Shaft horsepower. Friction between the steam and the wheels absorbs 6 hp. Bearing and governors take 2 hp. Assume this 2 hp does not go into the steam.

i. Steam entering the turbine has an enthalpy of 1,200 Btu per lb. Neglecting radiation losses from the turbine, what should be the enthalpy of the steam leaving the turbine if the energy due to residual velocity goes to increasing the enthalpy of the steam.

6.16 A steam-electric generating station has four 10,000-kw turbogenerators. Steam is supplied at 250 psia and 600°F. Exhaust is at 2 in. Hg abs. Daily load factor is 75 per cent. Assuming average steam rates and efficiencies, determine:

a. Tons of coal required per day

b. Condenser cooling water in gallons per minute

6.17 A multistage impulse turbine running at 3,600 rpm and delivering 1,000 hp, receives steam at 180 psia and 600°F, and the exhaust is at 16 psia. It is desired to limit the peripheral velocity of the turbine blading to approximately 625 ft per sec. The stage efficiency is 75 per cent, all stages are of the same diameter, and the reheat factor may be initially assumed at 1.06. Making suitable allowances for various losses, determine the proper number of stages, the actual engine efficiency and the ideal and actual steam rates.

6.18 An ideal refrigeration cycle using ammonia as the refrigerant compresses isentropically from saturated vapor at 20 to 190 psi. The temperature at the high-pressure side of the expansion valve is 80°F. Draw the TS diagram. Considering 1 lb, find (*a*) the net work, (*b*) the heat rejected to the condenser water, (*c*) the refrigeration, (*d*) the coefficient of performance, and (*e*) the pounds of NH_3 circulating per ton of refrigeration.

6.19 A stoker-fired boiler plant having dutch-oven-type furnaces was operated for a year with an average flue-gas analysis of 13 per cent CO_2, 0 per cent CO, and 6.25 per cent O_2, and combustible matter to ash pit 10 per cent. The second year an effort was made to raise the efficiency and an average flue-gas analysis of 15 per cent CO_2, 0.1 per cent CO, and 3.9 per cent O_2 was maintained with 16 per cent combustible matter to ash pit. A high grade of Pennsylvania bituminous slack coal was burned both years, having 7 per cent ash and 14,200 Btu bone dry. At the end of the second year it was found that the efficiency was the same as for the first year of operation yet the cost per 1,000 lb of steam had increased 2 per cent. Explain.

6.20 A test of an oil-fired boiler indicated 12.77 lb of water evaporated per lb of oil. The boiler pressure was 200 psia, superheat was 87°F, feedwater temperature 93°F, and boiler and furnace efficiency 82.8 per cent. What was the calorific value of the oil?

6.21 A steam boiler on a test generates 885,000 lb of steam in a 4-hr period. The average steam pressure is 400 psia, the average steam temperature is 700°F, and the average temperature of the feedwater supplied to the boiler is 280°F. If the boiler efficiency for the period is 82.5 per cent, and if the coal has a heating value of 13,850 Btu per lb as fired, find the average amount of coal burned in short tons per hour.

6.22 *a.* Define a boiler horsepower.

b. What is the relationship between a boiler horsepower and a mechanical horsepower?

c. Calculate the horsepower of a boiler generating 4,000 lb of dry steam per hr, with a factor of evaporation of 1.06. The latent heat of steam at 212°F is 970 Btu per lb.

6.23 A fan whose static efficiency is 40 per cent has a capacity of 60,000 cu ft per hr at 60°F and a barometer of 30 in. Hg, and gives a static pressure of 2 in. of water column on full delivery. What size electric motor should be used to drive this fan?

6.24 A fan discharged 9,300 cfm of air through a duct 3 ft in diameter against a static pressure of 0.85 in. of water. The gage fluid density was 62.1 lb per cu ft, the air temperature 85°F and the barometric pressure 28.75 in. Hg. If the power input to the fan is measured as 3.55 hp, what is the over-all mechanical efficiency of the fan? What is the static efficiency of the fan?

6.25 What horsepower is supplied to air moving at 20 ft per min through a 2 by 3 ft duct under a pressure of 3 in. water gage?

6.26 A fan whose efficiency is 40 per cent has a capacity of 60,000 cu ft per hr at 60°, a barometer of 30 in., and gives a pressure of 2 in. of water column on full delivery. What size electric motor should be used to drive this fan? Give computations.

6.27 A refrigeration plant is rated at 20-ton capacity.

a. How many pounds of air per hour will it cool from 90 to 70°F at constant pressure?

b. What is the approximate engine horsepower required to operate the plant?

6.28 A single-acting twin-cylinder 12 by 12 in. compressor receives saturated ammonia vapor at 31.16 psia and discharges it at 180 psia. Saturated liquid enters the expansion valve. The water to be frozen is at 80°F and the manufactured ice is at 16°F. The speed of the compressor is 150 rpm and volumetric efficiency is 82 per cent. The ratio of the ideal to brake work of the compressor is 80 per cent. Assume specific heat of ice is 0.5 Btu per lb. Determine:

a. Capacity

b. Tons of ice per 24 hr, neglecting radiation losses

c. The brake horsepower of the compressor

6.29 In the two-speed planetary reduction gear shown schematically, either gear *A* or internal gear *C* can be clutched-in at will directly to a shaft rotating at constant speed, the other gear meanwhile being held stationary by means of a brake. (The clutch and brake mechanisms are not shown.) The gears

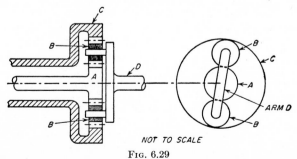

NOT TO SCALE

Fig. 6.29

are of 12 diametral pitch. Gear A has a pitch diameter of 3 in., and planet gears BB each have 18 teeth. Find the ratio between the high and low speeds of the follower shaft D.

6.30 An airplane engine having 14 cylinders and a $5\frac{3}{8}$-in. bore by 6-in. stroke develops a torque of 2,100 ft-lb at 2,500 rpm, peak power. The clearance volume of each cylinder is 23.4 cu in. Total engine weight is 1,450 lb. Calculate (a) the compression ratio, (b) the bmep at peak power, and (c) the bhp per lb.

6.31   a. A shaft made of SAE 1045 steel (ultimate = 97,000 psi and yield point = 58,000 psi) is subjected to a torque varying from 20,000 in-lb clockwise to 6,000 in-lb counterclockwise. What would be the diameter of the shaft if the factor of safety is 2 based on the yield point and the endurance strength in shear?

   b. Find the diameter of the shaft when it transmits a constant torque of 20,000 in-lb and the factor of safety is 2, as in a.

   c. Explain the reason for the difference in diameters obtained in a and b.

6.32   A Corliss engine 28 in. diameter by 54 in. stroke, driving a three-roller cane mill through double-reduction machine-cut spur-gearing runs at 60 rpm, receives live steam at the throttle at 150 psi saturated steam and exhausts against 15 psi pressure. Valve gear is set for $\frac{3}{8}$ cutoff. Piston rod is 5 in. diameter. Engine does not have tail rod. Assuming a transmission loss of 10 per cent in each gear reduction and an additional loss of 15 per cent for misalignments in the mill coupling and crownwheels, determine:

   a. Mean effective pressures at head and crank ends
   b. Horsepower delivered by engine at crankshaft
   c. Horsepower delivered to mill

6.33   A car engine rated at 12 hp gives a maximum torque of 6,516 in-lb. The clutch is of a single-plate type, and both sides of the plate are effective. The coefficient of friction is 0.3, the axial pressure is limited to 12 psi, and the external radius of friction surface is 1.25 times the internal radius. Find the dimensions of the plate and the total axial pressure which must be exerted by the springs.

6.34 A four-cylinder engine is equipped with a flywheel which may be considered to have an effective weight of 50 lb concentrated in the rim at a mean radius of 7 in. The excess energy delivered to the flywheel, which increases its speed from minimum to maximum, is estimated to be 20 per cent of the indicated work per revolution. At an output of 20 hp, the mean speed is 1,800 rpm and the mechanical efficiency of the engine is 80 per cent. Find the total speed variation.

6.35 A test on a 1-cylinder Otto 4-cycle (typical gasoline engine) internal-combustion engine yielded the following data: torque, 700 ft-lb; mean effective pressure, 110 psi; bore, 11 in.; stroke, 12 in.; speed, 300 rpm; fuel consumption, 24 lb per hr; and lower heating value of fuel, 18,000 Btu per lb. Find:

    a. The thermal efficiency of the engine.

    b. The mechanical efficiency.

    c. The fuel cost of horsepower per hour with fuel at 50 cents per gallon.

6.36 A built-up cylinder consists of an outer steel ring of thickness $\frac{1}{8}$ in. press-fitted on a copper ring of thickness $\frac{1}{16}$ in., as shown in Fig. 6.36. The copper ring has an outside diameter of 6 in. and after assembly is under a compressive stress of 2,000 psi. Steel density = 490 lb per cu ft and E = 30,000,000 psi. Copper density = 558 lb per cu ft and E = 16,000,000 psi. At what rpm will the stress in the copper ring become zero?

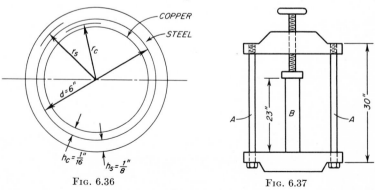

Fig. 6.36   Fig. 6.37

6.37 A screw press has the dimensions shown in Fig. 6.37. Two rods *A* are of steel 1.00 in. in diameter. A brass tube *B* is 2.50 in. in outside diameter and 1.50 in. inside diameter. The screw is turned until the unit stress in the brass tube is 5,000 psi at 70°F. Assume no deformation of the heads of the press, or of the screw and bolts. The coefficients of thermal expansion for brass and steel are, respectively, 0.0000105 and 0.0000065 per unit per degree F. At what temperature will the stress in the brass tube be 10,000 psi? $E_{steel} = 30,000,000$ psi, and $E_{brass} = 14,500,000$ psi.

6.38 A steam engine develops 80 hp at 150 rpm against a steady load. The flywheel weighs 3 tons and its radius of gyration is 5 ft. If the load suddenly changes to 0.1 of the initial value and there is no change in steam supply for 5 revolutions after the reduction of the load, calculate the change in speed from beginning to end of this period.

6.39 *a.* An automobile motor has a torque of 230 ft-lb, driving the rear axles through a differential having a ratio of 4.11, and an efficiency of 97 per cent. Calculate the tractive effort at the rear wheels, their diameters being 30 in.

b. What rear-wheel horsepower is developed when the speed is 40 mph?

6.40 Calculate the bore and stroke of a 6-cylinder automobile motor to deliver 30 bhp at 1,800 rpm, the ratio of stroke to bore being 1.4. Assume the mean effective pressure in the cylinder to be 90 psi and the mechanical efficiency 85 per cent.

6.41 A composite shaft is made of solid-steel rod 1.50 in. of outside diameter on which is shrunk an aluminum tube of 2.25 in. outside diameter. E_{shear} for steel = 12,000,000 psi, and E_{shear} for aluminum = 3,500,000 psi. If the shaft is 5 ft long and carries a torque of 15,000 in.-lb, determine:

a. The total angle of twist of the shaft

b. The maximum shearing stress in the steel and in the aluminum

6.42 A screw press, having two screws of $\frac{1}{2}$ in. pitch and exerting a pressure of 50 tons, is to be driven by an electric motor running at 1,700 rpm and connected to the screws by a suitable train of gears, the successive ratios of which are not to exceed 6 to 1. The pressure head is moved at the rate of 1 in. per min. The over-all efficiency is assumed to be 40 per cent. Sketch a suitable gear train and calculate the horsepower delivered by the motor when the press is operating at 50 tons pressure.

6.43 Find suitable pitch and width of face for a pair of $14\frac{1}{2}°$ involute spur gears to transmit 8 hp. The velocity ratio is to be 2 to 3 and the distance between centers is 10 in. The pinion is to be of mild steel and is to run at 150 rpm. The gear is to be bronze.

6.44 A slugging or striking wrench is one designed for tightening bolted connections in restricted quarters with the aid of a sledge or heavy hammer. The free end of this wrench is hit with a hammer in a tangential direction while the socket engages the nut. Determine the diameter of circular section required at the root of an alloy-steel wrench if designed to accommodate the impact caused by a 5-lb hammer striking the free end of the wrench at a velocity of 12 fps. The wrench handle is 18 in. long and of uniform diameter.

6.45 A single-stage double-acting reciprocating air compressor is guaranteed to deliver 500 cu ft per min of free air with clearance of 3 per cent and suction conditions of 14.7 psia and temperature of 70°F and discharge pressure of 105 psia. When tested under these conditions, it satisfied the manufacturer's guarantee, and test results show that compression and reexpansion curves follow pv^n = constant with $n = 1.34$.

  a. What is the piston displacement of the unit in cubic feet per minute?

  b. If the compression ratio is held constant and the unit is operated at an altitude of 6,000 ft where barometric pressure and temperature are 23.8 in. Hg and 70°F, respectively, what will be its capacity and discharge pressure?

  c. Assuming that the bore, stroke, and clearance remain unchanged, what would you suggest doing to the compressor under (b) to make it deliver the same weight of air per minute at the same discharge pressure as under guaranteed sea-level test conditions?

6.46 A d-c motor-driven pump running at 100 rpm delivers 500 gal per min of water against a total pumping head of 90 ft with a pump efficiency of 60 per cent.

    a. What motor horsepower is required?

    b. What speed and capacity would result if the pump rpm was increased to produce a pumping head of 120 ft, assuming no change in efficiency?

    c. Can a 25-hp motor be used under conditions indicated by (*b*)?

6.47 What weight can be lifted by a screw jack that has an efficiency of 80 per cent, if it is operated by a 50-lb force at the end of a 30-in. lever and the pitch of the screw is $\frac{1}{2}$ in.?

6.48 A hoist consists of a drum 24 in. in diameter fastened to a gear wheel with 80 teeth. The gear wheel is turned by a double-threaded worm. If the hoist can raise a 4,000-lb weight on a $\frac{7}{8}$-in. rope at the rate of 100 fpm, what torque and horsepower are being supplied at the shaft of the worm gear? Assume a mechanical efficiency of 60 per cent.

6.49 A safety valve spring having $9\frac{1}{2}$ coils has the ends squared and ground. The outside diameter of the coil is $4\frac{1}{2}$ in. and the wire diameter is $\frac{1}{2}$ in. It has a free length of 8 in. Determine the length to which this spring must be initially compressed to hold a boiler pressure of 205 psig on a $1\frac{1}{4}$-in. diameter valve seat.

    Determine the stress if the spring is compressed to its solid length.

$$E_s = 11.5 \times 10^6 \text{ psi}$$

Calculations and formula must be shown.

6.50 Construct the outline of a disk cam to give uniform acceleration and deceleration to a pointed follower which rises 3 in. while the cam turns through 180°, dwells for 60° cam rotation, falls with uniform velocity while the cam turns through 60°, and dwells the remaining 60°.

6.51 A reverted gear train is required to have an output-to-input speed ratio of exactly 1,131 to 2,000. The input and output gear shafts are colinear. Make a sketch of the gear train and specify the number of teeth on each gear. Use gears of not less than 20 nor more than 50 teeth.

1. Will the number of teeth of the input gear be: (1a) 39, (1b) 40, (1c) 41, or (1d) 42 teeth?

2. Will the number of teeth of the output gear be: (2a) 40, (2b) 45, (2c) 50, or (2d) 55 teeth?

6.52 Two helical gears of the same hand are used to connect two shafts that are 90° apart. The smaller gear has 24 teeth, and a helix angle of 35°. The nominal circular pitch is 0.7854 in. If the speed ratio is $\frac{1}{2}$, will the center distance between the shafts be: (a) 14.000, (b) 14.125, (c) 14.375, or (d) 14.750 in.?

6.53 Body 1, a 2-in. diameter disk, is driven by body 2. Body 2 is a 2-in. square rotating clockwise at a constant speed of 10

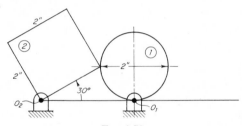

Fig. 6.53

radians per sec. as shown in Fig. 6.53. When the center of the circle is vertically upward,

1. Will the angular velocity of the body be: (1a) 8.5, (1b) 9.0, (1c) 9.5, or (1d) 10.0 radians per sec?

2. Will the rotational direction be: (2a) clockwise or (2b) counterclockwise?

3. Will the angular acceleration of the body be: (3a) 471.7, (3b) 472.2, (3c) 473.2, or (3d) 475.0 radians per sec^2?

4. Will the rotational direction be: (4a) clockwise or (4b) counterclockwise?

6.54 A rotor has two unbalances as shown in Fig. 6.54. Determine the necessary ounce-inch corrections and their angular positions in the end planes if the rotor is to be dynamically balanced.

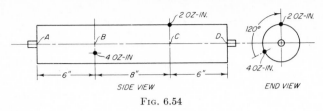

Fig. 6.54

6.55 Figure 6.55 is a schematic diagram of a viscous-damped phonograph arm. What is the value of the damping constant C if the arm is to be critically damped?

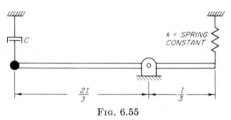

Fig. 6.55

6.56 A rectangular steel beam is subjected to a bending moment of 240,000 lb-ft. The yield point of the steel is equal to 35,000 psi. If the beam is 6 in. wide and 8 in. deep, determine to what distance from the outer surfaces a plastic state exists in the beam. Assume a theoretical stress-strain diagram as shown in Fig. 6.56.

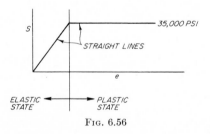

Fig. 6.56

6.57 4,000 gal per min of oil having a specific gravity of 0.85 flows through a clear wrought-iron pipe for a distance of 10,000 ft with an energy loss of 75 ft-lb per lb. Upstream and downstream elevations are 200 and 50 ft, respectively. The pressure upstream is 40 psi. The oil has a kinematic viscosity of 0.0001 sq ft per sec.

1. Will the theoretical inside diameter of the pipe be: (1*a*) 16.0, (1*b*) 16.6, (1*c*) 17.0, or (1*d*) 17.2 in.?

2. Will the downstream pressure be: (2a) 60.0, (2b) 62.5, (2c) 65.0, or (2d) 67.5 psi?

6.58   Tests of airfoils in a wind tunnel are to be used to determine the lift and drag of hydrofoils with a 6-in. chord length for a boat designed to travel at 60 fps in water at 70°F. In order to avoid compressibility effects in the wind tunnel, assume standard air is to be used at a maximum velocity of 200 fps. What chord length should be used for the model airfoils? The hydrofoils are to run deep, so that only viscous effects need be considered.

6.59   The curved beam shown in Fig. 6.59 has a uniform thickness of 1 in. Determine the value of the vertical force F in order that the resultant stresses normal to the vertical cross section A-B be numerically equal at points A and B.

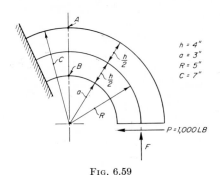

$$h = 4''$$
$$a = 3''$$
$$R = 5''$$
$$C = 7''$$

$P = 1,000\,LB$

F

Fig. 6.59

6.60   For the welded connection in Fig. 6.60, find the permissible static load P. An E6010 electrode is used and the factor of safety by the maximum shear theory of failure is to be equal to 2.

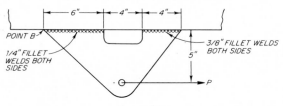

POINT B

1/4" FILLET WELDS BOTH SIDES

3/8" FILLET WELDS BOTH SIDES

P

Fig. 6.60

STRUCTURAL ENGINEERING

7.01 A steel hanger consists of a plate 5 by $\frac{1}{4}$ in. held at the top by four $\frac{3}{4}$-in. rivets in single shear. The rivets are arranged in diamond form so that the first net section has one hole and the second net section has two holes. In which way is the joint most likely to fail? What is the efficiency of the joint? Use unit stresses of 16,000 psi for tension, 12,000 psi for shear, and 24,000 psi for bearing.

7.02 A tension member is made up of $\frac{3}{8}$-in. thick steel plates 9 in. wide and spliced by a lap joint made with three rows of $\frac{3}{4}$-in. bolts snug fit in drilled holes and arranged in the pattern shown. The permissible stresses in the bolts and the plates are: 20,000 psi for tension in the net section, 16,000 psi for shear, and 32,000 psi for bearing. (a) What is the maximum permissible tension in the plates in pounds? (b) What is the efficiency of the joint in per cent?

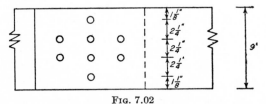

Fig. 7.02

7.03 A double-covered riveted butt seam has main plates $\frac{1}{2}$ in. thick and $\frac{5}{16}$-in. cover plates. The rivets are $\frac{3}{4}$ in. in diameter (the holes $\frac{13}{16}$ in.) and are arranged in four rows, two

146

on each side of the seam. The pitch in the rows next to the seam
is three rivet diameters, and in the outer rows six rivet diameters.
Find the allowable load per inch of seam and the working efficiency. The permissible working stresses in the rivets and plates
are: 16,000 psi for tension, 10,000 psi for shear, and 20,000 psi
for bearing.

7.04 A 5-ft diameter tank is to be designed to withstand an
internal working pressure of 500 psi. All rivets are to be the same
size for both the transverse and longitudinal double cover-
plated butt joints. Rivet holes are drilled $\frac{1}{16}$ in. larger diameter
than the rivets. Working stresses for plates and rivets are:
tension, 20,000 psi; shear, 15,000 psi; and bearing, 24,000 s.s.,
30,000 d.s. If the longitudinal rivet pattern is as shown, what
is the rivet pitch to the nearest inch, the efficiency of the joint,
and the thickness of the main plate and cover plates?

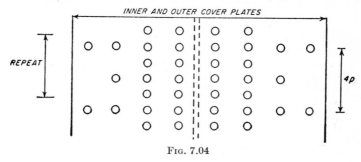

INNER AND OUTER COVER PLATES

REPEAT

4p

FIG. 7.04

7.05 What will be the rivet pitch and cover-plate thickness
for the transverse plate splice of question 7.05 for the rivet
pattern shown?

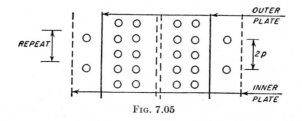

OUTER PLATE

REPEAT

2p

INNER PLATE

FIG. 7.05

7.06 The longitudinal quadruple-riveted butt joint shown has ¾-in. thick main plates and ½ in. splice plates. Rivets are 1⅛ in. diameter and holes, 1³⁄₁₆ in. If this splice was designed by using boiler code unit stresses of 11,000 psi in tension, 8,800 psi in shear, and 19,000 psi in bearing, what is the efficiency of the splice and in what way is the splice likely to fail?

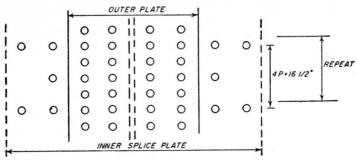

FIG. 7.06

7.07   Design a longitudinal lap joint for a tank 72 in. in diameter carrying 125 psi internal pressure. The allowable working stresses are: 16,000 psi in tension, 12,000 psi in shear, and 24,000 psi in bearing.

7.08   Design a welded joint as shown to develop the full strength of the angle in tension. The allowable stress in the throat of the fillet welds is 15,000 psi and the tensile stress in the angle is 20,000 psi.

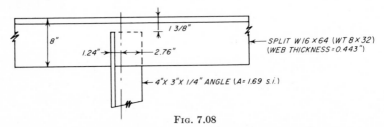

FIG. 7.08

7.09   Design a welded joint to develop the full strength of the T in tension. The allowable stresses in fillet welds are: 2,000 lb per in. for ¼-in. fillet weld; 2,500 lb per in. for ⁵⁄₁₆ in.; 3,000 lb per in. for ⅜ in.; and 3,500 lb per in. for ⁷⁄₁₆ in. Tensile stress in the T is 20,000 psi.

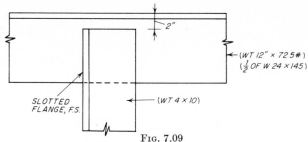

FIG. 7.09

7.10 A single plate 9 by 26 in. is fastened to the flange of a column with six bolts snug fit in drilled holes and arranged as shown. The working stresses in the bolts and the plate may be: 16,000 psi in shear, 32,000 psi in bearing, and 20,000 psi in tension. What diameter of bolts and thickness of plate should be used and what is the size of the load P if the maximum load on a corner rivet shall not exceed 9,500 lb?

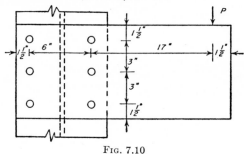

FIG. 7.10

7.11 For the bracket shown in Fig. 7.11,

1. Will the force in the outermost rivet under combined shear and torsion be: (1*a*) 8,200, (1*b*) 8,700, (1*c*) 9,100, or (1*d*) 9,400 lb?

2. Will the maximum stress in the plates be: (2*a*) 16,850, (2*b*) 17,300, (2*c*) 18,200, or (2*d*) 20,000 psi?

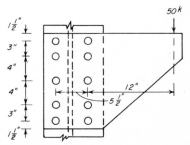

FIG. 7.11. Two ½-in. plates, one on each flange of column. All rivets are 1 in. diameter.

7.12 Find the force per inch in the outermost inch of fillet weld under combined shear and torsion of the following bracket of one plate and 32 in. of ¼-in. fillet weld.

Is this a good arrangement of welds?

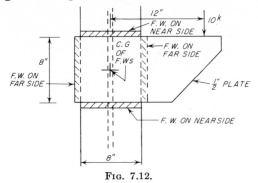

Fig. 7.12.

7.13 An A4 standard beam connection is made up of two angles 4 by 3½ by ⅜ in. with 4-in. legs outstanding. Rivets are ⅞ in. diameter. What is the tension in the top rivet under a maximum shear loading?

7.14 An A10 standard beam connection is made up of two angles 4 by 3½ by 7/16 in. with 4-in. legs outstanding. Rivets are ⅞ in. diameter. What is the tension in the top rivet under a maximum shear loading?

7.15 A bracket in the form of a plate 12 by ⅝ in. by 1 ft 6 in. is to be welded to the 12-by 1-in. flange of a WF column section. The plate is to be placed with the 18-in. side horizontal and symmetrically located with respect to the center line of the flange. The bracket is to support a pair of vertical forces of 100,000 lb each spaced 16 in. apart and symmetrically located with respect to the flange. Design the weld for this connection if the allowable shearing stress through the throat of the weld is 13,000 psi.

7.16 The main material of a plate girder consists of a web plate 42 by ⅜ in., four angles 6 by 4 by ⅝ in. placed 42½ in. back to back of 6-in. legs, and two cover plates 14 by ½ in. What total uniform load may be placed on this girder if the span is 30 ft? What is the maximum permissible spacing of ¾-in. rivets in the vertical legs of the flange angles at a point where the total shear is 75,000 lb?

7.17 A simply supported 50-ft-long plate girder of 36½ in. depth back to back of angles and without cover plates is to be designed for a single moving concentrated load of 70 kips (including impact). The top flange is considered to be laterally supported. The web plate has already been selected at 36 by ⅜ in. Select proper flange angles and compute the maximum end of girder flange to web rivet pitch for ¾-in. rivets. The bending stress is 18,000 psi, shear is 15,000 psi, and bearing is 40,000 psi.

7.18 A plate girder having a simple span of 60 ft carries a total uniform load of 5,000 lb per lin ft. Design a section to take the maximum moment.

7.19 A channel and a W beam form a T beam of 30 ft simple span. What concentrated load P may be supported at the center of the T beam if the shear in the rivets is not to exceed 15,000 psi, the fiber stress in the W is not to exceed 20,000 psi in tension, the fiber stress in the channel is not to exceed 17,050 psi, and the deflection for the concen-

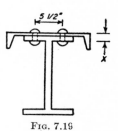

Fig. 7.19

trated load is not to exceed ⅟₃₆₀ of the span? Channel: weight = 33.9 lb, d = 15 in., a = 9.90 in.², I = 312.6 and 8.2 in.⁴, x = 0.79 in., web t = 0.400 in.; W section: d = 18.16 in., a = 28.22 in.², I = 1,674.7 and 206.8 in.⁴, weight = 96 lb. Rivets: diameter = ⅝ in., spacing 6-in. centers longitudinally.

7.20 A crane-runway girder of 21 ft span is simply supported at the ends. The girder carries a pair of moving wheel loads with axles 11 ft 4 in. on centers. Each wheel load carries 38,500 lb. Determine the maximum bending stress in the girder if the section is made up of a W 24 × 80 on top of which is fastened a 12-in. by 20.7-lb channel with the flanges turned down, and indicate the reasonableness of the bending stresses. Also compute the maximum permissible pitch for the two rows of ¾-in.-diameter rivets, one on each side of the top beam flange, holding the two sections together.

Allowable rivet shearing stress = 15,000 psi, and bearing stress = 32,000 psi. W section: depth = 24 in., area = 23.54 in.², I = 2,230 and 82.4 in.⁴, flange thickness = 0.727 in., and web thickness = 0.455 in. Channel: d = 12 in., a = 6.03 in.²,

$I = 128.1$ and 3.9 in.^{4}, $x = 0.70$ in., flange $t = 0.501$ in., and web $t = 0.280$ in.

7.21 A steel beam 24 ft long carrying a total uniformly distributed load of 10,000 lb per ft is supported at the two ends and at the center. Compute the change in bending moment at the center of the span when the center reaction settles ½ in. I of the beam $= 1,373$ in.^{4}, and E of the beam $= 30,000,000$ psi.

7.22 A simply supported beam has a rectangular cross section 1.0 in. wide and 6.0 in. deep and is 6 ft long. It carries loads of 4,800 lb and 2,400 lb located 2 ft from the left and right ends respectively. At a section 1.5 ft from the left end and 1.5 in. from the bottom face,

 1. Will the maximum normal stress be: (1a) 5,910, (1b) 5,970, (1c) 6,030, or (1d) 6,090 psi?

 2. Will the maximum normal stress be: (2a) tension or (2b) compression?

 3. Will the minimum normal stress be: (3a) 90, (3b) 100, (3c) 110, or (3d) 120 psi?

 4. Will the minimum normal stress be: (4a) tension or (4b) compression?

 5. Will the maximum shearing stress be: (5a) 2,960, (5b) 3,030, (5c) 3,090, or (5d) 3,140 psi?

7.23 What would be the sizes required for the three W steel beams of the horizontal frame in question 1.618, considering only the tank loads? What AISC standard end connections are needed? Neglect deflection considerations.

7.24 Find the amount and direction of the deflection (in terms of EI) at the 4P load.

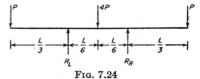

Fig. 7.24

7.25 Find the location and amount of the maximum deflection (in terms of EI) for this beam.

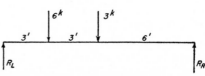

Fig. 7.25

7.26 A 20-ft beam of I moment of inertia has top and bottom cover plates over the center 10 ft of the beam giving a moment of inertia of $2I$ for the built-up section. The beam carries a uniform load of 1 kip per ft over the full length of the beam and 10 kips of load concentrated at the center. Find the deflection at the center in terms of EI.

7.27 An elevated water tank has legs consisting of two channels, one cover plate, and one open side adequately latticed. These legs are continuous for a height of 100 ft, but they are fixed at their ends and braced both ways at each 20-ft interval. What is the maximum capacity of one leg if the allowable unit stress is given by $f_s = 17,000 - 0.485(L/r)^2$?

Fig. 7.27

One plate 13 by $\frac{3}{8}$ in.
Two 8-in. channels at 11.5 lb each.
Properties of one channel are: $A = 3.36$ in.^{2}, $I_{x-x} = 32.3$ in.^{4}, $I_{y-y} = 1.3$ in.^{4}, $x = 0.58$ in.

7.28 A top chord member of a bridge truss is made up of two 15-in. 50 lb channels (placed 12 in. clear distance with flanges turned out), with one 22 by $\frac{3}{4}$-in. cover plate, and with one open side adequately latticed. The member is supported in either a horizontal or a vertical plane by members or bracing at panel points that are 25 ft on centers, and the chord may be considered fixed at these points. What is the safe axial load, applied at the center of gravity of the cross section, that the chord may carry (neglecting deflection), if the unit stress is governed by the formula $15,000 - L^2/4r^2$?

7.29 A 15-ft H column is made up of four angles 4 by 3 by $\frac{3}{8}$ in. and a web plate 8 by $\frac{3}{8}$ in. The angles are $8\frac{1}{2}$ in. back to back with long legs outstanding. What is the permissible concentric load? Use AISC column formula. $F_y = 36,000$ psi.

7.30 A steel column unbraced for a length of 20 ft is made of a W 14 × 287 section and two cover plates 22 by 2 in. The column is braced against sidesway and its ends are unrestrained. Will this column carry a concentrated load of 1,250 kips placed on the axis of the web of the W section but acting on the inner face of one of the cover plates? Use the AISC specifications. For the W 14 × 287, $A = 84.37$ in.^{2}, $d = 16.81$ in., $I_{1-1} = 3,912$ in.^{4}, $I_{2-2} = 1,466.5$ in.^{4}.

7.31   A W 14 × 167 column is subjected to a total axial load of 500 kips, a moment of 220 ft-kips at the top and a moment of 200 ft-kips at the bottom. The column is 15'-0" long and is subject to sidesway. The effective length in the plane of loading has been found to be $K_x = 1.80$. In the perpendicular plane there is bracing and $K_y = 0.75$. Determine if the column is adequate. Use A36 steel and the AISC code.

7.32   The bottom chord of a roof truss is to be made of two steel angles with vertical legs ½ in. clear back to back. The angles are 10 ft long center to center of truss panel points and carry a load of 65,000 lb in tension. The angles are connected to a ½-in. thick gusset plate at each end by a single row of ⅞-in. rivets passing through the vertical legs. What size angles will be required, taking into consideration both the eccentricity of the connection and bending due to the weight of the angles, if the maximum tensile stress is not to exceed 22,000 psi?

7.33   The bottom chord of a bridge truss is to be made of two channels with flanges outstanding and with webs vertical and 12 in. clear distance apart. If the tensile stress is not to exceed 22,000 psi and rivets are ⅞ in. in diameter, what size channels will be required (if they are adequately laced one to the other) to carry a tension of 350,000 lb and bending due to their own weight? The length of the chord is 25 ft center to center of panel points.

7.34   The top chord of a roof truss, having a very slight slope, is to be made of a structural T (WT) section and is continuous over panel points 8 ft on centers. Purlins bring 4,000-lb concentrations of load to the roof truss at the panel points and halfway between these points. This is to be a welded truss with center of gravity of members matching the joint-line diagram of the truss. Design the chord member, by AISC standards, to carry an axial compressive load of 60,000 lb.

7.35   If the top chord in question 7.34 had carried a compressive load of 75,000 lb and been made up of an angle and plate set symmetrically on each side of ½-in. gusset plates and fastened with ⅞-in. rivets, what would be a satisfactory cross section for

7.36   The top chord of a roof truss has a slope of 1 vertical to 2 horizontal. What size standard I-beam purlin is required to satisfy the following conditions if the combined bending stress is not to exceed 20,000 psi?

Span of purlin is 20 ft. Purlin is simply supported at its ends with the bottom flange resting on top chord of roof truss.

The purlin is supported in the plane of the top chord of the roof truss by sag rods at the third points of the purlins running to the ridge purlins. There is a load normal to the top flange of the purlin of 90 lb per ft of purlin, and a vertical load of 160 lb per ft including the weight of the purlin.

7.37   Design a beam 24'-0" long to support a uniformly distributed load of 5 kips per ft. Lateral support is provided only at the ends. Use A36 steel and the AISC code. Will the beam be: (*a*) W 14 × 136, (*b*) W 14 × 142, (*c*) W 14 × 150, or (*d*) W 14 × 158?

7.38   What diameter bolt is required if the bolt is subjected to a tension of 20,000 lb and a torque (for tightening) of 6,000 in-lb and the maximum normal stress is not to exceed 25,000 psi?

7.39   A 4- by 6-in. timber beam has a 4- by ½-in. steel plate connected to the lower surface. Find the neutral axis of the combination. What is the maximum fiber stress in the steel when that in the wood is 600 psi? $E_S = 30,000,000$ psi, and $E_W = 1,500,000$ psi. Use nominal sizes for the timber.

7.40   W24 × 160 stringers are spaced 5 ft on centers and have a 7-in. concrete slab bonded to the top flange of the steel stringers by means of Z-bar shear adapters. The top of the concrete is 5½ in. above the top of the W. If the stringers make 40-ft simple spans and carry the full dead weight, what static concentrated load at the center of each stringer may the composite section carry if the stress in the steel is not to exceed 18,000 psi and the stress in the concrete 1,350 psi?

7.41   In a composite beam of question 7.40, if the end shear were 33,000 lb, what would be the spacing of the Z-bar shear adapters at the end of the beam if they are fastened to the top flange of the W stringer with two ⅞-in. rivets?

7.42 A vertical retaining wall made of reinforced concrete is 18 ft high and is rigidly supported at the base. The wall is loaded by earth pressure which varies from 0 at the top to 540 psf at the base. For a 1-ft wall length, determine the wall thickness at the base and design the steel needed at the base. Use $f_c' = 3,000$ psi, $f_y = 50,000$ psi, and ACI specifications. Use ultimate strength method.

7.43 A reinforced concrete slab is simply supported on two parallel supports 10 ft center to center and carries a uniform live load of 400 psf. Design the slab by ACI specifications if $f_c' = 3,000$ psi and $f_y = 50,000$ psi. Use ultimate strength method.

7.44 A concrete beam as shown has a simple span of 15 ft. It was designed according to ACI specifications with $f_c' = 3,000$ psi and $f_y = 40,000$ psi. Assuming that the web reinforcement is satisfactory, what concentrated live load may be carried at the center of the span if the deflection due to this load is not to exceed $\frac{1}{800}$ of the span? Use ultimate strength method and the ACI specifications.

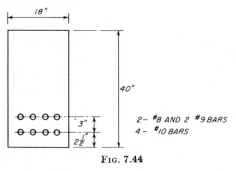

18″

40″

3″

2½″

2 – #8 AND 2 #9 BARS
4 – #10 BARS

Fig. 7.44

7.45 Design the section of a singly reinforced concrete beam that is simply supported on a 20′-0″ span and carries a dead load of 0.60 kips per ft (including an estimated value for its own weight) and a live load of 1.2 kips per ft. Use the ACI code. Use $f_c' = 4,000$ psi and $f_y = 50,000$ psi. Use ultimate strength method.

7.46 Design a singly reinforced, three-span continuous beam, each span 20 ft long. The beam carries a dead load of 0.60 kips per ft (including an estimated value for its own weight) and a live load of 1.2 kips per ft. Use the ACI code. Use $f'_c = 4,000$ psi and $f_y = 50,000$ psi. Use ultimate strength method.

7.47 This double-reinforced concrete beam is now to carry a maximum bending moment of 50,000 ft-lb. The concrete tested better than 3,000 psi. Steel may be stressed to 20,000 psi. Assume adequate web reinforcement. Is this bending moment permissible? Use the alternate method of ACI.

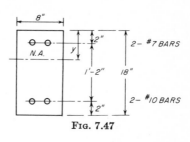

FIG. 7.47

7.48 A simply supported concrete beam 12 in. wide by 15.5 in. deep to the centroid of two rows of tension steel is to carry a dead load moment of 75 ft-kips, including an estimate of its own weight, and a live load moment of 120 ft-kips. If $f'_c = 5,000$ psi and $f_y = 50,000$ psi, design the required flexural steel reinforcement using ultimate strength methods and the ACI specifications.

7.49 The dimensions for the concrete beam designed in question 7.45 must be retained but the beam must now carry an additional live load of 600 lb per ft. What will the new schedule of reinforcing steel be?

7.50 A 4-in. slab is supported by T beams 45 in. on centers. The T beams have a 20'-0'' span, 12-in. web and an over-all depth of 24 in. What tensile reinforcement is required so that the beam can resist a usable moment of 985 ft-kips? Use $f'_c = 4$ ksi and $f_y = 60$ ksi. Use the ACI code.

7.51 A T beam has a 40-ft span and is simply supported. The slab has a thickness of 6 in. and the T beam stem is 24 in. wide. The stems of the T beams are 6 ft 0 in. on centers. The live loading is a truck which runs with one front wheel and one rear wheel on the longitudinal axis of the T beam, and produces a moment at the midpoint of the T beam of 172 ft-kips. Impact adds moment of 52 ft-kips. If d is 24 in. from top of slab to bottom of stem, design the necessary steel, in tension and/or compression, to carry the live + impact + dead moment and in the web to carry the shear (live + impact is 22 kips). Check design by using $f = Mc/I$. Use $f'_c = 3,000$ psi and $f_s = 20,000$ psi.

7.52 A floor is composed of T beams, 8 ft on centers, of 16-ft span center-to-center bearings, with a partial restraint at the ends equal to 20 per cent of the simple T beam moment. The live loading is 1,000 psf. If the slab is 4.5 in. of total thickness, design the T beam, showing sizes of all steel required work out a suitable stirrup spacing. Use $f'_c = 4,000$ psi and $f_s = 20,000$ psi.

7.53 A T beam has a flange 25 in. wide and 4 in. thick, an effective depth of 24 in., and a web width of 10 in. Tensile reinforcement consists of six #9 bars placed in two rows. Using $f'_c = 4$ ksi and $f_y = 60$ ksi and the ACI specifications for ultimate strength design, will the moment capacity of the beam be: (a) 569.5, (b) 578.0, (c) 585.0, or (d) 590.5 ft-kips?

7.54 Design a 14-ft long round spirally reinforced concrete column to carry an axial 266,000-lb dead load and 530,000-lb live load. Use the ACI specifications with $f'_c = 4,000$ psi and $f_y = 60,000$ psi. Use ultimate strength design.

7.55 Redesign the column in question 7.54, using a square section but with longitudinal steel circularly arranged within a spiral. Use same loads, stresses, and ultimate strength design.

7.56 A rectangular tied column is to support a dead load of 210 kips and a live load of 140 kips. Dead and live load moments of 63 and 42 ft-kips, respectively, are applied normal to the major axis. The column is braced against sidesway. The unsupported length of the column is 8'-0". The total moment at the top of the column is 2.5 times that at the bottom. The column bends in double curvature along the major axis. Architectural considerations limit the size of the column to 14 by 20 in. Using $f'_c = 4,000$ psi, $f_y = 60,000$ psi, and the ACI specifications, design the required reinforcement. Use ultimate strength method.

7.57 Given a 14 by 22 in. short rectangular tied column with a longitudinal reinforcement of four #10 bars placed 2.5 in. from the faces of the column to the center of the steel. The eccentricity of the compressive force is 16 in. from the center line of the column along the long axis. Use $f'_c = 4{,}000$ psi and $f_y = 60{,}000$ psi, and the ACI specifications for ultimate strength design. Will the ultimate permissible load be: (a) 207, (b) 219, (c) 234, or (d) 250 kips?

7.58 A 24 by 24 in. column reinforced with eight #10 bars, four in two opposite faces, has an unsupported length of 17'-0''. The column is subjected to the following loads:

Dead load	350 kips
Live load	153 kips
Dead load moment at top	50 ft-kips
Live load moment at top	50 ft-kips
Dead load moment at bottom	20 ft-kips
Live load moment at bottom	15 ft-kips

The column bends in double curvature and is not braced against sidesway. Determine whether the column is adequate. Use $f'_c = 5{,}000$ psi, $f_y = 60{,}000$ psi, the ACI specifications, and the ultimate strength method.

7.59 A reinforced concrete tied column 15 by 24 in. has a normal load N of 150,000 lb applied 5 in. eccentrically along the short axis. If $f'_c = 3{,}000$ psi and $f_y = 50{,}000$ psi, check the adequacy of the column on the basis of a cracked section using elestic relationships. There are three #9 bars centered 2 in. from each of the 24 in. long faces.

7.60 Determine the section of a composite column to carry an axial load of 1,000 kips. Use alternate method, the ACI specifications, with $f'_c = 3{,}000$ psi, $f_y = 40{,}000$, and A36 structural steel.

7.61 A steel W 12 × 65 column is encased in an 18 by 18-in. concrete section and is unsupported for its 20-ft height. Determine the axial load this column may carry. Use alternate method, the ACI specifications, $f'_c = 3{,}000$ psi, and A36 structural steel.

7.62 A pipe is filled with concrete and is unsupported for its 20-ft height. Determine the pipe required to support an axial load of 350 kips. Use alternate method, the ACI specifications, $f'_c = 3{,}000$ psi, and A36 structural steel.

7.63 Design a spread footing for a 20 by 20-in. square column which carries a dead load of 350 kips and a live load of 125 kips. The bearing capacity of the soil is 2.75 tons per sq ft. The bottom of the footing is 4'-6'' below grade. The weight of the soil is 120 pcf. Use $f'_c = 4,000$ psi, $f_y = 50,000$ psi, and the ACI specifications. Use ultimate strength method.

7.64 Design a rectangular concrete spread footing for a centrally placed column load of 700,000 lb, where the short dimension of the footing is 9 ft. The column is 28 by 28 in. in cross section. $f'_c = 3,000$ psi and $f_y = 50,000$ psi. Soil bearing = 3 tons per sq ft. Use working stress method.

7.65 A 12 by 12-in. concrete column carrying a load of 75,000 lb and a 15 by 15-in. concrete column carrying a load of 120,000 lb are 12 ft center to center. If $f'_c = 3,000$ psi and $f_y = 50,000$ psi, design a combined footing to support these loads if the soil bearing = 2 tons per sq ft. Use working stress method.

7.66 If the timbers are short-leaf southern pine with working stresses of $P = 1,200$ psi compression parallel to the grain, $Q = 380$ psi compression perpendicular to the grain, and N psi compression at an angle θ with the grain, is this joint satisfactory when $N = \dfrac{PQ}{P \sin^2 \theta + Q \cos^2 \theta}$?

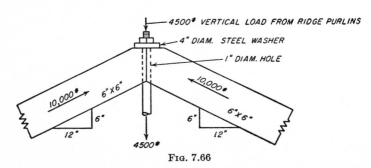

FIG. 7.66

7.67 A 4 by 6-in. timber diagonal of a timber roof truss is cut square ended and notched into a 6 by 6-in. bottom chord timber. If the unit stresses are the same as in the preceding question is this notch satisfactory in bearing?

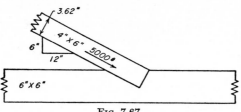

Fig. 7.67

7.68 Timber joists of 3 by 12 in. nominal dimension are on 16-in. centers and their ends rest upon steel beams of 6-in. flange width and on 14-ft centers. Including the weight of the joist and floor, what is the maximum allowable working load per square foot of floor if dressed timbers are used and the following stresses are not to be exceeded: shear at 120 psi; bending at 1,600 psi; bearing at 380 psi; and deflection not to exceed $\frac{1}{360}$ span with $E = 1,600,000$ psi?

7.69 A wood floor with a 14-ft span is to carry 250 psf including its weight. An opening 5 ft square is to be left in the center of the floor. Sketch a plan showing how the floor joists should be placed and specify the size of the floor joists if the stress conditions in the preceding question are not to be exceeded.

7.70 How should four 2- by 8-in. dressed planks be framed to give the maximum stiffness as a beam?

7.71 How should four 2- by 8-in. dressed planks be framed to carry the maximum bending shear?

7.72 The top chord of a wooden roof truss is continuous, supported, and braced at 6-ft centers. The chord carries a maximum axial load of 60,000 lb and purlins bring loads to the truss only at panel points. What 6-in. dressed section of dense southern yellow pine is required?

7.73 A timber column 12 ft long carries a compressive axial load of 27,500 lb and is to be of one piece. What size dressed dense southern yellow pine is required?

7.74 A timber column 12 ft long carries a compressive axial load of 75,000 lb. The column is to be a hollow box made of four 2-in. planks. What size dressed planks of dense southern yellow pine will be required?

7.75 Two 2- by 4-in. and one 4- by 8-in. dressed dense southern yellow pine timbers are nailed together to form an approximate 8- by 8-in. cross (in cross section). If all three timbers are 10 ft long, what safe axial load may this composite column carry?

LAND SURVEYING

8.01 Compute the missing dimensions of Lots 1, 3, 7, and 8 of block *A* from the information shown in the sketch below:

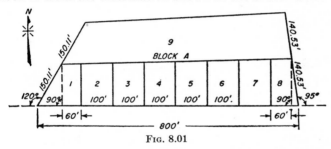

Fig. 8.01

8.02 Write a metes and bounds deed description for the house and lot where you live, assuming courses and distances where necessary. Draw a sketch map of the property and buildings, showing all information which should appear on a plan to be filed with the deed description at the registry of deeds.

8.03 In addition to the usual information shown on an ordinary property-survey plan, what pertinent information or data should appear on a plan of survey of property abutting on tidal waters or on an inland river where a riparian grant or lease is involved.

8.04 One side of a piece of property is defined in a deed in the following forms: "Beginning at a point in the northerly line of First Street in the town of London, N.J., at a drill hole in a stone monument sunk in the ground at the southeast corner of land now or formerly belonging to Joe Doe, and running N2°10′E a distance of 156.74 ft to a chisel cut in a traprock ledge . . . all bearings being true . . . ." In resurveying this property you find without any reasonable doubt both the drill hole and the chisel cut. The distance measures 144.74 ft and the true bearing is found to be N1°20′E. Is the owner of the property entitled to 144.74 ft or to 156.74 ft along this line? Explain your answer. How should the distance and bearing be shown on a map which you are to prepare for the owner?

8.05   A curved street, 60 ft between property lines, is to be laid out with a center-line radius of 400 ft. The intersection angle of the two tangents is 40°00′. Compute the deflection angles and chords necessary to lay out stakes on the inner property line with the transit at the point of curvature, the arc distance between all stakes except the last pair to be 25.00 ft.

8.06   Compute the area (in acres) enclosed by the following compass and chain traverse by the DMD method.

Point	Bearing	Distance (chains) (1 chain = 66 ft)
1	N09°15′W	6.22
2	N66 30 E	4.88
3	S88 30 E	4.32
4	S61 30 E	3.72
5	S12 30 W	4.34
6	S79 15 W	10.30

8.07   List five or more methods of describing a parcel of land for conveyance by deed.

8.08   Plot the boundaries of a parcel of land on a scale of 1 in. to 200 ft, the parcel to be approximately 600 ft wide by 1,000 ft long; show on the plat the following items:

   a. A line bearing N45°E.

   b. A line having a length of 4.25 chains.

   c. Two curves, reverse to each other, with radii of 200 ft and 500 ft, respectively.

   d. Lines tangent to the curves in *c*.

   e. A boundary street having an angle which is an angle of the parcel boundary also.

f. A side tie of 175 ft being measured on the prolongation of the north line of lot *A.*

g. Any additional item you wish to add to produce a closed figure.

h. Show all dimensions you deem necessary for a surveyor to make a ground location, in addition to those required above.

8.09 Write a proper legal description of the parcel you have platted in the above question; omit the caption of county, tract, and place of record. The description must be sufficiently detailed to enable a surveyor to stake the boundaries without other information.

8.10 Compute the latitudes and departures for the closed field bounded by the following bearings and distances. Balance, compute the closure error, and compute the area of the enclosed field in square feet by coordinates.

Line	Bearing	Distance (ft)
AB	N30°E	420
BC	S20°E	610
CD	S46½°W	535
DA	N3°W	578

8.11 An angle is to be measured with a 10-sec. transit with as great accuracy as the instrument is capable of giving. State the complete procedure which you would use to measure this angle, eliminating such instrumental and personal errors as possible. Indicate which steps will eliminate each type of instrumental or personal error. How closely would you estimate the angle could be measured with this instrument in 1 hr devoted to fieldwork?

8.12 A series of perpendicular offsets are taken from a transit line to a curved boundary line. These offsets were taken 25 ft apart and were taken in the following order: 0, 4.6, 8.9, 15.3, 10.4, 13.2, 18.5, 23.4, 17.6, 9.3, 3.2, 0. Find the area included between the transit line and the curved line by Simpson's one-third rule and check by the trapezoidal rule.

1. Will the area included between the transit line and the curved line by Simpson's one-third rule be: (1a) 3,110, (1b) 3,140, (13c) 3,170, or (1d) 3,200 sq ft?

2. Will the area by the trapezoidal rule be: (2a) 3,110, (2b) 3,140, (2c) 3,170, or (2d) 3,200 sq ft?

8.13 In testing a dumpy level for adjustment by the peg method the following rod readings were taken. Instrument at A: B.S. on A, 3.148 ft; F.S. on B, 3.398 ft. Instrument at B: B.S. on B, 4.465 ft; F.S. on A, 4.035 ft.

1. Will the true difference in elevation between A and B be: (1a) 0.330, (1b) 0.340, (1c) 0.350, or (1d) 0.360 ft?

2. Should the rod reading on A, to give a level sight with the instrument 4.465 ft above B, be: (2a) 3.875, (2b) 4.000, (2c) 4.125, or (2d) 4.250 ft?

3. Should the cross hair be: (3a) moved up, (3b) moved down, or (3c) not moved?

8.14 You have been retained by a client to survey a city lot, with the buildings on it, which he contemplates purchasing. In the course of the survey you discover that the principal building, constructed in 1900, projects 0.2 ft beyond the street line. What information would you show on your plan of the lot? How would you advise your client in this matter? On what reasoning would your advice be based?

8.15 In running differential levels, what are the reasons for insisting that the aggregate of the backsight *distances* should very nearly equal the aggregate of the foresight *distances?* If the sums of these distances are equal, is it essential that each backsight distance should equal that of the corresponding foresight? Explain.

8.16 *a.* A line is described in a deed as running a definite distance to a definite known object of certain position. Will the line be construed as running to that object or the prescribed distance if the two do not agree?

b. What is the term "northerly" construed to mean in the absence of other information? With other information?

c. A piece of land described by "metes and bounds" and also by area is sold. Which description will stand if the two do not agree?

8.17 *a.* List five or more methods of describing a parcel of land for conveyance by deed.

b. What are the necessary elements that should be shown upon a completed survey plan of a piece of property?

c. The precedence of certain types of evidence of original boundary locations is known as the order of calls. List these in their order of precedence.

d. Define the term party wall. Explain the significance of a party wall to a surveyor.

8.18 In 1760, a line AB had a magnetic bearing S88°30′E. The declination of the needle at the time was 2°00′W. At present the declination is 1°30′E. In running a line from A in a resurvey to attempt to find the buried monument at B, should the present magnetic bearing be: (*a*) S85°00′E, (*b*) S87°00′E, (*c*) N90°00′E, or (*d*) N88°00′E?

8.19 On Jan. 10 of a certain year a transit was set over point A with a backsight reading of 0°00′00″ on B. A horizontal angle of 45°15′30″ was turned clockwise to Polaris at the instant the star was at western elongation. Point A was in latitude 40°40′00″N and longitude 74°10′40″W. The declination of Polaris was +89°04′26″ and its right ascension was 1ʰ 55ᵐ 09ˢ. Find the true bearing of the line AB.

8.20 Complete the following level notes for a double-rodded line. Show the usual arithmetical check.

Point	B.S.	H.I.	F.S.	Elevation, ft
B.M. 713	10.736			740.874
T.P.1 H	6.943		3.916	
T.P.1 L	7.897		4.873	
T.P.2 H	8.337		2.431	
T.P.2 L	9.746		3.842	
T.P.3 H	5.173		5.469	
T.P.3 L	7.549		7.842	
T.P.4 H	3.411		8.188	
T.P.4 L	4.963		9.742	
B.M. 714			7.238	

8.21   The curved line in Fig. 8.21 represents a stream forming the boundary between two pieces of property. It is proposed to place this stream in a storm drain and to straighten the boundary.

Starting at a point A, a random line AB was run and the areas a, b, c, d, and e were determined from offset measurements, using Simpson's rule, to equal 3,040, 926, 2,384, 1,592, and 68 sq ft, respectively. Find the distance BH (along the boundary EF) such that the straight line AH will provide the same areas in the

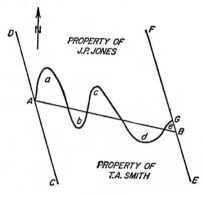

Fig. 8.21

properties of Smith and Jones as was the case when the boundary was curved. Compute the length and bearing of the line AH. $AB = 375.00$ ft, and its bearing is S77°40'E. Bearing $CD = EF = $ N15°30'W.

8.22   A church steeple had been located by triangulation. It was desired to determine the elevation of the cross bar of the weather vane on the steeple for use in connection with plane-table surveys in the vicinity. The conical spire rose centrally from a tower 35 ft square set above and with the two outside faces in the northeast walls of the church. A transit was set over a stake 100.00 ft away from the north wall of the church and on a line parallel to and 1.45 ft distance from the east wall. A backsight of 6.82 ft was taken with the telescope level on a bench mark having an elevation of 246.18 ft. A vertical angle of 40°10' was then turned to the bar of the weather vane. The telescope was then reversed and a second sight taken to the vane, the vertical angle reading 40°12′.

1. Would the horizontal distance between the transit and the spire be: (1a) 117.50, (1b) 119.02, (1c) 135.00, or (1d) 136.45 ft?

2. Would the altitude of the vane above the telescope be: (2a) 100.22, (2b) 100.32, (2c) 100.42, or (2d) 100.52 ft?

3. Would the elevation of the vane in reference to the bench mark be: (3a) 353.52, (3b) 354.62, (3c) 355.72, or (3d) 356.82 ft?

8.23 *a.* As transitman on a survey party you have a transit and also a watch which has stopped but which needs only winding to restart. State the manner in which you could determine the direction of the true meridian. The latitude, longitude, and correct time are unknown. You do not have access to an ephemeris, almanac, or star catalogue.

b. A survey party took equal altitude observations on Polaris for azimuth in the early evening of a January night. The pointer stars of the Big Dipper were obscured by clouds but Polaris was tentatively identified as the brightest star about halfway along a roughly horizontal line from Mizar (the second star in the handle of the Big Dipper) to the star δ of Cassiopeia (at the bottom of the first stroke of the W which is the constellation of Cassiopeia). The chief engineer rejected the data without even looking at the computed answer. What was the reason for his action?

8.24 Subdivide the area shown in Fig. 8.24 into five lots as indicated. Calculate all the information necessary to lay out the

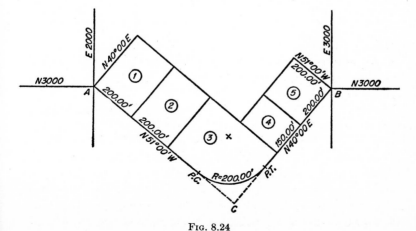

FIG. 8.24

lot corners in the field, with the exception of points A and B, which are existing monuments having the coordinates shown. The bearings of lot lines not marked are either N40°00′E or N51°00′W.

1. Is the length of line AC: (1a) 762.50, (1b) 764.00, (1c) 766.15, or (1d) 768.25 ft?

2. Is the length along the northeasterly line of lot 3 from the northerly corner of lot 3 to the westerly corner of lot 4: (2a) 164.00, (2b) 166.15, (2c) 168.30, or (2d) 170.45 ft?

3. Is the tangential distance of the curve on lot 3: (3a) 193.84, (3b) 194.74, (3c) 195.64, or (3d) 196.54 ft?

8.25   The line of a curved highway crosses a straight railroad siding as shown in Fig. 8.25. Determine the station along the highway of its intersection with the railroad. State how you would stake out the point in the field from the P.C.

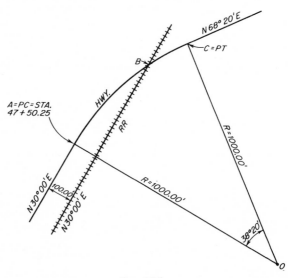

Fɪɢ. 8.25

PART THREE

Answers to Questions from Professional
Engineer's Examinations

BASIC FUNDAMENTALS

CHEMISTRY, MATERIALS SCIENCE, AND NUCLEONICS

1.101 A liter of hydrochloric acid (density 1.201) contains $1,000 \times 1.201 \times 0.4009 = $ **481.5** grams HCl.

Molecular weights: $HCl = 1.008 + 35.46 = 36.47$
$$NaCl = 23 + 35.46 = 58.46$$
$$H_2SO_4 = 2 \times 1.008 + 32 + 4 \times 16 = 98.02$$
$$Na_2SO_4 = 2 \times 23 + 32 + 4 \times 16 = 142.00$$

Check equation: $2 \times 58.46 + 98.02 = 214.92$
$$= 142.00 + 2 \times 36.47$$

Weight of NaCl required $= 481.5 \times 58.46/36.47 = $ **771.7** grams

Weight of (100 per cent) H_2SO_4 required

$$= \frac{481.5}{36.47} \times \frac{98.02}{2} = \textbf{646.8} \text{ grams} \qquad \textbf{b}$$

Basis: 1 mole $NaCl \rightarrow$ 1 mole HCl
   ½ mole $H_2SO_4 \rightarrow$ 1 mole HCl

1.102 *a.* Oxidation and reduction occur simultaneously in a chemical reaction. In the process of oxidation, one or more electrons are lost by an atom or an ion, while in reduction one or more electrons are gained by an atom or an ion. The oxidizing agent is reduced by the gain of electrons and the reducing agent is oxidized by the loss of electrons.

b. The observations of Newlands, Lotharmeyer, and Mendeleev indicated to Mendeleev that when the elements were

arranged in the order of increasing atomic weights, they could
be divided on the basis of chemical behavior into series or periods,
not all of which, however, contained the same number of elements. The properties of the elements were thought to be a
periodic function of their atomic weights. Later investigations
have shown that most of the physical and chemical properties
of the elements are periodic functions of their atomic numbers.

c. Radioactivity has been defined as the spontaneous dis-
integration of atoms accompanied by the emission of "rays."
In the case of radium, the rays have been identified as alpha
(helium with a double positive charge), beta (high-velocity
electrons), and gamma (X-rays with wavelengths from 10^{-9}
to 10^{-8} cm).

d. Acidity is customarily considered to be a function of the
concentration of the hydrogen ion (H^+) or oxonium ion (H_3O^+).
The more general definition of an acid, however, is that the ion
or molecule may act as a proton donor.

The hydrogen-ion concentration in any solution is a function
of the nature of the acid under consideration and of the nature
of the solvent used.

In aqueous solutions, the greater the concentration of hydrogen
ion (H^+)—more recent concept, oxonium ion (H_3O^+)—the
greater the acidity of the solution.

The concentration of the hydrogen ion is usually expressed
as pH, where $pH = 1/H^+$ per liter. That is, $pH = 1$ when
$H^+ = 10^{-1}$ moles per liter.

1.103 One ton ore that shows 60 per cent copper contains
1,200 lb copper per ton ore. One pound CuS contains 0.666 lb
copper. Therefore, 1 ton ore contains $1,200/0.666 - 1,200$, or
600 lb sulfur.

Sulfur available for H_2SO_4 production equals $600 - 60$, or
540 lb. Atomic weight of $H_2SO_4 = 2 \times 1 + 32 + 4 \times 16 = 98$.
One pound 93 per cent H_2SO_4 contains $0.93 \times {}^{32}\!/_{98}$ or 0.304 lb
sulfur. Therefore, 1 ton ore will produce $540/0.304 = 1,775$ lb
93 per cent H_2SO_4.   c

1.104 The production of sodium hydroxide (caustic soda) from
soda ash (Na_2CO_3) is known as the lime-soda process. Caustic
soda prepared from soda ash as manufactured by the Solvay
process is characterized by its low chlorine content.

The reaction depends on the low solubility product of calcium carbonate.

$$Na_2CO_3 + Ca(OH)_2 \rightleftarrows 2NaOH + CaCO_3\downarrow$$

The degree of conversion is a function of the initial concentration of soda ash, a 12 per cent solution giving approximately 95 per cent conversion and a 15 per cent solution giving only 90 per cent conversion.

The reaction equilibrium is insensitive to temperature changes, but the reaction rate and rate of settling of $CaCO_3$ are favored

Step	Equipment	Engineering problems
1. Preparation of soda-ash solution	Tanks with agitation	Economic consideration of cost of processing dilute solutions with high degree of conversion as compared to cost of processing more concentrated solution with a lower degree of conversion
2. Blending of soda solution with reburned and make-up lime	Classifier shaker	Feed rates, screen size, etc.
3. Soda solution causticized with slight excess of lime	Tanks with agitation	Corrosion, agitation, retention time, temperature of reaction
4. Separation of liquids and solids in first of two thickeners	Dorr thickener sludge pump	Settling rates, retention time, etc.
5. Overflow from 4 containing 10 to 11% NaOH is fed to evaporator	Evaporator and auxiliary equipment	Materials of construction, heat transfer, economic operating conditions
6. Underflow from 4 pumped to second thickener. Filtrate from 7 and water is added	Same as 4	Same as 4
7. Underflow from 6 filtered on continuous filter	Continuous filter	Optimum operating conditions; that is, vacuum, filter media, etc.
8. Overflow from 6 used as weak liquor in 1		
9. Filter cake from 7 returned to lime kiln		

by increased temperatures. The commercial process may be carried out as a batch or as a continuous operation. The process, equipment, and engineering problems are shown in the table on the preceding page.

1.105 Equal numbers of molecules of different (nonvolatile and nonelectrolyte) solutes dissolved in equal weights of the same solvent normally depress the vapor pressure, raise the boiling point, and lower the freezing point by constant amounts.

When 1 gram molecular weight nonvolatile nonelectrolyte is dissolved in 1,000 grams benzene, the freezing point of benzene is lowered 5.12°C.

Twenty grams unknown substance in 1,000 grams benzene lowers the freezing point (5.49 − 4.82) or 0.67°C.

Therefore, (5.12/0.67) × 20, or 152.8, grams would be required to lower the freezing point by 5.12°C. The experimental value of the molecular weight is 152.8.   d

$$ M = \frac{1,000 \times K_f \times W}{W_0 \times \Delta T} $$

where M = the molecular weight
K_f = the molal lowering of freezing point
W = weight of the unknown substance
W_0 = weight of solvent used
ΔT = the experimental freezing point lowering

1.106 The freezing point-solubility curve of binary systems (H_2O–Na_2SO_4 in this instance) should be studied carefully in problems of this type. The transition temperature between Na_2SO_4 and H_2O lies at 32.38°C. At temperatures above this point, the decahydrate becomes unstable with respect to the anhydrous salt. In this problem, the solution was saturated at 32°C so that only the decahydrate need be considered. Analysis of solubility data indicates that Na_2SO_4 is soluble to the extent of 46.25 grams per 100 grams water when equilibrium is attained with the decahydrate as the solid phase.

Basis of solution: 46.25 grams Na_2SO_4 and 100 grams H_2O at 32°C

Molecular weight: $Na_2SO_4 = 142$

$ H_2O \ = \ 18$

$ Na_2SO_4 + 10H_2O = 142 + 180 = 322$

Total weight of solution $= 146.25$ gram

Weight to be removed $= 146.25 \times 0.6 = 87.75$ gram

Weight of solution crystallized per gram $Na_2SO_4 = {}^{322}\!/_{142}$

$ = 2.27$

Weight of Na_2SO_4 crystallized $= 87.75/2.27 = 38.6$

Weight of H_2O removed from solution $= 38.6 \times {}^{180}\!/_{142} = 49.0$

Concentration of liquor

$$= \frac{\text{grams } Na_2SO_4}{100 \text{ grams } H_2O} \times (46.25 - 38.6) \times \frac{100}{(100 - 49)}$$

$$= \frac{15 \text{ grams } Na_2SO_4}{100 \text{ grams } H_2O}$$

Inspection of solubility curve indicates a temperature of 16.6°C.

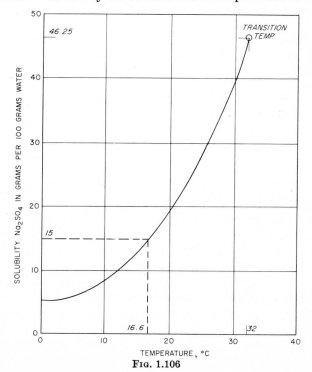

b

Fɪɢ. 1.106

1.107   1 barrel = 4 sacks cement = 4 cu ft
1 gal water = 0.134 cu ft

1 sack cement	= $(1 - 0.513)$	= 0.487 cu ft
$1\frac{1}{2}$ cu ft sand	= $1.5(1 - 0.360)$	= 0.960 cu ft
3 cu ft stone	= $3(1 - 0.330)$	= 2.010 cu ft
5 gal water	= 5×0.134	= 0.670 cu ft
		4.127 cu ft

One barrel cement will produce 16.51 cu ft of $1:1\frac{1}{2}:3$ concrete with 20 gal water.

1.108 $Fe_3O_4 + CO \rightarrow 3FeO + CO_2$
                 $3FeO + 3CO \rightarrow 3Fe + 3CO_2$

Each lb-mole Fe_3O_4 requires 4 lb-moles CO for the reduction of Fe_3O_4 to Fe.
Atomic weight $Fe_3O_4 = 3 \times 55.84 + 4 \times 16 = 231.52$

$3FeO = 215.5$ and $CO = 28$
$2,000/231.52 = 8.63$ lb-moles in 1 ton Fe_3O_4
$8.63 \times 0.84 = 7.25$ lb-moles Fe_3O_4 in 1 ton ore
$n = 7.25 \times 4 = 29.00$ lb-moles CO required to reduce Fe_3O_4 in 1 ton ore

Using the perfect-gas law, which states that 1 lb-mole perfect gas occupies 359 cu ft at atmospheric pressure (14.7 psi) and 492°R (Rankin or absolute temperature) which are standard conditions and correcting for temperature and pressure by

$$VP = nRT$$
$$V(\text{vol}) = \frac{29.00 \times 10.71 \times (460 + 85)}{(100 + 14.7)} = \frac{1472 \text{ cu ft CO}}{1 \text{ ton ore}}$$

where $R = 10.71$ = the gas constant or unit conversion factor $\dfrac{(14.7 \times 359)}{1 \times 492}$

[To change from $\dfrac{\text{psi} \times \text{cu ft}}{°R}$ to $\dfrac{\text{psf} \times \text{cu ft}}{°R}$ multiply $R(10.71)$ by 144, and the new $R = 1,544$.]

Since the gas used in the reduction is only 28 per cent CO, the total volume of gas required is $1,472 \div 0.28 = 5,230$ cu ft gas at 85°F and 100 psi.   **d**

1.109 $Al = 27$, $2Al$ or $Al_2 = 54$
                $Fe = 56$, $2Fe$ or $Fe_2 = 112$
                $SO_4 = 32 + 4 \times 16 = 96$, $(SO_4)_3 = 288$

$$?Al + ?Fe_2(SO_4)_3 \rightarrow Al_2(SO_4)_3 + ?FeSO_4$$
$$2Al + 3Fe_2(SO_4)_3 \rightarrow Al_2(SO_4)_3 + 6FeSO_4$$
$$2 \times 27 + 3 \times (112 + 288) \rightarrow (54 + 288) + 6 \times (56 + 96)$$
$$54 + 1{,}200 \rightarrow 342 + 912$$

Valences: $Al = 0$, $Fe_2(SO_4)_3 = (+3) \times 2$, $(-2) \times 3$
                  $Al_2(SO_4)_3 = (+3) \times 2$, $(-2) \times 3$
                      $FeSO_4 = (+2)(-2)$
Valence change: $Al + 0 \rightarrow +3 = $ loss of 3 electrons
                $Fe + 3 \rightarrow +2 = $ gain of 1 electron
Therefore: $2Al$ loses 6 electrons when $1Al_2(SO_4)_3$ is formed.
            $3Fe_2(SO_4)_3$ gains 6 electrons when $6FeSO_4$ is formed.

1.110   3. Lime

1.111   5. Methane (approximately 60 to 75 per cent)

1.112   1. 0 to 25

1.113   3. Relative stability

1.114   2. Saprophytic

1.115   5. The maximum velocity of flow is 1 ft per sec.

1.116   4. The sludge and raw sewage are not mixed.

1.117   2. 0.1 per cent

1.118   3. Oxidize putrescible matter

1.119   5. A leaping weir

1.120   7.5 gal water (per cu ft) weighs 62.4 lb, sp gr $= 1$. **One** bag cement (1 cu ft) weighs 94 lb.

Material	Weight (in lb)	Divided by sp gr $\times$ wt (1 cu ft water)	Absolute vol (in cu ft)
Cement	94	$3.10 \times 62.4 =$	0.49
Sand	200	$2.65 \times 62.4 =$	1.21
Gravel	350	$2.65 \times 62.4 =$	2.11
5 gal water	41.6	$1 \times 62.4 =$	0.67

a. Total weight concrete = 685.6 lb; total volume = 4.48 cu ft
b. One cu ft concrete weighs 685.6/4.48 or 153 lb
c. 27 ÷ 4.48 = 6.026 (of the above) batches per cu yd
Weight of materials per cu yd of concrete:

Cement:	94 × 6.026 =	567 lb
Sand:	200 × 6.026 =	1,205 lb
Gravel:	350 × 6.026 =	2,109 lb
Water:	41.6 × 6.026 =	250 lb
Weight 1 cu yd concrete =	153 × 27	= 4,131 lb

1.121 **d** Composition.
1.122 **b** Coarse-grained.
1.123 **d** Inhibits further deterioration.
1.124 **b** $2m_0C^2$.
1.125 **c** Gamma radiation.

ELECTRICITY

1.201
$$5 \text{ mph} = \frac{5 \times 5{,}280}{3{,}600} = 7.33 \text{ fps}$$

$$\text{Horsepower} = \frac{200 \times 7.33}{550} = 2.67 \text{ hp}$$

$$\text{Current} = \frac{2.67 \times 746}{0.70 \times 110} = 25.8 \text{ amp} \qquad \mathbf{a}$$

1.202 12 and 20 ohms in parallel.

$$\frac{1}{R_1} = \frac{1}{12} + \frac{1}{20} = \frac{32}{240} = \frac{1}{7.5}$$
$$R_1 = 7.5$$

Resistance A to B = 16 ohms = R_{AB}

$$\frac{1}{R_{AB}} = \frac{1}{16} = \frac{1}{(R_2 + 7.5)} + \frac{1}{30} = \frac{(30 + R_2 + 7.5)}{30(R_2 + 7.5)}$$
$$1.87R_2 + 14.08 = R_2 + 37.5$$
$$0.87R_2 = 23.42, \ R_2 = \mathbf{26.8 \text{ ohms}}$$

1.203 Three resistors in parallel:
$$\frac{1}{R_1} = \frac{1}{5} + \frac{1}{2} + \frac{1}{10} = \frac{8}{10}$$
$$R_1 = 1.25$$

Two resistors in parallel:
$$\frac{1}{R_2} = \frac{1}{4} + \frac{1}{4} = \frac{2}{4}$$
$$R_2 = 2.00$$

Two lines of resistors: Line 1. $6.75 + 1.25 = 8$ ohms

Line 2. $2.00 + 6.00 = 8$ ohms

$$\frac{1}{R_3} = \frac{1}{8} + \frac{1}{8} = \frac{2}{8}$$
$$R_3 = 4 \text{ ohms}$$

$$\text{Current} = \frac{V}{R} = \frac{12}{(4+1)} = \mathbf{2.4} \text{ amp from battery}$$

1.204 Two resistors in parallel:
$$\frac{1}{R} = \frac{1}{5} + \frac{1}{17} = \frac{22}{85}$$
$$R = 3.86$$

$$\text{Current} = \frac{V}{R} = \frac{6}{3.86} = 1.553 \text{ amp}$$

Each coil has 6 volts.

Coil *1* takes $\frac{6}{5}$ $= 1.2$ amp

Coil *2* takes $\frac{6}{17} = 0.353$ amp

$$\Sigma = 1.553 \text{ amp (check)}$$

1.205 Increase in cathode weight $= 300 - 15 = 285$ lb. At 382.52 amp-hr per lb this requires $285 \times 382.52 = \mathbf{109{,}000}$ amp-hr.

$$\text{Current} = \frac{109{,}000}{15 \times 24} = \mathbf{303} \text{ amp}$$

$$\text{Plate area} = \frac{40 \times 36}{144} = \mathbf{10} \text{ sq ft}$$

Current density $= 303/10 =$ **30.3** amp per sq ft

Reference: "Standard Handbook for Electrical Engineers."

1.206   Fifty kva at 90 per cent pf $= 0.9 \times 50 =$ **45 kw**

$$\text{Input} = 45 + \frac{400 + 600}{1,000} = 46 \text{ kw}$$

$$\text{Efficiency} = {}^{45}\!/_{\!46} = 97.83 \text{ per cent}$$

Maximum efficiency occurs when copper loss is equal to core loss. To reduce the full-load copper loss to equal the core loss requires that the load be reduced to

$$\sqrt[2]{{}^{400}\!/_{\!600}} = \sqrt[2]{0.667} = 0.816 \text{ or } 81.6 \text{ per cent of full load}$$

1.207   The equivalent resistance of similar lamps in parallel varies inversely as the number of lamps in the group. The voltage across groups in series varies directly as the equivalent resistance, by Ohm's law. Therefore, the voltage across each group will vary inversely as the number of lamps in the group.

$$\text{Voltage across 25-lamp group} = 220 \times \frac{15}{(15 + 25)} = 82.5 \text{ volts}$$

$$\text{Voltage across 15-lamp group} = 220 \times \frac{25}{(15 + 25)} = 137.5 \text{ volts}$$

Assuming constant resistance, the current through each **lamp** varies directly as the voltage.

Current through each lamp in 25 group $= (82.5/110) \times 0.9$
$=$ **0.675** amp

Current through each lamp in 15 group $= (137.5/110) \times 0.9$
$=$ **1.12** amp

1.208   Wire tables in handbook show that copper wire of 0.204-in. diameter has a resistance of 0.253 ohms per 1,000 ft at 25°C. The two wires from the generator to the load 500 ft away will then have a resistance of 0.253 ohms. The load requires a current of $10,000/120 = 83.3$ amp. Generator voltage must be **greater** than the load voltage by the IR drop in the line:

$$V_g = 120 + 83.3 \times 0.253 = 120 + 21.1 = 141.1 \text{ volts} \qquad \textbf{1a}$$

The power loss in the line will be I^2R, or

$$P_L = (83.3)^2 \times 0.253 = \textbf{1,755} \text{ watts} \quad \textbf{2c}$$

1.209 Charge is the product of capacitance and voltage, or $Q = CE$. The voltage across capacitors in series varies inversely as the capacitance:

Voltage across 8 μf $= 120 \dfrac{12}{(8 + 12)} = \textbf{72}$ volts

Voltage across 2 μf and 10 μf $= 120 \dfrac{8}{(8 + 20)} = \textbf{48}$ volts

The charge of the 8 μf $= 8 \times 10^{-6} \times 72 = \textbf{576} \times \textbf{10}^{-6}$ coulombs

The charge of the 2 μf $= 2 \times 10^{-6} \times 48 = \textbf{96} \times \textbf{10}^{-6}$ coulombs

The charge of the 10 μf $= 10 \times 10^{-6} \times 48 = \textbf{480} \times \textbf{10}^{-6}$ coulombs

Energy stored $= \frac{1}{2}$ charge $\times$ voltage

Total energy $= \frac{1}{2} \times 576 \times 10^{-6} \times 72 + \frac{1}{2} \times (96 + 480)10^{-6} \times 48 = 0.0207 + 0.0138 = \textbf{0.0345}$ joule

1.210 *a.* By Ohm's law and Kirchhoff's law:

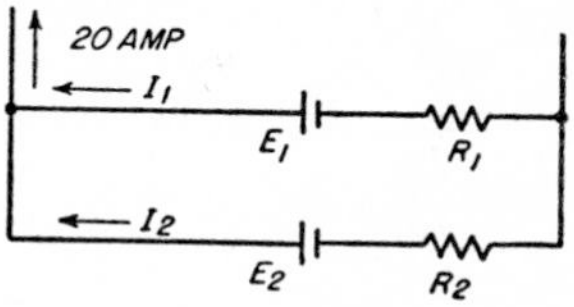

$$E_1 - I_1R_1 = E_2 - I_2R_2 \quad (1)$$
$$I_1 + I_2 = 20 \quad (2)$$

Substituting (2) in (1)

Fig. 1.210

$$E_1 - I_1R_1 = E_2 - (20 - I_1)R_2$$

$$I_1 = \frac{E_1 - E_2 + 20R_2}{R_1 + R_2} = \frac{(2.1 - 2.0 + 20 \times 0.05)}{(0.05 + 0.05)} = \frac{1.10}{0.10}$$
$$= \textbf{11} \text{ amp}$$

$$I_2 = 20 - I_1 = 20 - 11 = \textbf{9} \text{ amp}$$

b. As a lead storage cell discharges, the following reaction takes place:

$$\underset{\substack{\text{(positive} \\ \text{plate)}}}{PbO_2} + \underset{\substack{\text{(negative} \\ \text{plate)}}}{2H_2SO_4} + \underset{\substack{\text{(positive} \\ \text{plate)}}}{Pb} \rightarrow \underset{}{PbSO_4} + \underset{\substack{\text{(negative} \\ \text{plate)}}}{PbSO_4} + 2H_2O$$

The sulfuric acid combines with the lead peroxide of the positive plate and the lead of the negative plate to form lead sulfate and water. The specific gravity of the sulfuric acid electrolyte therefore decreases as this reaction takes place.

1.211 The equation for emf of a d-c generator may be found in any handbook: $E = \dfrac{ZPNBA}{60a10^8}$ volts

where Z = number of armature conductors

P = number of poles

N = speed in rpm

B = average flux density under the poles in lines per sq in.

A = area of each pole face in sq in.

a = number of parallel paths through the armature

Substituting the values given:

$$E = \frac{360 \times 4 \times 1200 \times 40{,}000 \times (6 \times 6)}{60 \times 4 \times 10^8} = 103.7 \text{ volts} \qquad \mathbf{d}$$

1.212 By Ohm's law and Kirchhoff's law:

FIG. 1.212

$$IR = 15 - 2I_2 = 10 - 1I_1 \qquad (1)$$
$$I_1 + I_2 = I \qquad (2)$$

If $I_1 = I_2$, then $I_1 = \dfrac{I}{2}$ and $I_2 = \dfrac{I}{2}$ [from (2)]

Substituting in (1)

$$IR = 15 - 2\frac{I}{2} \qquad \text{or } I(1 + R) = 15 \qquad (3)$$

and $\qquad IR = 10 - \dfrac{I}{2} \qquad$ or $I(0.5 + R) = 10 \qquad (4)$

Dividing (3) by (4)

$$\frac{(1 + R)}{(0.5 + R)} = \frac{15}{10} \qquad \text{and } 10 + 10R = 7.5 + 15R$$
$$5R = 2.5 \qquad \text{and } R = 0.5 \text{ ohm}$$

1.213 The 60-watt lamp could be connected in series with the parallel combination of the two 30-watt lamps across 240 volts for normal operation as shown. Since resistance

= (volts)^2/watts, the resistance of two 30-watt 120-volt lamps in parallel would be equal to the resistance of the 60-watt 120-volt lamp and the voltage across each lamp would be normal.

1.214 Total equivalent resistance:

$$R = 15 + \frac{7 \times 11}{7 + 11} = 15 + 4.28 = 19.28 \text{ ohms}$$

The total current, by Ohm's law of $I = V/R$:

$$I = 110/19.28 = 5.7 \text{ amp}$$

This is the current through the 15-ohm resistance. The total current divides inversely as the resistances through each branch of the parallel combination:

$$\text{Current through the 7-ohm branch} = 5.7 \times \frac{11}{7 + 11}$$
$$= 3.48 \text{ amp}$$

$$\text{Current through the 11-ohm branch} = 5.7 \times \frac{7}{7 + 11}$$
$$= 2.22 \text{ amp}$$

1.215 $I = \dfrac{\text{hp} \times 746}{V \times \text{efficiency}} = \dfrac{5 \times 746}{220 \times 0.85} = 19.9 \text{ amp}$

IR permitted $= 0.05 \times 220 = 11$ volts
R for 300 ft $= 11/19.9 = 0.553$ ohm
$R/1,000$ ft $= (1,000/300) \times 0.553 = 1.84$ ohms
Size wire $= $ #12 AWG
Current capacity check—25 amp

1.216 $P = V \times I = 115 \times 15 = 1,725$ watts

Therefore, **17** 100-watt lamps can be used

1.217 $V_2 = \dfrac{20,000}{16,000} V_1 = 1.25 \times 60 = 75$ volts

$$V = V_1 + V_2 = 60 + 75 = \mathbf{135} \text{ volts}$$

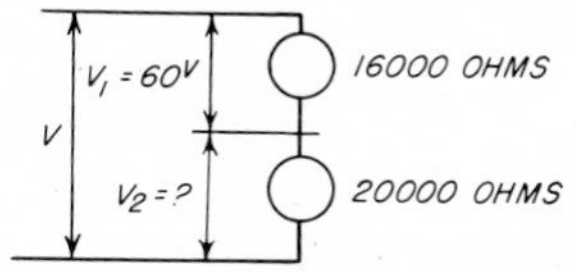

Fig. 1.217

1.218 $R_{\text{total}} = 1 + 7 + \dfrac{3 \times 6}{3 + 6} = 10$ ohms

$I_{\text{total}} = {}^{30}\!\!/_{10} = 3.0$ amp

a. Voltage across 7-ohm device $\quad = 3 \times 7 = 21$ volts

b. Voltage across 3-ohm device $\left.\vphantom{\begin{array}{c}a\\b\end{array}}\right\}$ and voltage across 6-ohm device $\Big\} = 3 \times 2 = 6$ volts

Power expended on 3-ohm device $= I^2R = 6 \times {}^{6}\!/_{3} = 12$ watts

c. Current in 6-ohm device $\quad\quad = \dfrac{V}{R} = \dfrac{6}{6} = 1.0$ amp

1.219 $I_{\text{line}} = \dfrac{\text{hp} \times 746}{V \times \text{efficiency}} = \dfrac{10 \times 746}{220 \times 0.83} = 40.8$ amp

$I_{\text{starting}} = 2 \times 40.8 = 81.6$ amp

$I_a = 81.6 - {}^{220}\!\!/_{100} = 79.4$ amp

$I_a = \dfrac{V}{R_1 + Ra} = 79.4 = \dfrac{220}{R_1 + 0.25}$

$R_1 + 0.25 = 2.77, \; R_1 = 2.5$ ohms   b

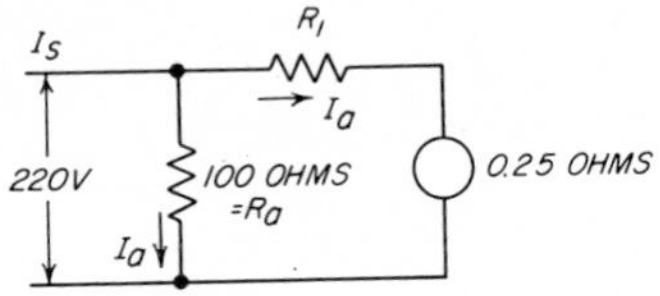

Fig. 1.219

1.220 $I = \dfrac{V}{Z}$ and $Z = R + jX_L - jX_c$

$Z = 63 + j377 \times 0.15 - j\dfrac{10^6}{377 \times 15}$

$= 63 + j56.6 - j177$

$= 63 - j120 = 135\underline{/-62.2^\circ}$

$I = {}^{110}\!\!/_{135}\underline{/-62.2} = 0.815\underline{/62.2^\circ}$ amp

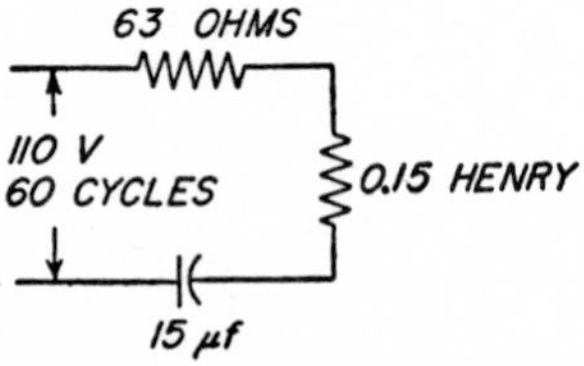

Fig. 1.220

1.221   Power input $= \dfrac{\text{hp} \times 746}{\text{efficiency}} = \dfrac{8 \times 746}{0.82} = 7{,}280$ watts

The apparent power = power divided by power factor and $\text{va}_{\text{input}} = 7{,}280/0.70 = 10{,}400$ va. As the motor is a three-phase balanced load, one-third or 3,460 va is applied to each phase. The three phases are Y connected with a line-to-line voltage of 220 volts. Hence each phase voltage is $220/\sqrt{3} = 127$ volts. Since the apparent power per phase is 3,460 va and the phase voltage is 127 volts, $I\varphi = 3{,}460/127 = 27.3$ amp current per phase and also per line in a Y connection. Voltage **1a**, current **2d**.

1.222   The plate characteristic curves, commonly found in tube manuals, are a family of graphs, each drawn for a constant value of control grid bias voltage, E_c, in which the vertical axis is plate current, I_p, and the horizontal axis is plate voltage, E_p. Suppose the wish is to find r_p, g_m, and μ for a 6J5 triode at the point $E_c = -9$ volts, $I_p = 5.5$ ma, and $E_p = 240$ volts. To find r_p, find the change in E_p for a small change in I_p, with E_c held constant at -9 volts. Find that at $E_c = -9$ and $I_p = 6$, $E_p = 246$, and that at $E_c = -9$ and $I_p = 5$, $E_p = 236$. Hence:

$$r_p = \frac{\Delta E_p}{\Delta I_p} = \frac{(246 - 236)}{(6 - 5)10^{-3}} = 10{,}000 \text{ ohms}$$

To find g_m, find the change in I_p, occasioned by a small change in E_c with E_p held constant at 240 volts. Find that at $E_p = 240$ and $E_c = -10$, $I_p = 3.5$, and that at $E_p = 240$ and $E_c = -8$, $I_p = 7.5$. Hence:

$$g_m = \frac{\Delta I_p}{\Delta E_c} = \frac{(7.5 - 3.5)10^{-3}}{-8 - (-10)} = 2 \times 10^{-3} \text{ or } 2{,}000 \text{ micromhos}$$

To compute μ, find the change in E_p caused by a small change in E_c with I_p held constant at 5.5 ma. Find that at $I_p = 5.5$ and $E_c = -8$, $E_p = 222$, and that at $I_p = 5.5$ and $E_c = -10$, $E_p = 262$. Hence:

$$\mu = -\frac{\Delta E_p}{\Delta E_c} = -\frac{(262 - 222)}{-10 - (-8)} = 20$$

As a check, $\mu = g_m \times r_p = 2 \times 10^{-3} \times 10^4 = 20$

1.223 The internal resistance of the cell on short circuit is found from Ohm's law.

$$R_i = \frac{E_i}{I_{\text{short-circuit}}} = \frac{1.5}{25} = 0.06 \text{ ohms}$$

When connected to an external resistance, R_e, and using Ohm's law again

$$I = \frac{E_i}{R_i + R_e} = \frac{1.5}{0.06 + 1.00} = 1.414 \text{ amp}$$

The voltage across the battery terminals when connected to the 1-ohm load will be the internal voltage minus the IR drop,

$$E_t = E_i - IR_i = 1.5 - 1.414 \times 0.06 = 1.414 \text{ volts}$$

This voltage is also that across the 1-ohm load,

$$E_t = IR_e = 1.414 \times 1 = 1.414 \text{ volts (check)}$$

1.224 *a.* The period of any oscillation is the time required per cycle. Hence the period

$$T = \frac{1}{f} = \frac{1}{5} = 0.2 \text{ sec}$$

b. In the absence of resistance, the natural undamped frequency of transient oscillations is the same as the a-c resonant frequency; that is, the angular frequency is

$$\omega_0 = \frac{1}{\sqrt{LC}}, \text{ whence } C = \frac{1}{\omega_0^2 L} = \frac{1}{(5 \times 2\pi)^2 \times 100}$$
$$C = 10.1 \times 10^{-6} \text{ f or } 10.1 \ \mu\text{f}$$

1.225 *a.* The differential equation for a series L-C-R circuit during discharge is, by Kirchhoff's voltage law,

$$Ri + L\frac{di}{dt} + \frac{1}{C}\int_{-\infty}^{t} i \, dt = 0$$

In this problem the initial charge Q is the value of the last term at $t = 0$.

The equation then becomes $Ri + L\dfrac{di}{dt} + \dfrac{1}{C}\displaystyle\int_0^t i\,dt - \dfrac{Q}{C} = 0$

b. In the equation as first written, each term represents a voltage drop. Thus $R \times i$ is the instantaneous resistance drop and $L\,di/dt$ is the instantaneous inductive voltage drop. The voltage drop across the capacitance, because of the voltage due to the initial charge, is

$$e_c = \frac{1}{C}\int_0^t i\,dt - \frac{Q}{C}$$

FLUID MECHANICS

1.301 Depth cylinder = 60 ft
Diameter cylinder = 40 ft
Weight water = 62.5 lb per cu ft
Intensity water pressure at base = $60 \times 62.5 = 3{,}750$ psf
Total pressure on base = $\pi\,(20)^2 \times 3{,}750 = 4{,}710{,}000$ lb
Total pressure on side = $\pi \times 40 \times 60 \times 3{,}750/2$
$= 14{,}130{,}000$ lb
Total normal pressure on base and sides = **18,840,000 lb**

1.302

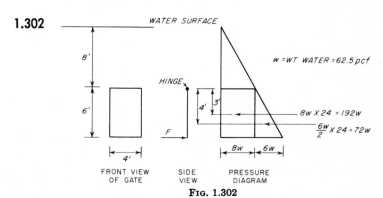

Fig. 1.302

Sum of moments of forces about hinge = 0
$+ 3 \times 192w + 4 \times 72w - 6F = 0$

$$F = 96w + 48w = 144 \times 62.5 = \textbf{9,000 lb}$$

1.303 $x_c - x_0 =$

$$\frac{l_1{}^2}{12x_0} \quad \text{or} \quad \frac{1}{12} = \frac{9}{12x_0}$$

$$x_0 = 9 \text{ ft}$$

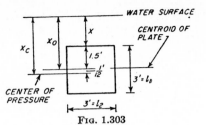

FIG. 1.303

$x = 9 \text{ ft} - 1.5 \text{ ft} = 7.5 =$ distance top edge of plate is below water surface.

1.304 $v_{10} \times d_{10} = v_{12} \times d_{12}$

$$v_{10} = 6 \times {}^{12}\!/_{10} = 7.2 \text{ fps} \qquad \textbf{1d}$$
$$w = 62.22 \text{ lb per cu ft (at } 80°\text{F)}$$
$$\mu = \frac{0.00003716}{0.4712 + 0.01435T + 0.0000682T^2}$$

For $T = 80°$F, $\mu = 0.000018$ lb-sec/sq ft

$$\rho = \frac{w}{g} = \frac{62.22}{32.17} = 1.934 \frac{\text{lb per cu ft}}{\text{ft per sec}^2} \left(\frac{\text{lb-sec}^2}{\text{ft}^4}\right)$$

$$\textbf{R} = \text{Reynolds number} = \frac{6 \times 1 \times 1.934}{0.000018} = \textbf{648,000} \qquad \textbf{2c}$$

1.305 For the oil, $\mu = \dfrac{0.80 \times 62.4}{32.17} = 1.55$

$$\frac{v \times 0.833 \times 1.55}{0.000042} = 648,000, \; v = \textbf{21.0 fps} \qquad \textbf{a}$$

1.306 Width barge
    = 20 ft
Weight water displaced
    = weight barge = 250T
d = draft in feet

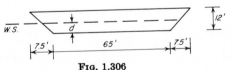

FIG. 1.306

$$\frac{250 \times 2000}{62.4} = \left(65 + \frac{7.5d}{12}\right)d \times 20$$
$$d^2 + 104d = 640, \quad d = \textbf{5.8 ft}$$

1.307   Wood cube $2 \times 2 \times 2$ ft will displace $2 \times 2 \times 1$ ft or 4 cu ft of water.

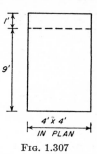

Water will rise in tank h ft.

$$h \times 4 \times 4 = 4, \qquad h = 0.25 \text{ ft.}$$

Original water pressure on one side of tank
$= 4 \times 9 \times \frac{1}{2}(0 + 62.5 \times 9) = 10{,}100$ lb
New water pressure $=$
$4 \times 9.25 \times \frac{1}{2}(0 + 62.5 \times 9.25) = 10{,}675$ lb

FIG. 1.307

Increase in water pressure per side = **575** lb

1.308   Pressure head + velocity head + elevation + lost head (if any) = constant = energy equation. All heads and elevations are in feet.

$$Q = v \times a, \quad 49.5 \text{ cfs} = v \text{ fps} \times 7.07 \text{ sq ft}$$

$$v = 7 \text{ fps and } \frac{v^2}{2g} = \frac{49}{64.4} = 0.7 \text{ ft}$$

P_R = pressure head at summit = $2.31 \times 30 = 69.3$ ft
Total head of pump = $69.3 + 0.7 + 300 + 10 = 380$ ft

$$\text{Energy of pump} = \frac{380 \text{ ft} \times 49.5 \text{ cfs} \times 62.4 \text{ lb/cu ft}}{550 \text{ fps}} = \textbf{2,130 hp}$$

1.309 $V_A = \dfrac{\text{discharge}}{A \text{ of 12-in. pipe}} = \dfrac{2}{0.785} = 2.55$ fps

$$V_D = \frac{2}{0.196} = 10.2 \text{ fps}$$

Equate energy of points A and B.

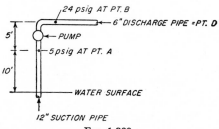

FIG. 1.309

$$-5 \times 2.31 + \frac{2.55 \times 2.55}{2 \times 32.2} + 0 + \text{pump head}$$

$$= +24 \times 2.31 + \frac{10.2 \times 10.2}{64.4} + 5$$

Pump head $= +55.4 + 11.6 + 1.60 - 0.10 + 5 = 73.5$ ft

Horsepower output of pump $= 73.5 \times 2 \times 62.4 \div 550$
$$= \textbf{16.7 hp}$$

1.310   This is the case of water passing through an orifice under constant head. The barge sinks as fast as the water rises within the barge.

$$v = \sqrt{2gh} = \sqrt{64.4 \times 4} = 16.06 \text{ fps}$$

The rate water enters the barge $= 0.60 \times 0.196 \times 16.06$
$$= 1.88 \text{ cfs}$$

Time needed to sink barge $= \dfrac{20 \times 10 \times 2}{1.88} = \textbf{213 sec}$

1.311   Francis's formula for a contracted weir with no velocity of approach is given by

$$Q = 3.33(b - 2H/10)H^{3/2} \text{ with } b = 10 \text{ ft and } H = 0.875 \text{ ft}$$

$$Q(\text{discharge of cfs}) = 3.33 \left(10 - \frac{2 \times 0.875}{10} \right) \sqrt{0.875^3}$$

$$Q = 3.33(10 - 0.175)\sqrt{0.67} = 3.33 \times 9.825 \times 0.819 =$$
$$\textbf{26.8 cfs}$$

1.312   Francis's formula for a suppressed weir with no velocity of approach is given by

$$Q = 3.33 \times bH^{3/2} = 3.33 \times 7 \times 0.819 = \textbf{18.8 cfs}$$

1.313   Use energy equation (all items in feet).
Elevation $=$ entrance loss $+$ discharge loss $+$ friction loss

$$100 = 0.5 \frac{v^2}{2g} + \frac{v^2}{2g} + f \frac{l}{d} \frac{v^2}{2g}$$

f for clear smooth pipe and an assumed velocity of 5 to 6 fps from tables = 0.0182

$$l = 12,000 \text{ ft}, \; d(\text{diameter of pipe in feet}) = 1 \text{ ft}$$

$$100 = \left(0.5 + 1.0 + \frac{0.0182 \times 12,000}{1}\right) \times \frac{v^2}{64.4}$$

$$v^2 = \frac{6,440}{220} = 29.2, \; v = 5.41 \text{ fps}$$

$$Q = \text{average} = 0.785 \times 5.41 = 4.25 \text{ cfs}$$

1.314 $7,500$ gal per min $= \dfrac{7,500}{7.5 \times 60} = 16.7$ cfs

$$Q = \text{average}, \; v = \frac{16.7}{1.77} = 9.45 \text{ fps}$$

Elevation = entrance loss + velocity loss + pipe loss + pressure head

$$180 = \left(0.5 + 1.0 + 0.0160 \times \frac{4,000}{1.5}\right) \frac{v^2}{2g} + h(\text{pressure})$$

$$h(\text{pressure}) = 180 - 44.2 \times \frac{9.45 \times 9.45}{64.4} = 180 - 61.3$$

$$= 118.7 \text{ ft}$$

$$\text{Pressure in pipe} = \frac{118.7}{231} = 51.4 \text{ psi}$$

1.315   Pressure head (pump) = total elevation pumped + total pipe losses.

$$Q = \text{average}, \; v = \frac{2}{0.196} = 10.2 \text{ fps}$$

$$h(\text{pump}) = 1,250 - 1,000 + \frac{0.0225 \times 8,000}{0.5} \times \frac{v^2}{2g} + \frac{1.5v^2}{2g}$$

$$= 250 + 361.5 \times \frac{10.2 \times 10.2}{64.4} = 833 \text{ ft}$$

$$833 = 1,100 - 1,000 + \frac{0.0225 \times 3,200}{0.5} \times \frac{10.2 \times 10.2}{64.4}$$

$$+ \text{ pressure head}_{3,200}$$

Pressure head at $3,200 = 833 - 100 - 233 = 500$ ft

Pressure at $3,200 = 500/2.31 = \textbf{216 psi}$

1.316 Equate energy between A and B and between A and M.
(A and B):

$$h(\text{pump}) = 900 - 800 + 0.02 \times \frac{4,000}{1} \times \frac{v^2}{2g} + \frac{1.5v^2}{2g}$$

$$= \left(100 + 81.5 \frac{v^2}{2g}\right)$$

(A and M):

$$\left(100 + 81.5 \frac{v^2}{2g}\right) + 100 = 100 \times 2.31 + 0.02 \times \frac{2,000}{1} \frac{v^2}{2g} + \frac{1.5\,v^2}{2g}$$

$$40 \frac{v^2}{2g} = 31, \ v^2 = \frac{31 \times 64.4}{40} = 50, \ v = 7.07 \text{ fps}$$

$Q = \text{average} = 0.785 \times 7.07 = \textbf{5.55} \text{ cfs}$
Pump horsepower output

$$= \frac{5.55 \times 62.4}{550}\left(100 + 81.5 \times \frac{31}{41.5}\right)$$

$$= 0.629(100 + 61) = \textbf{101} \text{ hp}$$

1.317 Write energy relationship between points A, B, C, and Y. Hydraulic gradient at point Y will be called Y.

1. (A and Y): $1,000 - Y = \dfrac{0.02 \times 3,000}{0.67} \dfrac{V_a^2}{2g} = 90 \dfrac{V_a^2}{2g}$

2. (Y and B): $Y - 850 = \dfrac{0.02 \times 2,000}{0.5} \dfrac{V_b^2}{2g} = 80 \dfrac{V_b^2}{2g}$

3. (Y and C): $Y - 875 = \dfrac{0.02 \times 1,000}{0.5} \dfrac{V_c^2}{2g} = 40 \dfrac{V_c^2}{2g}$

   a. (1) + (2): $150 = 90 \dfrac{V_a^2}{2g} + 80 \dfrac{V_b^2}{2g}$

   b. (1) + (3): $125 = 90 \dfrac{V_a^2}{2g} + 40 \dfrac{V_c^2}{2g}$

   c. $Q(\text{pipe } A) = Q(\text{pipe } B) + Q(\text{pipe } C)$
   $0.349 V_a = 0.196 V_b + 0.196 V_c$

Equations a, b, c can best be solved by assuming values of V_a, and then V_b and V_c can be found from equations a and b. If these values satisfy equation c, the assumed value of V_a was good. Thus, if V_a is taken as 8.03 fps, then $V_b = 6.9$ fps and $V_c = 7.4$ fps.

$$Q_a = 2.80 \text{ cfs} = Q_b + Q_c = 1.35 \text{ cfs} + 1.45 \text{ cfs}$$

1.318 $Q_R = aV_R, \quad 7 = 1.767V_R, \quad V_R = 3.96 \text{ fps}, \quad \dfrac{V_R{}^2}{2g} = 0.244$

Let elevation of hydraulic gradient at Y be Y. Equate energy between R and Y, Y and A, Y and B.

1. $400 - Y = 0.02 \times (2{,}000/1.5) \times 0.244 = 6.5$
    $$Y = 393.5$$

2. $393.5 - 250 = 143.5 = 0.02 \times \dfrac{13{,}000}{1} \times \dfrac{V_a{}^2}{2g} = 4.04V_a{}^2$

$$V_a{}^2 = 35.6, \ V_a = 5.96 \text{ fps}$$

$$Q_R - Q_A = Q_B = 7 - 0.785 \times 5.96 = 2.32 = \dfrac{\pi d^2}{4} \times V_b$$

$$d^2 = 2.95 \div V_b$$

3. $393.5 - 50 = 343.5 = 0.02 \times \dfrac{4{,}000}{2g} \times \dfrac{V_b{}^2}{d}$

$$\dfrac{V_b{}^2}{d} = 276, \quad \dfrac{V_b{}^4}{d^2} = 76{,}200$$

$$V_b{}^5 = 2.95 \times 76{,}200 = 225{,}000$$

by trial: $11^5 = 161{,}000, \ 12^5 = 248{,}500$

$11.9^5 = 239{,}000, \ 11.8^5 = 228{,}500, \ 11.75^5 = 224{,}000$

Now $d^2 = \dfrac{2.95}{11.75} = 0.251$

$$d = 0.5 \text{ ft} = \mathbf{6} \text{ in.}$$

1.319   The loss in head between points A and B in either the 36- or the 24-in. pipe must be the same.

$$0.022 \times \dfrac{2{,}000}{3} \times \dfrac{V_3{}^2}{2g} = 0.023 \times \dfrac{6{,}000}{2} \times \dfrac{V_2{}^2}{2g}$$

$$\dfrac{44}{3}V_3{}^2 = 69V_2{}^2, \quad V_3 = V_2\sqrt{\dfrac{3 \times 69}{44}} = 2.17V_2$$

$$Q_4 = Q_2 + Q_3$$
$$75.4 = 3.14V_2 + 7.07V_3 = 3.14V_2 + 7.07 \times 2.17V_2$$
$$V_2 = \frac{75.4}{3.14 + 15.33} = 4.08 \text{ fps}$$

Now $\quad V_3 = 8.85$ fps

$$Q_2 = 3.14 \times 4.08 = \underline{12.9 \text{ cfs}}$$
$$Q_3 = 7.07 \times 8.85 = \underline{62.5 \text{ cfs}}$$
$$\Sigma = Q_4 \qquad\qquad = \overline{75.4 \text{ cfs}}$$

1.320 Let p stand for pipe and n for nozzle.

$$A_p \times V_p = 8 \times A_n \times V_n$$
$$19.63 \times V_p = 8 \times 0.049 \times V_n = 0.392V_n$$
$$V_n = 50V_p$$

Equate energy of reservoir and nozzles. $\quad 600 = 0.5 \times \dfrac{V_p{}^2}{2g}$

$$+ 0.017 \times \frac{25,000 \times V_p{}^2}{5 \times 2g} + \left[1 + \left(\frac{1}{(0.95)^2} - 1 \right) \right] \times \frac{2,500V_p{}^2}{2g}$$

$$38,600 = 85.5V_p{}^2 + 2,775V_p{}^2 = 2,860.5V_p{}^2$$
$$V_p{}^2 = 13.5, \; V_p = 3.68 \text{ fps}, \; V_n = 183.0 \text{ fps}$$
$$Q_p = 3.68 \times 19.63 = 72.3 \text{ cfs}$$

Horsepower of eight jets $= \dfrac{72.3 \times 62.4}{550} \times \dfrac{183.0 \times 183.0}{64.4}$

$$= 4,300 \text{ hp}$$

1.321 Let t = throat and p = pipe.

$$\frac{A_t}{A_p} = \frac{1}{9} \qquad \frac{d_t}{d_p} = \frac{1}{3} = \frac{8 \text{ in.}}{24 \text{ in.}}$$

h_t and h_p = recorded pressure heads in feet.

$$Q = \frac{C \times A_t}{\sqrt{1 - (d_t/d_p)^4}} \sqrt{2g(h_t - h_p)}$$

$$= \frac{0.99 \times 0.349 \sqrt{64.4(36 - 9)}}{\sqrt{1 - \frac{1}{81}}} = \frac{0.346 \times 8.03 \times \sqrt{27}}{\sqrt{80/9}}$$

$$= \frac{9 \times 2.78 \times 5.2}{8.94} = 14.53 \text{ cfs}$$

1.322　Area $= \dfrac{2 \times 4 \times 12}{2} + 4 \times 30 = 168$ sq ft

Wetted perimeter $= 30 + 2\sqrt{16 + 144}$
$$= 30 + 2 \times 12.65 = 55.3 \text{ ft}$$

Hydraulic radius $= R = \dfrac{A}{P} = \dfrac{168}{55.3} = 3.04, \quad R^{\frac{1}{2}} = 1.743$

$S =$ slope of 1 in 1,000 $= 0.001$

$n = 0.02, \quad \dfrac{n}{R^{\frac{1}{2}}} = 0.01145$

Kutter's $C = \dfrac{41.65 + 0.00281/S + 1.811/n}{1 + (41.65 + 0.00281/S) \times \dfrac{n}{R^{\frac{1}{2}}}}$

$$= \dfrac{41.65 + 2.81 + 90.55}{1 + 44.46 \times 0.01145} = \dfrac{135.01}{1.51}$$

$$= 89.4$$

$v = C \times R^{\frac{1}{2}} \times S^{\frac{1}{2}} = 89.4 \times 1.743 \times 0.0316 = 4.92$ fps

$$Q = 168 \times 4.92 = \mathbf{826} \textbf{ cfs}$$

1.323　$y = \dfrac{\omega^2 x^2}{2g}$, where $\omega = \dfrac{2\pi(56)}{60}$

$$= 5.88 \text{ radians per sec}$$

$y_1 = \dfrac{(5.88)^2(3)^2}{64.4} = 4.82$ ft

$y_2 = \dfrac{(5.88)^2(2)^2}{64.4} = 2.14$ ft

$y_3 = \qquad\qquad \overline{2.68 \text{ ft}}$　　**1b**

(Check using parabola, $y_2 = \dfrac{4}{9}(4.82) = 2.14$.)

Volume of water lost $= \dfrac{Ay}{2} = \dfrac{\pi(2)^2(2.14)}{2} = \mathbf{13.45}$ **cu ft**　　**2d**

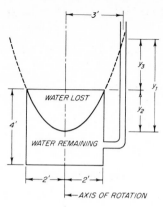

Fig. 1.323

1.324   40 mph = 58.67 fps

$$\text{Force} = C_D \times w \times A \times \frac{v^2}{2g}$$

$$= (1.12)(0.0807)(100\pi) \frac{(58.67)^2}{64.4} = \textbf{1,520 lb} \qquad \textbf{b}$$

1.325   Let point 1 be at the toe of the spillway and point 2 be at the point the jump has occurred.

$$d_1 d_2 \frac{(d_1 + d_2)}{2} = \frac{Q^2}{g}$$

$d_1 = 0.8$ ft, width $= 1.0$ ft, $a_1 = 0.8$ sq ft

$v_1 = 30$ fps, $Q = av = 0.8 \times 30 = 24$ cfs

$$0.8 d_2 \frac{0.8 + d_2}{2} = \frac{(24)^2}{32.2} = 17.9$$

$$d_2(0.8 + d_2) = d_2{}^2 + 0.8 d_2 = 44.8$$

$$d_2{}^2 + 0.8 d_2 + 0.16 = 44.8 + 0.16$$

$$d_2 + 0.4 = \sqrt{44.96} = 6.7$$

$$d_2 = \textbf{6.3 ft} \qquad \textbf{1a}$$

$$v_2 = \frac{24}{1 \times 6.3} = \textbf{3.81 fps} \qquad \textbf{2b}$$

$$E_{s1} = \frac{v_1{}^2}{2g} + d_1 = \frac{900}{64.4} + 0.8 = 14.75 \text{ ft-lb per lb}$$

$$E_{s2} = \frac{3.81^2}{64.4} + 6.3 = 6.52 \text{ ft-lb per lb}$$

$$E_{s1} - E_{s2} = 14.75 - 6.52 = 8.23 \text{ ft-lb per lb}$$

$$\text{Energy} = \frac{24 \times 200 \times 62.4 \times 8.23}{550} = \textbf{4,460 hp} \qquad \textbf{3c}$$

$$\frac{L}{d_2} = 4.8, \; L = 4.8 \times 6.3 = \textbf{30.24 ft} \qquad \textbf{4d}$$

MATHEMATICS AND MEASUREMENTS

1.401   By law of sines

$$\frac{BC}{AB} = \frac{\sin 30°}{\sin 10°}$$

$BC = 50 \times 0.5000/0.1737 = 143.97$ ft

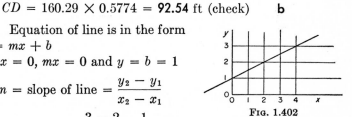

Fig. 1.401

$CD = 143.97 \times \sin 40°$
$\quad = 143.97 \times 0.6428 = 92.54 \text{ ft.}$

$BD = 143.97 \times \cos 40° = 143.97 \times 0.7660 = 110.29 \text{ ft}$

$$\tan 30° = \frac{CD}{AB + BD}$$

$CD = 160.29 \times 0.5774 = \textbf{92.54 ft (check)} \qquad \textbf{b}$

1.402   Equation of line is in the form of $y = mx + b$
when $x = 0$, $mx = 0$ and $y = b = 1$

$$m = \text{slope of line} = \frac{y_2 - y_1}{x_2 - x_1}$$

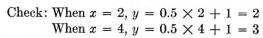

Fig. 1.402

$$= \frac{3 - 2}{4 - 2} = \frac{1}{2} = 0.5$$

$$y = \textbf{0.5}x + \textbf{1}$$

Check: When $x = 2$, $y = 0.5 \times 2 + 1 = 2$
        When $x = 4$, $y = 0.5 \times 4 + 1 = 3$

1.403   (1) $x^2 + y^2 = 5z$
         (2) $x^2 - y^2 = 3z$
        (1) + (2) $2x^2 = 8z$ and if $z = w^2$
             $x^2 = 4z = 4w^2$
            $x = 2w$ and $y = w$ or $x = 2y$

If $z = 1^2$ then $x = 2$ and $y = 1$
$z = 2^2$ then $x = 4$ and $y = 2$
$z = 3^2$ then $x = 6$ and $y = 3$
$z = n^2$ then $x = 2n$ and $y = n$
There are an infinite number of values that will satisfy.

1.404 *a.* $A = 89.42 = \pi d^2/4 = 0.785d^2$, $d = $ **10.67** in.

  b. Circumference $= \pi d = 3.14 \times 10.67 = 33.52$ in.

  c. Length of side of a regular inscribed hexagon $= d/2$
  $= $ **5.335** in.

1.405

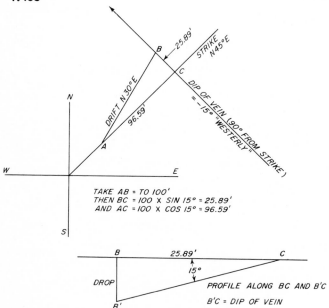

TAKE AB = TO 100'
THEN BC = 100 × SIN 15° = 25.89'
AND AC = 100 × COS 15° = 96.59'

PROFILE ALONG BC AND B'C
B'C = DIP OF VEIN

DROP BB' = 25.89 × TAN 15° = 25.89 × 0.2680 = 6.94'

PROFILE ALONG AB AND AB' (DRIFT OF MINE)

Fɪɢ. 1.405

Grade (per cent) $AB' = -6.94 \times 100/100 = $ **−6.94** per cent

1.406 Tape was used 9.17 times. If the tape is short, the distance measured is too long.

Total correction $= -0.10 \times 9.17 = -0.92$ ft

Corrected or true distance taped = 916.58 − 0.92 = **915.66** ft

1.407 $4 + \dfrac{(x + 3)}{(x - 3)} - \dfrac{4x^2}{(x^2 - 9)} = \dfrac{(x + 9)}{(x + 3)}$

To clear of fractions, multiply by $(x^2 - 9)$

$$4x^2 - 36 + x^2 + 6x + 9 - 4x^2 = x^2 + 6x - 27$$

Combining terms,

$$-27 = -27, \text{ or } 0 = 0$$

Since this is an identity, there are an infinite number of + or − values for x.

A special case is for $x = \pm 3$

$$x = +3, \quad +4 + \frac{6}{0} - \frac{36}{0} = \frac{12}{6}$$

or $+4 + \infty_1 - \infty_2 = 2$

$$x = -3, \quad +4 - \frac{0}{-6} - \frac{36}{0} = +\frac{6}{0}$$

or $+4 - 0 - \infty_2 = \infty_1$

If we approach ± 3 as a limit from 2 or 4 and use sufficient significant figures, the two very large values, here called infinity, will differ by enough to show that the equation is an identity.

1.408 The center of gravity of a circle is at its center and that of a triangle is one-third of the height above the base (or side).

Find center of gravity of 24-in. diameter circle with 9 by 12 triangular hole.

Above OX, take moments of areas about OX.

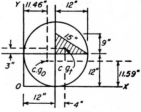

Fig. 1.408

Part	Area	×	Arm	=	Moment
Circle	452.4 sq in.	×	12 in.	=	+5,428.8 cu in.
Triangle	− 54.0	×	15	=	− 810.0
Net area	398.4 sq in.	×	(11.59 in.)	=	+4,618.8 cu in.

To right of OY take moments of areas about OY.

Circle	$+452.4$	$\times$	12	$=$	$+5,428.8$
Triangle	-54.0	$\times$	16	$=$	$-\ \ 864.0$
Net area	398.4 sq in.	$\times$	(11.46 in.)	$=$	$+4,564.8$ cu in.

1.409 The center of gravity of this area lies on the Y-Y axis. To find X, take moment of areas (of right half only because of symmetry) about top of figure.

Area no.	Area		$\times$	Arm	$=$	Moment
1	2×6	$= 12$ sq in.	$\times$	1 in.	$=$	12.00
2	1×10	$= 10$	$\times$	7	$=$	70.00
3	$\frac{1}{2} \times 1 \times 10 =$	5	$\times$	5.33	$=$	26.67
Total		27	$\times$	(4.02 in.)	$=$	108.67

The moment of inertia of an area about an axis other than that through its center of gravity $= I$ about center of gravity plus area times distance between center-of-gravity axis and new axis squared.

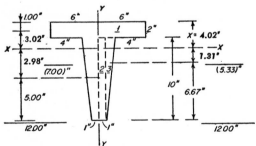

Fig. 1.409

$$I_{x\text{-}x} = 644.0 \text{ in.}^4$$

Double area	I_{cg}	$+$	Area $\times d^2$	$=$	New I
1	$\frac{1}{12} \times 12 \times 2 \times 2 \times 2$	$+ 12 \times 2 \times 3.02 \times 3.02$		$=$	$8.0 + 218.9$
2	$\frac{1}{12} \times 2 \times 10 \times 10 \times 10$	$+ 2 \times 10 \times 2.98 \times 2.98$		$=$	$166.7 + 177.6$
3	$\frac{1}{36} \times 2 \times 10 \times 10 \times 10$	$+ \frac{1}{2} \times 2 \times 10 \times 1.31 \times 1.31$		$=$	$55.6 + 17.2$
Total			644.0 in.^{4}	$=$	$230.3 + 413.7$

$$I_{y\text{-}y} = 313.0 \text{ in.}^4$$

Double area	I_{cg}	$+$	Area $\times d^2$	$=$	New I
1	$\frac{1}{12} \times 2 \times 12 \times 12 \times 12$	$+ 0$		$=$	288.00
2	$\frac{1}{12} \times 10 \times 2 \times 2 \times 2$	$+ 0$		$=$	6.67
3	$2 \times \frac{1}{36} \times 10 \times 1 \times 1 \times 1 + 2 \times \frac{1}{2} \times 1 \times 10 \times 1.33 \times 1.33$			$=$	0.55
					17.78
Total					313.00 in.^{4}

1.410 X distance to center of gravity = 2.14 in. to right. Moments about Y-Y axis:

Area no.	Area	$\times$	Arm	$=$	Moment
1	16 sq in.	$\times$	1 in.	$=$	16 cu in.
2	8	$\times$	4 in.	$=$	32
3	4	$\times$	3 in.	$=$	12
Total	28 sq in.	$\times$	(2.143 in.)	$=$	60 cu in.

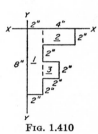

Fig. 1.410

Y distance to center of gravity = 3.29 in. down from x-x. Moments about x-x:

Area no.	Area	$\times$	Arm	$=$	Moment
1	16 sq in.	$\times$	4 in.	$=$	64 cu in.
2	8	$\times$	1	$=$	8
3	4	$\times$	5	$=$	20
Total	28 sq in.	$\times$	(3.286) in.	$=$	92 cu in.

$$I_{x\text{-}x} = 453.3 \text{ in.}^4$$

Area	I_{cg}	$+$	$A \times d^2$	$=$	New I
1	$\frac{1}{12} \times 2 \times 8 \times 8 \times 8$	$+$	$16 \times 4 \times 4$	$=$	$85.3 + 256$
2	$\frac{1}{12} \times 4 \times 2 \times 2 \times 2$	$+$	$8 \times 1 \times 1$	$=$	$2.7 + 8$
3	$\frac{1}{12} \times 2 \times 2 \times 2 \times 2$	$+$	$4 \times 5 \times 5$	$=$	$1.3 + 100$
				453.3 in.^{4} $=$	$89.3 + 364$

$$I_{v\text{-}v} = 197.3 \text{ in.}^4$$

Area	I_{og}	$+$	$A \times d^2$	$=$	New I
1	$\frac{1}{12} \times 8 \times 2 \times 2 \times 2$	$+$	$16 \times 1 \times 1$	$=$	$5.3 + 16$
2	$\frac{1}{12} \times 2 \times 4 \times 4 \times 4$	$+$	$8 \times 4 \times 4$	$=$	$10.7 + 128$
3	$\frac{1}{12} \times 2 \times 2 \times 2 \times 2$	$+$	$4 \times 3 \times 3$	$=$	$1.3 + 36$
			197.3 in.^4	$=$	$17.3 + 180$

1.411 Let x = rate (mph) of plane B and $x + 90$ = rate (mph) of plane A.

$$\frac{900}{(x + 90)} + 2.25 = \frac{900}{x}, \quad \text{(hours + hours = hours)}$$

Multiply through by $x(x + 90)$ to clear of fractions

$$900x + 2.25(x^2 + 90) = 900x + 81{,}000$$
$$x^2 + 90x = 36{,}000$$

Complete square

$$x^2 + 90x + 45 \times 45 = (x + 45)^2 = 36{,}000 + 2{,}025$$
$$x + 45 = 195, \quad x = 150, \quad x + 90 = 240$$

Check $\frac{900}{240} + 2.25 = 6.00 \text{ hr} = \frac{900}{150}$

1.412 Let A be one number and B the other. Given

$$A^2 + B^2 = 100 \text{ and } 2(A + B) = 28$$

This suggests a 3-4-5 triangle where

$$A = 6, \quad B = 8, \text{ and } C = 10$$

Thus $36 + 64 = 100$ and $2(6 + 8) = 28$
$$\text{and } AB = 48$$

By direct solution

$$A + B = 14 \text{ and } A^2 + 2AB + B^2 = 196$$

But $A^2 + B^2 = 100$ so $2AB + 100 = 196$ and $AB = 48$
For a check, solve for A and B.

$$A = 14 - B \text{ and } A^2 = 196 - 28B + B^2$$
$$196 - 28B + B^2 + B^2 = 100 \text{ and } B^2 - 14B + 48 = 0$$

By quadratic equation

$$B = \frac{-(-14) \pm \sqrt{+196 - 192}}{2} = \frac{+14 \pm 2}{2}$$

$$= +8 \text{ or } +6 \text{ and } A = +6 \text{ or } +8$$

The product of $A \times B = 48.$ **b**

1.413 Area of field $= xy$

(1) $+xy + 2,500 = (x + 100)(y - 25)$
(2) $+xy - 5,000 = (x - 100)(y + 50)$
            $xy + 2,500 = xy + 100y - 25x - 2,500$
(a) $+100y - 25x = +5,000$
            $xy - 5,000 = xy - 100y + 50x - 5,000$
(b) $-100y + 50x = 0$
(a) $\underline{+100y - 25x = +5,000}$
(a) $+$ (b) $+25x = +5,000$
                $x = +200$
                $y = +100$

Check: $xy = 20,000$
        $xy + 2,500 =$
        $22,500 = 300 \times 75$
        $xy - 5,000 =$
        $15,000 = 100 \times 150$

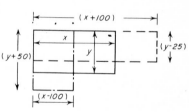

Fig. 1.413

1.414 Tire must be expanded

$$(62.378 - 62.263)\pi = 0.115\pi \text{ in.}$$

Coefficient of expansion of steel $= 0.0000065$ units per unit per degree F.

$t =$ degrees above 65°F that tire must be heated. Inside circumference of tire $= 62.263\pi$

$$0.115\pi = 62.263\pi \times 0.0000065 \times t$$

$$t = \frac{0.115}{0.405 \times 1/1,000} = 284°F$$

Temperature of tire must be raised to $284 + 65 = $ **349°F** or above.   **c**

1.415 A $-60°F$ change in temperature will shorten a 90.000-ft length of steel tape by $90 \times 60 \times 0.0000065 = 0.035$ ft.

90.035 ft must be read on the tape to lay out a 90.000-ft distance on the ground.

1.416 Ten men can dig $1\%_8 \times 150 = 187.5$ ft of trench in 7 hr. Ten men can dig 200 ft of trench in $(7 \times 200)/187.5 = 7.467$ hr.

Three men can backfill 200 ft of trench in 8 hr.

Ten men can backfill 200 ft of trench in $\%_{10} \times 8 = 2.4$ hr.

Ten men can dig and backfill 200 ft of trench in 9.867 hr =

9 hr 52 min   **a**

1.417 (1) $x^2 + y^2 = +15,025$

(2) $2x + 4y = +530$ or $x = +265 - 2y$

$x^2 = +70,225 - 1,060y + 4y^2$

(1) $y^2 + 4y^2 - 1,060y + 70,225 = +15,025$

$y^2 - 212y + 11,040 = 0$

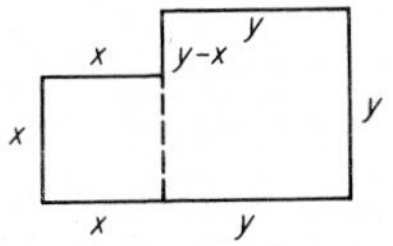

By quadratic equation

$$y = \frac{-(-212) \pm \sqrt{212 \times 212 - 44,160}}{2}$$

FIG. 1.417

$$y = \frac{+212 \pm \sqrt{784}}{2} = \frac{+212 \pm 28}{2} = +106 \pm 14 = 120 \text{ or } 92$$

When $y = 120$, $x = 25$ and $14,400 + 625 = 15,025$

When $y = 92$, $x = 81$ and $8,464 + 6,561 = 15,025$

Both solutions satisfy.

1.418 (a) $y^3 = x^3$ and $y = \sqrt[3]{x^3} = x$ or a straight line

Area $= \frac{1}{2} \times 4 \times 4 = $ **8**

(b)   Area $= \int_1^3 (x^3 + 3x^2)dx = \frac{x^4}{4} + \frac{3x^3}{3} \Big]_1^3$

$= 8\frac{1}{4} + 27 - \frac{1}{4} - 1 = 20 + 26 = $ **46**

(c) $\frac{d}{dx}(4x^2 + 17x) = $ **$8x + 17$**

Also $\frac{d}{dx}(ax^2 + b^{\frac{1}{2}}) = $ **$2ax$**

(d) $\int (7x^3 + 4x^2)dx = \dfrac{7x^4}{4} + \dfrac{4x^3}{3} + c_1$

Also $\int x \cos (2x^2 + 7)dx = \dfrac{\sin}{4} (2x^2 + 7) + c_1$

1.419   The increase in length of 100-ft tape from 10°C to 30°C = 100 × (30 − 10) × 0.000011 = 0.022 ft. The tape at 30°C = 100.042 + 0.022 = 100.064 ft. The measurement of 1,256.271 ft is short because it was made with a long tape.

The correction at +0.064 ft per 100 ft = 12.563 × 0.064 = +0.804 ft.

True length of line = 1,256.271 + 0.804 = 1,257.075 ft.   **a**

1.420   Arithmetical progression:

$$x + (x + a) + (x + 2a) = 45$$
$$3x + 3a = 45, \quad x + a = 15, \text{ and } a = 15 - x$$

If $x = 10$, then $a = 5$ and the three numbers are 10, 15, and 20. Geometrical progression:

$$10 + 2 = 12 = b \quad = 12 \text{ and } r = 1.5$$
$$15 + 3 = 18 = br \quad = 12 \times 1.5$$
$$20 + 7 = 27 = br^2 = 12 \times 1.5 \times 1.5$$

By direct solution:

(1) $x + a = +15 \text{ or } a = +15 - b + 2 = 17 - b$

(2) $x + 2 = \quad b \text{ or } x = b - 2$

(3) $x + a + 3 = br \text{ or } 18 = br \text{ and } r = \dfrac{18}{b}$

(4) $x + 2a + 7 = br^2 \text{ or } (b - 2) + 2(17 - b) + 7 = 18 \times \dfrac{18}{b}$

(4) becomes

$$b^2 - 39b + 324 = 0, \quad b = \dfrac{-(-39) \pm \sqrt{1,521 - 1,296}}{2}$$

$$b = \dfrac{+39 \pm 15}{2} = 12, \quad \text{and } x = 10, \quad a = 5, \quad r = 1.5$$

1.421   (a)   (1) $+4x^2 + 7y^2 = +32$, or $+12x^2 + 21y^2 = + 96$
(2) $-3x^2 + 11y^2 = +41$, or $\underline{-12x^2 + 44y^2 = +164}$
(1') + (2'). $+65y^2 = +260$

$$y^2 = 4, \quad y = +2, \quad x = \pm 1, \quad \text{and } y = -2, \quad x = \pm 1$$

(b)

(1) $+ x + 2y - z = + 6$
$2 \times$ (2) $+4x - 2y + 6z = -26$
(3) $\underline{+3x - 2y + 3z = -16}$
(a) = (1) + 2 × (2), $+ 5x + 5z = -20$, or $+x + z = -4$
(b) = (1) + (3) $+ 4x + 2z = -10$, or $\underline{+x + 0.5z = -2.5}$
(a') − (b') = $+0.5z = -1.5$

From (a'), $x = -1$; and from (1), $y = +2$ $z = -3$
Check using (2) and (3).

1.422   (a) $\cos 2A = 2\cos^2 A - 1$
$2\cos^2 75° - 1 = \cos 150° = -\cos 30°$
$$\cos 30° = \frac{\sqrt{3}}{2} = \frac{1.732}{2} = 0.866$$
$$\cos^2 75° = \frac{1 - 0.866}{2} = \frac{+0.133}{2} = 0.0666$$
$$\cos 75° = \sqrt{0.0666} = 0.258$$

(b) Angle $A = 75° = AoX$ (first quadrant)
$\sin A = Aa$
$\cos A = oa$
$\tan A = \dfrac{Aa}{oa}$

Angle $B = BoX = 105°$ (second quadrant)
$\cos B = bo = -oa$
$\sin B = Bb = Aa$
$\tan B = \dfrac{Bb}{bo} = -\dfrac{Aa}{oa}$

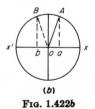

(b)

Fig. 1.422b

1.423   Volume $= x(20 - 2x)^2$

$$= 400x - 80x^2 + 4x^3$$

(for V max) $\dfrac{dV}{dx} = 0$

$$= +400 - 160x + 12x^2$$

or $+3x^2 - 40x + 100 = 0$

Fɪɢ. 1.423

Using quadratic equation

$$x = \frac{-(-40) \pm \sqrt{1,600 - 1,200}}{6} = \frac{+40 \pm 20}{6} = +10 \text{ or } 3.33$$

V min (0) when $x = 10$, so use $x = 3.33$

$V = 13.33 \times 13.33 \times 3.33 = 592.59$ cu in.

x may be found by trial (carry volume to $\frac{1}{100}$ s).

1. Try values of $x = 1, 2, 3$, and 4.
2. Try values of $x = 3.1$ to 3.5.
3. Try values of $x = 3.31$ to 3.35.

1.424 $s = \frac{1}{2}(a + b + c) = \frac{1}{2}(9 + 12 + 10) = 15.5$

$$r = \sqrt{\frac{(s - a)(s - b)(s - c)}{s}} = \sqrt{\frac{6.5 \times 3.5 \times 5.5}{15.5}}$$

$r = \sqrt{8.07} = 2.84$

$\tan \dfrac{1}{2} A = \dfrac{r}{(s - a)} = \dfrac{2.84}{6.5} = 0.437, \quad \dfrac{1}{2} A = 23.6°, \quad A = 47.2°$

$\tan \dfrac{1}{2} B = \dfrac{r}{(s - b)} = \dfrac{2.84}{3.5} = 0.812, \quad \dfrac{1}{2} B = 39.05°, \quad B = 78.1°$

$\tan \dfrac{1}{2} C = \dfrac{r}{(s - c)} = \dfrac{2.84}{5.5} = 0.517, \quad \dfrac{1}{2} C = 27.35°, \quad C = 54.7°$

Check: $\overline{180.0°}$

Area $ABC = \frac{1}{2} \times 12 \times 10 \times \sin 47.2° = 60 \times 0.7337 = 44.0$ ft^2

Volume prism $= \dfrac{(h_a + h_b + h_c)}{3} \times 44.0$

$$= \frac{(8.6 + 7.1 + 5.5)}{3} \times 44.0$$

$$= \frac{21.2}{3} \times 44.0 = 311 \text{ ft}^3$$

Fɪɢ. 1.424

1.425

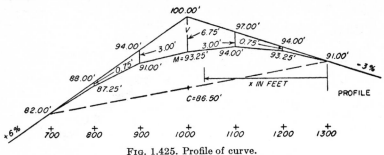

FIG. 1.425. Profile of curve.

'Elevations along Grade Tangents and Parabola

Station	Vertex Elevation	−	Drop to tangent	=	Elevation of tangent	−	Drop to parabola	=	Elevation of parabola
7 + 00	100.00	−	3 × 6	=	82.00	−	0.00	=	82.00
8 + 00	100.00	−	2 × 6	=	88.00	−	0.75	=	87.25
9 + 00	100.00	−	1 × 6	=	94.00	−	3.00	=	91.00
10 + 00				=	100.00	−	6.75	=	93.25
11 + 00	100.00	−	1 × 3	=	97.00	−	3.00	=	94.00
12 + 00	100.00	−	2 × 3	=	94.00	−	0.75	=	93.25
13 + 00	100.00	−	3 × 3	=	91.00	−	0.00	=	91.00

Point C has an elevation the mean of that at 7 + 00 and 13 + 00.

Point M has an elevation the mean of 100.00 and C. $V =$ 6.75 ft. Distance down (in feet) from tangent to parabola

$$= \left(\frac{x}{\frac{1}{2} \text{ length of parabola}} \right)^2 \times 6.75 = \left(\frac{x}{300} \right)^2 \times 6.75$$

For stations 8 + 00 and 12 + 00, drop = $(100/300)^2 \times 6.75$ = 0.75 ft

For stations 9 + 00 and 11 + 00, drop = $(200/300)^2 \times 6.75$ = 3.00 ft

From the table above it is fairly obvious that the high point of the parabola is at station 11 + 00. In other cases not so obvious, it is usually a matter of observation on which side of V

the high point lies. In this case, knowing the high point is to the right of V

Elevation of tangent $= 91.00 + 0.03x$

Elevation of curve $= y = 91.00 + 0.03x - \left(\dfrac{x}{300}\right)^2 \times 6.75$

When curve is at maximum elevation, the slope of the tangent is zero. Therefore

$$\frac{dy}{dx} = 0 = +0.03 - \frac{2x \times 6.75}{90,000}$$
$$13.5x = 2,700, \quad x = 200$$

or high point is at station $13 + 00 - 2 + 00 = 11 + 00$

The above work may be done by cut and try with more work and perhaps less accuracy.

MECHANICS (KINETICS)

1.501   Velocity = acceleration $\times$ time, or $v = at$

$$v = 16 \text{ ft per sec} = a \times 4, \quad a = 4 \text{ ft per sec}^2$$

Tension in cable $= Ma = \dfrac{W}{g}(g + a)$

$$T = (2,000/32.2)(32.2 + 4) = \mathbf{2,250} \text{ lb} \qquad \mathbf{b}$$

1.502   Final velocity squared = initial velocity squared plus twice acceleration times distance body moves.

$$v^2 = 0 = 16 \times 16 + 2 \times a \times 5, \quad a = -25.6 \text{ ft per sec}^2$$

Tension in cable $= Ma = \dfrac{W}{g}(g + a)$

$$= (2,000/32.2)(32.2 - 25.6) = \mathbf{410} \text{ lb}$$

1.503   Force normal to plane $= 40 \times 0.866$
         $= 34.67$ lb

Force parallel to plane $= 40 \times 0.500$
         $= 20.00$ lb

Friction force $= 34.67 \times 0.3 = 10.40$ lb

Resultant force parallel to plane
         $= 20.00 - 10.40 = 9.60$ lb

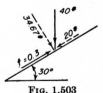

Fig. 1.503

$$\text{Force} = \text{mass} \times \text{acceleration or } 9.60 = \frac{40}{32.2} \times a$$

$$a = 7.73 \text{ ft per sec}^2$$

Distance weight moves $= \frac{1}{2} \times$ acceleration $\times$ time squared

$$D = \frac{1}{2}at^2 \text{ and } 60 = \frac{1}{2} \times 7.73t^2$$
$$t = 3.94 \text{ sec}$$

Distance weight moves during third second

$$= (7.73/2)(3^2 - 2^2) = 19.33 \text{ ft}$$

1.504   Initial velocity $= \dfrac{30 \times 5{,}280}{60 \times 60} = 44$ ft per sec

Final velocity = initial velocity minus the acceleration $\times$ time

$$0 = 44 - a \times 5, \quad a = 8.8 \text{ ft per sec}^2$$

Force = mass $\times$ acceleration, $F = (6{,}000/32.2) \times 8.8 = 1{,}640$ lb
Taking moments about bottom of rear wheel

$$\Sigma M_{W_R} = 0 = +10 \times 2 \times W_F - 6{,}000 \times 5 - 1{,}640 \times 2$$
$$W_F = 1{,}500 + 164 = 1{,}664 \text{ lb}$$

Normal pressure on each front wheel = 1,664 lb
Normal pressure on each rear wheel = 1,336 lb

1.505   The weight $\times$ distance of free fall + the weight $\times$ displacement of spring = work of spring = average force of spring $\times$ displacement of spring squared.

$$1{,}000 \times 6 + 1{,}000 \times d_s = (2000/2) \times d_s^2$$
$$d_s^2 - d_s - 6 = 0 = (d_s - 3)(d_s + 2)$$
$$d_s = +3 \text{ in.} \quad \textbf{a}$$

(The spring does not elongate, hence $d_s = -2$ is meaningless.)

1.506   Work of spring = kinetic energy of car
Let R = resistance of spring in pound per inch.

$$\frac{R}{2} \times d_s^2 = \frac{1}{2} Mv^2$$

$$\frac{R}{2} \times 2.5 \times 2.5 = \frac{1}{2} \times \frac{100{,}000}{32.2} \times 2 \times 2 \times 12$$

$$R = 23{,}850 \text{ lb per in.}$$

1.507 Block M has a 50-lb component parallel to the 30° plane and an 86.67-lb component normal to the 30° plane.

Friction force block $M = 86.67 \times 0.2 = 17.33$ lb

Block N has an 86.67-lb component parallel to the 60° plane and a 50-lb component normal to the 60° plane.

Friction force block $N = 50 \times 0.2 = 10$ lb

Net force block $M = 50 + 17.33 = 67.33$ lb

Net force block $N = 86.67 - 10 = 76.67$ lb

Resultant force $= 76.67 - 67.33 = 9.33$ lb down 60° plane

Force = mass $\times$ acceleration, $9.33 = \dfrac{(100 + 100)}{32.2} \times a$

$$a = 1.505 \text{ ft per sec}^2$$

Distance weights move $= \frac{1}{2} \times$ acceleration $\times$ time squared

$$15 = \tfrac{1}{2} \times 1.505 \times t^2, \quad t = 4.47 \text{ sec}$$

Tension in the connecting cord $= T$

(On weight M side) $T = 67.33 + (100/32.2) \times 1.505 = 72.00$ lb

(On weight N side) $T = 76.67 - (100/32.2) \times 1.505 = 72.00$ lb

1.508 $T_x = T_y$ and $T_x + T_y = T_z = 2T_x = 2T_y$ (where $T =$ tension in cords carrying weights)

Weights y and z will fall, weight x will rise.

Say weight x rose 10 ft and weight y fell 8 ft, then the pulley carrying their joint cord must rise one-half the difference or 1 ft in order for the cord to remain taut. That gives $d_z = \dfrac{d_x - d_y}{2}$

$$\therefore (1) \quad a_z = \frac{a_x - a_y}{2} = 0.5(a_x - a_y)$$

$$(2) \quad T_x = \frac{10g}{g} + \frac{10}{g}a_x = +10 + 0.31a_a$$

$$(3) \quad T_y = \frac{20g}{g} - \frac{20}{g}a_y = +20 - 0.62a_y$$

$$(4) \quad T_z = \frac{30g}{g} - \frac{30}{g}a_z = +30 - 0.93a_z$$

$$(2') \quad T_z = \qquad\qquad +20 + 0.62a_x$$
$$(3') \quad T_z = \qquad\qquad +40 - 1.24a_y$$
$$(5) = (3') - (2'), \quad 0 = +20 - 0.62a_x - 1.24a_y$$
$$(6) = (3') - (4), \quad 0 = +10 - 1.24a_y + 0.93a_z$$

but $(1') = 0.93 \times (1) + 0.93a_z = +0.465a_x - 0.465a_y$

$(6')$　Substituting $(1')$ in (6)
$$-0.465a_x + 0.465a_y + 1.24a_y = +10$$

$(7) = \frac{4}{3} \times (6')$
$$-0.62a_x + 2.27a_y = +13.33$$
(5) $\quad\underline{+0.62a_x + 1.24a_y = +20.00}$
$(8) = (5) + (7)$ $\quad 3.51a_y = +33.33$
$(8')$ $\qquad\qquad\qquad a_y = +\ 9.5$ ft per sec²
From (5) $\qquad\qquad a_x = +13.3$ ft per sec²
From (1) $\qquad\qquad a_z = +\ 1.9$ ft per sec²

Now $d = \frac{1}{2}at^2$ and $v = at$

At the end of 1 sec $d_x = 6.70$ ft up, $\quad v_x = 13.3$ fps
$\qquad\qquad\qquad d_y = 4.75$ ft down, $v_y = \ 9.5$ fps
$\qquad\qquad\qquad d_z = 0.85$ ft down, $v_z = \ 1.9$ fps

1.509　Positive work − negative work = final kinetic energy.
Positive work = foot-pounds of drop
$$= \frac{1 \times 100 \text{ ft}}{100} \times 120,000 \text{ lb} = 120,000 \text{ ft-lb}$$
Negative work = friction loss
$$= 10 \text{ lb per ton} \times 60^T \times 100 = \underline{-60,000 \text{ ft-lb}}$$
Final kinetic energy $= \dfrac{1}{2} \times \dfrac{120,000}{32.2} \times v^2 = 60,000$ ft-lb
$$v^2 = 32.2$$

Striking velocity = **5.67** fps

Final kinetic energy $= 60,000 \times 12 = \dfrac{100,000}{2} \times d_s{}^2$

Spring will shorten **3.795** in.

1.510　Initial kinetic energy = uphill work + friction

$$30 \text{ mph} = \frac{30 \times 5,280}{60 \times 60} = 44 \text{ fps}$$

(1)

$$\frac{1}{2} \times \frac{80,000}{32.2} \times (44)^2 = \frac{80,000}{2,000} \times 10 \times d_1 + \frac{2 \times d_1}{100} \times 80,000$$

divide (1) by 800

$$3,010 = 0.5d_1 + 2d_1 = 2.5d_1 \text{ or } d_1 = \textbf{1,204} \text{ ft on slope}$$

Downhill work $-$ friction $=$ kinetic energy

$$+\frac{2 \times 1,204}{100} \times 80,000 - 10 \times 1,204 \times \frac{80,000}{2,000} = \frac{80,000 \times v^2}{64.4}$$

$$+24.08 - 6.02 = 18.06 = \frac{v^2}{64.4}, \quad v = 34 \text{ fps}$$

Kinetic energy $=$ friction

$$80,000 \times \frac{v^2}{64.4} = \frac{80,000}{2,000} \times 10 \times d_2$$

$$18.06 = \frac{d_2}{200}$$

$$d_2 = \textbf{3,612} \text{ ft on level track}$$

1.511 $\qquad \frac{50}{r} = \sin 5° = 0.0872, \ r = 574 \text{ ft}$

$$45 \text{ mph} = 45 \times 5,280/3,600 = 66 \text{ fps}$$

Gage of a standard railroad track $=$ 4 ft 8½ in. or 4.9 ft center-to-center rails.

    a. Superelevation for equal wheel pressures

$$= \frac{G \times v^2}{g \times r} = \frac{4.9 \times 66 \times 66}{32.2 \times 574} = \textbf{1.15} \text{ ft}$$

    b. Horizontal force 5 ft. above rail for 60 mph speed $=$

$$\frac{Wv^2}{g \times r} = \frac{100,000 \times 88 \times 88}{32.2 \times 574} = 41,950 \text{ lb}$$

Superelevation $=$ 8 in. in 4.9 ft or 0.67/4.9, and position of vertical force (weight) off center at top of rails $=$

$$x = 5 \times 0.67/4.9 = 0.683 \text{ ft}$$

Distant vertical force from center inner rail $=$

$$4.9/2 - 0.683 = 1.767 \text{ ft}$$

$\Sigma M_{\text{inner rail}}$

$$0 = +5 \times 41,950 + 1.767 \times 100,000 - 4.9 \times R_{\text{outer}}$$

Normal pressure outer rail $= R_{outer} =$

$$42{,}750 + 36{,}000 = \mathbf{78{,}750}\ \text{lb}$$

1.512   60 mph = 88 fps

a. $\tan A = 0.15$
           $A = 8.53° =$ angle of superelevation

b. Now

$$\tan (A + B) = \frac{88 \times 88}{32.2 \times 500} = 0.482$$

$$A + B = 25.73°$$
$$A = \mathbf{8.53°}\ \text{(angle of superelevation)}$$
$$B = 17.20°\ \text{and}\ \tan B = 0.310$$

Skidding will impend if f is below **0.310**

1.513   Vertical is shorter than 100 ft by $\dfrac{5 \times 5}{200} = 0.125\ \text{ft} = \boldsymbol{h}$

$$\text{Tension in cable} = \frac{100}{99.875} \times 20{,}000 = \mathbf{20{,}025}\ \text{lb}$$

$$\text{Horizontal pull} = \frac{5}{99.875} \times 20{,}000 = \mathbf{1{,}001}\ \text{lb}$$

$$v^2 = 2gh = 2 \times 32.2 \times 0.125\ \text{or}\ 0.25g$$

Tension of cable when swinging through vertical position

$$= 20{,}000 + \frac{20{,}000 \times 0.25g}{g \times 100} = \mathbf{20{,}050}\ \text{lb}$$

1.514 $\sin 45° = \cos 45° = 0.707$

The vertical velocity at the high point of path $= 0 =$ initial velocity $\times \sin 45° - gt$

$$300 \times 0.707 = 32.2 \times t\ \text{and}\ t = 6.6\ \text{sec}$$

Maximum height projectile will reach
$$= 300 \times t \times \sin 45° - \tfrac{1}{2}gt^2$$
$$= 300 \times 6.6 \times 0.707 - \tfrac{1}{2} \times 32.2 \times 6.6 \times 6.6$$
$$= 1{,}400 - 700 = \mathbf{700}\ \text{ft}$$

For 45° angle, $h_{max} =$ initial velocity squared divided by $4g$ or

$$\frac{300 \times 300}{4g} = 700\ \text{ft}$$

Range = twice the initial velocity $\times$ cos 45° $\times$ t

$$= 2 \times 300 \times 0.707 \times 6.6 = \textbf{2,800 ft}$$

$$\left(\text{Also max range} = \frac{v^2}{g} = \frac{300 \times 300}{g} = 2,800 \text{ ft}\right)$$

1.515 (See the special formulas in answer 1.514.)

Maximum range = $40 \times 5280 = \dfrac{v^2}{32.2}$ or $v^2 = 6,800,000$ and

$v = \textbf{2,600 fps}$

Maximum height = $\dfrac{v^2}{4g} = \frac{1}{4} \times$ max range = $^{40}\!\!/_4 = \textbf{10 miles}$

1.516   18 in. = 1.5 ft

sin 60° = $0.866 \times 1.50 = 1.30$ ft (radius)
cos 60° = $0.500 \times 1.50 = 0.75$ ft (center ball below top end
of rod)

tan 60° = $1.732 = \dfrac{v^2}{gr}$, $\quad v^2 = 1.732 \times 32.2 \times 1.30$

$$v = \textbf{8.52 fps} = \textbf{62.6 rpm}$$

Weight of 3-in. cast-iron ball = $\frac{1}{6} \times \pi \times (\frac{1}{4})^3 \times 450 = 3.68$ lb

Moment (about point where arm connects to vertical shaft)
of horizontal effective force and weight (vertical)

$$\Sigma M = 0 = -0.75 \times \left(\frac{3.68 \times v^2}{gr}\right) + 1.30 \times (3.68)$$

Check: $0.75 \times 1.732 = 1.30$

Tension in arm = $3.68 \sqrt{(1)^2 + (1.732)^2} = \textbf{7.36 lb}$

1.517   60 mph = 88 fps

Radius = 1,000 ft, very nearly.
Let A = the angle the cord makes with the vertical

$$\tan A = \frac{v^2}{gr} = \frac{88 \times 88}{32.2 \times 1000} = 0.2735, \quad A = \textbf{15.3°}$$

The horizontal displacement of the weight =

$$6 \times \sin 15.3° = 6 \times 0.264 = \textbf{1.58 ft}$$

The tension in the cord $= 25 \sqrt{(1)^2 + (0.274)^2}$
$$= 25 \times 1.036 = \textbf{25.9 lb}$$

1.518 Distance projectile must travel
$$= \text{initial velocity squared} \times \frac{\sin 2A}{g} = \left[2(v \cos A) \left(\frac{v \sin A}{g} \right) \right]$$

$$10 \times 5{,}280 = \frac{1{,}500 \times 1{,}500}{32.2} \times \sin 2A$$

$$\sin 2A = \frac{322 \times 0.528}{225} = 0.756, \quad 2A = 49.1°$$

$$A = 24° \ 33' \quad \textbf{c}$$

1.519 $I = \frac{1}{2} M r^2$ for a solid cylinder about geometric axis

M(for 24-in. diameter cylinder)
$$= \pi \times \left(\frac{12}{12} \right)^2 \times \frac{4}{3} \times \frac{450}{32.2} = 58.5$$

M(for 22-in. diameter cylinder)
$$= \pi \times \left(\frac{11}{12} \right)^2 \times \frac{4}{3} \times \frac{450}{32.2} = 49.2$$

I(24-in. cylinder) $= \frac{1}{2} \times 58.5 \times 1^2 \qquad = 29.3$
I(22-in. cylinder) $= \frac{1}{2} \times 49.2 \times (11/12)^2 = 20.6$
$$I(\text{net}) \ (\text{lb} \times \text{ft} \times \sec^2) = \quad \textbf{8.7} \qquad \textbf{d}$$

1.520 $M = (490/32.2) \times \pi \times r^2 \times L = 47.8 \times r^2 \times L$
$I = \frac{1}{2} M r^2 = 23.9 \times r^4 \times L$
I(about axis of 40-in. disk) $= I_{cg} + M \times d^2$
where d is the distance between axes
I (40 in.) $= 23.9 \times \frac{1}{3} \times (20/12)^4 = +61.50$
I (10 in.) $= 7.97 \times (5/12)^4 \qquad = -0.24$
I (12 in.) $= 7.97 \times (6/12)^4 + 15.93 \times (6/12)^2 \times (12/12)^2$
$\qquad = -0.50 - 3.98 = -4.48$

$$I(\text{net}) = +61.50 - 4.72 = \textbf{56.78 lb} \times \text{ft} \times \sec^2$$

1.521   Angular velocity $= 1{,}500$ rpm $= 1{,}500 \times \dfrac{2\pi}{60}$

$\qquad\qquad\qquad\qquad\qquad\qquad\qquad = 157$ radians per sec

$\qquad 157 = \alpha t = \alpha 60, \quad \alpha = 2.617$ radians per sec^2

Number of revolutions until wheel stops

$\qquad\qquad = $ mean angular velocity $\times$ time

$\qquad\qquad = \dfrac{1{,}500 + 0}{2} \times 1 \text{(minute)} = 750$ revolutions

Angular velocity at end of 40 sec $= 157 - 2.617 \times 40$

$\qquad = 52.3$ radians per sec $= 500$ rpm

1.522   Angular velocity $= 60$ mph $= 88$ fps at rim

88 fps $= \dfrac{88}{3\pi}$ rps $= \left(\dfrac{88}{3\pi}\right) \times 2\pi = 58.7$ radians per sec

Number of revolutions to stop wheel $= \dfrac{500}{3\pi}$ or $\left(\dfrac{500}{3\pi}\right) 2\pi$

$\qquad = 333$ radians

Angular velocity squared $=$ twice the total number of revolutions to stop times the angular acceleration.

$\qquad (58.7)^2 = 2 \times 333 \times \alpha, \alpha = 5.17$ radians per sec^2

1.523 M(hemisphere) $= \dfrac{490}{32.2} \times \dfrac{1}{6} \times \pi \times \left(\dfrac{6}{12}\right)^3 \times \dfrac{1}{2}$

$\qquad\qquad\qquad\qquad\qquad = 0.498$ lb $\times$ sec^2 per ft

Center of gravity hemisphere $= \frac{3}{8}r = \frac{3}{8} \times 3 = 1.125$ in. above diametral plane

Angular velocity $= 300$ rpm $= 31.4$ radians per sec

Outward (horizontal) force $=$ mass $\times$ radius distance (axis to center of gravity of weight) $\times$ angular velocity squared

$\qquad$ ohf $= 0.498 \times 12\frac{2}{12} \times 31.4 \times 31.4 = 493$ lb

Moment in rod $= \dfrac{(1.125 - 0.500)}{12} \times 493 = 25.6$ ft-lb

For zero moment in rod, place rod at center of gravity of weight or 1.125 in. above diametral plane.

1.524

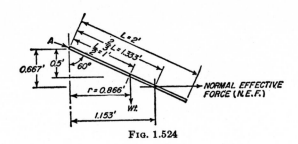

FIG. 1.524

$$\Sigma M_A = 0 = -0.667 \times \text{nef} + 0.866 \times \text{weight}$$
$$0 = -0.667 \times M \times 0.866 \times V_a{}^2 + 0.866 \times 32.2M$$
$$V_a{}^2 = \frac{32.2}{0.667} = 47.6$$

V_a (angular velocity) $= 6.9$ radians per sec $= 66$ rpm   **C**

1.525   Angular velocity $= V_a = 240$ rpm $= 25.1$ radians per sec

$$2T = \frac{M}{2} r_{c.g} \times V_a{}^2 \quad \text{where} \quad r_{c.g} = \frac{2}{\pi} \times \text{(mean radius of rim)}$$

$$\frac{M}{2} = \frac{1}{2} \times \frac{450}{32.2} \times \frac{18}{12} \times \pi \left[\left(\frac{72}{12}\right)^2 - \left(\frac{68}{12}\right)^2 \right] = 125$$

$$T = \frac{125}{2} \times \frac{2}{\pi} \left(\frac{72+68}{2 \times 12}\right) \times 25.1 \times 25.1 = 146{,}000 \text{ lb}$$

$$\text{Tension in rim} = \frac{146{,}000}{4 \times 18} = 2{,}032 \text{ psi}$$

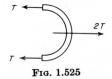

FIG. 1.525

MECHANICS (STATICS)

1.601 $\Sigma V = +10 - 10 - 10 + 10 = 0$
$\Sigma M(\text{about } F_1) = +5 \times 10 + 9 \times 10 - 12 \times 10$
$$= +20 \text{ ft-lb (clockwise)}$$

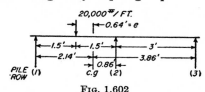

Fɪɢ. 1.601

The resultant is equal to a clockwise couple of 20 ft-lb. Had the right hand force (F_4) acted downward, then

$$\Sigma V = +10 - 10 - 10 - 10 = -20 \text{ lb acting downward}$$
and $\quad \Sigma M(F_1) = +5 \times 10 + 9 \times 10 + 12 \times 10 = +260 \text{ ft-lb}$

The resultant would be a 20-lb force acting downward and $260/20 = 13$ ft to right of F_1.

1.602　Find center of gravity of pile group.

Fɪɢ. 1.602

M about Pile Row 1

Pile row	Pile spacing center to center	Area piles per ft	Distance from row 1.	Area × distance	x = distance center of gravity pile group to pile	I = area × x^2
1.	2.5 ft	0.400 sq ft	0	0	+2.14 ft	1.835
2.	4.0 ft	0.250	3 ft	0.750	−0.86	0.185
3.	6.0 ft	0.167	6 ft	1.000	−3.86	2.485
Sum		0.817	(2.14) ft	1.750		4.505

Center of gravity of pile group lies 1.750/0.817 or 2.14 ft to right of row 1. The 20,000 lb per ft load P is 0.64 ft eccentric e to center of gravity of piles.

$$\text{Load on any piles} = \frac{P}{A} + \frac{Pex}{I} \text{ and } \frac{P}{A} = \frac{20,000}{0.817} = 24,500 \text{ lb}$$

$$\frac{Pex}{I} = \frac{20,000(0.64)x}{4.505} = 2840x$$

Pile row	$\dfrac{P}{A}$	$\dfrac{Pex}{I}$	Stress in pile
1.	$+24,500$	$+ 6,080$	$= 30,580$ lb per pile
2.	$+24,500$	$- 2,440$	$= 22,060$ lb per pile
3.	$+24,500$	$-10,960$	$= 13,540$ lb per pile

Check: (1)　$30,580 \div 2.5 = 12,210$ lb per ft
　　　　(2)　$22,060 \div 4.0 = 5,515$ lb per ft
　　　　(3)　$13,540 \div 6.0 = \underline{2,258}$ lb per ft
　　　　　　　　　　　　　　$19,983$ lb per ft　(say OK)

1.603　$\Sigma M_{R_L} = 0 = +4 \times 1,000 \times 2 + 10,000 - 12R_R,$
　　　　　　　　　　　　　　　　　　$R_R = +1,500$ lb
　　　　$\Sigma M_{R_R} = 0 = +12R_L + 10,000 - 4 \times 1,000 \times 10,$
　　　　　　　　　　　　　　　　　　$R_L = +2,500$ lb

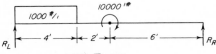

FIG. 1.603. Forces.

Check:

$$V = 0 = +R_L + R_R - 4 \times 1,000 = +2,500 + 1,500 - 4,000$$

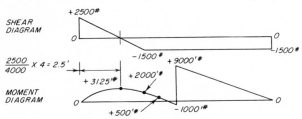

FIG. 1.603. Shear and moment.

M at point 2.5 ft to right of R_L
　　　$= +2,500 \times 2.5 - 1,000 \times 2.5 \times 2.5/2 = +3,125$ ft-lb
M at point 4.0 ft to right of R_L
　　　　$= +2,500 \times 4 - 1,000 \times 4 \times 2 = +2,000$ ft-lb
M at point 5.0 ft to right of R_L
　　　　$= +2,500 \times 5 - 1,000 \times 4 \times 3 = +500$ ft-lb
M at point 6.0 ft to right of R_L but to left of couple
　　　　$= +2,500 \times 6 - 1,000 \times 4 \times 4 = -1,000$ ft-lb
M at point 6.0 ft to right of R_L but to right of couple $= 9,000$ ft-lb

1.604 $\Sigma M_{R_L} = 0$

$$= +5 \times 2{,}000 + 10 \times 4{,}000 + 15 \times 6{,}000$$
$$+ 20 \times 1{,}000 \times 10 - 20 R_R$$
$$R_R = 17{,}000 \text{ lb}$$

$\Sigma M_{R_R} = 0$

$$= -5 \times 6{,}000 - 10 \times 4{,}000 - 15 \times 2{,}000$$
$$- 20 \times 1{,}000 \times 10 + 20 R_L$$
$$R_L = 15{,}000 \text{ lb}$$

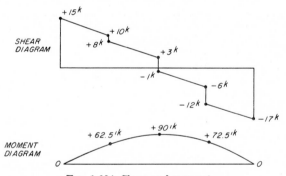

Fig. 1.604. Forces.

Check: $\Sigma V = 0$

$$= +15{,}000 - 2{,}000 - 4{,}000 - 6{,}000 - 20{,}000 + 17{,}000$$

SHEAR
DIAGRAM

$+15^k$ $+10^k$ $+8^k$ $+3^k$ -1^k -6^k -12^k -17^k

MOMENT
DIAGRAM

$+62.5'^k$ $+90'^k$ $+72.5'^k$ 0 0

Fig. 1.604. Shear and moment.

Forces to left point of moments:

M at left quarter point
$$= +15 \times 5 - 1 \times 5 \times \tfrac{5}{2} = 62.5 \text{ ft-kips}$$

M at center line
$$= +15 \times 10 - 2 \times 5 - 1 \times 10 \times \tfrac{10}{2} = 90 \text{ ft-kips}$$

M at right quarter point
$$= +15 \times 15 - 2 \times 10 - 4 \times 5 - 1 \times \frac{15 \times 15}{2} = 72.5 \text{ ft-kips}$$

Check: M at right quarter point, forces to right of point of moments $= +17 \times 5 - 1 \times 5 \times \tfrac{5}{2} = +72.5 \text{ ft-kips}$

1.605 ΣM_{R_L}
$$0 = +2 \times 20 \times 6 - 14R_R, \quad R_R = 17.14 \text{ kips}$$
ΣM_{R_R}
$$0 = -2 \times 20 \times 8 + 14R_L, \quad \underline{R_L = 22.86 \text{ kips}}$$
Check: $\Sigma V = 0 = +40.00 \text{ kips} - 2 \times 20$

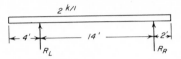

Fig. 1.605. Forces.

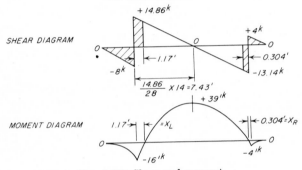

Fig. 1.605. Shear and moment.

$$M \text{ at } R_L = -2 \times 4 \times 2 = -16 \text{ ft-kips}$$
$$M \text{ at } R_R = -2 \times 2 \times 1 = - 4 \text{ ft-kips}$$

$$M \text{ at } 11.43 \text{ ft } (M \text{ max}) = -2 \times 11.43 \times \frac{11.43}{2} + 22.86 \times 7.43$$
$$= +39 \text{ ft-kips}$$

$$M = 0 \text{ when } x_L = 1.17, \text{ and } \frac{4 \times 8}{2} = 16 = 14.86x - \frac{2x^2}{2}.$$
$$\text{Similarly } M = 0 \text{ when } x_R = 0.304 \text{ ft}$$

1.606

$$\Sigma M_{R_L} = 0 = +1 \times 24 \times 6 + 15 \times 9 - 18R_R, \quad R_R = 15.5 \text{ kips}$$
$$\Sigma M_{R_R} = 0 = -1 \times 24 \times 12 - 15 \times 9 + 18R_L, \quad \underline{R_L = 23.5 \text{ kips}}$$
Check: $\Sigma V = 0 = -15 - 24 + \overline{39.0 \text{ kips}}$

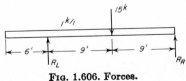

Fig. 1.606. Forces.

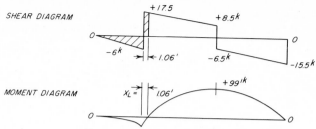

SHEAR DIAGRAM

MOMENT DIAGRAM

Fig. 1.606. Shear and moment.

M at $R_L = -1 \times 6 \times 3 = -18$ ft-kips

M_{max} at 15 kips load $= -1 \times 15 \times 15\!/\!2 + 23.5 \times 9 = 99$ ft-kips

$M = 0$ when $x_L = 1.06$ ft, and $18 = 17.5x - \dfrac{x^2}{2}$

1.607 ΣM_{R_L}

$0 = +4 \times 15 \times 15\!/\!2 + 30 \times 20 - 30 \times R_R,$ $R_R = +35$ kips

ΣM_{R_R}

$0 = +30R_L - 4 \times 15 \times 22.5 - 30 \times 10,$ $\underline{R_L = +55 \text{ kips}}$

Check: $\Sigma V = 0 = \overline{-60 - 30 + 90 \text{ kips}}$

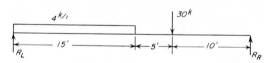

Fig. 1.607. Forces.

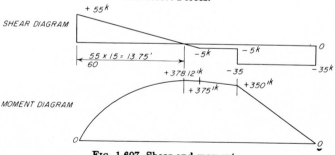

SHEAR DIAGRAM

MOMENT DIAGRAM

Fig. 1.607. Shear and moment.

Maximum moment at point 13.75 ft to right of R_L

$$= +55 \times 13.75 - 4 \times \frac{13.75^2}{2} = 378.12 \text{ ft-kips}$$

M at 15-ft point $= +55 \times 15 - 4 \times \dfrac{15^2}{2} = +375$ ft-kips

M at 20-ft point $= +35 \times 10 = 350$ ft-kips

1.608   It is given that $M_L = PL/16 = 16P/16 = P$ ft-lb.

$$\Sigma M_L = 0 = -P - 16R_R + 10P, \ R_R = \frac{9P}{16} \text{ lb}$$

$$M_P = 6 \times \frac{9P}{16} = \frac{54P}{16} \text{ ft-lb} > M_L$$

$$I_{xx} = \frac{bh^3}{12} = 4 \times \frac{16^3}{12} = 1{,}365 \text{ in.}^4$$

$$f = \frac{Mc}{I} = \frac{12 \times 54P \times 8}{16/1{,}365} = 1{,}350 \text{ and } P = \textbf{5,687} \text{ lb} \qquad \textbf{1d}$$

For deflection Y_P, use tables for the deflection of each item of loading and add the deflections together. Referring to Merritt, "Structural Steel Designers' Handbook" pp. 3-433 and 3-435, McGraw-Hill, 1972, find the deflection at load P for the load P on a simple beam and the deflection at load P for the moment P on the same simple beam.

$$Y_{P_1} \text{ (due to load } P) = \frac{+Pa^2b^2}{3EIL}$$

$$= \frac{+(5687)(100)(36)(1728)}{(3)(16)(10)^5(1365)(16)} = 0.338 \text{ in. (down)}$$

Y_{P_2} (due to negative moment P at the left end)

$$= \frac{-P}{6EIL}(2L^2a - 3La^2 + a^3)$$

$$= \frac{-(10)(5687)(1728)}{(6)(16)(10)^5(1365)(16)}(2 \times 256 - 3 \times 16 \times 10 + 100)$$

$$= \frac{-4.72(132)}{10^4} = -0.062 \text{ in. (up)}$$

Net $Y_P = +0.338 - 0.062 = \textbf{0.276}$ in. (down)   **2a**

If tables for deflections are not available, a double integration of $EI\ d^2y/dx^2$ may take too much time on an examination and a graphical integration using $Mm\ dx/EI$ is usually faster. Referring to Fig. 1.608a

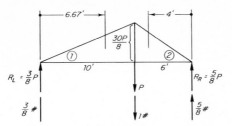

Fig. 1.608a

Area no.	Area	Dummy force	Arm	Product
1	$\dfrac{+(10)(30P)}{(2)(8)} = \dfrac{+300P}{16}$	$\dfrac{+3\#}{8}$	6.67	$+46.9P$
2	$\dfrac{+(6)(30P)}{(2)(8)} = \dfrac{+180P}{16}$	$\dfrac{+5\#}{8}$	4.00	$+28.1P$
				$+75.0P$

$$Y_{P_1} = + \frac{(75)(5,687)(1,728)}{(16)(10)^5(1,365)} = +0.338 \text{ in. (down)}$$

Referring to Fig. 1.608b

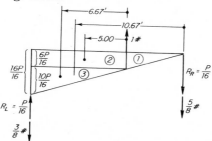

Fig. 1.608b

Area	Area	Dummy force	Arm	+ Prod.	− Prod.
1 + 2 + 3	$\dfrac{-P(16)}{2} = -8P$	$\dfrac{-5\#}{8}$	10.67	$+53.33$	
2	$\dfrac{-(6P)(10)}{16} = \dfrac{-60P}{16}$	$+1\#$	5.00		$-18.75P$
3	$\dfrac{-(10)(10)}{(16)(2)} = \dfrac{-100P}{16}$	$+1\#$	6.67		$-20.83P$
					$-39.58P$
				$-39.58P$	
				$+13.75P$	

$$Y_{P_2} = \frac{+(13.75)(5,687)(1,728)}{(16)(10)^5(1,365)} = 0.062 \text{ in. (up)}$$

Net $Y_P = +0.338 - 0.062 = +0.276$ in. (down)

FIG. 1.608c

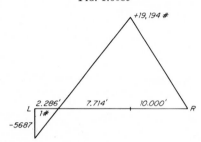

FIG. 1.608d

1.609 *Using moment distribution:*
Fixed end moments (clockwise $\curvearrowright$ = +, counterclockwise $\curvearrowleft$ = −)

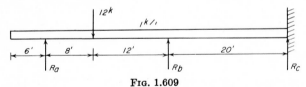

FIG. 1.609

Uniform load: To left of $R_a = 1 \times 6 \times 3 = +18$ ft-kips

To right of R_a (and elsewhere) $= \dfrac{wl^2}{12} =$

$$\frac{1 \times \overline{20}^2}{12} = \pm 33.33 \text{ ft-kips}$$

Concentrated load: To right of $R_a = \dfrac{Pab^2}{l^2} = \dfrac{12 \times 8 \times 12 \times 12}{20 \times 20}$

$$= -34.56 \text{ ft-kips}$$

To left of $R_b = \dfrac{Pa^2b}{l^2} = \dfrac{12 \times 8 \times 8 \times 12}{20 \times 20}$

$$= +23.04 \text{ ft-kips}$$

Distribution factors	1.0		0.5	0.5		0
Fixed end +18	−34.56		+23.04			
moments	−33.33		+33.33	−33.33		+33.33
Net fixed end M +18	−67.89		+56.37	−33.33		+33.33
	+49.89	→	+24.94			
	−11.99	←	−23.99	−23.99	→	−11.99
	+11.99	→	+ 6.00			
	− 1.50	←	− 3.00	− 3.00	→	− 1.50
	+ 1.50	→	+ 0.75			
	− 0.19	←	− 0.38	− 0.38	→	− 0.19
	+ 0.19	→	+ 0.10			
			− 0.05	− 0.05	→	− 0.02
Final moment +18	−18 ft-kips		+60.75	−60.75 ft-kips		+19.63 ft-kips

(To left) M at R_b
$$0 = -1 \times 26 \times 13 - 12 \times 12 + 20R_a + 60.75,$$
$$R_a = +21.06 \text{ kips}$$

(Left and right) M at R_b
$$0 = -1 \times 46 \times 3 - 12 \times 12 + 20 \times 21.06 + 19.63 - 20R_c,$$
$$R_c = + 7.94 \text{ kips}$$

M at R_c
$$0 = -1 \times 46 \times 23 - 12 \times 32 + 40 \times 21.06 + 20R_b + 19.63,$$
$$R_b = +29.00 \text{ kips}$$

Check: $\Sigma V = 0 = +21.06 + 29.00 + 7.94 - 46 - 12$

M at 12 kips load $= -1 \times 14 \times 7 + 21.06 \times 8 = 70.48$ ft-kips **la**

Using three moment equations (signs follow fiber stress convention):

$$M_1 l_1 + 2M_2(l_1 + l_2) + M_3 l_2$$
$$= -\tfrac{1}{4} w_1 l_1^3 - \tfrac{1}{4} w_2 l_2^3 - P_1 l_1^2 (k_1 - k_1^3)$$

1. Take two spans, R_a to R_c and $k_1 = \tfrac{8}{20} = 0.4$

$$-18 \times 20 + 2M_b(20 + 20) + M_c \times 20$$
$$= -\frac{1 \times \overline{20}^3}{4} - \frac{1 \times \overline{20}^3}{4} - 12 \times (20^2)(0.4 - 0.064).$$

$$-360 + 80M_b + 20M_c = -2{,}000 - 2{,}000 - 1{,}614.4$$
$$= -5{,}254.4$$

(a) $+4M_b + M_c = -262.70$

2. Take two spans R_b to R_d where $R_d = 0$ ft to right of R_c

$$M_b(20) + 2M_c(20 + 0) + M_d(0) = -2,000$$
(b) $$M_b + 2M_c = -100$$
(b') $$4M_b + 8M_c = -400$$

subtracting (a) from (b') $+7M_c = -137.30$
$$M_c = -19.62 \text{ ft-kips}$$

and the minus sign indicates tension in top fiber over R_c.

Substituting in (b), $M_b = -100 + 39.25$
$$= -60.75 \text{ ft-kips} \qquad \textbf{2c}$$

Reactions are found as before.

1.610 The first step is to cut pulley ropes and place rope forces at axles of pulleys or on beam.

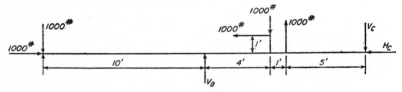

FIG. 1.610. Forces.

$\Sigma H = 0 = +1,000 - 1,000 - H_c, \quad H_c = 0$
$M_B = 0$
$= -10 \times 1,000 - 1 \times 1,000 - 5 \times 1,000 + 4 \times 1,000 + 10V_c$
$$V_c = 1,200 \text{ lb}\downarrow$$

$M_C = 0$
$= -20 \times 1,000 - 1 \times 1,000 - 6 \times 1,000 + 5 \times 1,000 + 10V_B$
$$V_B = 2,200 \text{ lb}\uparrow$$

Check: $\Sigma V = 0 = -1,000 + 2,200 - 1,000 + 1,000 - 1,200$

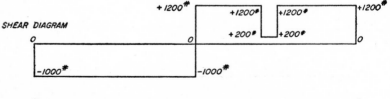

FIG. 1.610. Shear and moment.

1.611 $0.75H + 1.333H = 2.083H = 1{,}000$ lb,

$$H = 480 \text{ lb and } 360 + 640 = 1{,}000$$

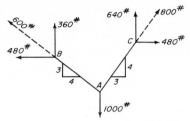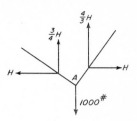

Fɪɢ. 1.611

Check: $\Sigma M_B = 0 = +4 \times 1{,}000 + 1 \times 480 - 7 \times 640$

$$= +4{,}480 - 4{,}480$$

1.612 $\Sigma M(\text{left rope}) = 0 = -200x + 150(10 - x)$

$$x = 4.29 \text{ ft}, \quad \text{and} \quad 10 - x = 5.71 \text{ ft}$$

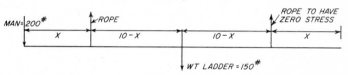

Fɪɢ. 1.612

Check: $\Sigma_M(\text{right rope})$

$$0 = -200 \times 15.71 - 150 \times 5.71 + \text{rope}_L \times 11.42$$

$$\text{Rope}_L = 350 \text{ lb}. \quad \Sigma V = 0 = +350 - 200 - 150$$

1.613

$$h = \sqrt{400 - 64} = 18.32 \text{ ft}$$

$$\Sigma M_a = 0 = +4 \times 150 - 18.32H$$

$$H = 32.75 \text{ lb} \quad \textbf{b}$$

Frictional force at A must at least $=$
H.

Fɪɢ. 1.613

1.614

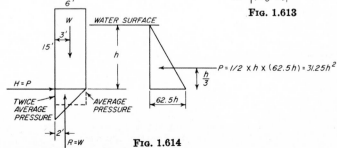

Fɪɢ. 1.614

Length is 1 ft perpendicular to the cross section.

$$W = 1 \times 6 \times 15 \times 150 = 13,500 \text{ lb}$$

$$\text{Overturning couple} = P \times \frac{h}{3} = 10.417h^3$$

$$\text{Righting couple} = W \times 1 = 13,500 \text{ lb-ft}$$

$$h^3 = \frac{13,500}{10.417} = 1,296, \; h = \mathbf{10.9} \text{ ft} \qquad \mathbf{a}$$

1.615   Length is 1 ft along wall.

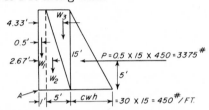

FIG. 1.615. Vertical and horizontal forces.

$$W_1 = 1 \times 1 \times 15 \times 150 \; = \; 2,250 \text{ lb}$$
$$W_2 = 1 \times \tfrac{5}{2} \times 15 \times 150 = \; 5,625 \text{ lb}$$
$$W_3 = 1 \times \tfrac{5}{2} \times 15 \times 100 = \; 3,750 \text{ lb}$$
$$\Sigma W = R = \qquad\qquad\qquad\quad 11,625 \text{ lb}$$

At the moment of overturning, R would act at A.

For M(about A)

	F	$\times$	a	$=$	M
$W_1 =$	2,250 lb	$\times$	0.5 ft	$=$	$+ \; 1,125$ ft-lb
$W_2 =$	5,625	$\times$	2.67	$=$	$+15,000$
$W_3 =$	3,750	$\times$	4.33	$=$	$+16,250$
$(P =$	3,375$)$	$\times$	5.00	$=$	$-16,875$
$R = \Sigma W =$	11,625	$\times$	(1.33)	$=$	$+15,500$

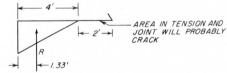

FIG. 1.615. Force on base.

The wall is safe against overturning.

$$\Sigma W \times f = 3,375, \; f = \frac{3,375}{11,625} = 0.29$$

The wall is probably safe against sliding.

1.616

To locate R between front and rear axles.

ΣM(16-ton axle) $= 0 = +20 \times d - 14 \times 4$

$$d = 2.8 \text{ ft}, \quad \frac{d}{2} = 1.4 \text{ ft}$$

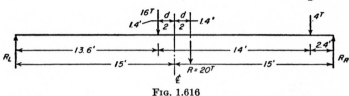

FIG. 1.616

To place a truck on a span so as to produce maximum moment, the rear axle is placed as far to one side of the center line as the resultant of the front and rear axles lays on the other side of the center line. If the span is too short to get both front and rear axles on the span, place rear axle at the center line.

$$R_L = \frac{13.6}{30} \times 20 = 9.067 \text{ tons}, \quad R_R = \frac{16.8}{30} \times 20 = 10.933 \text{ tons}$$

Maximum moment (about 16-ton axle) $= 9.067 \times 13.6$
$$= 123.3 \text{ ft-tons} \qquad \textbf{1d}$$

If 16-ton axle is placed at the center line of the span

$$R_L = \frac{12.2}{30} \times 20 = 8.133 \text{ tons and } M = 8.133 \times 15 = 122 \text{ ft-tons}$$

Maximum shear occurs when the 16-ton axle is just to the right of $R_L = \dfrac{27.2}{30} \times 20 = \textbf{18.13 tons} \qquad \textbf{2a}$

1.617   First, read the answer to 1.616. Then locate the resultant of the tractor and semitrailer axle loads by taking moments about the 8-kip axle.

$$14 \times 32 + 28 \times 32 = x(8 + 32 + 32), \quad x = \frac{42 \times 32}{72} = 18.67 \text{ ft}$$

$$d = 18.67 - 14 = 4.67 \text{ ft}, \frac{d}{2} = 2.33 \text{ ft}$$

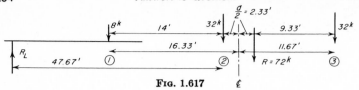

FIG. 1.617

a. The 8-kip axle (①) is (50 − 16.33 ft) or **33.67** ft to right of R_L.

b. The maximum moment (about axle ②)

$$R_L \times 47.67 - 8 \times 14 = 0$$

$$R_L = \frac{47.67}{100} \times 72 = 34.4 \text{ kips}$$

∴ Maximum $M = 34.4 \times 47.67 - 112$
$$= 1{,}640 - 112 = \textbf{1,528} \text{ ft-kips}$$

c. The maximum shear is R_R with truck and trailer backed up so that axle ③ is at R_R.

$$R_R = \frac{(100 - 9.33)}{100} \times 72 = \textbf{65.3} \text{ kips} = Ma \times \text{shear}$$

1.618 M(about line TR)

$$0 = 100 \times 7.07 - V_s \times 17.07, \qquad V_s = 41.4 \text{ kips}$$

M (about line ESF)

$$0 = -100 \times 10 + 2V_T \times 17.07, \qquad V_T = 29.3 \text{ kips}$$
$$V_R = 29.3 \text{ kips}$$
$$\overline{ 100.0 \text{ kips}}$$

Check: $\Sigma V = 0 = -100 + V_R + V_s + V_T$

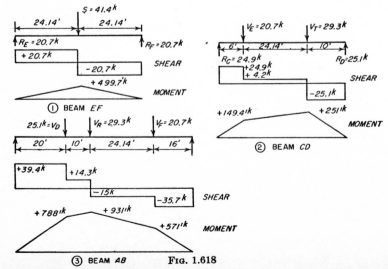

FIG. 1.618

1.619 By symmetry and inspection, $R_L = R_R = 5,000$ lb.

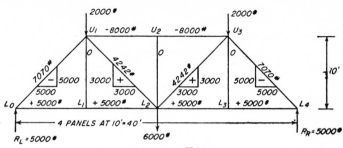

FIG. 1.619. Truss.

By analyzing joints L_1, U_2, and L_3, the stress in the verticals U_1L_1, U_2L_2, and U_3L_3 must be zero, as ΣV must equal zero and there is no outside vertical force or other vertical member or member having a vertical component to counteract any stress in these three members.

Use the "method of joints" ($\Sigma H = 0$ and $\Sigma V = 0$ for all stresses or components of stresses of all members meeting at a joint), letting $+$ indicate tension and $-$ indicate compression, but noting that for any joint the sum of the forces acting in one direction must equal the sum of the forces acting in a 180° direction regardless of signs.

Also note that no joint can be analyzed if the stresses in more than two members meeting at that joint are unknown.

FIG. 1.619. Joints.

1.620 To solve for the stress in L_0L_1, pass the section 1-1, cutting members L_0U_1 and L_0L_1. Take moments about L_0U_1 extended to U_1. Assume stress of L_0L_1 acts out of L_0.

$$\Sigma M_{U_1} = 0 = +(4,000 - 1,000) \times 10 - 5 \times S_{L_0L_1}$$
$$S_{L_0L_1} = +6,000 \text{ lb (tension)}$$

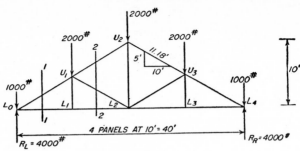

FIG. 1.620. Forces on truss.

To solve for the stress in U_1U_2 and U_1L_2, pass the section 2-2, cutting members U_1U_2, U_1L_2, and L_1L_2.

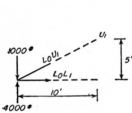

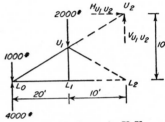

FIG. 1.620. Stress in L_0L_1.   FIG. 1.620. Stress in U_1U_2.

Take moments at L_2 and assume U_1U_2 to be a tension member (acting out of joint U_1) with components at U_2.

$$\Sigma M_{L_2} = 0 = +20 \times 3{,}000 - 10 \times 2{,}000 + 10 H_{U_1U_2}$$
$$H_{U_1U_2} = -4{,}000 \text{ lb (compression)}$$

(− sign indicates compression)

and
$$S_{U_1U_2} = \frac{11.18}{10} \times (-4{,}000)$$
$$= -4{,}472 \text{ lb (compression)}$$

Now take moments at L_0 and assume U_1L_2 to be taken with components acting at L_2. Forces U_1U_2 and L_1L_2 and $H_{U_1}L_2$ or their lines of action pass through L_0.

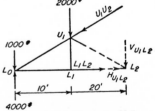

FIG. 1.620. Stress in U_1L_2.

$$\Sigma M_{L_0} = 0 = +10 \times 2{,}000 + 20 \times V_{U_1L_2}$$
$$V_{U_1L_2} = -1{,}000 \text{ lb (compression)}$$
$$S_{U_1L_2} = \frac{11.18}{5} \times 1{,}000 = -2{,}236 \text{ lb (compression)}$$

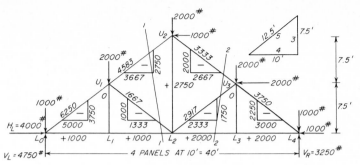

FIG. 1.621

$$\Sigma M_{L_0} = 0 = +10 \times 2{,}000 + 20 \times 2{,}000 + 30 \times 2{,}000 +$$
$$40 \times 1{,}000 - 7.5 \times 2{,}000 - 15 \times 1{,}000 - 40V_R$$
$$V_R = \textbf{3,250} \text{ lb}$$
$$\Sigma M_{L_4} = 0 = -10 \times 2{,}000 - 20 \times 2{,}000 - 30 \times 2{,}000 -$$
$$40 \cdot \times 1{,}000 - 7.5 \times 2{,}000 - 15 \times 1{,}000 + 40V_L$$
$$V_L = \textbf{4,750} \text{ lb}$$

Check: $\Sigma V = 0 = -1{,}000 - 2{,}000 - 2{,}000 - 2{,}000 - 1{,}000$
$$+ 8{,}000 \text{ lb}$$

Pass vertical sections through second and third panels and find $H_{U_1U_2}$, $V_{U_1L_2}$, $H_{U_2U_3}$, and $V_{L_2U_3}$ as in the answer to 1.620.

Section 1-1:

$$\Sigma M_{L_2} =$$
$$0 = +(4{,}750 - 1{,}000) \times 20 - 2{,}000 \times 10 + H_{U_1U_2} \times 15$$
$$H_{U_1U_2} = \textbf{−3,667} \text{ lb}$$
$$\Sigma M_{L_0} = 0 = +2{,}000 \times 10 + V_{U_1L_2} \times 20$$
$$V_{U_1L_2} = \textbf{−1,000} \text{ lb}$$

Section 2-2:

$$\Sigma M_{L_4} = 0 = -2{,}000 \times 10 - 2{,}000 \times 7.5 - V_{L_2U_3} \times 20$$
$$V_{L_2U_3} = \textbf{−1,750} \text{ lb}$$
$$\Sigma M_{L_2} = 0$$
$$= +2{,}000 \times 10 - 2{,}000 \times 7.5 - (3{,}250 - 1{,}000)20 - H_{U_2U_3} \times 15$$
$$H_{U_2U_3} = \textbf{−2,667} \text{ lb}$$

Stresses or components of stress in other members and checks are obtained by applying $\Sigma H = 0$ and $\Sigma V = 0$ at each joint.

1.622   Section 1-1 and truss to right of section (assume U_2U_3 as tension):

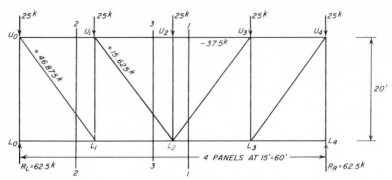

FIG. 1.622

$$\Sigma M_{L_3} = 0 = +15 \times 25 + 30 \times 25 - 30 \times 62.5 - 20 S_{U_2 U_3}$$
$$S_{U_2 U_3} = -37.5 \text{ kips (compression)}$$

Section 2-2 and truss to left of section (assume $U_0 L_1$ as tension):

$$\Sigma V = 0 = +62.5 - 25 - V_{U_0 L_1}, \quad V_{U_0 L_1} = +37.5 \text{ kips}$$
$$S_{U_0 L_1} = +46.875 \text{ kips (tension)}$$

Section 3-3 and truss to left of section (assume $U_1 L_2$ as tension):

$$\Sigma V = 0 = +62.5 - 25 - 25 - V_{U_1 L_2}, \quad V_{U_1 L_2} = +12.5 \text{ kips}$$
$$S_{U_1 L_2} = +15.625 \text{ kips (tension)}$$

1.623

$$\Sigma M_{L_6} = 0$$
$$= -1 \times 48 - 2(40 + 32 + 24 + 16 + 8) - 6 \times 32 + 40 H_{U_6}$$
$$H_{U_6} = +12 \text{ kips (to right)}$$
$$H_{L_6} = -12 \text{ kips (to left)} \quad \text{and} \quad V_{L_6} = +12 \text{ kips (up)}$$
$$\Sigma V = 0 = -17 + 12 + V_{U_6}, \quad V_{U_6} = +5 \text{ kips (up)}$$

a. Vertical section in second panel from left end and components of $U_1 U_2$ acting at U_2

$$\Sigma M_{L_3} = 0 = -1 \times 16 - 2 \times 8 + 8 H_{U_1 U_2}$$
$$H_{U_1 U_2} = +4 \text{ kips (tension)}$$

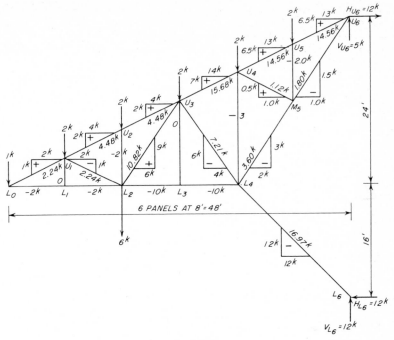

Fig. 1.623

b. Vertical section in fourth panel from left end and components of U_3U_4 acting at U_4

$$\Sigma M_{L_4} = 0 = -1 \times 32 - 2 \times 24 - 8 \times 16 - 2 \times 8 + 16H_{U_3U_4}$$

$$H_{U_3U_4} = +14 \text{ kips}$$

c. Vertical section in sixth panel from left end and components of U_5U_6 acting at U_4

$$\Sigma M_{L_4} = 0 = -1 \times 32 - 2 \times 24 - 8 \times 16 - 2 \times 8 + 2 \times 8$$
$$+ 16H_{U_5U_6}$$

$$H_{U_5U_6} = +13 \text{ kips}$$

d. The remaining components and the necessary checks are easily obtained by applying $\Sigma H = 0$ and $\Sigma V = 0$ at each joint. Having one component of stress, the other component of stress and the actual stress in the members are found from simple proportion.

1.624   Use moments to find $H_{U_1U_2}$, L_2L_3, and $H_{U_4U_5}$

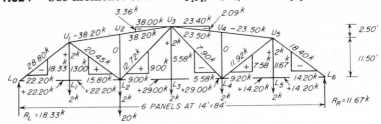

FIG. 1.624

$\Sigma M_{L_2} = 0$ (truss to left of L_2), section through second panel.

$$0 = +18.33 \times 28 - 2 \times 14 + 12.75 \times H_{U_1U_2}$$

$$H_{U_1U_2} = -38.00 \text{ kips (compression)}$$

$\Sigma M_{L_4} = 0$ (truss to right of L_4), section through fifth panel.

$$0 = -11.67 \times 28 + 2 \times 14 - 12.75 \times H_{U_4U_5}$$

$$H_{U_4U_5} = -23.40 \text{ kips}$$

$\Sigma M_{U_3} = 0$ (truss to right of U_3), section through fourth panel.

$$0 = -11.67 \times 42 + 2 \times 28 + 2 \times 14 + 14 \times L_3L_4$$

$$L_3L_4 = +29.0 \text{ kips (tension)}$$

The remaining components and the necessary check are easily obtained by applying $\Sigma H = 0$ and $\Sigma V = 0$ at each joint. Having one component of stress, the other component of stress and the actual stress in the members are found from simple proportion.

1.625 $\Sigma M_{(B-C)}$

$$0 = +40 \times 2{,}400 + 30V_{AD}$$

$$V_{AD} = -3{,}200 \text{ lb}$$

$$H_{AD} = -1{,}600 \text{ lb}$$

$$S_{AD} = -3{,}580 \text{ lb (compression)}$$

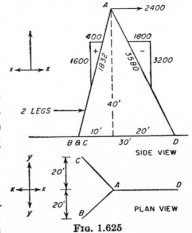

In both legs AB and AC there are horizontal components parallel to BC equal to $\frac{20}{10} \times 400$ = 800 lb. Thus each leg, AC and AB, has three components:

$$x = 400, \; y = 800, \; V = 1{,}600$$

Stress in AB and AC =

$$400 \sqrt{1^2 + 2^2 + 4^2} = 400 \sqrt{21}$$

$$= +1{,}832 \text{ lb (tension)}$$

FIG. 1.625

THERMODYNAMICS

1.701 *a. Latent heat* (latent heat of evaporation): The quantity of heat required to evaporate 1 lb saturated liquid.

b. Specific heat: Amount of transferred heat required to change the temperature of one unit weight of a substance one degree unit of temperature.

c. Heat of fusion: The quantity of heat required to change the state of a substance from solid to liquid without change of temperature.

d. Btu: $\frac{1}{180}$ the quantity of heat required to raise the temperature of 1 lb pure water from 32 to 212°F under standard atmospheric pressure.

e. Absolute temperature: A measure of the ability to transfer heat to other bodies based on a reference temperature (absolute zero) where a body has given up all the thermal energy it possibly can. Absolute zero = -273.16°C = -459.69°F.

f. Calorie: The quantity of heat required to raise the temperature of 1 gram water from 14.5°C to 15.5°C.

g. Watt: Power expended when 1 amp flows between two points having a potential difference of 1 volt.

1 watt = 1 joule per sec = 10^7 ergs per sec = 10^7 dyne-cm per sec

1.702 T = final temperature (°F)
T_I = initial temperature of ice (°F)
T_w = initial temperature of water (°F)
W_I = weight of ice (lb)
W_W = weight of water (lb)
HF = heat of fusion = 144 Btu per lb

Heat lost by water = heat gained by ice

$$W_W(T_w - T) = W_I(\text{HF}) + W_I(T - T_I)$$
$$100(100 - T) = 30(144) + 30(T - 32)$$
$$10,000 - 100T = 4,320 + 30T - 960$$
$$130T = 6,640$$
$$T = 51°\text{F} \quad \textbf{d}$$

1.703   Initial temperature of cold water $= T_C$ (°F)
Initial temperature of hot water $= T_H$ (°F)
Volume of cold water $= V_C$ (cu ft)
Volume of hot water $= V_H$ (cu ft)
Total volume $= V_T$ (cu ft)
Weight of cold water $= W_C$ (lb)
Weight of hot water $= W_H$ (lb)
Total weight $= W_T$ (lb)
Specific weight $= w$ (lb/cu ft)
Heat lost by hot water = heat gained by cold water

$$W_H(T_H - T) = W_C(T - T_C)$$
$$V_H(w)(T_H - T) = V_C(w)(T - T_C)$$
$$V_H(62.5)(160 - 70) = V_C(62.5)(70 - 40)$$
$$90V_H = 30V_C \text{ and } V_C = 3V_H$$
$$V_C + V_H = V_T = (25)(75)(5) = 9375 = 3V_H + V_H = 4V_H$$
$$V_H = 2{,}344 \text{ cu ft}$$

$$V_C = V_T - V_H = 9{,}375 - 2{,}344 = \mathbf{7{,}031} \text{ cu ft}$$

1.704   Weight of water $= W_w$ (lb)
Weight of ice $= W_I$ (lb)
Weight of steam $= W_s$ (lb)
HV = heat of vaporization = 970 Btu per lb
HF = heat of fusion   = 144 Btu per lb
Heat lost by steam in condensing $= (HV)(W_s) =$
970W_s (Btu)
Heat lost by condensed steam $= (T_s - T)W_s$
$= (212 - 80)W_s = 132W_s$ (Btu)
Heat gained by ice in melting $= (HF)W_1 = 144 \times 100$
$= 14{,}400$ (Btu)
Heat gained by ice and cold water $=$

$$(T - T_w)(W_I + W_w) = (80 - 32)(100 + 300) = 19{,}200 \text{ (Btu)}$$

Heat lost by steam = heat gained by ice and water

$$970W_s + 132W_s = 1{,}102W_s = 14{,}400 + 19{,}200 = 33{,}600$$
$$W_s = \frac{33{,}600}{1{,}102} = \mathbf{30.5} \text{ lb}$$

1.705 V_1 = initial volume, p_1 = initial pressure
            V_2 = final volume, p_2 = final pressure
               T_1 = initial temperature
               T_2 = final temperature
               $V_1 = 12V_2$, $K = 1.4$ for air

For adiabatic process $\dfrac{p_2}{p_1} = \left(\dfrac{V_1}{V_2}\right)^k$ and $\dfrac{p_2}{13} = \left(\dfrac{12V_2}{V_2}\right)^{1.4}$

$$p_2 = 13(12)^{1.4} = 13 \times 32.4 = \textbf{421 psi}$$
$$\frac{T_2}{T_1} = \left(\frac{V_1}{V_2}\right)^{k-1} \text{ and } \frac{T_2}{580} = \left(\frac{12V_2}{V_2}\right)^{1.4-1} = (12)^{0.4}$$
$$(°R = 120°F + 460° = 580°R)$$
$$T_2 = 580(12)^{0.4} = 580 \times 2.71 = \textbf{1570°R} = \textbf{1110°F}$$

1.706 $\dfrac{p_2}{p_1} = \dfrac{T_2}{T_1}$ (constant-volume process)

$$p_1 = 32 + 14.7 = 46.7 \text{ psia}$$
$$T_2 = 75 + 460 = 535°R$$
$$T_1 = 50 + 460 = 510°R$$
$$p_2 = 46.7 \times {}^{535}\!/_{510} = 49.1 \text{ psia}$$
$$= 49.1 - 14.7 = \textbf{34.4 psig} \quad \textbf{b}$$

1.707 $R = \dfrac{pv}{T} = \dfrac{p_1 V_1}{W T_1} = \dfrac{p_2 V_2}{W T_2}$ where $v = \dfrac{V}{W}$

v = specific volume, R = gas constant,
                    V = total volume, W = total weight

   $0°F = 460°R$ and $200°F = 660°R$
$$V_1 = 100V_2, \quad \frac{14.7 \times 100V_2}{460} = \frac{p_2 V_2}{660}$$
$$p_2 = {}^{660}\!/_{460} \times 14.7 \times 100 = \textbf{2,110 psia}$$

1.708   *a.* The heating value of a fuel is the amount of heat that must be removed from the products of complete combustion of a unit amount of fuel to cool the products down to the temperature of the original air-fuel mixture. Most common fuels contain hydrogen, which when burned forms water vapor or steam. The higher heating value is obtained when all the water vapor of combustion condenses, thus giving up its latent heat of evaporation. The lower heating value will be obtained when sufficient excess air is used to prevent any condensation of water vapor.

    b. If the final volume of the products of a constant-pressure combustion exceeds the original volume, part of the fuel energy must be used to push the atmosphere out of the way. In this case, the heating value at constant pressure is less than that at constant volume by a quantity equal to the work done in pushing the atmosphere out of the way. If the final volume of the products is less than the original volume, the constant-pressure heating value exceeds the constant-volume value by a quantity equal to the work done on the gas by the atmosphere during the volume contraction.

1.709 When a gas or a gaseous mixture remains in contact with a liquid surface, it will acquire vapor from the liquid until the partial pressure due to the water vapor in the air is equal to the saturated vapor pressure of the water at the existing temperature. When the vapor concentration reaches the equilibrium value, the gas is said to be saturated with vapor. It is not possible for the gas to contain a greater concentration of vapor, for as soon as the vapor pressure of the liquid is exceeded by the partial pressure of the vapor, condensation takes place.

    Vapor pressure of the pond water at 70°F = 0.363 psia. Partial pressure of air at 90°F dry bulb and 65°F wet bulb = 0.170 psia.

    Since 0.363 is greater than 0.170, water will be evaporated into the air stream.

1.710 *a.* Near the floor. Since the specific weight of the hot air is less than that of the cold air, the warm air will rise, thus setting up convection currents.

    b. e = emissivity factor, expressing the degree to which the source surface approaches an "ideal black body." An "ideal black body" is one which could absorb (or emit) all the radiant energy that falls on it.

    Most nonmetallic substances regardless of color have e equal to or greater than 0.8.

    Aluminum paint has an e = 0.3 to 0.5. Therefore, to heat a room most efficiently, the radiators should be painted with black rather than aluminum paint.

1.711
$$1 \text{ kw} = 3{,}413 \text{ Btu per hr}$$
$$50{,}000 \text{ kw} = 170{,}650{,}000 \text{ Btu per hr}$$
$$\frac{\text{Btu per hr}}{\text{lb per hr}} = \frac{\text{Btu}}{\text{lb}} = \frac{170{,}650{,}000}{569{,}000}$$
$$= h_1 - h_2' = 300 \text{ Btu per lb}$$
$$\eta_e = \text{engine efficiency}$$
$$h_1 - h_2 = \frac{h_1 - h_2'}{\eta_e} = \frac{300}{0.75} = 400 \text{ Btu per lb}$$
$$h_1 = 950 + 300 = 1{,}250 \text{ Btu per lb}$$
$$h_2 = 1{,}250 - 400 = 850 \text{ Btu per lb}$$

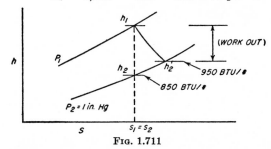

FIG. 1.711

Using a Mollier chart, find s_2 by following constant pressure line (1 in. Hg) down to $h_2 = 850$ Btu per lb and then project vertically up to $h_1 = 1{,}250$ Btu per lb.

Condition at $h_1 = T_1 = 476°F$, $P_1 = $ **243** psia

1.712   Total weight of snow = volume × specific weight
$$= (10 \times 50 \times \tfrac{1}{2}) \times 10 = 2{,}500 \text{ lb}$$
$$\text{Btu necessary to melt snow} = \frac{\text{weight} \times (\text{heat of fusion})}{\text{efficiency}}$$
$$= \frac{2{,}500 \times 144}{0.50} = 720{,}000 \text{ Btu}$$
$$3{,}415 \text{ Btu} = 1 \text{ kwhr and kwhr needed} = \frac{720{,}000}{3{,}415} = 211$$
$$\text{Cost} = 0.02 \times 211 = \textbf{\$4.22}$$

1.713 W_{bo} = lb steam per hr = 885,000 ÷ 4 = 221,250 lb per hr
h_1 = enthalpy of water (280°F at 400 psia) reaching boiler = 248.4 Btu per lb (see steam tables)
h_2 = enthalpy of steam delivered by boiler (700°F **at** 400 psia) = 1,362.7 Btu per lb

Boiler output per hr $= W_{bo} \times (h_2 - h_1)$

$\qquad = 221{,}250\ (1{,}362.7 - 248.4) = 247 \times 10^6$ Btu per hr

Pounds coal necessary per hr

$$= \frac{\text{Btu per hr output}}{(\text{heating value per lb})(\text{efficiency})}$$

$$= \frac{247 \times 10^6}{13{,}850 \times 0.825} = 21{,}600 \text{ lb per hr}$$

Coal burned per hour $= \dfrac{21{,}600}{2{,}000} = 10.8$ tons per hr

1.714

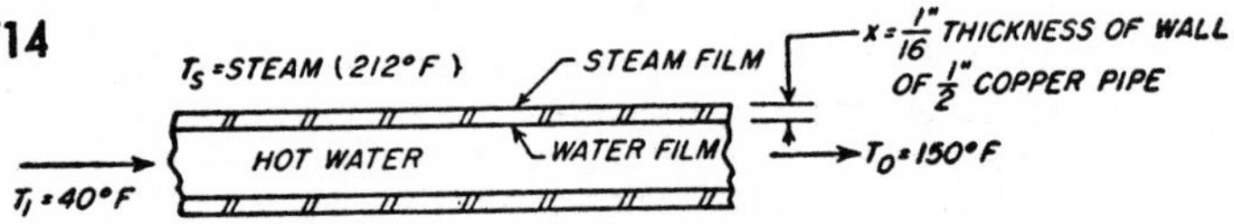

Fig. 1.714

$Q = UA\Delta t_m$ where Q = heat transferred in Btu per hr

$\qquad\qquad\qquad\qquad U$ = Transmittance in
$\qquad\qquad\qquad\qquad\qquad$ Btu/(sq ft) (°F)(hr)

$\qquad\qquad\qquad \Delta t_m$ = logarithmic mean temperature
$\qquad\qquad\qquad\qquad$ difference per °F

Surface area $= \pi \left(\dfrac{D}{12}\right) \times L = \pi \left(\dfrac{0.5}{12}\right) \times 20 = 2.62$ sq ft

$$\Delta t_m = \frac{\Delta t_a - \Delta t_B}{\log_e (\Delta t_a / \Delta t_B)}$$

where $\Delta t_a = T_s - T_1 = 212 - 40 = 172°F$

$\qquad\ \Delta t_B = T_s - T_0 = 212 - 150 = 62°F$

$\qquad\ \Delta t_m = \dfrac{172 - 62}{\log_e (172/62)} = \dfrac{110}{\log_e 2.77} = \dfrac{110}{1.02} = 108°F$

$$\frac{1}{U} = \frac{1}{h_w} + \frac{x}{K} + \frac{1}{h_s}$$

where K = thermal conductivity of copper in

$\qquad\qquad \dfrac{(\text{Btu per in. of thickness})}{\text{hr}/(\text{sq ft})(°F)}$

h_w = surface coefficient of heat transfer (water film) in Btu/(hr)(sq ft)(°F)

h_s = surface coefficient of heat transfer (steam film) in Btu/(hr)(sq ft)(°F)

(Values of K, h_w, and h_s obtained from Kent's "Mechanical Engineers' Handbook," Power, Wiley, 1950.)

$$\frac{1}{U} = \frac{1}{600} + \frac{0.0625}{2{,}100} + \frac{1}{1{,}000}$$
$$= 0.00167 + 0.00003 + 0.001 = 0.00270$$

$U = 370$ Btu/(sq ft)(°F)(hr)

$Q = 370 \times 2.62 \times 108 = 105{,}000$ Btu per hr

w = weight of water (lb per hr)

c_p = mean specific heat coefficient = 1

$Q = Wc_p(T_0 - T_1)$

and $105{,}000 = W(1)(150 - 40) = W(110)$

$$W = \frac{105{,}000}{110} = \textbf{956} \text{ lb per hr}$$

$$\frac{W}{500} = \frac{956}{500} = \textbf{1.91} \text{ gal water per min}$$

1.715 Btu output per lb of coal = (heat of combustion per lb coal)(Eff.) = $(14{,}000)(0.50) = 7{,}000$ Btu/lb

h_2 = enthalpy of steam (212°F and 50 per cent quality)

$h_2 = (h_f + x_2 h_{fg})$ [from vapor tables]
$$= 180 + 0.50 \times 970 = 665 \text{ Btu/lb}$$

h_{fg} = enthalpy of a fluid and vapor (gas).

h_1 = enthalpy of water (70°F) = 38 Btu/lb

Btu output per lb coal = $w_{bo}(h_2 - h_1)$

where w_{bo} = lb steam per lb coal

$7{,}000 = w_{bo}(665 - 38) = 627w_{bo}$

$w_{bo} = 7{,}000 \div 627 = \textbf{11.2}$ lb steam per lb coal

1.716 Indicated horsepower = $I_{hp} = \dfrac{P_m LAN}{33{,}000}$

P_m = mean effective pressure, psia

L = stroke, ft

A = net area of piston, sq in.

N = power cycles per minute per cylinder end

D = piston diameter, in.

D_R = piston rod diameter, in.

Area of head end of piston $= \dfrac{\pi D^2}{4} = \dfrac{\pi}{4} \times (6)^2 = 28.4$ sq in.

Area of crank end of piston $= \dfrac{\pi}{4}\,(D^2 - D_R{}^2)$

$$= \frac{\pi}{4}\,(36 - 1.56) = 27.0 \text{ sq in.}$$

I_{hp}, head end $= 62 \times 0.67 \times 28.4 \times 300 \div 33{,}000$
$$= 10.65 \text{ hp}$$

I_{hp}, crank end $= 62 \times 0.67 \times 27 \times 300 \div 33{,}000$
$$= 10.15 \text{ hp}$$

Total $I_{hp} = 10.65 + 10.15 = 20.8$ hp

Brake hp $=$ (total I_{hp})(mean efficiency)
$$= 20.8 \times 0.83 = 17.25 \text{ hp}$$

1 hp $= 0.746$ kw

Generator output $=$ (brake hp)(0.746)(generator efficiency)
$$= 17.25 \times 0.746 \times 0.92 = 11.85 \text{ kw}$$

1.717 $I_{hp} = \dfrac{P_m LAN}{33{,}000}$ (see answer to 1.716)

Work per revolution $= 2P_m LA = 2 \times 120 \times 1 \times \dfrac{\pi}{4}(10)$

$$= 18{,}850 \text{ ft-lb}$$

$$I_{hp} = 18{,}850 \times 300 \div 33{,}000 = 171.5 \text{ hp}$$

1.718 $\dfrac{P_1 V_1}{T_1} = \dfrac{P_2 V_2}{T_2}$ and $\dfrac{740 \times 400}{291} = \dfrac{760 \times V_2}{273}$

where T is in degrees absolute $= {}^\circ\text{C} + 273$

$$V_2 = {}^{273}\!/_{291} \times {}^{740}\!/_{760} \times 400 = 366 \text{ cc} \qquad c$$

1.719 1 liter water $= 1{,}000$ cc $= 1{,}000$ grams

$$Q = WC(T_2 - T_1)$$

where Q = heat added (calories)

W = weight of water (grams)

C = thermal capacity $\left(\dfrac{\text{calories}}{{}^\circ\text{C per gram}}\right)$

$$500 = 2,000 (1)(T_2 - 30)$$
$$T_2 - 30 = 0.25$$
$$T_2 = 30.25°C$$

1.720 For a Carnot cycle the $T - S$ diagram is a rectangle.

$$T_1 = 273 + 157 = 430°K$$
$$T_2 = 273 + 100 = 373°K$$

Thermal efficiency $= \dfrac{(T_1 - T_2)}{T_1} \times 100$

$$= \dfrac{(430 - 373)}{430} \times 100 = \textbf{13.25 per cent}   \qquad \textbf{b}$$

1.721 1 watt $= 5.689 \times 10^{-2}$ Btu per min

Weight of 1 gal water $= 62.5/7.5 = 8.33$ lb

Btu per min effective in heating water

$= $ (watts input)(efficiency)(5.689×10^{-2})

$= (300)(0.75)(5.689)(10)^{-2} = 12.8$ **Btu per min**

Btu per gal necessary to heat water

$= $ (weight of water in lb) $\times$ (temperature change in °F)

$= 8.33(120 - 60) = 500$ Btu per gal

Time to heat 1 gal water $= \dfrac{\text{Btu per gal}}{\text{Btu per min}} = \dfrac{500}{12.8}$

$= \textbf{39.1}$ min per gal or 0.65 hr per gal

Cost to heat 1 gal of water $= \dfrac{\text{cost}}{\text{kwhr}} \times \text{kw} \times \dfrac{\text{hr}}{\text{gal}}$

$= 0.05 \times 0.300 \times 0.65 = \textbf{\$0.01 or 1 cent}$

1.722 Saving in Btu

$= \left[\dfrac{\text{(Btu saved)}}{\text{(hr)(°F)(area)}} \right] \times \text{(hr)}(\Delta T)\text{(area in sq ft)}$

$= \dfrac{(0.40 - 0.18)}{1 \times °F \times 1} \times 5,000 \times 35 \times 10,000 = 385 \times 10^6 \textbf{ Btu}$

Pounds of coal saved $= \dfrac{\text{saving in Btu}}{\text{(Btu/lb coal)} \times \text{efficiency}}$

$= \dfrac{385 \times 10^6}{13 \times 10^3 \times 0.60} = \textbf{49,400 lb}$

Saving in dollars $= $ (cost/ton) (tons saved) $= 15 \times 49,400/2,000$

$= \textbf{\$371}$

1.723   psia $= 14.7 + $ psig $= 14.7 + 20 = $ **34.7**

$h_2 = h_f + xh_{fg}$ and $h_1 = h_f$

where $x = $ quality of the steam

$h_1 = $ enthalpy of condensed water

$h_2 = $ enthalpy of original mixture per lb

$h_f = $ enthalpy of 1 lb water

$h_{fg} = $ change of enthalpy during vaporization

$$\Delta(\text{Btu per lb}) = h_2 - h_1 = h_f + xh_{fg} - h_f = xh_{fg}$$

$$x = \frac{\Delta(\text{Btu per lb})}{h_{fg}} = \frac{475}{939.5} = 0.507 = \textbf{50.7} \text{ per cent} \qquad \textsf{a}$$

1.724   (See answer to 1.716 for symbols.)

$I_{hp} = P_m LAN/33{,}000$

Area piston, head end $= \dfrac{\pi}{4} \times 144 = 113.1$ sq in.

Area piston, crank end $= \dfrac{\pi}{4}(12^2 - 2^2) = 110.0$ sq in.

$$I_{hp} \text{ head end} = \frac{70 \times 1.5 \times 113.1 \times 350}{33{,}000} = 126.0$$

$$I_{hp} \text{ crank end} = \frac{70 \times 1.5 \times 110.0 \times 350}{33{,}000} = 122.5$$

$$I_{hp(total)} = \overline{248.5} \text{ hp}$$

Brake hp $= (I_{hp})(\text{mechanical efficiency}) = 248.5 \times 0.92$

$$= \textbf{228.5} \text{ hp}$$

1.725   1 hp $= 42.41$ Btu per min

Btu per min dissipated as heat $= 42.41 \times $ hp

$= 42.41 \times 300 = 12{,}720$

Btu per min necessary to heat water from 80°F to 180°F

$= 12{,}720 = $ weight water per min $\times 100$

Weight water per min $= 127.2$ lb per min

$$= \frac{127.2 \times 7.5}{62.5} \text{ or } \textbf{15.3} \text{ gal per min}$$

GENERAL, MULTIPLE CHOICE

1.801 Assuming a current exists in the direction as shown in Fig. 1.801 of the question, Kirchhoff's voltage law states that the sum of the E's is zero. Therefore, the vector sum

$$\mathbf{E}_{CB} + \mathbf{E}_{BA} + \mathbf{E}_{AD} + \mathbf{E}_{DC} = 0$$

1. Assuming $\mathbf{E}_{CB}$ as reference, the voltage vector diagram is

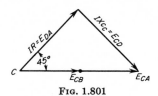

FIG. 1.801

shown in Fig. 1.801. For a 45° angle

$|IR| = |IX_{c_c}|$ and $R = X_{c_c}$
$R = 1/(2\pi fC_c) = 1/(2\pi 60 \times 10^{-6}) = \mathbf{2{,}650}$ ohms   **1c**

2. $|E_{CB}| = |E_{BA}| = |E| = |E_{CA}/2|$
$|E_{CA}| = 1.414|E_{AD}|$
and $|E_{AD}| = 2|E|/1.414 = \mathbf{1.414|E|}$ **2a**

1.802 1. At resonant frequency in a series R-L-C circuit, the supply voltage equals the voltage across the resistance.

Therefore, $P = E_R^2/R$ or $R = E^2/P$
and $R = (100)^2/10 = \mathbf{1{,}000}$ ohms   **1a**

2. The Q at the resonant frequency equals the resonant frequency divided by the band width.

$$Q_0 = \frac{\omega_0}{\Delta\omega} = \frac{2\pi \times 10^6}{0.1(2\pi \times 10^6)} = 10$$

$$Q_0 = \omega_0 L/R \text{ or } L = Q_0 R/\omega_0$$

Therefore,

$L = (10)(1{,}000)/(2\pi \times 10^6) = \mathbf{1.59 \times 10^{-3}}$ henrys   **2b**

3. $Q_0 = 1/(\omega_0 RC)$ or $C = 1/(Q_0\omega_0 R)$

Therefore,

$C = 1/[(10)(2\pi \times 10^6)(1{,}000)] = \mathbf{15.9 \times 10^{-12}}$ farad   **3d**

To check:

$$\omega_0 = \frac{1}{\sqrt{LC}} = \frac{1}{\sqrt{1.59 \times 10^{-3} \times 15.9 \times 10^{-12}}} = 2\pi \times 10^6$$

radians per sec

1.803   1. The induction motor kva load $= \dfrac{80}{0.8}\ \underline{/\cos^{-1} 0.8}$

$= 100\underline{/36.8^\circ}$ kva

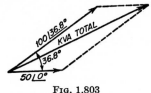

FIG. 1.803

The total kva load is the vector sum of the loads as shown in Fig. 1.803.

$$\text{Kva}_{\text{total}} = 100\underline{/36.8^\circ} + 50\underline{/0^\circ} = 80 + j60 + 50$$
$$= 130 + j60 = 143\underline{/24.7^\circ}\ \text{kva} \qquad \textbf{1d}$$

2. Power factor $= \cos 24.7^\circ = \textbf{0.910} \qquad \textbf{2d}$
3. Current per phase $= \text{kva} \times 1{,}000/(1.732 \times \text{voltage})$
$$= 143 \times 1{,}000/(1.732 \times 2{,}200)$$
$$= \textbf{37.6 amp} \qquad \textbf{3c}$$

1.804   1. If $R = 0$ and $G = 0$, the resonant angular frequency would be $\omega_0 = 1/\sqrt{LC} = 10^6$ radians per sec. By definition of the Q of a coil, $Q = \omega L/R$. The frequency at which the Q was measured should have been stated in the question. Since it was not, assume that it was measured at ω_0 and that the resonant frequency will not differ greatly from ω_0. Then, approximately,

$$R = \omega_0 L/Q = 10^6 \times 10^{-3}/2 = \textbf{500 ohms} \qquad \textbf{1a}$$

From the LC product

$$C = LC/L = 10^{-12}/10^{-3} = \textbf{1} \times \textbf{10}^{-9}\,\textbf{farad} \qquad \textbf{1a}$$

From the formula for the time constant of the series $G\text{-}C$ circuit comprising the capacitor alone

$$T = C/G \text{ and } G = C/T = 10^{-9}/(2 \times 10^{-6})$$
$$= 5 \times 10^{-4} \text{ mho} \qquad \textbf{1a}$$

2. The equivalent impedance of the given circuit is given by

$$Z_e = R + j\omega L + \frac{1}{G + j\omega C}$$

Series resonance may be defined as either of two conditions as ω is varied: (1) the imaginary (reactive) part of Z_e equal to zero, or (2) the magnitude of Z_e equal to a maximum. In general, for low-Q circuits, these two definitions will yield slightly differing values of the resonant frequency. Choosing here to use the first definition,

$$Z_e = R_e + jX_e = R + j\omega L + \frac{1}{G + j\omega C}$$

Multiplying both sides by $G + j\omega C$

$$GR_e - \omega C X_e + j(X_e G + \omega C R_e) = RG - \omega^2 LC$$
$$+ j(\omega LG + \omega C R) + 1$$

Equating the real terms,

$$GR_e - \omega C X_e = RG - \omega^2 LC + 1$$

Equating the imaginary terms,

$$X_e G + \omega C R_e = \omega LG + \omega C R$$

From the latter, if $X_e = 0$, $R_e = LG/C + R = \dfrac{10^{-3} \times 5 \times 10^{-4}}{10^{-9}}$
$+ 500 = 1{,}000$ ohms. From the former, if $X_e = 0$, $GR_e = RG - \omega^2 LC + 1$.

Whence $\omega = \sqrt{\dfrac{1 + G(R - R_e)}{LC}} = \omega_0 \sqrt{1 - G(R_e - R)}$

$$= \omega_0 \sqrt{1 - \frac{LG^2}{C}}$$

$$= \omega_0 \sqrt{1 - \frac{10^{-3} \times 25 \times 10^{-8}}{10^{-9}}} = \omega_0 \sqrt{1 - 0.25}$$

$$= 0.865\omega_0 = \textbf{0.865} \times \textbf{10}^6 \textbf{ radians per sec} \qquad \textbf{2c}$$

Thus the series-resonant frequency is lower than the L-C resonance by the factor 0.865. At this frequency, the value of R is $0.865 \times 500 = 433$ ohms, but this correction does not change the resonant frequency.

   3. $R_e = LG/C + R = 500 + 433 = \mathbf{933}$ ohms   **3d**

1.805 The kva per phase is

$$\text{kva}_\phi = \text{kw}/(3 \times \text{pf}) = 1{,}000/(3 \times 0.85) = 392$$

The voltage per phase is

$$|E_R| = \text{kva}_\phi \times 10^3/I_\phi = 392 \times 10^3/56.4 = 6{,}960 \text{ volts to neutral}$$

If I_ϕ is chosen as reference, then for $\cos\theta = 0.850$ or $\theta = 31.6°$ and $\sin\theta = 0.524$,

$$I_\phi = 56.4\underline{/0°} \text{ amp per wire}$$
$$E_R = 6{,}960\underline{/31.6°} \text{ volts to neutral}$$

In each phase the line impedance, $Z_\phi = 1 + j10$, and for a balanced load there is no current, and hence no voltage drop, in the neutral. Hence, the sending-end voltage is simply $E_s = E_R + T_\phi Z_\phi = 5{,}930 + j3{,}650 + 56.4 + j564 = 5{,}990 + j4{,}214 = 7{,}344\underline{/35.3°}$ volts to neutral.

   1. The line voltage required at the sending end will be

$$7{,}344 \times 1.732 = \mathbf{12{,}700} \text{ volts line to line}   \mathbf{1c}$$

   2. The regulation is

$$\frac{|E_S|}{|E_R|} - 1 = \frac{7{,}344}{6{,}960} - 1 = 0.052 = \mathbf{5.2} \text{ per cent}   \mathbf{2b}$$

1.806 $R \times \tan 24° = 0.445R$

$$(0.445R - 10)^2 + (R - 300)^2 = (R - 150)^2 \quad \text{(Fig. 1.806)}$$
$$0.198R^2 - 8.9R + 100 + R^2 - 600R + 90{,}000 = R^2 - 300R + 22{,}500$$
$$0.198R^2 - 308.9R = -67{,}600$$
$$R^2 - 1{,}560R = -341{,}000$$
$$(R - 780)^2 = (-780)^2 - 341{,}000 = 608{,}400 - 341{,}000 = 267{,}400 = (\pm517)^2$$
$$R = +780 + 517 = \mathbf{1{,}297} \text{ ft}   \mathbf{d}$$

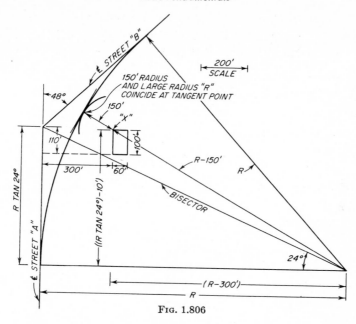

Fɪɢ. 1.806

Check: $0.445 \times 1,298 = 577$, and $577 - 10 = 567$, $1,298 - 300 = 998$, $1,298 - 150 = 1,148$.

$$567^2 = \quad 321,000$$
$$998^2 = \quad 996,000$$
$$1,148^2 = \overline{1,317,000}$$

1.807 $1,100 \times \sin 5° = 1,100 \times 0.0872 = 95.9$
$1,100 \times \cos 5° = 1,100 \times 0.996 \ = 1,095.8$

Coordinates of the radius center are (Fig. 1.807)

$$1,570.0 - 1,095.8 = 474.2 \text{ N}$$
$$420.0 + \quad 95.9 = 515.9 \text{ E}$$

To find coordinates of center line of road (N85°E tangent) and intersection of property line bearing N9°E through coordinates 280 N and 990 E:

(1) $420 + 0.996X = 990 + 0.1564Y$
$0.996X - 0.1564Y = 570$
$1,570 + 0.0872X = 280 + 0.988Y$

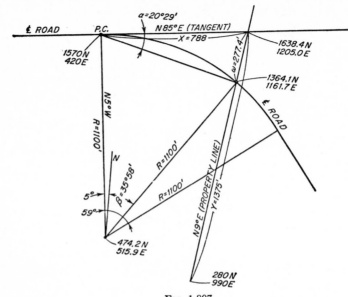

Fig. 1.807

(2) $-0.0872X + 0.988Y = 1{,}290$
(3) = 6.32 × (1) $+6.3000X - 0.988Y = 3{,}600$
(2) + (3) $+6.213X = 4{,}890$
$$X = 788 \text{ ft}$$
From (2) $0.988Y = 1{,}290 + 0.0872 \times 788$
$$= 1{,}290 + 68.6 = 1{,}358.6$$
$$Y = 1{,}375 \text{ ft}$$

Check coordinates:

$$1{,}570 + 788 \times 0.0872 = 1{,}570 + 68.6 = 1{,}638.4 \text{ N}$$
$$280 + 1{,}375 \times 0.998 = 280 + 1{,}358.6 = 1{,}638.4 \text{ N}$$
$$420 + 788 \times 0.996 = 420 + 785 = 1{,}205 \text{ E}$$
$$990 + 1{,}375 \times 0.1564 = 990 + 215 = 1{,}205 \text{ E}$$

To find coordinates of center line of road on the curve and the
intersection of line bearing N9°E, solve for angle and distance.

(4) $1{,}100 \sin \beta + 0.1564 \text{ W} = 1{,}205.0 - 515.9 = 689.1$
(5) $1{,}100 \cos \beta + 0.988 \text{ W} = 1{,}638.4 - 474.2 = 1{,}164.2$

(6) = 6.32 × (4) $6{,}952 \sin \beta + 0.988\,W = 4{,}355.1$
(7) = (6) − (5) $6{,}952 \sin \beta - 1{,}100 \cos \beta = 3{,}190.9$
(7a) = (7)/1,100 $6.32 \sin \beta = \cos \beta + 2.90$
$39.94 \sin^2 \beta = 39.94 - 39.94 \cos^2 \beta = \cos^2 \beta + 5.80 \cos \beta + 8.41$
$40.94 \cos^2 \beta + 5.80 \cos \beta = 31.53$
$\quad \cos^2 \beta + 0.1416 \cos \beta = 0.770$
$\quad\quad (\cos \beta + 0.0708)^2 = 0.770 + 0.005 = 0.775 = (\pm 0.8800)^2$
$\quad\quad\quad\quad \cos \beta = +0.8800 - 0.0708 = 0.8092$
$\quad\quad\quad\quad\quad \beta = 35°58'$
$\quad\quad\quad\quad\quad \alpha = 0.5\,(5° + 35°58') = 20°29'$ **b**
$\quad\quad 1{,}100 \sin \beta = 1{,}100 \times 0.587 = 645.8$
$\quad\quad 1{,}100 \cos \beta = 1{,}100 \times 0.809 = 889.9$
from (5) $\quad 889.9 + 0.988\,W = 1{,}164.2$
$\quad\quad\quad\quad 0.988\,W = 274.3$
$\quad\quad\quad\quad\quad\quad W = 277.6$

Check coordinates:

$\quad\quad 0.1564\,W = 0.1564 \times 277.6 = 43.3$
$\quad 474.2 + 889.9 = 1{,}364.1\ N + 274.3 = 1{,}638.4\ N$
$\quad 515.9 + 645.8 = 1{,}161.7\ E + 43.3 = 1{,}205.0\ E$

1.808 $\mu = 0.005/32$ slugs per ft, and $F = 25$ lb

$$V_{\text{transverse}} = \sqrt{F/\mu} = \sqrt{25 \times 32/0.005} = \sqrt{160{,}000}$$
$$= \textbf{400 fps}\quad \textbf{d}$$

1.809 $V = \sqrt{\gamma RT/M}$ (ideal gas)

where γ = ratio of the heat capacity at constant pressure to the
heat capacity at constant volume
R = universal gas constant
T = temperature in °K
M = mean molecular weight of air
$V = \sqrt{1.40 \times 8.3 \times 10^7 \times {}^{300}\!/_{29}} = 34{,}600$ cm per sec
$\quad = \textbf{1,130 fps}$ **b**

1.810   1. $\lambda = (1{,}130 - 88)/500 = \textbf{2.09 ft}$ **1c**
2. $\lambda = (1{,}130 + 88)/500 = \textbf{2.44 ft}$ **2d**
3. $f = f_0 \times V/(V - v_s) = 500 \times 1{,}130/(1{,}130 - 88)$
$\quad\quad = \textbf{5.42 cps}$ **3b**
4. $f = f_0 \times V/(V + v_s) = 500 \times 1{,}130/(1{,}130 + 88)$
$\quad\quad = \textbf{4.64 cps}$ **4c**

5. $f = f_0(V + v_L)/(V - v_s)$
     $= 500(1,130 + 44)/(1,130 - 88)$
     $= 5.62$ cps   **5d**

6. $f = f_0(V - v_L)/(V + v_s)$
     $= 500(1,130 - 44)/(1,130 + 88)$
     $= 4.46$ cps   **6a**

1.811   1. $B_2 - B_1 = 10 = 10 \log I_2/I_1$
                    $\log I_2/I_1 = 1 = \log 10$
                    therefore, $I_2/I_1 = 10$ **1a**

2. $B_2 - B_1 = 20 = 10 \log I_2/I_1$
                    $\log I_2/I_1 = 2 = \log 100$
                    therefore, $I_2/I_1 = 100$ **2d**

3. $B_2 - B_1 = 10 \log I_2/I_1 = 10 \log 4$
     $= 10 \times 0.6 = 6$ db   **3b**

4. $B_2 - B_1 = 20 \log p_2/p_1 = 20 \log 6 = 20 \times 0.778$
     $= 15.56$ db   **4c**

1.812   1. The potential horsepower available is

$$300 \times 62.5 \times {}^{125}\!/\!_{550} = 4,260 \text{ hp}$$

The efficiency of the plant is

$$100 \times 3,500/4,260 = \mathbf{82.2} \text{ per cent}   \mathbf{1c}$$

2. Area of pipe $= 0.785 \times 25 = 19.63$ sq ft
Velocity of water in pipe $= 300/19.63 = 15.3$ fps
Friction factor for pipe (from table) $= f = 0.01276$

and $f \times l/d = 0.01276 \times 1,000/5 = 2.552$

Head loss in pipe $= (1.03 + 2.552) \times 15.3^2/64.4 = 13$ ft
Head available for turbine $= 125 - 13 = 112$ ft

The potential horsepower of the turbine is

$$300 \times 62.5 \times {}^{112}\!/\!_{550} = 3,820 \text{ hp}$$

The efficiency of the turbine is

$$100 \times 3,500/3,820 = \mathbf{91.5} \text{ per cent}   \mathbf{2d}$$

1.813   Area of 12-in.-diameter pipe $= 0.785$ sq ft
2,000 gpm $= 2,000/(60 \times 7.48) = 4.45$ cu ft per sec
$v = Q/A = 4.45/0.785 = 5.68$ fps

6.5 psi $= 6.5/0.434 = 15$ ft of head

Pump develops a net head of $1,000 - 15 = 985$ ft

$\mu =$ coefficient of viscosity (lb-sec per sq ft)

$\rho =$ mass per unit volume

$\nu =$ kinematic viscosity $= \mu/\rho$

The API specific gravity relative to water at 60°F

$\qquad = 141.5/(131.5 + \text{deg API}) = 141.5/(131.5 + 40)$

$\qquad = 141.5/171.5 = 0.825$

$\mu = (0.0022 \times 400 - 1.30/400) \times 0.825$

$\quad = (0.88 - 0.0032) \times 0.825 = 0.724$ poise

$\quad = 0.724/478.69 = 0.00151$ lb-sec per sq ft

$\rho = 62.4 \times 0.825/32.2 = 1.60$ slugs per cu ft

$\nu = \mu/\rho = 0.00151/1.60 = 0.000944$ sq ft per sec

$\text{Re} = \nu d/\nu = 15.3 \times 1.0/0.000944 = 16,200$ (turbulent

$\qquad\qquad\qquad\qquad\qquad\qquad\qquad\qquad\qquad$ flow)

from a table of friction factors vs. Reynolds numbers

$f_{\max} = 0.039$

$985 = (f \times l \times v^2)/(d \times 2g) = 0.039 \times l \times 5.68^2/64.4$

$l = (985 \times 64.4)/(0.039 \times 32.2) =$ **50,500 ft**   **b**

1.814   Area of 4-in. pipe $= 0.0872$ sq ft $= 12.56$ sq in.

μ (viscosity) at 100°F $= 0.397 \times 10^{-6}$ (from tables)

$w_1 = p/RT$

$\quad = [(100 + 14.7) \times 144]/[53.3 \times (100 + 459.4)]$

$\quad = 0.554$ lb per cu ft

$v_1 = W/(0.554 \times 0.0872) = 20.7W$

$\nu = \mu/\rho = 0.397 \times 10^{-6} \times 32.2/0.554 = 23.1 \times 10^{-6}$

$\text{Re} = v_1 d/\nu = (20.7W \times 0.33)/(23.1 \times 10^{-6})$

$\qquad\qquad\qquad\qquad\qquad\qquad\qquad\qquad = 295,000W$

$(p_1{}^2 - p_2{}^2) \times 144^2 = [W^2 RT(f \times l/d)]/(ga^2)$

$$W^2 f = \frac{[(114.7)^2 - (74.7)^2] \times (12.56)^2 \times 32.2}{53.3 \times 559.4 \times 3,000}$$

$$= \frac{(13,140 - 5,580) \times 5,070}{89,400 \times 1,000} = \frac{7,560 \times 5.07}{89,400}$$

$\qquad\qquad\qquad\qquad\qquad\qquad\qquad\qquad\qquad = 0.428$

Assume a value for W, as 5.00.

Then $\text{Re} = 1,475,000$ and $f = 0.0151$

$$W^2 = 0.428/0.0151 = 28.4$$

$$W = 5.33$$

Then for $W = 5.33$, Re $= 1,570,000$, and $f = 0.0149$

$$W^2 = 0.428/0.0149 = 28.7$$
$$W = 5.36 \text{ say OK} \qquad \mathbf{d}$$

1.815 $T_2/T_1 = e^{f\theta} = 2.718^{0.12 \times 2\pi n}$
where θ is the angle in radians and n is the number of turns about the capstan

$$4,000/50 = 80 = 2.718^{0.755n}$$
$$\log 80 = n \times 0.755 \log 2.718$$
$$1.9031 = n \times 0.755 \times 0.4343 = n \times 0.327$$
$$n = 1.9031/0.327 = 5.81, \text{ or use } \mathbf{6} \text{ turns} \qquad \mathbf{a}$$

1.816 $P = (2\pi nT)/(33,000 \times 12)$ hp

Torque $= (33,000 \times 12 \times 80)/(6.28 \times 300) = 16,800$
$$\text{in.-lb}$$

Moment $= 30 \times 1,000 = 30,000$ in.-lb

Shear due to torque $= s = \dfrac{Tr}{\pi r^4/2} = \dfrac{16,800}{\pi r^3/2}$

Fiber stress due to moment $= f = \dfrac{30,000r}{\pi r^4/4} = \dfrac{60,000}{\pi r^3/2}$

Shear stress due to torque and moment $= s'$
$$= (\tfrac{1}{2}) \sqrt{4s^2 + f^2}$$

$$10,000 = \frac{1 \times 2}{2 \times \pi r^3} \sqrt{4(16,800)^2 + (60,000)^2}$$

$$\pi r^3 = \frac{1,000}{10,000} \sqrt{1,130 + 3,600} = 6.88$$

$$r^3 = 2.196$$
$$r = 1.30 \text{ in.}$$
$$d = 2.60 \text{ in., or use } \mathbf{2.625} \text{ in.} \qquad \mathbf{d}$$

Check bending stress due to torque and moment

$$f' = (f/2) + s'$$
$$= 120,000/(2 \times 6.88) + 10,000$$
$$= 8,725 + 10,000 = 18,725 \text{ (less than 20,000)}$$

1.817 Neglect fillets.

Then stem of $T = 0.885 \times 6.5 \quad = \quad 5.75$ sq in.
Net flange of $T = 1.570 \times 15.71 = \underline{24.70}$ sq in.
$$\text{Area of } T = \overline{30.45} \text{ sq in.}$$

c.g. distance from back of flange

$$x = (3.25 \times 5.75 + 0.785 \times 24.70)/30.45$$
$$= (18.70 + 19.40)/30.45 = 1.25 \text{ in.}$$

I of T

Stem $5.75 \times 6.5^2/12 + 5.75 \times 2^2 = 20.20 + 23.00 \qquad = 43.20$
Flange $24.70 \times 1.57^2/12 + 24.70 \times 0.465^2 = 5.07 + 5.35$
$$= \underline{10.42}$$
$$53.62$$

Use $I = 54$ in.^{4}

$$I_2 = 2(54 + 30.45 \times 28.75^2) = 2(54 + 25{,}200) = 50{,}500$$
$$S_2 = 50{,}500/30 = 1{,}683 \text{ in.}^3$$
$$S_2/S_1 = 1{,}683/1{,}031 = \mathbf{1.63} \quad \mathbf{c}$$

1.818 1. $E = PL/A\Delta = (10{,}000 \times 600)/(1 \times 0.222)$
$$= 27 \times 10^6 \text{ psi} \quad \mathbf{1a}$$
2. $\Delta = (3.4 \times 600 \times 0.5 \times 0.222)/10{,}000 = \mathbf{0.0226}$ ft $\quad \mathbf{2b}$

1.819 1. $\Delta = \tfrac{3}{32} = 36 \times 65 \times 10^{-7}t$
$$t = (3 \times 10{,}000{,}000)/(32 \times 36 \times 65) = \mathbf{400°F} \quad \mathbf{1d}$$

2. Tension in the ring $= \Delta \times A \times E/L$
$$= (\tfrac{1}{32}) \times (\tfrac{4}{4}) \times (30{,}000{,}000) \times (\tfrac{1}{36})$$
$$= \mathbf{26{,}000} \text{ lb} \quad \mathbf{2c}$$

3. $2T = 4 \times 36 \times p$
$$p = (2 \times 26{,}000)/(4 \times 36) = \mathbf{361} \text{ psi} \quad \mathbf{3a}$$

1.820 $\Delta_s = (P_s \times L)/(1 \times 30 \times 10^6) = \Delta_c$
$$= (P_c \times L)/(1 \times 17 \times 10^6)$$

$$17P_s = 30P_c$$
$$17P_s - 30P_c = 0$$
(1) $\qquad P_s - 1.765P_c = 0$
(2) $\qquad P_s + P_c = 30{,}000$
(2) − (1) $\qquad +2.765P_c = 30{,}000$
$$P_c = 10{,}830 \text{ lb}$$
$$P_s = \mathbf{19{,}170} \text{ lb} \quad \mathbf{1c}$$

$$\Delta_s = \Delta_c + 60 \times (93 - 65) \times 10^{-7} \times L$$

$$\frac{P_s \times L \times 10^{-6}}{1 \times 30} = \frac{P_c \times L \times 10^{-6}}{1 \times 17} + \frac{6 \times 28 \times 10^{-6} \times L}{1}$$

$$17P_s = 30P_c + 17 \times 30 \times 168$$

(3) $P_s - 1.765P_c = 5{,}040$

(2) $P_s + P_c = 30{,}000$

(2) − (3) $+2.765P_c = 24{,}960$

$$P_c = \textbf{9,030} \text{ lb} \qquad \textbf{2b}$$

$$P_s = 20{,}970 \text{ lb}$$

1.821

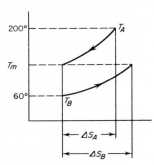

Fig. 1.821

1.821   lb water × temp. = product

$$4 \times 200.0 = 800$$
$$2 \times 60.0 = 120$$
$$\overline{6 \times (153.3) = 920}$$

$T_m = 153.3°\text{F} = 153.3 + 460 = 613.3°$ Rankine;

$\qquad\qquad T_A = 660°$ Rankine; $T_B = 520°$ Rankine

By Fig. 1.821

$$\Delta S_A = wc_p \log_e (T_m/T_A) = (4)(1) \log_e (613.3/660)$$
$$= 4 \log_e 0.93 = -0.290 \text{ units (drop)}$$
$$\Delta S_B = wc_p \log_e (T_m/T_B) = (2)(1) \log_e (613.3/520)$$
$$= 2 \log_e 1.18 = +0.331 \text{ units (rise)}$$
$$\Delta S_{A-B} = 0.331 - 0.290 = \textbf{0.041} \text{ units} \qquad \textbf{a}$$

1.822   1. For ideal turbine

$$W = \frac{KwRT_1}{1 - K}\left[\left(\frac{p_2}{p_1}\right)^{\frac{K-1}{K}} - 1\right]$$
$$= \frac{(1.4)(1)(53.3)(1{,}660)}{1 - 1.4}\left[\left(\frac{15}{60}\right)^{\frac{1.4-1}{1.4}} - 1\right]$$
$$= 103{,}800 \text{ ft-lb per sec}$$
$$= 103{,}800/550 = \textbf{188.5} \text{ hp} \qquad \textbf{1b}$$

2. For turbine developing 150 hp

Temperature of water entering = 50°F
Temperature of water leaving = 100°F
Each pound of water picks up $(100 - 50) \times 1 = 50$ Btu

The general expression for steady flow (neglecting the change in kinetic energy) is

$$H_1 + Q = H_2 + W$$

$$
\begin{aligned}
Q &= W + (H_2 - H_1) = W + wc_p\,\Delta T \\
&= (150 \times 550)/778 + (1)(0.24)(300 - 1{,}200) \\
&= 106 - 216 = -110 \text{ Btu per sec (to be carried away by}
\end{aligned}
$$
cooling water)

$(^{110}\!/_{50})(60/8.34) = \mathbf{15.85}$ gal per min   **2c**

1.823

Steam	Ice	Calorimeter	Water
$w_s = 7$ lb	$w_i = 8$ lb	$w_c = 5$ lb	$w_w = 50$ lb
$t_s = 242°F$	$t_i = 25°F$	$t_c = 60°F$	$t_w = 60°F$
$c_s = 0.48$	$c_i = 0.50$	$c_c = 0.093$	
$p_s = 14.7$ psia			

Let t_m = resulting temperature of mixture

Heat given up by steam = heat gained by ice + heat gained by calorimeter + heat gained by water

$$
\begin{aligned}
&[w_s c_s(t_s - 212) + 970 w_s + w_s(212 - t_m)] \\
&\quad = [w_i c_i(32 - 25) + 144 w_i + w_i(t_m - 32)] \\
&\qquad\qquad + [w_c c_c(t_m - 60)] + [w_w(t_m - 60)] \\
&[7 \times 0.48(242 - 212) + 970 \times 7 + 7(212 - t_m)] \\
&\quad = [8 \times 0.50 \times 7 + 144 \times 8 + 8(t_m - 32)] \\
&\qquad\qquad + [5 \times 0.093(t_m - 60)] + [50(t_m - 60)] \\
&(101 + 6{,}790 + 1{,}484 - 7t_m) = (28 + 1{,}152 + 8t_m - 256) \\
&\qquad\qquad + (0.465t_m - 27.9) + (50t_m - 3{,}000) \\
&\qquad 65.465 t_m = +11{,}659 - 1{,}180 = 10{,}479 \\
&\qquad\qquad t_m = \mathbf{160°F} \quad \mathbf{d}
\end{aligned}
$$

1.824 Heat loss without insulation

$$Q = U_1 A\, \Delta t = 0.40 \times 10,000 \times 35 = 140,000 \text{ Btu}$$
$$\text{per hr}$$

Heat loss with insulation
$$Q = U_2 A\, \Delta t = 0.18 \times 10,000 \times 35 = \underline{\quad 63,000} \text{ Btu}$$
$$\text{per hr}$$
$$\text{Saving} = 77,000 \text{ Btu}$$
$$\text{per hr}$$

Total saving for heating season
$$Q = 77,000 \times 5,000 = 385 \times 10^6 \text{ Btu}$$
Coal saved $= (385 \times 10^6)/(13,000 \times 0.60 \times 2,000)$
$$= 24.7 \text{ tons}$$

Dollar saving $= \$15 \times 24.7 = \$370$ **b**

1.825 1. $Q_A = 10$ tons $= 10 \times 200$ Btu per min per ton
$$= 2,000 \text{ Btu per min}$$
Input to compressor $= 15 \times (2,545/60) = 637$ Btu
$$\text{per min}$$
Condensing water must carry away $2,000 + 637$
$$= 2,637 \text{ Btu per min}$$

$Q_{\text{water}} = wc_p\, \Delta t$
$2,637 = w \times 1 \times (70 - 50)$
$w = 2,637/20 = 131.85$ lb per min $= 131.85/8.34$
$\quad = \textbf{15.8}$ gal per min   **1a**

2. C.O.P. $= \dfrac{T_1}{(T_2 - T_1)} = \dfrac{(0 + 460)}{(70 + 460) - (0 + 460)} = \dfrac{460}{70}$
$$= \textbf{6.57} \quad \textbf{2a}$$

3. C.O.P. $= 4.71$ per hp per ton $= 4.71/(^{15}\!/_{10}) = \textbf{3.14}$ **3d**

CHEMICAL ENGINEERING

Reference will be made a number of times in this section to J. H. Perry, "Chemical Engineers' Handbook," 3d ed., McGraw-Hill, 1950, simply by stating "Perry, p.—."

2.01

Coal analysis (dry)	Refuse	Flue gas (volatile)
C = 75% Ash = 8% $\left.\begin{array}{l}H_2 \\ O_2\end{array}\right\} = 17\%$	C = 0	CO_2 = 12.6% CO = 1.0% O_2 = 6.2% N_2 = 80.2%
100%		= 100.0%

Basis will be 100 lb fuel.

100 lb fuel contains 75 lb carbon or $^{75}\!/_{12}$ (atomic weight)
$$= 6.25 \text{ lb-atoms}$$

1 lb-mole of flue gas contains 0.136 lb atoms of carbon, i.e.,

$$CO_2 = 0.126 \text{ lb-mole}$$
$$CO = \underline{0.010} \text{ lb-mole}$$
$$0.136 \text{ lb-mole}$$

Therefore $\dfrac{6.25}{0.136} = \dfrac{46 \text{ lb-moles dry stack gas}}{100 \text{ lb fuel}}$

The stack gas (dry) contains $\dfrac{0.062 \text{ lb-mole oxygen}}{\text{lb-mole stack gas}}$ and for each 100 lb fuel, there is $46 \times 0.062 = 2.86$ lb-moles of oxygen in the stack gas.

265

Since this oxygen came from the air (21 per cent oxygen and 79 per cent nitrogen) there is $2.86 \times \dfrac{0.79}{0.21}$ or 10.75 lb-moles of N_2 in stack gas from excess air.

The nitrogen in the stack gas that comes from the air supplied for combustion is

$$46 - (5.79 + 0.46 + 2.86 + 10.75) = 26 \text{ lb-moles } N_2$$

where

$$\frac{0.126}{0.136} \times 6.25 = 5.79 \text{ lb-moles } CO_2$$

and

$$\frac{0.010}{0.136} \times 6.25 = 0.46 \text{ lb-mole } CO$$

The oxygen supplied for combustion by the air is

$$26 \times \frac{0.21}{0.79} = 6.90 \text{ lb-moles}$$

Oxygen required for combustion of carbon is

$$C + O_2 \rightarrow CO_2 \text{ or } \frac{0.126}{0.136} \times 6.25 = 5.79 \text{ lb-moles}$$

$$C + \tfrac{1}{2} \times O_2 \rightarrow CO \text{ or } \tfrac{1}{2} \times \frac{0.010}{0.136} \times 6.25 = 0.23 \text{ lb-mole}$$

$$\overline{ 6.02 \text{ lb-moles}}$$

Oxygen supplied by air for combustion of H in fuel is $6.90 - 6.02 = 0.88$ lb-mole per 100 lb fuel

Available O_2 and H_2 in fuel per 100 lb fuel is

$H_2 + \tfrac{1}{2} \times O_2 \rightarrow H_2O$

$2X + (\tfrac{1}{2}X - 0.88) \, 32 = 17$ (from fuel analysis)

$X = 2.55$ lb-moles hydrogen or 5.10 lb per 100 lb fuel

$\tfrac{1}{2}X = 1.275$ lb-moles total oxygen required.

$(\tfrac{1}{2}X - 0.88) = 0.395$ lb-mole available oxygen in fuel

$\phantom{(\tfrac{1}{2}X - 0.88)} = 12.65$ lb per 100 lb fuel

a. *Per cent oxygen*

lb-mole of O_2 for combustion of 100 lb fuel

CO_2 = 5.79 lb-moles
CO = 0.23 lb-mole
H_2 = 1.27 lb-moles $(= \frac{1}{2}X)$
      7.29 lb-moles

But there are 2.86 lb-moles oxygen in stack gas

$$\therefore \text{ per cent oxygen } = \frac{2.86}{(7.29 - 0.39)} \times 100 = \textbf{41.5 per cent}$$

b. *Complete analysis of fuel*

   C = 75.0 per cent
ash = 8.0 per cent
 H_2 = 5.1 per cent ($\times$ above)
 O_2 = 11.9 per cent $(17.0 - 5.1)$

c. *Cubic feet stack gas per pound fuel* (at 680°F)

Basis—100 lb fuel
H_2O in air (80 per cent relative humidity and 36 mm vapor

$$\text{pressure)} = \frac{36 \times 0.80}{760 - 36 \times 0.80} = 0.0393 \text{ mole } H_2O \text{ per}$$

mole air

$$\therefore \frac{46 \times 0.802}{0.79} = 46.75 \text{ lb-mole air (dry) and } 46.75 \times$$

$0.0393 = 1.84$ lb-mole water vapor in air. H_2O from combustion of $H = 5.1/2 = 2.55$ lb-mole
Total lb-moles stack gas per 100 lb fuel
                = $46 + 1.84 + 2.55 = 50.39$ lb-moles
Total lb-moles stack gas per 1 lb fuel = 0.504 lb-mole = n

$$\therefore V = \frac{n \cdot RT}{P} = \frac{0.504 \times 10.71 \times (460 + 680)}{14.7} = \textbf{418.0 cu ft}$$

d. *Cubic feet of air per pound fuel* (at 90°F)

$$n = 0.4675 + 0.0184 = 0.486 \text{ lb-mole per lb fuel}$$
$$V = \frac{0.486 \times 10.71 \times (460 + 90)}{14.7} = \textbf{195.0 cu ft}$$

2.02 At top, 740 mm, 20°C, 0.13 per cent NH_3 by volume
At base, 745 mm, 40°C, 4.90 per cent NH_3 by volume
   100 cfm gas
100 per cent $-$ 4.90 per cent $=$ 95.1 per cent $=$ 0.951.
Air, cfm $=$ 100 $\times$ 0.951 $=$ 95.1

$$(mV_1) = (mV_0) \times \frac{T_1}{T_0} \times \frac{P_0}{P_1} = \text{mole-volume at other than}$$

standard condition
(mV_1) for air at 745 mm and 40°C

$$= 359 \times \frac{(273 + 40)}{273} \times \frac{760}{745} = 421 \text{ cu ft per mole}$$

$\dfrac{95.1}{421.0} = 0.226$ mole of air flowing per min

$\dfrac{\text{Mole } NH_3}{\text{Mole air}}$ at entrance $= \dfrac{0.049}{0.951} = 0.0515$

$\dfrac{\text{Mole } NH_3}{\text{Mole air}}$ at exit $= \dfrac{0.0013}{0.9987} = 0.0013$

Mole ammonia absorbed per mole air flowing
$$= 0.0515 - 0.0013 = 0.0502$$

a. Rate of flow of gas at exit in cfm

$$V = 0.226 \times 359 \times \frac{(273 + 20)}{273} \times \frac{760}{740} = 89.5 \text{ cfm}$$

b. Weight of ammonia absorbed per min
   $= 0.226 \times 0.0502 \times 17 = 0.193$ lb NH_3 per min
where atomic weight $NH_3 = 17$

2.03 1. Look up x-y data (Perry, p. 574).
   2. Construct x-y diagram.
   3. Locate top product and feed, and bottom product on x-y diagram. x-y data indicates that a top product boiling at 78.41°C would have a vapor composition of 0.7815 mole fraction of ethanol.
   4. Material balance per 44 moles feed
      a. $0.17F = 0.7815D + 0.01W$
   or $17 \times 44 = 78.15D + W$
      b. $44 = D + W$
         $77.15D = 16 \times 44$
            $D = 9.12$ moles per hr
            $W = 34.88$ moles per hr

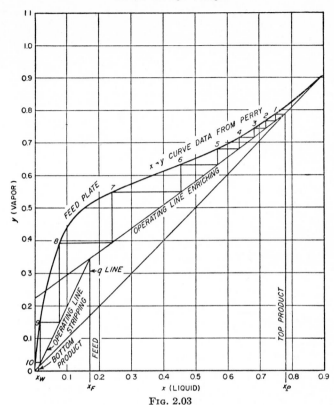

FIG. 2.03

5. Intercept of operating line $= \dfrac{9.12}{9.12 + 31} \times 0.7815 = 0.177.$

6. Construct operating line for enriching section.

7. Since the feed is a saturated liquid q (where $c_p\Delta t = 0$) $= \dfrac{c_p\Delta t + \lambda}{\lambda} = 1$ and the slope of the q line $= \dfrac{q}{q-1} = \dfrac{1}{1-1}$ $= \dfrac{1}{0} = \infty$. Construct q line.

8. Construct operating line in stripping section.

9. Construct McCabe-Thiele stepwise representation of theoretical plates.

   a. Ten theoretical plates are required. Therefore 10/0.60 or 17 actual plates are required. If reboiler is considered to be 100 per cent efficient then $9/0.60 + 1$ or 16 plates are required.

   b. The condenser must remove the latent heat from $9.12 + 31$ or 40.12 moles of top product per hr. The latent heat of a vapor (78.15 moles per cent ethanol) is 420 Btu per lb. Hence $40.12 \times 39.84 \times 420 = $ **672,000** Btu per hr where $0.7815 \times 46 + 0.2185 \times 18 = 39.84$. The reboiler duty is calculated as follows:

$$\bar{V} = F (q - 1) + V$$

where $F = 44$ moles per hr
   $(q - 1) = (1 - 1) = 0$
     $\bar{V} = V = 40.12$ moles per hr
     $\bar{V} = $ vapor volume in the stripping section
     $V = $ vapor volume in the enriching section.

The vapor in equilibrium with a 1 mole per cent ethanol solution contains approximately 0.10 mole per cent ethanol and its latent heat of vaporization is 850 Btu per lb. Therefore $q = 40.12 \times 18.28 \times 850 = $ **623,000** Btu per hr where $0.01 \times 46 + 0.99 \times 18 = 18.28$.

   c. Obtained from x-y diagram.

2.04   This problem could be solved directly by using the graph in Perry, p. 682, where ΔP is shown as a function of L and G/ϕ for a variety of packing material.

   Solution: 5 gal water per min is equivalent to

$$\frac{5}{7.5} \times 62.5 \times 60 = 2{,}500 \text{ per hr}$$

$$L = 2{,}500 \div \frac{\pi}{4} \times 1 \times 1 = 3{,}180 \text{ lb/(hr)(sq ft)}$$

$$G/\phi = (1.5 \times 3{,}600) \times \left(\frac{\pi}{4} \times 1^2\right) \times \left(\frac{492}{530} \times 0.0807\right)$$

$$= 318 \text{ lb per hr}$$
$$= (\text{ft per sec} \times \text{sec per hr}) \times (\text{sq ft}) \times (\text{weight air in lb per cu ft at } 70°\text{F})$$

From graph, for ½-in. Berl saddles, $\Delta p = 0.20$ in. per ft
$\therefore$ $7 \times 0.20 = 1.40$ in. water = the drop of pressure through the tower

2.05   Perry, p. 674.

1. For 25 gram of NH_3 per 100 gram of H_2O, the partial pressure of NH_3 = 227 mm.

2. For 2.04 gram of NH_3 per 100 gram of H_2O, the partial pressure of NH_3 = 12.2 mm.

Assume that the total pressure of the exit gas is 760 mm and that the gas leaves saturated with water at 20°C.

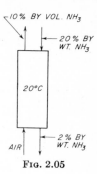

FIG. 2.05

p (water) = 17.5 mm (from steam tables)

p (air + NH_3) = 760 − 17.5 = 742.5 mm

p (NH_3) = 0.10 × 742.5 = 74.25 mm

Top: $Pai - Pag$ = 227 − 74.2 = 152.8 mm

Bottom: $Pai - Pag$ = 12.2 − 0 = 12.2 mm

$$(\Delta p)_{lm} = \frac{152.8 - 12.2}{2.3 \times \log_{10}(152.8/12.2)} = 55.6 \text{ mm}$$

$$= 0.0732 \text{ atm}$$

N = lb-moles NH_3 stripper per hour = pV/RT

$$= \frac{74.25 \times 8{,}000}{760 \times 0.730 \times 293 \times 1.8} = 2.03$$

$$V = \frac{N}{K_g a (\Delta p)_{lm}} = \frac{2.03}{7 \times 0.0732} = \mathbf{3.96} \text{ cu ft}$$

When gas rate is decreased 25 per cent,

$$V_2 = (\tfrac{4}{3})^{0.8} \times V_1 = 1.259 \times 3.96 = \mathbf{5.00} \text{ cu ft}$$

2.06   It is assumed that the alcohol to be cooled is 95 per cent alcohol. The arithmetic mean temperature of each fluid will be used in evaluating fluid properties:

$$T_{\text{alc.}} = \frac{172 + 70}{2} = 121°F$$

$$T_{\text{water}} = \frac{80 + 50}{2} = 65°F$$

The volume of alcohol to be handled is assumed to have been measured at room temperature. Hence

Alcohol, lb per hr = $200 \times 8.33 \times 0.804 = 1{,}340$ lb per hr
Specific heat of 95 per cent alcohol = 0.742
Heat to be transferred from alcohol to water = q
$$= 1{,}340 \times 0.742 \times (172 - 70) = 101{,}400 \text{ Btu per hr}$$

Calculation of film coefficient, h_i, for alcohol:

$$\text{Mass velocity} = G = \frac{1{,}340 \times 4 \times (16)^2 \times (12)^2}{60 \times 60 \times \pi \times (19)^2}$$

$$= 48.4 \text{ lb per sec} \times \text{sq ft}$$

$$R_e = \frac{DG}{\mu} = \frac{(19)}{16 \times 12} (48.4) \frac{1}{(0.77)(0.000672)}$$

$$= 9{,}250 \text{ (no units)}$$

where μ = lb per sec $\times$ ft = centipoise $\times$ 0.000672
$k = 0.087$ Btu/(hr)(sq ft)(°F)

$$h_i = 0.0225 \frac{k}{D} (R_e)^{0.8} \left[\frac{C\mu}{k} \right]^{0.3}$$

$$= 0.0225 \frac{(0.087)(16)(12)}{19} (9{,}250)^{0.8} \left[\frac{0.742 \times 0.77 \times 2.42}{0.087} \right]^{0.3}$$

$$= 67.5 \text{ Btu/(hr)(sq ft)(°F)}$$
where $2.42 = 0.000672 \times 3{,}600$

Calculation of h_0 for water:

$$\text{Rate of flow} = \frac{101{,}400}{(80 - 50) \times 3{,}600} = 0.939 \text{ lb per sec}$$

$$\text{Area of annulus} = \frac{\pi}{4 \times 144} \left[(2.067)^2 - \left(\frac{21}{16} \right)^2 \right] = 0.01395 \text{ sq ft}$$

$$\text{Mass velocity} = G = \frac{0.939}{0.01395} = 67.3 \text{ lb per sec} \times \text{sq ft}$$

$$\text{Equivalent diameter} = D_e = \frac{4 \times \text{area annulus}}{\text{area heating surface}}$$

$$D_e = \frac{4 \times 0.01395}{\pi \times 2\frac{1}{16} \times \frac{1}{12}} = 0.162 \text{ ft}$$

$$R_e = \frac{D_e G}{\mu} = \frac{0.162 \times 67.3}{1.052 \times 0.000672} = 15{,}420 \text{ (no units)}$$

By McAdams equation (modified D and B equation).

$$h_0 = 160 \, (1 + 0.01 \times 65) \left(\frac{67.3}{62.3}\right)^{0.8} \times \frac{1}{(0.162 \times 12)^{0.2}}$$

$$h_0 = 246 \text{ Btu/(hr)(sq ft)(°F)}$$

Over-all coefficient, V_0, based on outside surface of copper tube is given by:

$$\frac{1}{U_0} = \frac{1}{h_0} + \frac{LA_0}{KA_{av}} + \frac{A_0}{h_i A_i}$$

$$\frac{1}{U_0} = \frac{1}{246} + \frac{(\tfrac{1}{16} \times \tfrac{1}{12})(2\tfrac{1}{16}\pi \times 1)}{222 \times (2\tfrac{0}{16}\pi \times 1)} + \frac{(2\tfrac{1}{16}\pi \times 1)}{67.5 \times (1\tfrac{9}{16}\pi \times 1)}$$

$$\frac{1}{U_{0_c}} = 0.02047 \quad \text{and} \quad U_{0_c} = 48.9 = \text{the clean heat-transfer}$$

      coefficient.

Fouling resistances of at least 0.001 should be assumed on the inside and on the outside of the tubing.

$$\frac{1}{U_{0_D}} = 0.02047 + 0.002 = 0.02247$$

$$U_{0_D} = 44.5 = \text{the heat-transfer design coefficient}$$

The logarithmic mean temperature difference from alcohol to water:

$$(\Delta t)_{lm} = [(172 - 80) - (70 - 50)] \div 2.3 \log_{10} \frac{(172 - 80)}{(70 - 50)}$$

$$= 47.2\text{°F}$$

Required heat-transfer area:

$$A = \frac{q}{U_0(\Delta t)_{lm}} = \frac{101{,}400}{44.5 \times 47.2} = 48.3 \text{ sq ft}$$

Area per pass, based on outside surface of copper tube:

$$a = (\pi 2\tfrac{1}{16} \times \tfrac{1}{12}) \times 24 = 8.25 \text{ sq ft}$$

$$\text{Number of passes in series} = \frac{48.3}{8.25} = 5.8$$

Therefore, use **6** passes.

2.07 X-Y data from Perry, p. 574

Boiling point of 50 mole per cent solution = 73.1°C

$\Delta t = 73.1 - 20 = 53.1°C$

c_p = specific heat of feed = 0.81

λ = latent heat of feed = 16,300 Btu per lb-mole

$$q = \frac{\lambda + c_p \times \Delta t}{\lambda} = \frac{16{,}300 + 0.81 \times 25 \times 53.1 \times 1.8}{16{,}300}$$

$$= 1.12$$

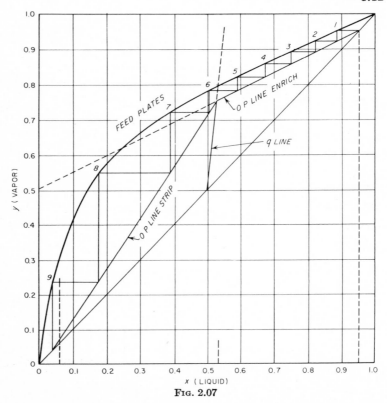

Fɪɢ. 2.07

where 25 = average mole weight of feed (lb per lb-mole)

Slope of q-line = $q/(q-1) = 1.12/0.12 = 9.33$

a. $R_{\min} = (0.95 - 0.79)/(0.79 - 0.53) = 0.615$

b. $R = \left(\dfrac{100 + 40}{100} \right) \times 0.615 = 0.861$

$$Y\text{-intercept} = \frac{X_D}{R+1} = \frac{0.95}{1.861} = 0.51$$

From the diagram, number of perfect plates is 8.85.
Therefore, use **9** plates.

   c. 2,000 lb per hr feed = 80 lb-moles per hr.
      Product = 40 lb-moles per hr
        = $(0.95 \times 40 \times 32 + 0.05 \times 40 \times 18)$ = 1,252 lb per hr
      Residue = 40 lb-moles per hr
        = $(0.95 \times 40 \times 18 + 0.05 \times 40 \times 32)$ =   748 lb per hr
                                                  2,000 lb per hr

   d. $V = RD = 0.861 \times 40 = 34.44$ lb-moles per hr
      $\bar{V} = F(q-1) + V = 80(1.12 - 1) + 34.44$
                             = 44.0 lb-moles per hr
    Pounds steam per hr = $^{970}\!\!/_{960} \times 18 \times 44.0$ = **800**

2.08   Heat of this reaction *at* 25°C is:

$$\Delta H = 369.4 - 356.2 + 10.5 = 23.7 \text{ kcal per gram-mole} = 23,700 \text{ cal per gram-mole}$$

Standard entropy change for this reaction at 25°C is:

$$\Delta S = 35.0 - 42.0 + 45.1 = 38.1 \text{ cal/(gram-mole)(°K)}$$
$$\Delta c_p = c_p(H_2O) = 6.89 + 3.283 \times 10^{-3}T - 0.343 \times 10^{-6}T^2$$

At 500°K:

$$\Delta H = 23,700 + \int_{298}^{500} (6.89 + 3.283 \times 10^{-3}T$$
$$- 0.343 \times 10^{-6}T^2)dT$$
$$= 23,700 + 6.89(500 - 298)$$
$$+ 1.641 \times 10^{-3}(500^2 - 298^2) - 0.114(500^3 - 298^3) \times 10^{-6}$$
$$= 23,700 + 1,393 + 264 - 11$$
$$= 25,346 \text{ cal per gram-mole}$$

$$\Delta S = 38.1 + \int_{298}^{500} \left(\frac{6.89}{T} + 3.283 \times 10^{-3} + 0.343 \times 10^{-6}T \right) dT$$

$$= 38.1 + 6.89 \times 2.3 \log_{10}(^{500}\!\!/_{298})$$
$$+ 3.283 \times 10^{-3}(500 - 298) - 0.172 \times 10^{-6}(500^2 - 298^2)$$
$$= 38.1 + 3.57 + 0.66 - 0.03$$
$$= 42.3 \text{ cal/(gram-mole)(°K)}$$

$$\Delta G^c = 25{,}346 - 500 \times 42.3 = +4{,}196 \text{ cal/(gram-mole)}$$

$$\log K = \log P_{(\text{H}_2\text{O})} = -\frac{\Delta G^\circ}{2.3RT} = -\frac{4{,}196}{2.3 \times 1.987 \times 500}$$

$$= -1.835 = \frac{1}{1.835}$$

$$P_{(\text{H}_2\text{O})} = \frac{1}{68.4} = 0.0146 \text{ atm} = 11.1 \text{ mm Hg}$$

At 700°K:

$$\Delta H = 23{,}700 + \int_{298}^{700} (6.89 + 3.283 \times 10^{-3}T - 0.343 \times 10^{-6}T^2)dT$$

$$= 27{,}093 \text{ cal per gram-mole}$$

$$\Delta S = 38.1 + \int_{298}^{700} (6.89/T + 3.283 \times 10^{-3} - 0.343 \times 10^{-6}T)dT$$

$$= 45.2 \text{ cal/(gram-mole)(°K)}$$

$$\Delta G^\circ = 27{,}093 - 700 \times 45.2 = -4{,}547 \text{ cal per gram-mole}$$

$$\log P_{(\text{H}_2\text{O})} = +\frac{4{,}547}{2.3 \times 1.98 \times 700} = 1.420$$

$$P_{(\text{H}_2\text{O})} = 26.3 \text{ atm} = 386 \text{ psi}$$

The thermal dehydration of the acid appears to be commercially feasible in the temperature range of 500 to 700°K without the use of any powerful desiccating agents.

2.09 *Reference:* Hougen, Watson, and Ragatz, "Chemical Process Principles," Part 3, pp. 834–835, Wiley, 1936.

$$\text{Eq. } (b) = \frac{Vr}{F} = \frac{n_0{}^2}{K\pi^2}\left[Hx_A + J\frac{x_A}{n_{A_0}(n_{A_0} - x_A)} \right.$$
$$\left. + M \ln\left(\frac{n_{A_0}}{n_{A_0} - x_A}\right) + N(\cdots) \right]$$

V_r = volume of reactor = 5 cu ft
F = feed rate = 1 lb-mole per hr
$n_0 = 1$ (assumed), $n_{A_0} = 1$, $w = -\frac{1}{2}$
$x_A = \frac{1}{2}$, $\pi = 1$ atm
$H = \omega^2 = \frac{1}{4}$, $J = (1 + \omega n_{A_0})^2 = \frac{1}{4}$
$M = -2\omega(1 + \omega n_{A_0}) = \frac{1}{2}$
$N = 0$

$$\frac{5}{1} = \frac{(1)^2}{K \times (1)^2} \left[\left(\frac{1}{4}\right)\left(\frac{1}{2}\right) + \left(\frac{1}{4}\right) \frac{\frac{1}{2}}{1(1 - \frac{1}{2})} \right.$$
$$\left. + \frac{1}{2} \, ln \left(\frac{1}{(1 - \frac{1}{2})}\right) \right]$$

$$K = \frac{1}{5} \left[\frac{1}{8} + \frac{1}{4} + \frac{2.3}{2} \, log_{10} 2 \right]$$
$$= 0.1442 \text{ lb-mole/(cu ft) (hr)(atm}^2\text{)}$$

Let rate constant at 300°C = K'

$$log \left(\frac{K}{K'}\right) = \frac{30,000}{4.576} \left(\frac{773 - 573}{773 \times 573}\right) = 2.97$$

$$K' = \frac{0.1442}{932} = 1.55 \times 10^{-4} \text{ lb-mole/(cu ft)(hr)(atm}^2\text{)}$$

Feed rate $= \dfrac{PV}{RT} = \dfrac{5 \times 10,000}{0.730 \times 573 \times 1.8} = 66.5$ lb-mole per hr

$H = J = \frac{1}{4}, \quad M = \frac{1}{2}, \quad N = 0$ as before but $x_A = \frac{1}{4}$
and $\pi = 5$ atm

$$\frac{V_r}{66.5} = \frac{1}{(1.55 \times 10^{-4})(5)^2} \left\{ \left(\frac{1}{4}\right)\left(\frac{1}{4}\right) + \left(\frac{1}{4}\right)\left[\frac{\frac{1}{4}}{1(1 - \frac{1}{4})} \right] \right.$$
$$\left. + \frac{1}{2} \, ln \left[\frac{1}{(1 - \frac{1}{4})}\right] \right\}$$

$$V_r = \frac{66.5 \times 0.2886}{(1.55 \times 10^{-4})(25)} = \mathbf{4,960} \text{ cu ft} = \text{volume of reactor}$$

2.10 Since the vapor pressures of the two pure components are given, this question will be solved on the assumption that Raoult's law and Dalton's law hold.

Total pressure = $P = 0.50 \times 3.0 + 0.50 \times 0.9 = 1.95$ atm

Composition of the vapor phase:

Yn-butane = $1.50/1.95 = 0.769$ mole fraction
Yn-pentane = $0.45/1.95 = 0.231$ mole fraction

This question may also be solved using K constants (Thermodynamics books give p-x data).

Assume $P = 2$ atm:

	x	K	K_z	$Y = K_z/\Sigma K_z$
n-Butane	0.50	1.8	0.90	0.773
n-Pentane	0.50	0.53	0.265	0.227
			1.165	

Assume $P = 3$ atm:

	x	K	K_z	$Y = K_z/\Sigma K_z$
n-Butane	0.50	1.2	0.60	0.769
n-Pentane	0.50	0.36	0.18	0.231
			0.78	

Interpolating for $\Sigma K_z = 1.00$

$$P = 2 + (3 - 2)\ \frac{(1.165 - 1.000)}{(1.165 - 0.780)} = 2.43$$

Yn-butane $= 0.771$ mole fraction
Yn-pentane $= 0.229$ mole fraction

2.11 The conversion of SO_2 to SO_3 is a heterogeneous catalytic reaction: $SO_2 + \frac{1}{2}O_2 = SO_3$.

Since the reaction is exothermic, the percentage conversion of SO_2 to SO_3 decreases with increasing temperature and for this reason the process should be operated at the lowest temperature which gives a satisfactory rate of conversion. In order to take advantage of the high rates of conversion obtainable at high temperatures and at the same time to attain a high equilibrium conversion, which is only possible at lower temperatures, the reaction is usually carried out in two stages. The rectants enter the first converter at about 400°C and leave it at about 600°C. The reaction mixture is then cooled, entering the second converter at about 400°C and leaving it at about 460°C. Conversions of about 80 per cent are obtained in the first converter and of about 96 to 97 per cent after passage through the second converter. First-class commercial plants attain within 1 to 2 per cent of the equilibrium conversion.

The percentage conversion of SO_2 to SO_3 is favored by the presence of excess oxygen, as may be seen from the equation. Feed gases may be prepared by burning sulfur (brimstone) in air. The higher the SO_2 content of the burner gas, the lower the O_2 content. Thus, with brimstone-burner gas containing 10 per cent SO_2, the equilibrium conversion to SO_3 is 96.8 per cent at 460°C, while with 5 per cent SO_2 gas the equilibrium conversion is 97.8 per cent at this temperature. However, gas rich in SO_2 requires less gas-handling equipment and less catalyst and is therefore used in preference to the weaker gas in spite of slightly smaller conversions.

While an increase in the total pressure of the system should result in increased equilibrium conversion of SO_2 to SO_3, as may be seen by applying Le Châtelier's principle to this reaction, the effect would be so small that it would not compensate for additional costs of operation at high pressure. Hence, operation at atmospheric pressure is practiced.

Reference: Rogers, "Manual of Industrial Chemistry," Vol. I, Chap. 7, Van Nostrand, 1942.

2.12 *Reference:* Hougen and Watson, "Chemical Process Principles," 2d ed., Part 3, p. 824, Wiley, 1936.

$$kt = ln\left(\frac{n_{A0}}{n_{A0} - x_A}\right) = 2.3 \log_{10}\left(\frac{100}{(100 - 99)}\right) = 4.606$$

$$t = 4.606/0.0098 = 470 \text{ min}$$

2.13 Synthetic methods of preparation have been chosen as answers to this question in order to emphasize that the attempt to utilize elementary carbon in the production of organic compounds is an active and successful field. The hydrocarbons are still produced from petroleum and alcohol from grains and other materials, but the production of these materials from carbon is ever increasing.

1. During World War II, a large part of Germany's liquid fuels were produced by the hydrogenation of low-grade oils and tars and by the reduction of carbon monoxide by hydrogen in the presence of a catalyst. The latter process, known as the Fischer-Tropsch synthesis is an excellent example of hydrocarbon synthesis.

The raw materials are carbon monoxide produced from coal and hydrogen. Successful catalysts employed in the process are: cobalt, thoria, and kieselguhr (1:0.08:1); and iron, copper, and alkali (5:1:0.1). For a complete discussion of the process variables, refer to the current literature (e.g., Schlesinger, Cromwell, Leva, and Storch, *Ind Eng Chem*, Vol. 43, p. 1474, 1951; and Groggins, "Unit Processes in Organic Synthesis," McGraw-Hill, 1952). The products are used as liquid and gaseous fuels.

2. Synthetic methanol. The synthesis of methanol from carbon monoxide by means of the Fischer-Tropsch synthesis has been carried out in this country for 25 years. An excellent review of synthetic methanol production is to be found in *Ind Eng Chem*, Vol. 40, p. 2230, 1948. About one-half of the methanol produced is oxidized to formaldehyde. The remainder is used as antifreeze solvent, etc. The raw materials are coke or natural gas, and the catalysts consist of metal-metallic oxide mixtures. A typical catalyst is copper with oxides of zinc, chromium, manganese, and aluminum. When contact with iron is excluded, methanol of 99 per cent purity can be produced. For a discussion of process variables see Groggins, "Unit Processes in Organic Synthesis," McGraw-Hill, 1952.

3. Ketone. Acetone is a typical example of a ketone produced on a large scale. It may be produced from starch or sugar by fermentation and the products of butanol, ethanol, and acetone are separated by distillation. Acetone is also prepared from propylene obtained from cracking still gas. The propylene is treated with sulfuric acid in a packed tower. The product of the tower is heated in a lead lined tank and isopropyl alcohol is distilled. The isopropyl alcohol in the presence of a copper catalyst yields acetone.

4. Ester.

Reference: Groggins, "Unit Processes in Organic Synthesis," 4th ed., McGraw-Hill, 1952, for a complete discussion of the large-scale production of esters.

5. Metallic organic compounds. Lead tetraethyl is perhaps one of the best known organo-metallic compounds and is used in the prevention of "knocking" in internal-combustion engines.

Tetraethyl lead is produced by the interaction of ethyl chloride with a lead-sodium alloy in a dry and inert atmosphere.

2.14 *Reference:* Perry, pp. 1870–1872.

It is assumed that the plant has been properly designed so as to include the proper materials of construction, venting and ventilating systems, fire equipment, etc.

(1) An active safety- and fire-protection organization should exist in the plant. (2) Special safety-protection equipment should be available and used. (3) Special fire-protection equipment should be available and its proper use should be common knowledge in the plant. (4) Preventive maintenance should be practiced continuously.

2.15   *a.* Fats are glycerol esters of saturated and unsaturated fatty acids. Hydrolysis or splitting of the fat may be accomplished in either alkaline or acid media. The alkaline media method is described here.

$$
\begin{array}{ll}
C_zH_yCOO-CH_2 & CH_2OH \\
\quad | & \quad | \\
C_zH_yCOO-CH + 3NaOH \rightarrow 3C_zH_yCOONa + CHOH \\
\quad | & \quad\quad\quad\quad\; \text{Soap} \quad\quad\quad | \\
C_zH_yCOO-CH_2 & CH_2OH \\
\quad\; \text{Fat} & \quad\;\; \text{Glycerol}
\end{array}
$$

The fat is heated to approximately 80°C with live steam, and sodium hydroxide is added. The removal of the acid radical in the fat is a stepwise procedure. The partially hydrolyzed fat is "salted out" with sodium chloride and the procedure is repeated until the fat is completely hydrolyzed, the soap appearing in the top layer and the glycerol in the bottom layer.

b. Sodium chloride is added to "salt out" the soap.

Reference: Riegel, "Industrial Chemistry," 4th Ed., Reinhold, 1942, or Groggins, "Unit Processes in Organic Synthesis," 4th Ed., McGraw-Hill, 1952, for a description of acid hydrolysis.

2.16 $\dfrac{50,000 \text{ units} \times 0.020/100}{1,000,000 \text{ units}} = \dfrac{10 \text{ units}}{1,000,000 \text{ units}}$

Use 50/1,000 or 5 per cent of 0.020 per cent manganese by weight.

2.17 (a) Formic acid. (b) Alcohol. (c) Acid. (d) Glycerol and salts of higher fatty acids. (e) Glucose and fructose.

2.18 When resistance offered by press and cloth is negligible, $V^2 = K\theta$, where V = volume of filtrate collected in time θ and K = constant for constant pressure filtration. The volume of filtrate to be collected is directly proportional to the thickness of the cake. Hence, new time of filtration is $\theta_2 = \theta_1(V_2/V_1)^2 = 2(\frac{3}{2})^2 = 4.50$ hr. The rates of washing are in each case inversely proportional to the final volumes of filtrate. Hence, the wash rate, $R_2 = R_1(V_1/V_2) = R_1(\frac{2}{3})$. The time for washing is proportional to the volume of wash water and inversely proportional to the wash rate. Hence, $(\theta_{wash})_2 = (\theta_{wash})_1(V_2/V_1)(R_1/R_2) = (30)(\frac{3}{2})(\frac{3}{2}) = 67.5$ min $= 1.125$ hr. The new total time per cycle $= 4.50 + 1.125 + 0.500 = 6.125$ hr. The new gal per cycle $= \frac{3}{2}$(old gal per cycle). Therefore,

$$\frac{\text{New output of press, gal per hr}}{\text{Old output of press, gal per hr}} = \frac{1.5(3.00)}{1\,(6.125)} = 0.735$$

Hence, average output from press is reduced by 26.5 per cent.

2.19

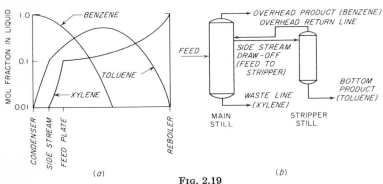

Fig. 2.19

2.20 Modern industrial chemical operations in many cases require water which is purer than the naturally occurring supply. Since in most chemical plants large quantities of process steam are used, requirements for boiler feedwater are large. Treatment of boiler water includes, when necessary, removal of solids by sedimentation, coagulation, settling, and filtration, and removal of dissolved salts by lime-soda, zeolite, and hydrogen-cation exchange methods. Large amounts of water are used in chemical process plants for cooling purposes in condensers and heat exchangers. Such waters, particularly when recirculated through cooling towers or spray ponds, may need to be treated for prevention of scale by the lime-soda, zeolite, or hydrogen-cation exchange methods and may have to be chlorinated to inhibit organic growths. Water used directly in the manufacture of chemicals should in general be clear, colorless, and free from iron, manganese, hydrogen sulfide, and organic growths. Hardness in water is generally objectionable in such industrial uses as dyeing, laundering, bleaching, paper making, tanning, etc. The lime-soda or zeolite process is usually used to remove hardness caused by calcium and magnesium salts. In the manufacture of fine chemicals and drugs distilled water has been used in the past, but the present trend is toward the use of demineralized water obtained by ion-exchange methods.

So-called temporary hardness in water is caused by dissolved calcium bicarbonate, while permanent hardness is due to dissolved noncarbonate calcium salts such as $CaCl_2$ and to magnesium salts such as $MgHCO_3$ and $MgCl_2$.

1. In the lime-soda process, slaked lime and soda ash are used to remove calcium and magnesium by precipitation:

$$Ca(HCO_3)_2 + Ca(OH)_2 = 2CaCO_3{\downarrow} + 2H_2O$$
$$Mg(HCO_3)_2 + 2Ca(OH)_2 = Mg(OH)_2{\downarrow} + 2CaCO_3{\downarrow}$$
$$+ 2H_2O$$
$$CaCl_2 + NaCO_3 = CaCO_3{\downarrow} + 2NaCl$$
$$MgCl_2 + Na_2CO_3 + Ca(OH)_2 = Mg(OH)_2{\downarrow} + CaCO_3{\downarrow}$$
$$+ 2NaCl$$

2. The zeolite softening process is a cation-exchange process in which the zeolite reacts chemically with the hard water, removing Ca and Mg ions and liberating Na ions in their stead. The process may be represented by the equation

$$Na_2Ze + CaCl_2 = CaZe + 2NaCl$$

When the zeolite is spent, it may be regenerated by treatment with brine:

$$CaZe + 2NaCl = Na_2Ze + CaCl_2$$

Zeolite-treated waters tend to be alkaline because bicarbonate hardness is transformed by the treatment to dissolved sodium bicarbonate.

3. The demineralizing process involves the use of cation and anion exchange resins which remove dissolved salts by first replacing cations with hydrogen ions, then replacing the anions with hydroxyl ion:

$$CaCl_2 + 2HR = 2HCl + CaR_2$$
$$2HCl + 2R'OH = 2R'Cl + 2H_2O$$

The demineralizing process yields the purest product but is also the most expensive method of the three. The zeolite treatment is convenient and works with a minimum amount of attention or supervision. The zeolite treatment is cheaper than the cold lime-soda process for removing noncarbonate hardness while the reverse is true in the case of bicarbonate hardness. Zeolite-treated water may in some cases be objectionable because of its usually high alkalinity and will also have a tendency to foam because of the presence of dissolved sodium salts.

The lime-soda process is not as effective in removing all hardness as the zeolite process but is cheaper with waters high in bicarbonate hardness. A combination lime-zeolite process is sometimes used, the lime to remove the bicarbonate hardness and the zeolite to remove all remaining hardness. The cost of equipment in this combination process is, however, high.

Reference: Nordell, "Water Treatment for Industrial and Other Uses," Reinhold, 1951.

2.21 *a.* The equation that expresses the liquid level Y as a function of the controller set point R is the following:

$$Y(s) = \frac{GG_vG_c}{1 + GG_vG_cH} R(s) = \frac{(1/4\pi s)(3)(6)}{1 + (1/4\pi s)(3)(6)(0.5)} R(s)$$
$$= \frac{9/2\pi}{s + (9/4\pi)} R(s)$$

b. For this case the above equation becomes

$$Y = \frac{(9/2\pi)}{s + (9/4\pi)} \left(\frac{\gamma_0}{s}\right) = \frac{9}{2\pi} \gamma_0 \left\{\frac{1}{s[s + (9/4\pi)]}\right\}$$
$$= \gamma_0 \left[\frac{1}{s} - \frac{1}{s + (9/4\pi)}\right]$$

The time-domain response $y(t) = 2\gamma_0[1 - e^{-(9/4\pi)t}]$
If $t \to \infty$, $y(t) = 2\gamma_0 = 6$ or $\gamma_0 = 3$ psi

2.22 Both columns operate with overhead vapors at the azeotrope; condensation yielding two liquid phases one butanol-rich (upper layer) and one water-rich phase (lower layer).
Heterogeneous azeotrope $\pi = 760$ mm Hg
Vapor: 25 per cent n-butanol/75 per cent water (mole basis)
Liquid butanol-rich: 40 per cent butanol/60 per cent water (mole basis)
Water-rich: 2.5 per cent butanol/97.5 per cent water (mole basis), liquid is at $\sim$200° no subcooling.
Operation of the recovery plant depends on a steady condition existing. Therefore, check separation in the decanter.
Butanol-rich layer; essentially temperature independent:

	Moles	mw	Weight	Wt per cent
Butanol	40	74	2,960	73.27
Water	60	18	1,080	26.73
			4,040	

Butanol-rich storage-tank composition will vary with time.
Water-rich layer; temperature dependent:

	Moles	mw	Weight	Wt per cent
Butanol	2.5	74	185	9.54
Water	97.5	18	1,755	90.46
			1,940	

Water-rich storage-tank composition will vary with time.
At 90°F ($\sim$110° subcooling) the 92.3 wt per cent water is a possibility. Since indicated flow sketch is inconsistent with separation achieved in the decanter, the entire flow arrangement needs to be modified. Suggested modification is to return decanted streams directly to the respective columns onto the top stage along with the fresh feeds from the storage/holdup tanks.

2.23   The reactants in the Fischer-Tropsch synthesis of motor fuels are carbon monoxide and hydrogen. This synthesis gas may be prepared from coal, coke, natural gas (i.e., CH_4), etc.
The principle reactions, which are a function of
  1. Temperature (200–300°C)
  2. Pressure (1–25 atm)
  3. Type of catalyst (CO, Ni, Fe, or Cu with alkali oxide promotors)
  4. Carbon monoxide–hydrogen ratio
  5. Purity of synthesis gas
are of two general types:

$$a\mathrm{CO} + \frac{(2a + b)\mathrm{H}_2}{2} \rightarrow \mathrm{C}_a\mathrm{H}_b + a\mathrm{H}_2\mathrm{O}$$

$$2a\mathrm{CO} + \frac{b\mathrm{H}_2}{2} \rightarrow \mathrm{C}_a\mathrm{H}_b + a\mathrm{CO}_2$$

The products of the reaction vary in chain length from 1 to 40 or more carbon atoms and a variety of saturated and unsaturated products. The principle by-products are alcohols and ketones.

2.24   *a.* In order to concentrate a solution of a material whose solubility decreases with increase in temperature, the latent heat of vaporization for the solvent must be supplied at reduced temperatures. This can be done by operating at reduced pressures, i.e., vacuum evaporators. Care must be taken in designing the heating unit so that there is not an accumulation of solute at this point. A vertical-tube, forced-circulation evaporator is indicated (see Perry, p. 507). If multiple effects are used, forward feed should be used.
  b. (See Perry, p. 1061) In cases where solutes have a large positive temperature coefficient of solubility, cooling without evaporation can produce enough supersaturation to yield crystallization. Typical equipment for this type of operation are Wulft-Bock crystallizers, Swenson-Walker crystallizers, agitated batch crystallizers, etc.
  c. Sodium chloride has a very small positive temperature coefficient of solubility and is typical of solutes in this class. In this case cooling has very little influence and crystallization

is dependent on evaporation. Crystallizing evaporators are used.

d. Use a selfpriming centrifugal pump. If temperature is high, consideration of cooling the packing is important.

e. Use vacuum or vacuum freeze driers and spray driers.

f. Use an exchanger which produces high linear rates of flow on the viscous fluid side and which permits easy cleaning.

g. Use a vacuum still which permits distillation at "safe" temperatures.

h. The drier must be supplied with adequate means for controlling drier temperature and humidity. A tunnel-vertical-turbo type is suitable for lumber. A recirculating drier control system is described in Perry, p. 1335.

2.25 *a.* The weight of sulfur conveyed from the storage pile to the melting tank may be controlled automatically by a Harding weigh feeder, a Schaeffer belt system, or a Jeffrey vibrating feeder if the process is continuous. An automatic hopper scale or conveyor-type batching scale may be used in a batch operation. (See Perry, p. 1293.)

b. A two-control valve system for the heating tank is described in Perry, p. 1330. A control system for a continuous process is described in Perry, p. 1334.

c. A ratio flow-control system is described in Perry, pp. 1333 and 1334.

2.26 *a.* Iron and aluminum removal is costly since the precipitate of the iron and aluminum ion drags down, by adsorption, appreciable quantities of phosphate which must be reclaimed. Finely ground limestone is added to the phosphoric acid. Soluble CO_2HPO_4, CaH_2PO_4, and a precipitate of iron and aluminum phosphate and calcium fluoride are formed. The precipitate must be washed thoroughly to remove the soluble phosphates. The wash water is added to the acid solution of CaH_2PO_4 and pure sulfuric acid is added with the resulting precipitate of $CaSO_4$. After the removal of the $CaSO_4$, the acid solution is concentrated in a vacuum evaporator or a spray chamber.

b. There are three general methods of defluorination of phosphate rock to produce tricalcium phosphate, which is used as an ingredient in animal food products. The chief difference in the two methods which employ SiO_2 and steam is in the heat treatment. One process involves fusion and the other does not. The principle reaction is $Ca_{10}F_2(PO_4)_6 + H_2O + SiO_2 \rightarrow 3Ca_3(PO_4)_2 + CaSiO_3 + 2HF$. For a complete discussion, refer to Wagga-man, "Phosphoric Acid, Phosphates, and Phosphate Fertilizers," p. 390, Reinhold.

   c. $Na_3PO_4 \cdot 12(H_2O)$ is used as an internal "softening agent" in cases where the boiler acidity and/or the natural hardness of the water is relatively low. Phosphates when added internally produce a flocculant precipitate of the scale-producing ions, which is easily removed in boiler blowdown. The principle reactions for the removal of scale-producing ions are:

$$3Ca(HCO_3)_2 + 3Na_3PO_4 \rightarrow Ca_3(PO_4)_2\downarrow + 6NaHCO_3$$
$$3MgSO_4 + 2Na_3PO_4 \rightarrow Mg_3(PO_4)_2\downarrow + 3Na_2SO_4$$
$$FeCl_3 + Na_3PO_4 \rightarrow FePO_4\downarrow + 3NaCl$$

The phosphates have an added advantage in that they control the pH of the boiler water.

2.27   24,000 gal water is equivalent to $24,000 \times \dfrac{62.4}{7.5} \times \dfrac{1}{2,000}$ or 100 tons water.

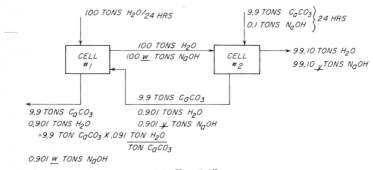

Fig. 2.27

Let w = equilibrium concentration of NaOH in cell #1 in tons NaOH per ton H_2O
Let y = equilibrium concentration of NaOH in cell #2 in tons NaOH per ton H_2O
Material balance of NaOH in cell #1

$$0.901w + 100w = 0.901y$$

Material balance of NaOH in cell #2

$$0.901y + 99.10y = 0.1 + 100w$$

(1) $100.901w = 0.901y$ or $100w = 0.89y$
(2) $100.001y = 0.1 + 100w = 0.1 + 0.89y$
(3) $98.91y = 0.1$ and $y = 1.01 \times 10^{-3}$ tons NaOH per ton H_2O

Therefore $w = 9 \times 10^{-6}$ tons NaOH per ton H_2O

Therefore, since each ton of $CaCO_3$ carries 0.091 tons of water and each ton of water carries 9×10^{-6} ton NaOH, the product will contain per ton $CaCO_3$

1 (ton $CaCO_3$) $\times$ 0.091 (ton H_2O per ton $CaCO_3$) $\times$ 9×10^{-6} ton NaOH per ton H_2O = 8.19×10^{-7} ton NaOH

2.28 The following symbols will be used for the solution of the problem. (*Reference:* William H. McAdams, "Heat Transmission," 3d ed., chap. 9, McGraw-Hill, 1954.

S = cross section of stream in a tube, sq ft
G = mass velocity, lb/(hr)(sq ft)
q = heat-transfer rate, Btu per hr
U_0 = over-all coefficient of heat transfer between two streams based on outside area, Btu/(hr)(sq ft)
A_0 = area for heat transfer at outside radius
N_{Re} = Reynolds number = DG/μ, μ = velocity of fluid
k = thermal conductivity of fluid
h = coefficient of heat transfer between fluid and surface, Btu/(hr)(sq ft)
T = absolute temperature
t_1, t_2 = temperature at stations 1 and 2, respectively
t_w = temperature of wall
i = as subscript means isothermal
w = mass rate of flow per tube, lb fluid per hr
c_p = specific heat at constant pressure, Btu/(lb)(°F)
S = 0.00413(13) = 0.0537 sq ft, $G = 60,000/0.0537 = 1.118 \times 10^6$ lb/(hr)(sq ft)
$q = wc_p(t_2 - t_1) = UA\Delta T_L$ $A_0 = 0.2618(52)(6) = 81.6$ sq ft

$$U_0 = \cfrac{1}{\cfrac{1}{2,000} + \cfrac{1}{0.935} \cfrac{(0.065/12)}{220} + \cfrac{1}{0.87 h_i}} = \cfrac{1}{0.000526 + \cfrac{1.15}{h_i}}$$

$$N_{\mathrm{Re_{min}}} = \frac{(0.87/12)(1.118 \times 10^6)}{2.42} = 33,500, \text{ so turbulent.}$$

Assume $t_2 = 100$, $t_{av} = 80°$, $\mu = 0.862$ cp, $k = 0.35$.

Neglecting the μ/μ_w term,

$$h_{i(\text{approx.})} = \frac{0.023(0.35)}{0.87/12} \left(\frac{33,500}{0.862}\right)^{0.8} \left(\frac{2.42 \times 0.862}{0.35}\right)^{0.33} = 944$$

Therefore $U_0 = 573$.

$$q = 60,000(t_2 - 60) = 573(81.6) \left\{\frac{t_2 - 60}{\ln[(227 - 60)/(227 - t_2)]}\right\};$$
$$t_2 = 151°, \text{ so no good.}$$

Try $t_2 = 160°$, $t_{av} = 110°$, $\mu = 0.62$, $k = 0.367$.
Now include $(\mu/\mu_w)^{0.14}$; $t_w \approx 192°F$, $\mu_w = 0.32$.

$$h_i = \frac{0.023(0.367)}{0.87/12} \left(\frac{33,500}{0.62}\right)^{0.8} \left(\frac{2.42 \times 0.62}{0.367}\right)^{1/3} \left(\frac{0.62}{0.32}\right)^{0.14} = 1,267$$
$$U_0 = 697, \quad q = 60,000(t_2 - 60)$$
$$= 697(81.6) \left\{\frac{t_2 - 60}{\ln[(227 - 60)/(227 - t_2)]}\right\}$$

The above equation yields $t_2 = 162°F$, close enough.

2.29   The following symbols will be used for the solution of the problem. (*Reference:* William H. McAdams, "Heat Transmission," 3d ed., chap. 6, McGraw-Hill, 1954.

P = absolute pressure (intensity of), lb$_F$ per sq ft
V = average velocity, fps
f = friction factor, dimensionless, in Fanning equation
N_{Re} = Reynolds number, dimensionless
ρ = absolute viscosity of fluid, lb/(sec)(ft)
H = skin friction

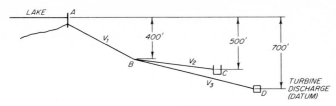

FIG. 2.29

$$P_D = (95.3)(144) = 13{,}720 \; \text{lb}_F/\text{sq ft}$$
$$P_A = \text{atm} = 2{,}115 \; \text{lb}_F/\text{sq ft}$$
$$V_1 = V_2 + V_3$$

$A \to B:$
$$\frac{2{,}115 - P_B}{62.3} = (300 - 700) + H_1$$

$B \to C:$
$$\frac{P_B - 2{,}115}{62.3} = (200 - 300) + H_2$$

$B \to D:$
$$\frac{P_B - 13{,}720}{62.3} = (0 - 300) + H_3$$

$$H_1 = \frac{4f_1(1{,}000)V_1^2}{(3.826/12)64.4}; \quad H_2 = \frac{4f_2(8{,}000)V_2^2}{(3.826/12)64.4}; \quad H_3 = \frac{4f_3(3{,}000)V_3^2}{(3.826/12)64.4}$$

f values are functions of N_{Re}, $N_{Re} = \dfrac{DV\rho}{\mu}$

Solve by trial and error using tables for values of f.

$$N_{Re} = \frac{(3.826/12)(62.3)V}{0.000672} = 29{,}500V$$

Solution gives $V_1 = 18$ fps, $V_2 = V_3 = 9$ fps

$N_{Re_1} = 511{,}000$, $f_1 = 0.004$
$N_{Re_2} = 268{,}500$, $f_2 = 0.0045 = f_3$, $P_B = 9{,}000 \; \text{lb}_F/\text{sq ft}$
gal per min $= 18(\pi/4)(3.826/12)^2(7.48)(60) = $ **645**   c

2.30   The following symbols will be used for the solution of the problem. (*Reference:* William H. McAdams, "Heat Transmission," 3d ed., chaps. 6 and 15, McGraw-Hill, 1954.)

$a_0 = $ constant, dimensionless
$B_0 = $ correction factor

D_i, D_o = inside and outside diameters of tube, ft

y_o = clearance, ft, between outer surfaces of tubes in a bundle, taken to correspond to the minimum free area

μ_o = viscosity for fluid outside tubes

m = exponent, dimensionless, in dimensionless equations

K = dimensional term

g_c = conversion factor in Newton's law equal to 4.17×10^8 (lb fluid)(ft)/(hr)(lb$_F$)

ρ_o = density of fluid for fluid outside tubes, lb per cu ft

G_{oo} = mass velocity for optimum velocity outside tubes, lb/(hr)(sq ft of cross section)

G = mass velocity, same units as G_{oo}

t = bulk temperature of fluid, °F

S = cross section normal to flow of fluid, sq ft

q = quantity of heat transfer, Btu per hr

A = area of heat-transfer surface, sq ft

$$a_0 = 0.25 + \frac{0.1175}{1.85^{1.08}} = 0.311$$

$$K_0' = K_0 \frac{D_i}{D_0} = \frac{2B_0 a_0 y_0 \mu_0{}^m}{\pi g_c \rho_0{}^2 D_0{}^{m+1}}$$

$$= \frac{(2)(0.311)(1.95/12)(0.0455)^{0.15}}{(3.14)(4.17 \times 10^8)(0.072)^2(1.05/12)^{1.15}}$$

$K_0' = 1.542 \times 10^{-7}$

$$G_{00} = \left[\frac{9 \times 10^{-5}}{(7.55 \times 10^{-9})(1.542 \times 10^{-7})} \times \frac{1}{3.75} \right]^{0.351}$$

$$= 4{,}225 \text{ lb/(hr)(sq ft)}$$

$\dfrac{\Delta t_m}{\Delta t_0} = 1.0$

Check on $\dfrac{D_0 G_{\max}}{\mu} = \dfrac{(1.05/12)(4{,}225)}{0.0455} = 8{,}120$ OK

$S_{\min} = \dfrac{w}{4{,}225} = \dfrac{64{,}000}{4{,}225} = 15.15$ sq ft.

If N = number of tubes in one row, then

$$S_{\min} = \left[\frac{(3N + 2.55) - 1.05N}{12} \right] 10 = 15.15, \text{ and } N = 8 \quad \textbf{1c}$$

$$h_0 = \frac{(0.33/1.25)(0.25)(4{,}225)}{(8{,}120)^{0.4}(0.74)^{2/3}} = 9.35 = U_0$$

$$\Delta t_L = 100/\ln\left(\frac{175}{75}\right) = 118.3°\text{F}$$

$q = 64,000(0.25)(100) = 9.35A\,(118.3)$, and $A = 1{,}440$ sq ft

Let N' equal total number of tubes.

$$1{,}440 = \pi\left(\frac{1.05}{12}\right)N'(10); \quad N' = 526$$

$$\frac{526}{8} = \mathbf{66}\text{ rows} \qquad \mathbf{2d}$$

2.31 *a.* Solvent recovery by adsorption of solvent vapors on activated adsorbing agents operates efficiently and safely for the recovery of solvent vapors over a wide range of concentrations of solvent in solvent-laden air.

The solvent-laden air is removed from the drying operation by means of a suction blower. The air is usually passed through a dust-removal device and then to adsorbers.

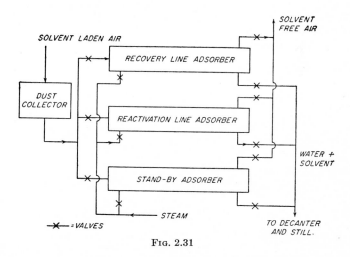

FIG. 2.31

The solvent-laden air then passes to an adsorber where the solvent is selectively adsorbed. When the adsorbing agent becomes "saturated" the solvent-laden gas is directed to a recently activated adsorber. Steam is directed to the "saturated adsorber" and the solvent and water are sent to decanters and stills.

b. Other methods of solvent recovery to be considered are:
1. Scrubbers
2. Compression and cooler
References: Perry, "Chemical Engineers' Handbook," 3d ed., McGraw-Hill, 1950.
Robinson, "The Recovery of Vapors," Reinhold, 1942.

2.32 1. Assume adiabatic operation.

$$Q = 0 = \Delta H;\ \text{energy balance around condenser}$$

Let t = temp. leaving condenser. Ref. 60°F.

$$0 = (100)(0.5)(t - 60) + 27.5(0.909)(0.4)(200 - 60)$$
$$+ 27.5(0.091)(1)(200 - 60) - 27.5(0.909)[0.3(300 - 260)$$
$$+ 100 + 0.4(260 - 60)] - 27.5(0.091)[0.5(300 - 212)$$
$$+ 970 + 1.0(212 - 60)]$$

The above equation yields t = **180°F**   **1b**

2. Energy balance around cooler. Ref. 65°F. Let w = lb per hr of water. The stream entering cooler is

$$(48{,}000\ \text{lb per day}) \left(\frac{27.5}{100}\right)\left(\frac{1}{24}\right) = 550\ \text{lb per hr}$$

$$Q = 0 = \Delta H$$

$$0 = 550(0.909)(0.4)(100 - 65) + 550(0.091)(1)(100 - 65)$$
$$+ w(1)(130 - 65) - 550(0.909)(0.4)(200 - 65)$$
$$- 550(0.091)(1)(200 - 65)$$

The above equation yields w = **385** lb per hr   **2d**

2.33 The following symbols will be used for the solution of the problem. (*Reference:* William H. McAdams, "Heat Transmission," 3d ed., chap. 6, McGraw-Hill, 1954.)

D_p = diameter of packing, average

G_o = The superficial mass velocity based on the cross section of the empty tube

ϵ = fraction voids, dimensionless

V = average velocity, fps

ρ = density, lb fluid per cu ft

f' = friction factor, dimensionless

$$D_p = \frac{6}{76} = 0.079\ \text{ft}$$

$$V_s = \frac{1,000 \text{ cfm}}{60 \text{ sec/min } (0.785)(4)^2} = 1.333 \text{ fps}$$

$$\rho_m = \left(\frac{29}{359}\right)\left(\frac{492}{530}\right) = 0.075 \text{ lb per cu ft, } \mu = 0.018 \text{ cp}$$

$G_o = V_s\rho_m$; substitute the above values in

$$\frac{D_pG_o}{\mu(1-\epsilon)} = \frac{(0.079)(1.333)(0.075)}{(0.018)(0.000672)(1-0.69)} = 2,110$$

$$f' = 1.75 + \frac{150}{2,110} = 1.82$$

$$\Delta P_f = \frac{f'LG_o^2(1-\epsilon)}{\epsilon^3 g_c D_p\rho_m} = \frac{1.82(30)[(1.333)(0.075)]^2(1-0.69)}{(0.69)^3(32.2)(0.079)(0.075)}$$

$$= 27.1 \text{ lb}_F \text{ per sq ft}$$

$[27.1 \text{ lb}_F \text{ per sq ft}/62.35 \text{ lb}_F \text{ per cu ft}]12 = $ **5.21 in. H_2O**   a

2.34   Flow rate, lb per hr $= \left(\dfrac{70,000 \times 60}{359}\right)\left(\dfrac{492}{710}\right)\left(\dfrac{1.1}{1.0}\right) 29.1 =$

260,000 lb per hr

$$G_s = \text{lb/(hr)(sq ft)} = \frac{260,000}{(\pi D^2)/4} = \frac{331,000}{D^2}$$

where D is the diameter in feet.

Tower height $= (0.32G_s^{0.18})(15)$ ft

Tower volume $=$ (height)(cross-sectional area) $=$ (height)$\left(\dfrac{\pi}{4}D^2\right)$

$$= 0.32G_s^{0.18}(15)\left(\frac{260,000}{G_s}\right) \text{ cu ft}$$

Tower cost dollars per hr (fixed cost) $=$ (volume)\$1.0(0.20)$\dfrac{1}{8,000}$

$$= \frac{0.32 \times 15 \times 260,000}{G_s^{0.82}}\left[(1)(0.2)\frac{1}{8,000}\right] = \frac{31.2}{G_s^{0.82}}$$

Total cost per hr $=$ operation cost per hr $+$ fixed cost per hr

$$= \left(1.8G_s^2 \times 10^{-8} + \frac{81}{G_s} + \frac{4.8}{G_s^{0.8}}\right) + \frac{31.2}{G_s^{0.82}}$$

$$\frac{d}{dG_s} \text{ (total cost)} = 0 = 3.6 \times 10^{-8}G_s - \frac{81}{G_s^2} - \frac{3.84}{G_s^{1.80}} - \frac{25.6}{G_s^{1.82}}$$

$$= 3.6 \times 10^{-8}G_s^3 - 3.84G_s^{0.20}$$

$$- 25.6G_s^{0.18} = 81$$

By trial and error G_s = 1,760 lb/(hr)(sq ft). Therefore, the desired values are

$$\text{Diameter} = \sqrt{\frac{331,000}{1,760}} = \textbf{13.8 ft} \qquad \textbf{1d}$$

$$\text{Height} = (0.32)(1,760^{0.18})(15) = \textbf{18.5 ft} \qquad \textbf{2a}$$

2.35 This a trial-and-error solution. As a first approximation, assume that the theoretical deliverabilities of the parallel lines are in the ratio to their diameters:

$$\frac{Q_{22}}{Q_{20}} = \left(\frac{22}{20}\right)^{2.67} = 1.29$$

The actual deliverabilities will be in the ratio:

$$\frac{Q'_{22}}{Q'_{20}} = 1.29 \left(\frac{0.88}{0.92}\right) = 1.233$$

This gives an estimated actual flow of

$$Q'_{22} = (1.233/2.233) \times 6 \times 10^6 \text{ per hr} = 3,310,000 \text{ cu ft per hr}$$

for the 22-in. line, and

$$Q'_{20} = (1.000/2.233) \times 6 \times 10^6 \text{ per hr} = 2,690,000 \text{ cu ft per hr}$$

for the 20-in. line. The theoretical flow, assuming smooth pipes, is then computed as

$$Q_{22} = 3,310,000/0.88 = 3,760,000 \text{ cu ft per hr, and}$$
$$Q_{20} = 2,690,000/0.92 = 2,930,000 \text{ cu ft per hr}$$

The Weymouth equation is

$$(P_1)^2 - (P_2)^2 = \frac{fLW^2zRT}{Dg_cMA^2}$$

where P_1 and P_2 are the inlet and outlet pressures in psf (abs), f is the friction factor for smooth pipe, L is the length of the pipe line in feet, W is the theoretical rate of flow in pounds per second, z is the compressibility factor, R is the universal gas constant, T is the absolute temperature, D is the internal diameter of the pipe in feet, g_c is 32.17 ft per sec², M is the molecular weight of the gas in pound-seconds per cubic feet, and in this case equals 0.65 × 29, and A is the cross section of the pipe in square feet. The Weymouth equation is used to compute the pressure drop in each line. If the calculated values of P_2 for the two lines agree within reason, an average P_2 value may be taken as the answer. If there is no close agreement of the calculated values of P_2, the assumed ratio of Q_{22}/Q_{20} is readjusted and the calculations repeated.

2.36 Approximate calculations for the amount of heat transfer: Temperature of steam = 228°F, and temperature in condenser = 101°F. Assume forward feed and boiling-point elevations of 3°F in the first effect and 6°F in the second effect, giving temperatures estimated to be 170°F for boiling point in first effect and 107°F for boiling point in second effect.

Pounds of water evaporated per pound of NaOH
$$= (^{90}\!/_{10} - ^{84}\!/_{16}) = 3.75$$

Heat transferred for vaporization in the two effects
$$= 3.75 \times \frac{(997 + 1,033)}{2} = 3,800 \text{ Btu per lb NaOH}$$

Heat required to preheat feed if admitted to first effect = $9(170 - 70) = 900$ Btu per lb NaOH

It is seen that approximately 24 per cent of the heat-transfer requirement is for preheating the feed to its boiling point. Accordingly, it is recommended that for a 10 per cent increase in production available heat exchangers be used to preheat the feed to about 160°F by means of steam bled from the vapor line leaving the first effect. The evaporators are to be operated as a double effect with forward feed. It is estimated that the resultant increase in capacity will be well over 10 per cent and that there will be no loss in steam economy as compared with normal operation with backward feed.

In order to handle a temporary 50 per cent overload, it will be necessary to operate the two evaporators in parallel, using the regular condenser on one of the effects and the spare condenser on the other one. High-pressure steam (5 psi) will be admitted to the steam chest of each evaporator and the feed will enter both effects in parallel. If the capacity of the spare condenser is insufficient to give a 50 per cent increase in production, then available heat exchangers may be used to preheat the feed, thus taking some load off the evaporators. High-pressure steam will have to be used for preheating the feed in the heat exchangers. Since high-pressure steam will be the heating medium in both evaporators and in the heat exchangers, there will be a resultant loss in steam economy which can be tolerated on a temporary overload basis.

2.37 *Reference:* Walker, Lewis, McAdams, and Gilliland, "Principles of Chemical Engineering," pp. 415–416, McGraw-Hill, 1937.

$$\frac{1}{V^2} = a + b\theta \text{ and } a = \frac{1}{(225)^2} \text{ and } \theta = 160$$

$$\frac{1}{(50)^2} = \frac{1}{(225)^2} + 160b \text{ and } b = 2.38 \times 10^{-6}$$

$$V_{av}\theta = \frac{Q}{A\Delta t} = \frac{2\,(\sqrt{a + b\theta} - \sqrt{a})}{b}$$

a. It is assumed that the evaporator is operated for 160 hr and cleaned for 8 hr, both each week. It follows that on substituting values for "*a*" and "*b*,"

$$V_{av} = \frac{Q}{A\Delta t(\theta - \theta_c)} = 1.27 \times \frac{10^4}{168} = 77.7 \text{ Btu/(hr)(sq ft)(°F)}$$

b. Let time of operation = θ
    Let time of cleaning = θ_c

    It is required that $\dfrac{Q}{(\theta + \theta_c)}$ be a maximum.

Hence $\dfrac{d}{d\theta}\left[\dfrac{Q}{(\theta - \theta_c)}\right] = \dfrac{d}{d\theta}\left[2A\Delta t\dfrac{(\sqrt{a + b\theta} - \sqrt{a})}{b(\theta + \theta_c)}\right] = 0$

When $\dfrac{d}{d\theta}\left[\dfrac{\sqrt{a + b\theta} - \sqrt{a})}{(\theta + \theta_c)}\right] = 0$, then $\theta = \theta_c + 2\sqrt{\dfrac{a\theta_c}{b}}$

Substituting values of a, b, and θ_c, $\theta = 24.3$ hr.

This time corresponds to a value for $V_{av} = \dfrac{2\sqrt{b\theta_c}}{b(\theta + \theta_c)}$
= 113 Btu/(hr)(sq ft)(°F) as compared with $V_{av} = 77.7$ in part *a.*

Accordingly, for maximum capacity the evaporator should be cleaned after 24.3 hr of operation, corresponding to a total time for the cycle of 32.3 hr. It is suggested that the evaporator be operated during three shifts and cleaned during the fourth shift, corresponding to a 32-hr cycle.

2.38 *a.* Plate and frame presses are suitable for batch operation with both compressible and noncompressible sludges. This press is the cheapest per unit filtering surface and requires the smallest floor space, but involves a large amount of hand labor. This press is best adapted to sludges which either contain small amounts of solids or in which the solids are highly valuable, as in the manufacture of dyes.

b. Rotary continuous filters are used where the process is continuous, where large amounts of solids must be handled, and where labor costs must be kept low. These filters are well adapted to homogeneous sludges of high solid concentrations and low compressibility, which require little washing and do not require high pressures.

c. The Sweetland press is a pressure leaf filter which involves batch operation and has high washing efficiency, the wash water displacing the filtrate when admitted into the press. This filter is suitable for handling slurries when the filtrate is valuable and dilution with wash water is to be held to a minimum.

d. The Moore filter is an intermittent vacuum filter adapted to batch operation. Vacuum is applied to the inside of the leaves during filtration, while compressed air is applied to discharging the cake. The filter basket is moved from filtering tank to washing tank and dump tank with a traveling crane. This filter is suitable for sludges which do not easily settle out and which form an adherent cake. It may be used with compressible sludges provided that a high rate of filtration is not required.

2.39 The problem is the application of the phase rule in metallurgy. Precipitation hardening is applied to alloy systems in which there is a decrease of solid solubility in solid solutions with decreasing temperature.

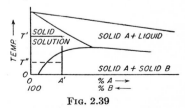

Fig. 2.39

If an alloy whose composition is represented by A' is heated to a temperature T', a single phase of solid solution results.

If this solution is cooled suddenly to T'', the solid solution will persist but will be unstable with respect to precipitation of component B (i.e., the solid solution is supersaturated with respect to component B). The precipitation of component B distorts the space lattice and interferes with the normal slip of the alloy. This in turn hardens and strengthens the alloy. There is an optimum value in this process, beyond which the strength and hardness of the alloy are decreased. If this process takes place at room temperature or is the result of cold working, the process is known as age hardening.

The two terms—precipitation hardening and age hardening— have currently become synonymous. The "Metals Handbook," ASM, however, still gives the following definitions: 1. "Precipitation hardening is a process which increases the hardness of an alloy by the controlled precipitation of a constituent from a supersaturated solid solution"; and 2. "Age hardening is a process that increases hardness and strength and usually decreases ductility. The process usually involves rapid cooling or cold work."

2.40   In checking the efficiency of combustion of flue gases it is desirable to ascertain the CO_2, CO, and O_2 content of the flue gases. All three can be determined by means of a manually operated Orsat analyzer which works by chemical absorption of each gas in a separate unit. A commercial automatic Orsat is manufactured which measures the CO_2 content only. Thermal-conductivity analyzers are in use which take advantage of the fact that the conductivity of CO_2 is considerably less than that of air. In one type of thermal-conductivity meter the CO_2 content is first determined. The sample is then passed over a heated carbon rod to burn CO to CO_2, and the CO_2 content is then redetermined. A thermal-flow meter is available which takes advantage of the magnetic property of oxygen and determines the oxygen by measuring the thermal conductivity of the gas sample in a magnetic field. Thermal conductivity methods are not suitable when components of unknown thermal conductivity are present in the gas mixture, and in any case require calibration for the type of gas mixture tested. They have the advantage of being fully automatic. The CO_2 content may also be determined by an automatic and recording instrument measuring the density of flue gases. Such instruments have no provision for the determination of the CO or O_2 content.

2.41 Instruments for the measurement of temperature are:

1. Liquid-in-glass thermometers, which depend on the expansion of the liquid with rise in temperature. Ranges covered are from $-40°F$ to $1000°F$. Use is limited by fragility. Exposed stem corrections and radiation corrections may be necessary to give accurate results. Accuracy obtainable may be as high as $\pm 0.01°F$ with calibrated instruments.

2. Pressure-type thermometers, in which the pressure of a liquid, gas, or vapor is used to distort a spiral or helix. These instruments are suitable in ranges from $-40°F$ to $1200°F$. These instruments are simple and easily installed, but accuracy is limited and varies from $\pm 1°F$ to $\pm 10°F$.

3. Thermocouples, which may be used for measuring temperatures in the range of $-250°F$ to $2800°F$. When used in conjunction with a millivoltmeter, precautions should be taken against changes in the resistance of the thermocouple wires with use. If the indicating instrument is a potentiometer, changes in the resistance of the thermocouple wires will not affect readings. Proper compensation must be made for any variation in the cold-junction temperatures. These instruments are capable of high accuracy when calibrated.

4. Resistance thermometers, in which the electrical resistance of a coil is measured by means of a wheatstone bridge. These instruments are useful in the range of from $-250°F$ to $1500°F$. They are fragile and may lose calibration as a result of contamination or volatilization of coil material. When properly calibrated, these instruments are capable of high accuracy, to better than $\pm 0.1°F$. Contact resistances must be avoided since they make readings unreliable.

5. Optical pyrometers, which are used for measurement of high temperatures by visually matching the intensity of monochromatic light emitted by the heat source against that from a calibrated constant source. They are useful for temperatures above $1200°F$. Commercial instruments are accurate to $\pm 10°F$. Correction for emissivity is necessary when sighting upon materials in the open. These instruments cannot be made to record temperature automatically.

6. Radiation pyrometers, which measure the total intensity of radiated light and heat. The radiation is absorbed by a battery of thermocouples in series and the electromotive force generated is a measure of the temperature of the heat source. They are useful for temperatures above 1200°F, are less accurate than optical pyrometers, and are more subject to error due to accumulation of dust and dirt and in the presence of cold strata of absorbing gases. To obtain correct readings with radiation pyrometers the diameter of the source must exceed the minimum required by the geometry of the instrument. Corrections for emissivity must be made when sighting on material in the open.

2.42 Four methods of removing dust from a gas:

1. Cyclone separators—most widely used dust-collection equipment. The dust-laden gas enters a cylindrical or conical chamber tangentially and leaves through a central opening. Cyclones are applicable for removal of solid or liquid particles of 5 to 200 microns in diameter. Cyclones are designed for an inlet velocity of about 50 ft per sec. Friction loss through cyclones may range from 1 to 20 inlet velocity heads. A cyclone may be placed on either the suction or pressure side of a fan.

2. Cloth collectors (bag filters)—dust-laden gases are passed through a woven fabric which filters out the dust. Such filters have a pressure drop of 2 to 6 in. of water and are rated at 1 to 8 cu ft/(min)(sq ft) of cloth area. In operation it is essential that the gas be kept above its dew point to avoid plugging of the bag pores.

3. Scrubbers—a liquid is employed in this equipment to assist in the removal of dust. The equipment may take the form of spray chambers, water jets, cyclones, packed towers, and mechanical liquid disintegrators. Dust particles down to diameters of 5 microns may be handled. Depending on the type of scrubber used, pressure drops may amount to from 0.1 to 8 in. of water. Water consumption may range between 3 and 100 gal/(min)(1,000 cu ft) of gas handled.

4. Electrical precipitators—best adapted to electrically conducting dusts. Electrical power consumption is in the range of 0.2 to 0.6 kw/(1,000 cu ft)/(min) of gas handled. Very high collection efficiencies may be obtained.

For the given case of a gas entering at 95°C and with a dew-point of 70°C, a chamber spray scrubber is recommended. The humidification of the gas by the spray will tend to flocculate the finer particles. Since the gas will be cooled below its dew point by the spray, condensation will take place on the dust particles, increasing their effective size. The dust is then eliminated by impingement against baffles. This method will involve low first cost and low operating cost.

2.43   See Perry, p. 844.

Initial mixture in pounds alcohol per pound air

$$0.05 \times 46 = 2.30 \text{ lb alcohol}$$
$$0.95 \times 29 = 27.60 \text{ lb air}$$
$$\therefore H = \frac{2.3}{27.6} = 0.0833$$

Hs for final air:

$$Hs = \frac{60}{\{[(114.7/14.7) \times 760] - 60\}} \times \frac{46}{29} = 0.0162 \frac{\text{lb alcohol}}{\text{lb air}}$$

and $\dfrac{0.0833 - 0.0162}{0.0833} \times 100 = 80.6$ per cent recovery of ethanol.

2.44 $6{,}000 \times 500 \times \dfrac{144}{163} \times \dfrac{1}{33{,}000 \times 60 \times 0.25} = 5.4 \text{ hp}$

$6{,}000 \times (4{,}200 - 500) \times \dfrac{144}{163} \times \dfrac{1}{33{,}000 \times 60 \times 0.30} = 33.0 \text{ hp}$

Total power $= \mathbf{38.4 \text{ hp}}$

2.45   *a.* Use aluminum or chromium alloys (over 18 per cent Cr) for cold acid where strength is over 68 per cent.

     b. Use iron or steel, or use monel or nickel when contamination must be avoided.

     c. Use iron or steel when the benzol is water and acid free.

     d. Use ordinary cast iron when slight contamination is permissible. Use monel metal or nickel when contamination must be avoided.

     e. Use iron or steel when there is less than 20 per cent water and when the H_2SO_4 is greater than 15 per cent. Use high-silica cast iron when the water content is between 20 and 30 per cent.

Use lead when the water content of the mixture is greater than 30 per cent.

2.46   Basis of solution: 1 sq ft and capacity of 1 in. cake on leaf.

1.
$$K'\theta = \frac{V^2}{A^2} = K' \times 5 = \frac{(3.8)^2}{(1)^2}$$
$$K' = 2.89$$

2. Differentiating equation 1, the rate of filtration at a given time is obtained

$$\frac{dV}{d\theta} = \frac{2.89A^2}{2V}$$

3. The cake thickness will be described in terms of t. When $t = 2$, then the cake thickness will be $1/t$ or one-half the thickness of the original cake and the volume of the filtrate delivered will be one-half the value delivered at $t = 1$.

Time Schedule for Leaf Press

Cycles, number	Cake thickness, in.	Time, hr				
		Filter cycle	Wash cycle	Clean	Total time	Time (n) cycles
1	1	5	5	0.5	10.5	10.5
2	$\frac{1}{2}$	1.25	1.25	0.5	3	6
3	$\frac{1}{3}$	0.549	0.549	0.5	1.6	4.8
4	$\frac{1}{4}$	0.312	0.319	0.5	1.12	4.48
5	$\frac{1}{5}$	0.200	0.200	0.5	0.9	4.5
6	$\frac{1}{6}$	0.139	0.139	0.5	0.78	4.68

Sample calculations:

1 cycle ($V = 3.8$ cu ft):

$$\text{Wash rate} = \frac{2.89 \times 1^2}{2 \times 3.8} = 0.38 \text{ cu ft per hr}$$
$$\text{Wash time} = \frac{3.8/2}{0.38} = 5 \text{ hr}$$

2 cycles ($V = 1.9$ cu ft):

$$\text{filter time} = \theta = \frac{(1.9)^2}{1^2 \times 2.89} = 1.25 \text{ hr}$$

$$\text{wash rate} = \frac{dv}{d\theta} = \frac{2.89 \times 1^2}{2 \times 1.9} = 0.76 \text{ cu ft per hr}$$

$$\text{wash time,} \ \theta = \frac{1.9/2}{0.76} = 1.25 \text{ hr}$$

2.47 $H_w - H = \dfrac{0.26}{\lambda} (t_g - t_w)$

where $t_g = 100°F$
$\quad\quad t_w = 75°F$
$\quad\quad H_w = 0.019$
$\quad\quad \lambda = 1,050$

$\quad\quad$ *Air in:* $H = 0.019 - \dfrac{0.26}{1,050} (100 - 75)$

$$H = 0.019 - 0.006 = 0.013 \ \frac{\text{lb water}}{\text{lb air}}$$

$\quad\quad$ *Air out:* $H = 0.059$ (psychometric chart)
$\quad\quad \therefore \ 0.059 - 0.019 = 0.040$ lb water removed per lb air

and $\dfrac{1}{0.04} = 25$ lb air circulated per lb water removed

$\quad q = 25 \times 0.238 \times 20 + 0.013 \times 25 \times 0.48 \times 20 + 1,032$
$\quad\quad$ (lb air) (c_p air) (Δt) (lb water) $\quad\quad$ (c_p) $\quad$ (Δt) $\quad$ (λ)
$\quad q = 1154$ Btu per lb of water removed

See Perry for experimental data, T-H charts, and steam tables.

2.48   Use the symbols given in the problem.

$\quad$ *a.* Tank 1: $VsC \dfrac{d\hat{T}_1(t)}{dt} = WC[\hat{T}_i(t) - \hat{T}_1(t)] + \hat{q}(t)$

or $\quad\quad 0.5s \dfrac{d\hat{T}_1(s)}{dt} = WC[\hat{T}_i(t) - \hat{T}_1(t)] + \dfrac{1}{400} \hat{q}(t)$

$\quad\quad\quad 0.5s\hat{T}_1(s) = \hat{T}_i(s) - \hat{T}_1(s) + \dfrac{1}{400} \hat{Q}(s)$

or
$$\hat{T}_1(s) = \left(\frac{1}{0.5s + 1}\right)\left[\frac{1}{400}\hat{Q}(s) + \hat{T}_i(s)\right]$$

Tank 2: $VsC\dfrac{dT(t)}{dt} = WC[T_1(t) - T(t)]$

or
$$0.5s\frac{d\hat{T}}{dt} = \hat{T}_1(t) - \hat{T}(t)$$

$$0.5s\hat{T}(s) = \hat{T}_1(s) + \hat{T}(s)$$

or $\hat{T}(s) = \dfrac{1}{0.5s + 1}\hat{T}_1(s) = \left(\dfrac{1}{0.5s + 1}\right)^2\left[\dfrac{1}{400}\hat{Q}(s) + \hat{T}_i(s)\right]$

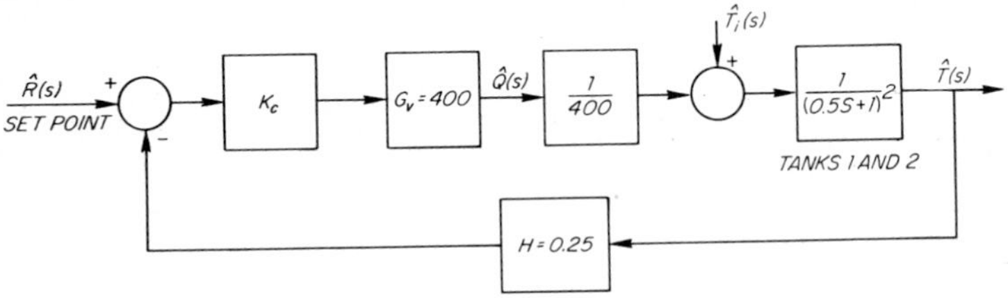

Fig. 2.48a

$b.\ \hat{T}(s) = \dfrac{[1/(0.5s + 1)^2](1/400)\,400\,K_c}{1 + [1/(0.5s + 1)^2](1/400)\,400(0.25K_c)}\hat{R}(s)$

$\qquad = \dfrac{K_c}{(0.5s + 1)^2 + 0.25K_c}\hat{R}(s)$

2.49 *a.* The packed volume is a function of the column diameter and the height of the column required to give the desired separation. The column diameter is determined on the basis of permissible vapor velocity. This information must be determined for individual types of packing and is usually presented under flooding data.

The height of the column is a function of:

1. The degree of removal of solute
2. The vapor-liquid equilibrium and solubility data for the solvent chosen
3. Height-of-transfer-unit, $K_g a$, or heat-equivalent-t-a-transfer-plate data for the system involved

b. 1. 50 to 75 per cent of the flooding velocity is chosen as vapor velocity.

2. An economic balance of cost of solute waste vs. cost of column is the ultimate aim of most design calculations. In order to calculate the column height, the following steps are used:

(a) Plot equation data

(b) choose liquid-gas ratio

(c) (1) Determine the number of transfer units by calculation or graphically, and the height equals the number of transfer units × HTU, or

(2) Evaluate $\dfrac{dx}{(x^z - x)}$ (liquid phase controlling)

or $\dfrac{dy}{(y^z - x)}$ (gas phase controlling) by graphical integration

and solve $\dfrac{dx}{(y^z - x)} = \dfrac{K_l a S \, dL}{L}$

or $\dfrac{dy}{(y^z - x)} = \dfrac{K_y a S \, dL}{L}$

when $K_y a$ or $K_l a$ are known and where y^z and x^z are equilibrium values.

c. The advantage of the HTU over the HETP is that the HTU is based on a true differential countercurrent process rather than a stepwise countercurrent process. The HTU method is essentially the same as capacity of $K_g a$ and $K_y a$ methods, but the HTU method has an advantage in that it involves only one dimension and does not vary appreciably with gas velocity. When the operating line and equilibrium line are parallel, HTU, HETP, and capacity methods are identical.

2.50 A multiple-effect, probably a triple-effect, evaporator would be used. If necessary, an economic balance could be carried out to determine the number of effects for which the sum of the fixed charges and operating costs is a minimum. To design the evaporator, it would be necessary to know the temperature of the steam available for heating in the first effect and the temperature of the cooling water for the condenser in the last effect. The boiling-point elevations of caustic solutions of various strengths between 8 and 50 per cent would be required.

For the purpose of making heat balances around the effects, it would be desirable to have an enthalpy concentration chart for caustic solutions or at least data on heats of mixing and specific heats of caustic solutions. The temperature of the feed to the evaporator would determine whether forward feed (for hot feed) or backward feed (for cold feed) should be used. The three effects would normally be designed for equal areas. In carrying out the calculations it would first be necessary to assume the concentrations of the solutions in the three effects. An approximation would be to assume equal evaporations. The over-all coefficients of heat transfer in each effect would then be estimated from some available correlation. The temperatures of the vapors leaving each effect would then have to be estimated, either assuming equal temperature drops across the heat-transfer surface in each effect or by assuming the temperature drops inversely proportional to the over-all coefficients. Allowance must be made for the boiling-point elevations of the solutions. A heat balance is then made around each effect and the amounts evaporated in each effect are calculated from these heat balances. The calculated amounts of heat transferred may now be used to readjust the assumed temperature drops across the heat-transfer surfaces, while the calculated evaporations can be used to correct the assumed concentrations and over-all coefficients in the effects. The calculations are then repeated until the calculated evaporations and temperature drops agree with the assumed ones.

Materials of construction will be determined by the degree of corrosiveness of the solutions which will depend on the concentrations and the temperatures. In the case of caustic solutions up to concentrations of 50 per cent and temperatures of 200°F, mild steel can be used. With forward feed, the effect containing the most concentrated solutions (50 per cent) could be operated below 200°F so that mild steel could be used for all effects. With backward feed, however, the first effect would contain the 50 per cent solution at some temperature above 200°F, so that stainless steel or nickel-clad steel or monel metal would then have to be used, at least in the first effect, and for any lines, valves, or pumps leading out of this effect. The higher cost of these materials may be a deciding factor in the choice between backward feed and forward feed.

2.51

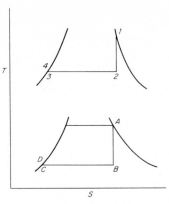

$$\text{Fig. } 2.51a$$

2.51   1. $-W_{\text{Hg}} = h_1 - h_2 = 148.6 - h_2$

$$S_1 = S_2 = 0.1277 = 0.02323 + x(0.1385)$$

or $\qquad\qquad x = 0.754$

$$h_2 = 15.85 + 0.754(126.95) = 111.45$$

$$-W_{\text{Hg}} = 148.6 - 111.45 = \mathbf{37.15} \text{ Btu per lb} \qquad \mathbf{1c}$$

2. $-W_{\text{steam}} = h_A - h_B = 1204.5 - h_B$

$$S_A = S_B = 1.4946 = 0.1151 + x(1.8867)$$

or $\qquad\qquad x = 0.731$

$$h_B = 59.99 + 0.731(1040.9) = 820.99$$

$$-W_{\text{steam}} = 1204.5 - 821 = \mathbf{383.5} \text{ Btu per lb} \qquad \mathbf{2b}$$

3. For heat balance around Hg condenser, let $y = $ lb Hg per lb steam.

$$y(h_2 - h_3) = (h_A - h_D)$$

$$y(111.45 - 15.85) = 1204.5 - 59.99)$$

$$y = \mathbf{11.98} \qquad \mathbf{3a}$$

4. Efficiency $= \dfrac{W_{\text{net}}}{Q} = \dfrac{11.98(37.15) + 383.5}{11.98(148.6 - 15.83)} = 0.521$ or $\mathbf{52.1}$

per cent   **4d**

2.52 $\dfrac{V}{v_0} = \dfrac{C_{A_0} - C_A}{kC_A}$ solving for v_0:

$$v_0 = \frac{kC_A}{C_{A_0} - C_A} V$$

$Q = (24)(60)$ min per day

Profit per day: $P_D = Q[(v_0 C_{A_0} x)2.5 - v_0(10)] - 10$

$$P_D = Q \frac{kC_A}{C_{A_0} - C_A} V(C_{A_0}x)(2.5) - Q \frac{kC_A}{C_{A_0} - C_A} V(10) - 10$$

Substitute $C_{A_0}(1 - X)$ for C_A:

$$P_D = Q \frac{kC_{A_0}(1 - x)}{C_{A_0} - C_{A_0}(1 - x)} V(C_{A_0}x)(2.5)$$

$$- Q \frac{kC_{A_0}(1 - x)}{C_{A_0} - C_{A_0}(1 - x)} V(10) - 10$$

$$= 2.5kQVC_{A_0}(1 - x) - \frac{10kQV(1 - x)}{x} - 10$$

$$\frac{\partial P_D}{\partial x} = -2.5kQVC_{A_0} - 10kQV \frac{-1}{x} - \frac{10kQV(1 - x)(-1)}{x^2} = 0$$

with $C_{A_0} = 5$, $-12.5x^2 + 10x + 10 - 10x = 0$ and $x = \mathbf{0.89}$

2.53 *a.* There are three general methods of demineralizing or desalting. All methods involve the replacement of metal cation by hydrogen ion and nonmetal anion by hydroxyl ion.

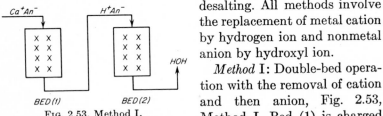

Method I: Double-bed operation with the removal of cation and then anion, Fig. 2.53, Method I. Bed (1) is charged with cation resin (H^+) form and bed (2) is charged with weakly basic anion resin (OH^-) form. The reaction in bed (1) is

$$RSO_3H + Ca^+An^- \leftrightarrows RSO_3Ca + H^+An^-$$

where $Ca^+ \approx Na^+$, K^+, Ca^{++}, Mg^{++}, Fe^{+++}, etc. The reaction in bed (2) is

$$RNH_3OH + H^+An^- \leftrightarrows RNH_3An + HOH$$

It should be noted that in this type of treatment the solution is acidic after treatment in bed (1) and until treatment in bed (2) is completed.

Method II: Double-bed operation with the removal of anion and then cation, Fig. 2.53, Method II. Bed (1) is charged with a strong base anion exchanger and the bed reaction may be represented by the equation

$$R_4NOH + Ca^+An^- \rightarrow R_4NAn + Ca^+OH^-$$

The solution in transit between bed (1) and bed (2) will be basic. The reaction in bed (2) is

$$RCOOH + Ca(OH)_2 \rightarrow RCOOCa + HOH$$

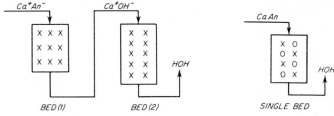

FIG. 2.53. Method II.   FIG. 2.53. Method III.

Method III: Simultaneous removal of cation and anion (mixed bed), Fig. 2.53, Method III. The simultaneous reactions in the mixed bed are

$$Ca^+An^- + RSO_3H + R_4NOH \rightarrow RSO_3Ca + R_4NAn + HOH$$

and conditions of pH deviating from 7 are localized.

b. The important factors to consider in comparing the methods described in (*a*) are:

1. Possible increased space requirements for Methods I and II due to multiple beds and increased rinse requirements in regeneration.

2. Influence of pH (less or greater then 7 in Methods I and II) where other materials present in the stream are susceptible to changes in pH.

3. The problem of "leakage" is not a major consideration in Method III.

4. Effluent qualities in Method III are more nearly independent of regenerant levels.

5. All methods may be operated to give high-quality effluent (0.1 to 1.0 ppm of iron in the case of water).

2.54 Corrosion in pipelines handling mixed wastes is best studied by installing in the line short sections of the materials to be tested. The test sections must be insulated from the existing pipe and from each other. Duplicate test sections of each material, with the outside surfaces painted to eliminate external sources of corrosion, should be installed.

The field test is preferable since

a. Corrosion products may contaminate and alter the corrosive nature of laboratory test solutions.

b. Trace materials in mixed wastes that enhance the corrosion process may not have been recognized. Further, if they are known, the small amounts may be depleted in a laboratory test so that erroneous results are indicated.

c. The laboratory test would fail to duplicate external conditions which may exist in actual practice.

d. Exact duplication of velocity, temperature gradients, and oxygen content are difficult in the laboratory. Also increasing flow rates tend to (1) reduce the thickness of stagnent films, (2) prevent the formation of protective coatings, (3) erode protective films already in existence, and (4) cause erosion corrosion.

The influence of velocity is known to be important. An example is to be found in tests on the effect of sea water on low-iron (0.06 per cent) cupron nickel (70-30). An increase in velocity from 15 to 25 fps increased the depth of attack (in mils) approximately sevenfold. Mantell, "Engineering Materials Handbook," McGraw-Hill, pp. 37–14, 1958.

After suitably cleaning the test specimens, corrected weight losses should be determined. Average and maximum depth of pits, if present, should be determined. The size, shape, depth, location, and distribution of pits should be noted. Bending tests to detect embrittlement should be carried out (not for Karbate, etc.).

The most economical installation is then determined on the basis of a comparative-cost calculation involving (1) cost of pipe, (2) cost of installation, (3) cost of maintenance, and (4) cost of pumping.

In instances where a new process is involved, tests should be carried out in the pilot plant.

2.55 a. Wrought-aluminum alloys are classified broadly into non-heat-treatable alloys and heat-treatable alloys.

1000 series alloys (old system 1S) are 99 per cent aluminum or higher. Iron and silicon are the major impurities. These alloys have superior corrosion resistance and below average mechanical properties.

2000 series alloys have copper as the major alloying material. This group, after proper heat-treatment, have mechanical properties approximately those of mild steel. Artificial aging may increase yield strength. The alloys have inferior corrosion resistance.

3000 series alloys have manganese as the major alloying element and are non-heat-treatable. Used where moderate strength and good workability are required.

4000 series alloys have silicon as the major alloying material. Most are non-heat-treatable and some have low melting points.

5000 series alloys contain magnesium as one of the major alloying elements. They are moderate to high-strength alloys. They are non-heat-treatable and have excellent resistance to corrosion in sea water.

6000 series alloys contain silicon and magnesium in a ratio approximately the same as in magnesium silicide. They are not so strong as the 2000 series or 7000 series but have good corrosion resistance.

7000 series alloys have zinc as the major alloying element. Magnesium, copper, and chromium are also used. 7075 is among the highest-strength alloys. It is heat-treatable.

One would, therefore, specify a 2000 series or 7000 series alloy, clad with a 6000 series alloy.

    b. 1. The iron fasteners will corrode. "Rusting" will be rapid.

       2. Copper fasteners will be protected at the expense of the aluminum. The aluminum will show cracks in the vicinity of the copper.

       3. The zinc will protect the aluminum. When the zinc coating fails, the results will be the same as in (1).

4. No evidence of corrosion except on portions of fasteners that may have been stressed during manufacture.

c. When loads below the proportional limit (conformity to Hooke's law—the ratio of stress to strain is proportional) are applied to most engineering materials, elastic deformation will occur. The deformation is constant irrespective of the total time that the material is under stress as long as the temperature T is not critical. At greater temperatures, a larger deformation may result, but a new phenomenon may occur in that there will be an increase in deformation with time while stress remains constant. This phenomenon is known as creep (see Fig. 2.55c1).

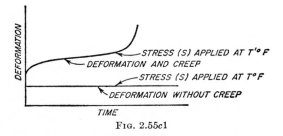

FIG. 2.55c1

Creep is a very sensitive physical property. Steels showing the same tensile properties and hardness at room temperature may show large differences in creep resistance at higher temperatures. Two steels of identical chemical composition may have the same properties at room temperature, yet show differences in creep strength at higher temperatures.

The stress-rupture data as a function of temperature and time are a very important consideration in the design of pressure vessels operating at elevated temperatures. When stainless steels are used as a fabricating material, for equipment to operate at temperatures exceeding 900°F, the yield point may be lower than that determined by the room-temperature short-time tensile test. Time is a major consideration. The most useful data concerning creep strength show the stress that will elongate steels 1 per cent in 10,000 hr. A typical stress to produce rupture curve in 1,000 hr would be similar to Fig. 2.55c2.

Reference: Mantell, "Engineering Materials Handbook," McGraw-Hill, 1958.

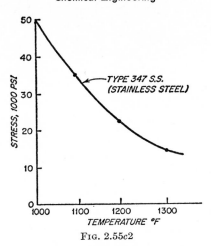

Fig. 2.55c2

2.56 Perry, p. 569, gives data on $K = Y/X$ as a function of temperature at 115 psia for hydrocarbons:

	1	2	3	4
	Mole fraction in feed, $L = 1$	K	X_L, when $L = 0.07$	Y_γ, when $\gamma = 0.93$
Methane	0.20	27	0.008	0.213
Ethane	0.20	4.7	0.045	0.212
Propane	0.30	1.6	0.192	0.307
n-Butane	0.10	0.51	0.184	0.094
Isobutane	0.10	0.71	0.137	0.097
n-Pentane	0.10	0.17	0.438	0.074
			1.004	0.997

Values of L_n (the remaining liquid) are now assumed. From a material balance for 1 mole of feed

$$X_m L_n + Y_m(1 - L_n) = 1X_{fm}$$
$$X_e L_n + Y_e(1 - L_n) = 1X_{fe}$$

where X_m and X_e refer to equilibrium values of mole fraction of methane and ethane, respectively. The Y values apply to equilibrium conditions in vapor and X_f values represent mole fraction of component in feed. A similar equation can be written for each component in the feed. Since $K_m = Y_m/X_m$, dividing both sides of the material balance by X_m and substituting for Y_m/X_m, the following equation is obtained:

$$L_n + K_m(1 - L_n) = \frac{1X_{fm}}{X_m} \text{ and } X_m = \frac{X_{fm}}{L_n + K_m(1 - L_n)}$$

The same type of equation can be written for each component. The proper solution is reached when

$$X_m + X_e + X_p + X_{n-b} + X_{i-b} + X_{n-p} = 1$$

The solution is trial and error. The values shown in the table are values of X (mole fraction in remaining liquid) when L (moles of liquid remaining) = 0.07 for 1 mole of feed. The vapor is therefore 0.93 moles and

$$X_m = \frac{0.20}{0.07 + 27(0.93)} = \frac{0.20}{25.2} = 0.008$$

$$X_e = \frac{0.20}{0.07 + 4.7(0.93)} = \frac{0.20}{4.44} = 0.045$$

$$X_p = \frac{0.30}{0.07 + 1.6(0.93)} = \frac{0.30}{1.56} = 0.192$$

$$X_{n-b} = \frac{0.10}{0.07 + 0.51(0.93)} = \frac{0.10}{0.544} = 0.184$$

$$X_{i-b} = \frac{0.10}{0.07 + 0.71(0.93)} = \frac{0.10}{0.730} = 0.137$$

$$X_{n-p} = \frac{0.10}{0.07 + 0.17(0.93)} = \frac{0.10}{0.228} = 0.438$$

$$\text{Summation } X = \overline{1.004}$$

The vapor compositions are now calculated from $Y = KX$

and $$Y_m = 27 \times 0.008 = 0.216$$

The composition of the remaining liquid (0.07 mole) is shown in column 3 of the table and that of the vapor (0.93 mole) is shown in column 4.

2.57 a. Figure 2.57 is a schematic flow diagram.

$$R_E = \frac{L_E}{P} = \text{external reflux ratio}$$

$$V_1 = L_E + P$$

= material balance top of column and reflux splitter

$$V_1 = R_E P + P = P(R_E + 1)$$
$$L_E + V_2 = L_1 + V_1 = \text{material balance top plate}$$
$$L_E H_{L_B} + V_2 H_{V_2} = L_1 H_{L_1} + V_1 H_{V_1} = \text{enthalpy balance}$$

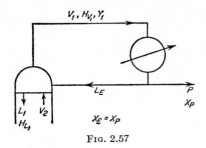

FIG. 2.57

Assuming that the enthalpies of vapors V_1 and V_2 are equal

$$L_E H_{L_B} + V_2 H_{V_1} = L_1 H_{L_1} + V_1 H_{V_1}$$
$$H_{V_1}(V_2 - V_1) = L_1 H_{L_1} - L_E H_{L_B}$$

and therefore

$$(L_1 - L_E)(H_{V_1}) = L_1 H_{L_1} - L_E H_{L_B}$$
$$L_1 H_{V_1} - L_1 H_{L_1} = L_E H_{V_1} - L_E H_{L_B}$$
$$L_1 = L_E \frac{H_{V_1} - H_{L_B}}{H_{V_1} - H_{L_1}}$$

The external reflux ratio is defined as $\dfrac{R_E}{R_E + 1} = \dfrac{L_E}{V_1}$, and the

internal reflux ration is defined as $\dfrac{R_I}{R_I + 1} = \dfrac{L_1}{V_2}$.

$$R_E V_1 = R_E L_E + L_E$$
$$V_1 = \frac{R_E L_E + L_E}{R_E}$$

and

$$\frac{R_I}{R_I + 1} = \frac{L_1}{V_2} = \frac{L_1}{L_1 + V_1 - L_E}$$
$$R_I L_1 + R_I V_1 - R_I L_E = R_I L_1 + L_1$$
$$V_1 = \frac{R_I L_E + L_1}{R_I}$$

and therefore $\dfrac{R_E L_E + L_E}{R_E} = \dfrac{R_I L_E + L_1}{R_I}$

$$R_I L_E = R_E L_I$$

$$R_I = \frac{R_E L_E}{L_E} \frac{H_{V_1} - H_{L_B}}{H_{V_1} - H_{L_1}} = R_E \frac{H_{V_1} - H_{L_B}}{H_{V_1} - H_{L_1}}$$

For a saturated vapor 0.8 mass fraction ethanol,

$$H_{V_1} = 595 \text{ Btu per lb}$$

For a saturated liquid 0.8 mass fraction ethanol

$$H_{L_B} = 103 \text{ Btu per lb}$$
$$H_{L_B} = 58 \text{ Btu per lb at } 120°F$$

Composition of liquid in equilibrium with 0.8 mass fraction vapor is 0.62 and

$$H_{L_1} = 115 \text{ Btu per lb}$$
$$R_E = 2 \frac{595 - 58}{595 - 115} = 2 \frac{537}{480} = 2.236$$

The slope of the operating line is 0.69.

b. For 1 lb of feed the amounts of top and bottom product are obtained by means of a material balance.

$$0.35 = 0.8V + 0.05B = 0.75V + (0.05V + 0.05B)$$
$$1.0 = V + B \text{ and } 0.05 = (0.05V + 0.05B)$$

therefore $$0.35 = 0.75V + 0.05$$

and $0.75V = 0.30$, whence $V = 0.40$ lb and $B = 0.60$ lb.

For every pound of feed 0.40 lb of top product are withdrawn and 0.80 lb of reflux are returned to the top plate. When the reflux is fed to the top plate as a saturated liquid, only the latent heat of vaporization which is removed in the condenser must be supplied in the reboiler or

$$0.80 \times (595 - 103) = 393.6 \text{ Btu per lb}$$

When the reflux is returned at 120°F, but with a composition of 0.8 mass fraction ethanol, its enthalpy is 58 Btu per lb. Therefore, $0.80 \times (595 - 58) = 429.6$ Btu per lb of feed must be supplied. When the "cold" liquid stream enters the top plate it removes latent heat from vapors on the top plate and internal or operating reflux is greater than external reflux. Data on enthalpies of ethanol-water mixtures are found in Brown and Associates, "Unit Operations," pp. 327 and 582, Wiley, 1950. Discussion on influence of cold reflux is found in Treybal, "Mass-transfer Operations," McGraw-Hill, p. 312, 1955.

2.58 At the point of incipient fluidization, the P due to the gas flow will be such that the forces tending to raise the particles of the catalyst will be equal to the weight of the particles.

a. $-\Delta P_f = L(1 - X)(\rho_s - \rho_g)$

$$\frac{-\Delta P_f}{L} = (1 - 0.4)(152 - 0.5) = 0.6(151.5) = 90.90 \text{ psf}$$

b. It is now necessary to determine the gas velocity that will give a pressure drop equal to that calculated in (a) using the method of Brownell and Katz explained in Brown and Associates, "Unit Operations," Wiley, 1950.

$$\text{Re} = \frac{D_P F_{\text{Re}} V \rho}{\mu} = \frac{0.174 \times 44 \times V \times 0.5}{12 \times 0.03 \times 6.72 \times 10^{-4}} = 1.59 \times 10^4 \times V \text{ fps}$$

$$f = \frac{2g_c D_P(-\Delta P)}{F_f L V^2 \rho}$$

$$\frac{\Delta P_f}{L} = \frac{f F_f V^2}{2g_c D} = \frac{f V^2 \times 1,400 \times 0.5 \times 12}{64.4 \times 0.174} = f V^2 \times 749$$

Assume $V = 2.6$ fps

$$\text{Re} = 1.59 \times 10^4 \times 2.6 = 4.13 \times 10^4$$
$$f = 0.0225$$

$$\frac{\Delta P_f}{L} = 0.0225 \times (2.6)^2 \times 749 = 112.35 \text{ psf}$$

Assume $V = 2.3$ fps

$$\text{Re} = 1.59 \times 10^4 \times 2.3 = 3.65 \times 10^4$$
$$f = 0.023$$

$$\frac{\Delta P_f}{L} = 0.023 \times (2.3)^2 \times 749 = 89.88 \text{ psf}$$

This velocity could be estimated as suggested by Leva in *Chemical Engineering*, November, 1957.

$$G_{mf} = 688D_P{}^{1.82} \frac{[\rho_f(\rho_s - \rho_F)]^{0.94}}{\mu^{0.88}}$$

$$= \frac{688(0.174)^{1.82}[0.5(151.5)]^{0.94}}{(0.03)^{0.88}} = 9.91 \text{ lb per ft}^2 \text{ sec}$$

$$\text{Re} = \frac{D_P G_{mf}}{\mu} = \frac{0.0145 \times 9.91}{0.03 \times 6.72 \times 10^{-4}} = 0.71 \times 10^4 = 7{,}100$$

Estimate the correction factor $= 0.2$ and

$$G_{mf} = \frac{9.91 \times 0.2}{0.5} = 4 \text{ fps}$$

c. Now assuming the porosity $= 1$, when the velocity of the gas is sufficient to elutriate particles whose diameter is 0.174 in.,

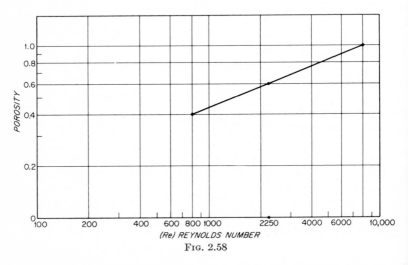

FIG. 2.58

determine the velocity of the fall of this particle. From Brown and Associates, "Unit Operations," p. 79, Wiley, 1950,

$$f_D = \frac{4g(\rho_s - \rho_f)D_P{}^3\rho_f}{3\mu^2} = \frac{128.8(151.5)(0.0145)^3(0.5)}{3(9 \times 10^{-4})(6.72 \times 10^{-4})^2} = 2.45 \times 10^7$$

When $\text{Re} = 1$, $f_D = 2.45 \times 10^7$. When $f_D = 1$, $\text{Re} = 4{,}950$. Therefore, $\text{Re} = 7{,}900$ and $V = 22.2$ fps. Richardson and Laki in *Transactions of the Institute of Chemical Engineers*, London,

Vol. 32, p. 35, 1954, have found that a straight line results when the logarithm of the porosity is plotted as a function of the logarithm of the Reynolds number.

$$\text{When porosity} = 0.4, \text{Re} = 816$$
$$\text{When porosity} = 1.0, \text{Re} = 7{,}900$$

Plot on loglog paper and draw a straight line as in Fig. 2.58. From this plot, when the porosity = 0.6, Re = 2,250, and $V = 6.35$ fps. There will be $10 \times 0.5 = 5$ ft available to accommodate the expanded bed. Each foot of static bed will occupy 1.25 ft in the fluidized state. Therefore, $5/1.25 = 4$ ft of static bed that can be accommodated.

2.59   *a.* The volumetric displacement of the piston is

$$\frac{\pi}{4} \times (0.667)^2 \left(\frac{10}{12}\right)(200) = 58.2 \text{ cu ft per min}$$

Since the mechanical efficiency is only 88 per cent, the operating volumetric displacement is

$$0.88 \times 58.2 = 51.2 \text{ cu ft per min}$$

The specific volume of saturated ammonia vapor at 0°F is 9.116 cu ft per lb (Perry, p. 250). And

$$51.2 \frac{(\text{cu ft})}{(\text{min})} \times \frac{1 \text{ (lb of ammonia)}}{9.116 \text{ (cu ft)}} = 5.61 \text{ lb ammonia per min}$$

b. The enthalpy of ammonia vapor at 0°F = 611.8 Btu per lb. The entropy of ammonia vapor at 0°F = 1.3352 Btu per lb per °F. The enthalpy of ammonia at 140 psia at h_B and $S = 1.3352$

T	h	S
190°	704.0	1.3328
	h_B	1.3352
200°	709.9	1.3418

(Table 204, Perry, p. 253).

$$h_B = 704 + 5.9 \times \frac{0.0024}{0.0090} = 705.6 \text{ Btu per lb}$$

The isentropic shaft work = 705.6 − 611.8 = 93.8 Btu per lb
The actual shaft work = 1.2 × 93.8 = 112.6 Btu per lb

$$\text{The horsepower input} = \frac{\text{actual shaft work} \times \text{lb per min}}{\text{efficiency} \times \text{Btu per hp}}$$

$$= \frac{112.6 \times 5.61}{0.85 \times 42.4} = 17.5$$

c. The enthalpy of the ammonia after compression = 611.8 + 112.6 = 724.4 Btu per lb, and as a saturated liquid at 70°F = 120.5 Btu per lb (Table 203, Perry, p. 250). The heat removed = (724.4 − 120.5)5.61 = 3380 Btu per min.

d. The enthalpy of ammonia (saturated liquid) at 70°F
= 120.5
The enthalpy of ammonia (liquid) at 30.42 psia = 42.9
The enthalpy of ammonia (vapor) at 30.42 psia = 611.8

Therefore, by means of a heat balance

$$120.5 = 42.9 + X(611.8 - 42.9)$$

$$X = \frac{120.5 - 42.9}{611.8 - 42.9} = \frac{77.6}{568.9} = 0.1364$$

or 13.64 per cent is vaporized.

2.60 a. The total weight of water in the feed stock is 10 × 2,000 × 0.1 = 2,000 lb per day. The weight of the salt is 20,000 − 2,000 = 18,000 lb per day. The weight of the salt leaving is $\frac{18,000}{0.99}$ = 18,181 lb = 9.09 tons per day.

b. The weight of the water removed is 2,000 − 181 = 1,819 lb per day, or 1,819/24 = 75.8 lb water per hr.

c_1. The humidity of the entering air is determined from the wet-bulb dry-bulb reading.

$$H_w - H_g = \frac{H_g}{29\lambda_w k_g}(T_g - T_w)$$

$$H_w - H_g = \frac{0.26}{\lambda_w}(T_g - T_w)$$

$$0.0595 - H_g = \frac{0.26}{962}(225 - 110)$$

Humidity of the air entering $= H_g = 0.0595 - 0.0310 =$
$$\frac{0.0285 \text{ lb water}}{\text{lb dry air}}.$$

c_2. When all the heat required for vaporizing water is supplied by air or other gases and the heat transfer by conduction through boundaries and by radiation is negligible, the drying operation may be termed adiabatic. The condition of the air with respect to humidity would therefore follow an adiabatic cooling line on the humidity chart. The humidity of the entering air is equal to the humidity of the exit air $= \dfrac{0.047 \text{ lb water}}{\text{lb dry air}}$.

2.61 The weight of copper is $3 \times 100 \times 0.5$, or 150 lb. The current requirements per pound of material produced are to be found in tables of electrochemical equivalents. See Mantell, "Industrial Electrochemistry," 3d ed., p. 741, McGraw-Hill, 1950. Cu^{++} requires 382.5 amp-hr per lb. Since the current efficiency is 85 per cent, the actual current requirement will be $382.5/0.85 = 450$ amp-hr per lb. Therefore, 150 lb of copper requires 450×150 or 67,500 amp-hr. The power requirement will be $67,500 \times 2.5/1000$, or 168.75 kwhr. Since conversion is 90 per cent and transformer loss is 3 per cent, the actual power will be $168.75/(0.90 \times 0.97)$ or 193.3 kwhr, and the cost will be $193.3 \times 0.01 = \$1.93$.

2.62 A 20,000-amp cell with a current efficiency of 92 per cent would effectively deliver 18,400 amp per hr. The table of electrochemical equivalents indicates that 342.9 amp-hr are required to produce 1 lb of chlorine when the change in valence is 1. Therefore, $18,400 \times 24/342.9 = 1,286$ lb of chlorine per day. The current efficiency includes undesirable anode products, current leaks, heat losses, etc.

2.63 *a.* The work of an adiabatic compression $(-w)$ is

$$-w = +\Delta E = + C_v\, dT$$

Since $C_p - C_v = R$, then $C_v = C_p - R$, and $-w = +\Delta E = \int (C_p - R)\, dT$, where $R = \dfrac{T_2}{T_1} = \dfrac{460 + 600}{460 + 80} = \dfrac{1060}{540} = 1.987$.

Therefore

$$-w = \frac{100}{28} \int \left(9.46 - \frac{3.29 \times 10^3}{T} + \frac{1.07 \times 10^6}{T^2} - 1.987 \right) dT$$

$$= 3.57 \int \left(7.47 - \frac{3.29 \times 10^3}{T} + \frac{1.07 \times 10^6}{T^2} \right) dT$$

$$= 3.57[(7.47)(1060 - 540) - 3.29 \times 10^3 \ln {}^{1060}\!\!/_{540}$$
$$+ 1.07 \times 10^6 ({}^1\!\!/_{540} - {}^1\!\!/_{1060})]$$

$$= 3.57[7.47 \times 520 - 3290(6.97 - 6.29)$$
$$+ 1070(1.85 - 0.95)]$$

$$= 3.57(3880 - 2240 + 960) = 3.57 \times 2600 = \mathbf{9300} \text{ Btu}$$

b. The general expression for the change in entropy of a system undergoing polytropic change is

$$\Delta S = R \ln \frac{T_2}{T_1} + R \ln \frac{p_1}{p_2} + \int_{T_1}^{T_2} (C_p - R) \frac{dT}{T}$$

$$= R \ln \frac{p_1}{p_2} + R \ln \frac{T_2}{T_1} - R \ln \frac{T_2}{T_1} + \int_{T_1}^{T_2} (C_p) \frac{dT}{T}$$

$$= R \ln \frac{p_1}{p_2} + \int_{T_1}^{T_2} (C_p) \frac{dT}{T}$$

Since the process is reversible and adiabatic, $\Delta S = 0$. Therefore,

$$0 = -R \ln \frac{p_2}{p_1} + \int_{T_1}^{T_2} (C_p) \frac{dT}{T}$$

or $$R \ln \frac{p_2}{p_1} = \int_{T_1}^{T_2} (C_p) \frac{dT}{T}$$

$$= \int_{540}^{1060} \left(\frac{9.46}{T} - \frac{3.29 \times 10^3}{T^2} + \frac{1.07 \times 10^6}{T^3} \right) dT$$

$$= 9.46 \ln \frac{1060}{540} - 3.29 \times 10^3 \left(\frac{1}{540} - \frac{1}{1060} \right)$$

$$+ 0.535 \times 10^6 \left[\frac{1}{(540)^2} - \frac{1}{(1060)^2} \right]$$

$$= 9.46 \times (6.97 - 6.29) - 3.29 \times (1.85 - 0.95)$$
$$+ 0.535 \times (3.42 - 0.89)$$

$$= +6.43 - 2.96 + 1.35 = 4.82$$

$$\log \frac{p_2}{p_1} = \frac{4.82}{2.3 \times 1.987} = 1.056$$
$$\log p_2 - \log p_1 = 1.056$$
$$\log p_2 = 1.056 + 1.176 = 2.232$$
$$p_2 = 170.6 \text{ psia}$$

2.64 A schematic diagram of the process is shown in Fig. 2.64a. A line connecting the feed and solvent is constructed on the

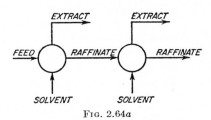

FIG. 2.64a

phase diagram in Fig. 2.64b. The gross composition of the mixture is calculated from the material balance.

$$1,200 X_{cm_1} = 1,000 \times 0.5 \qquad \text{component } C$$
$$X_{cm_1} = 500/1,200$$
$$= 0.416 \text{ weight fraction or } 41.6 \text{ per cent } C$$
$$1,200 X_{bm_1} = 200 \qquad \text{component } B$$
$$X_{bm_1} = 200/1,200$$
$$= 0.166 \text{ weight fraction or } 16.6 \text{ per cent } B$$

This falls on a tie line parallel to and a little above the tie line for point ①. A tie line is drawn through M_1 and the composition of the first raffinate (R_1) and extract (E_1) is determined.

R_1 weight per cent A 58.0 and E_1 weight per cent A 5.5
$\qquad\qquad\qquad\qquad B$ 6.2 $\qquad\qquad\qquad\qquad\qquad B$ 39.4
$\qquad\qquad\qquad\qquad C$ 35.8 $\qquad\qquad\qquad\qquad\qquad C$ 55.1

The weights of R_1 and E_1 are now determined by means of a material balance.

Component C: $1,200(0.416) = 0.358 R_1 + 0.551 E_1$
$\qquad\qquad 1,200 = R_1 + E_1 \text{ and } E_1 = 1,200 - R_1$
$\qquad\qquad R_1(0.551 - 0.358) = 1,200(0.551 - 0.416)$
$\qquad\qquad R_1 = 840 \text{ lb and } E_1 = 360 \text{ lb}$

Since the operation is to remove $500 \times 0.8 = 400$ lb, only $500 - 400 = 100$ lb is to remain in raffinate. One extraction with 200 lb of C leaves $840 \times 0.358 = 300.7$ lb of component C.

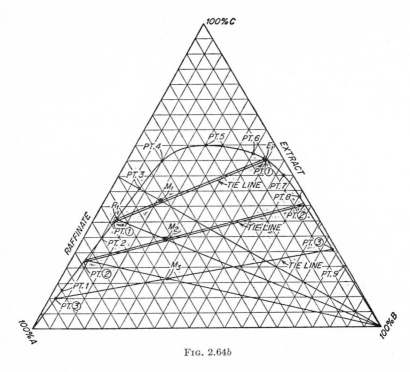

FIG. 2.64b

Hence one extraction is not sufficient. A new line is constructed from R_1 through C and a new mixture point M_2 is located.

$$1{,}040 X_{cm_2} = 840(0.358) \qquad \text{component } C$$
$$X_{cm_2} = 301/1{,}040 = 0.29$$
$$1{,}040 X_{am_2} = 840(0.58) \qquad \text{component } A$$
$$X_{am_2} = 488/1{,}040 = 0.469$$

This falls approximately on the tie-line data for point ②.

R_2 weight per cent A 73.7 and E_2 weight per cent A 1.8
$\phantom{R_2 \text{ weight per cent }}B$ 3.8 $\phantom{\text{ and } E_2 \text{ weight per cent }}B$ 57.7
$\phantom{R_2 \text{ weight per cent }}C$ 22.5 $\phantom{\text{ and } E_2 \text{ weight per cent }}C$ 40.5

The weight of R_2 is now determined from a material balance.

$$840(0.29) = R_2(0.225) + E_2(0.405)$$
$$840 = R_2 + E_2 \text{ and } E_2 = 840 - R_2$$
$$840(0.29) = R_2(0.225) + 840(0.405) - R_2(0.405)$$
$$R_2(0.405 - 0.225) = 840(0.405 - 0.290)$$
$$R_2 = 840(0.115)/(0.180) = 537 \text{ lb}$$
$$537(0.225) = 120.8 \text{ lb of } C \text{ left in raffinate}$$

Therefore, a third extraction is necessary if at least 80 per cent of the component C is to be removed. The new mixture point will be at:

$$737X_{am_3} = 537(0.737) \qquad \text{component } A$$
$$X_{am_3} = 0.537$$
$$737X_{bm_3} = 537(0.038) + 200 \qquad \text{component } B$$
$$X_{bm_3} = 220.4/737 = 0.304$$

This point is on tie-line data for point ③.

$$737(0.537) = R_3(0.984) + E_3(0.013) \qquad \text{component } A$$
$$737 = R_3 + E_3 \text{ and } E_3 = 737 - R_3$$
$$737(0.537) = R_3(0.984) + 737(0.013) - R_3(0.013)$$
$$R_3 = 737(0.537 - 0.013)/(0.984 - 0.013) = 398$$

Weight of component C in raffinate $= 398(0.10) = 39.8$ lb.

2.65   *a.* On-off control indicates opening or closing of a valve as the parameter to be controlled deviates from the desired value. For example, when the temperature falls below a selected point or the flow falls below a similar point, the valve is opened. This method may introduce large fluctuations unless the capacity of the system is large. This type of control may involve considerable transfer lag. In order to minimize this lag a proper supply-demand ratio should be selected and the supply valve should be throttled to a suitable fraction of its full operation. Throttling will vary with the supply-demand ratio.

b. Proportional control operates on the principle that the control will select a valve position for every temperature within the temperature range of the instrument. The range of variables of the measuring element (temperature or velocity) necessary to cause a complete cycle of the control valve (open-shut) is an important consideration in the operation of the instruments.

If the control is functioning properly at T_1, under demand D_1 and valve opening of 50 per cent of full capacity, and the demand is suddenly increased, (1) the temperature will drop, (2) the valve will open wider, (3) the control will set itself for a new temperature, and (4) this may introduce "droop" or temperature deviation from the desired value.

In order to minimize this temperature deviation a proportional band, which is related to the range of values of the measuring element (temperature) necessary to cause complete cycle of the value, must be chosen. This value or range is made smaller if considerable droop is indicated. A range that is too narrow, however, will cause oscillation in the control achievement.

c. Derivative or preact control operates on the principle that the correction is equal to some constant times the rate of change of the control variable (derivative dT/dt, dU/dT, etc.).

Derivative control is used in conjunction with some other method of control (proportional), since a large steady error would not be corrected by derivative controls. In this case its derivative would be zero. Derivative control does not operate on the deviation from a set point.

CIVIL ENGINEERING

3.01   *a.* To prevent the rapid drying of concrete the exposed surfaces should be covered. In some instances spraying with a resin-base solution or a covering of impregnated paper is used. For high early-strength concrete the curing period should be at least three days; and for ordinary concrete one week is required.
   b. In freezing weather fresh concrete should be at a temperature of at least 50°F at the time of pouring. This can be accomplished by heating the aggregates to 70 to 80°F. The concrete should be protected by covers and in the case of a building heaters should be installed to maintain the temperature at a minimum of 50°F for a period of one week for concrete with Type I cement and three days for concrete with Type III cement. If 2 per cent calcium chloride is added to the mix the above curing periods may be reduced to three days for Type I cement and two days for Type III cement.
   c.

Type of structure	*Length of time in days before forms and supports should be removed*
Arch centers	14–21
Centering under beams	14
Floor slabs	3–10
Walls	1–3
Columns	7–14
Sides of beams	3–10

d. Time limit between the addition of water at the mixer and the point of deposition may be 30 to 90 min depending on the degree of agitation of the mix during transit or during the waiting period.

e. Concrete may be admitted through temporary openings in the side of the forms, lowered into the form in bottom dump buckets, or allowed to slump down a pipe from the top, but concrete should never be dumped into the form from any height that will permit segregation of the mix. Filling should not be so rapid as to cause undue hydraulic pressure on the forms. Concrete should be sufficiently vibrated, as it is placed, to produce a dense concrete free of air pockets.

3.02 *a.* The way to prevent frost heave in a clay subgrade is to intercept the capillary rise. This can be done by installing at a depth of 3 ft (depth of frost penetration) an intercepting layer consisting of a water-tight membrane of asphalt and a sand cushion of 3 in. thickness or else a layer with low capillarity such as 18 in. of coarse-to-medium sand may be used. It is almost impossible to drain a subgrade containing large quantities of clay. The best thing to do is to seal the pavement and shoulders so that no moisture can enter the subgrade and drainage will not be necessary.

b. The strength of the concrete will be affected by the water-cement ratio, the length and type of curing, the quality of the aggregate, the thoroughness of mixing, any admixtures which may be used, and the type of cement used; the mix used may affect strength where the quantity of aggregate is excessive. To insure adequate strength a low water-cement ratio consistent with workability should be used. The curing should continue for at least 10 days. The aggregate should be clean and sound and should not include smooth or flat particles nor should the sand be excessively micaceous. The sand should not include large quantities of fine material. Mixing time should be around one minute. The water-cement ratio should be lowered to restore any strength lost by addition of admixtures. Try a mix $1:2:3\frac{1}{2}$ by volume and 5.75 gal water per sack cement.

c. Reinforcing steel is used in concrete pavements to control cracking. The steel should be placed about 2 in. below the surface of the slab. The steel is placed near the surface in order to better control surface cracks and prevent the entrance of water and other foreign matter. Reinforcing steel is not designed inasmuch as it serves not as full reinforcing, but rather as a crack-controlling medium. The size and spacing of the bar is chosen arbitrarily and varies widely from state to state.

d. Vibration makes the mixture more plastic and thus allows the use of a lower water-cement ratio thus increasing the strength of the concrete. In some cases of low slump concrete vibrating develops honeycombing. Also the concrete may become hard making it difficult to handtool around the edges and near the joints. High-spots may also prove difficult to reduce. The essential reason for vibrating concrete is to assure proper placement of concrete even where a stiff mix is used.

e. Concrete may be cured by:

1. Spraying a plastic coating on the surface of the concrete thus preventing loss of water through the surface.
2. Using calcium chloride in the concrete mixture thus retaining moisture in the concrete by the hygroscopic action of the material.
3. Periodically spraying the pavement with water providing excess moisture on the surface to replace any which may have been lost.
4. Ponding water and allowing it to stand to a depth of 2 in. or so on the surface.

Ponding probably provides the best method, where the water is prevented from leaking off the slab into the subgrade. The seal provided against moisture loss is probably the most secure and the presence of excess moisture is beneficial to the concrete.

3.03 Normal duration critical path requires 19 days (*A-H-I-J-K-M*). Reduce *A* to 2 days, *D* to 2 days, *H* to 3 days, and *L* to 1 day. Normal cost = \$12,750. Added cost = *A* + \$100, *D* + \$50, *H* + \$600, and *L* + \$500 for a sum of \$1,250. Total cost = \$14,000. *A* + *H* + *I* + *J* + *K* + *M* = 15 days. *A* + *B* + *C* + *D* + *L* + *M* = 15 days.

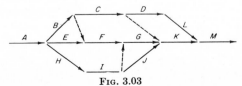

Fig. 3.03

3.04   *a.* The blast follows lines of least resistance. The effective depth will be 2 ft less.

   b. This is true.

   c. This is not so.

   d. No. A diesel is needed to drive the generator.

   e. This is true because there are fewer voids.

   f. $20T \times 12 = 0.75P$. $P = 320$ ft-tons, or 10.66 tons at a 30-ft radius.

   g. True. About 3 per cent per 1,000 ft for a four-cycle engine.

   h. This is so.

3.05   *a.* To avoid the possibility of a duplicate bid.

   b. Since it is difficult to allocate many items such as overhead, supervisory time, equipment use, etc., on each item, many contractors prefer to calculate a lump sum bid first and then break it down by proportion.

   c. A very low bid may be given to keep the organization together, or to keep equipment busy, or because the company is hungry for a job, or the contractor may be already on the job or nearby. A bid may be high because it is only a courtesy bid, or there may be uncertainties because of unusual specifications or because of type of inspection; and there are always variations in skill and equipment.

   d. Should be included in the contract if material is to be stored for any length of time.

   e. By leasing: cash is released as working capital, local taxes on personal property are avoided, specialized expensive equipment can be used and there may be tax advantages by charging against current year. Pride of ownership is missing, you may not be able to lease when you want the equipment, you may lose discounts available for paying cash, there is the possibility of an underequipped contractor bidding beyond his capabilities.

3.06   *a.* The original cracking of the concrete pavement is probably traceable to inadequate subgrade support. In order to reinforce the subgrade without removing the old pavement

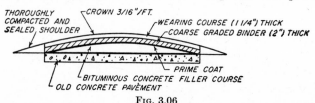

THOROUGHLY COMPACTED AND SEALED, SHOULDER
CROWN 3/16"/FT.
WEARING COURSE (1 1/4") THICK
COARSE GRADED BINDER (2") THICK
PRIME COAT
BITUMINOUS CONCRETE FILLER COURSE
OLD CONCRETE PAVEMENT

FIG. 3.06

it is necessary to drill a series of holes in the pavement and pump in a high-penetration slow-curing road oil under pressure. Thus we can realign the old slabs and seal and stabilize the subgrade. The shoulders should also be sealed against lateral encroachment of water into the subgrade.

b. The base should be thoroughly brushed clean and all joints and cracks should be sealed with a high-penetration rapid-curing asphalt. Then the old surface should receive a prime coat of low-penetration medium-curing road oil. A leveling course of coarse bituminous concrete may be used if the base requires to be brought to a true cross section.

c. The wearing course would consist of a coarse aggregate bituminous concrete binder, depth $1\frac{1}{2}$ to $2\frac{1}{2}$ in. and a wearing course 1 to $1\frac{3}{4}$ in. in depth. The binder course should have the aggregate distributed as follows: between $1\frac{1}{2}$-and $\frac{3}{4}$-in. screens, 20 to 30 per cent; $\frac{3}{4}$ to $\frac{1}{4}$ in., 25 to 40 per cent; $\frac{1}{4}$-in. to No. 10 sieve, 5 to 15 per cent; passing No. 10 sieve, 20 to 35 per cent; bitumen 4 to 7 per cent. All percentages are by weight. The asphalt should have a penetration of 60–70. For the surface course the proportions are: aggregates, between $1\frac{1}{4}$ and $\frac{3}{4}$ in., 15 to 25 per cent; $\frac{3}{4}$ to $\frac{1}{4}$ in., 20 to 35 per cent; $\frac{1}{4}$ in. to No. 10 sieve, 5 to 15 per cent; No. 10 to No. 200 sieve, 25 to 35 per cent; passing No. 200 sieve, 4 to 6 per cent; bitumen 5 to 8 per cent. The asphalt used is the same as for the binder course. The materials are mixed in a standard highway mixing plant using the hot-mix method at a temperature of 300°F. Application temperature should not fall below 225°F. The mixtures may be spread by hand using rakes and hoes although most jobs of any size use mechanical spreaders with a helical worm spreader and a strike-off screed. The mixture is then rolled longitudinally from the edges to the center with a small overlap on each subsequent trip. The roller used is a 10- to 12-ton three-wheel roller. A subsequent rolling in a diagonal direction by a tandem roller of 5 tons weight may be used for the wearing course.

3.07   a. See the answer to 3.63c.

b. See the answer to 3.63c.

c. This type of surface will require a surface treatment to be applied periodically. The surface treatment consists of a prime coat of emulsified or cut-back asphalt, a binder coat, cover stone, a seal coat, and screenings. From time to time patches may be required where weaknesses have developed. Where such patches are necessary the subgrade should be stabilized and compacted and then a base course and wearing course similar to the original should be placed. A stock of material from the original stockpiles should be kept for this purpose to ensure that the patch will become integral with the surrounding pavement and have the same composition.

3.08 To the local merchants it can be pointed out that studies have indicated that only a small percentage of the vehicles on the state route are headed for the central business district where a town of this size is involved. Most of their business comes from local sources, and this will be enhanced once local traffic is free to move on its own streets without the excessive burden of congestion brought on by through traffic.

The residents of the suburbs should be told that a fair, not to say generous, price will be paid for all parcels taken. The right of way will be fenced in through residential area to safeguard children. Pedestrian overpasses will be provided. Past experience has indicated that property appreciates, not depreciates, when a well-planned limited-access facility comes into an area.

The taxpayers' group should be told that the average loss in time through the center of town results in a monetary loss to 27,000 motorists in lost wages, wear and tear on vehicles, increased delivery costs, etc., which more than offsets the yearly costs of the new facility. Further, there are savings on intangibles such as decreased accidents and fatalities, decreased auto insurance rates for the town, and less noise in the center, which add to the arguments for the new facility.

3.09 a, 11; b, 3; c, 8; d, 5; e, 13; f, 4; g, 6; h, 12; i, 20; j, 19.

3.10 1. For the necessary thickness of pavement and base courses, see U.S. Corps of Engineers "Engineering Manual," Part XII, Chap. 2, p. 40, Fig. 3, of 1951, or similar charts in other publications.

Compacted subgrade CBR = 10, wheel load = 60,000 lb, tire pressure = 200 psi, combined thickness = 25.5 in., say 26 in.

Necessary pavement thickness ("Engineering Manual," p. 31), tire pressure = 200 psi, wheel load = 60,000 lb, thickness = 4 in. (for CBR in upper base = 80).

Thickness necessary above material (d) = 20 in. ("Engineering Manual," p. 40). Make material (d) 6 in. thick.

Thickness necessary above material (c) = 11 in. Make material (c) = 8 in. thick.

Thickness necessary above material (b) = 10 in. Thickness = 1 in. Omit material (b).

Make material (a) = 8 in. and pavement = 4 in.

Relative cost:
Pavement,	2 layers at	20.0 =	40.0
Material (a),	2 layers at	7.0 =	14.0
Material (c),	2 layers at	4.5 =	9.0
Material (d),	2 layers at	2.0 =	4.0
Subgrade,	2 layers at	1.0 =	2.0
		Total =	69.0

2. The pavement shall consist of a binder course 2.5 in. thick and a surface course 1.5 in. thick. The binder course shall have 5.5 per cent asphaltic cement and the surface 7 per cent asphaltic cement. The aggregate gradation for the surface course shall have the following characteristics:

Screen	Per cent passing
1½	100
1	84–100
½	66–84
No. 4	42–66
No. 10	31–55
No. 40	16–34
No. 80	10–22
No. 200	4–8

The binder course shall have the following aggregate gradation limits:

Screen	Per cent passing
2	100
1½	86–100
1	70–88
½	52–72
No. 4	32–52
No. 10	20–40
No. 40	10–22
No. 80	6–15
No. 200	3–7

A prime coat of cutback asphalt, grade MC-O at a rate of ½ gal per sq yd, will be applied to the base before the binder course is applied.

A heated tack coat of RC-3 should be applied to the binder course at the rate of 0.2 gal per sq yd before placing the surface course.

A seal coat of 0.2 gal per sq yd of hot cutback asphalt should be applied to the surface, followed by 8 to 15 lb of sand.

3.11   Elevation of low point must equal 98.42 ft.

$$\text{Station (L.P.)} = \frac{1.5L}{1.5 - (-2)} = 0.429L \text{ from P.V.T.}$$

Tangent offset at P.V.I. is T.O.

$$\frac{(0.429)^2}{(0.5L)^2} (\text{T.O.}) = 98.42 - (95.00 + 0.015 \times 0.071L)$$

$$\text{T.O.} = \frac{+3.42 - 0.00106L}{0.736} \tag{1}$$

$$2(\text{T.O.}) = \frac{(95.00 + 0.02 \times 0.5L) + (95.00 + 0.015 \times 0.5L)}{2}$$
$$- 95.00$$

$$\text{T.O.} = \frac{0.0100L + 0.0075L}{4} \tag{2}$$

By equating (1) and (2),

$$L = 3.42/0.00428 = \textbf{800} \text{ ft} \quad \text{c}$$

and T.O. = 3.50 ft.

$$\text{Station (L.P.)} = 100 + 0.071 \times 800 = 100 + 56.8$$

As an alternate, the length of curve could have been assumed as 800 ft. The T.O. is 3.50 ft. The elevation at the low point can be expressed as Y and the distance in stations from the P.V.T. as X. Then

$$Y = 101.00 - 1.5X + \frac{X^2}{4^2} (3.50)$$

At the low point the tangent will be horizontal and

$$\frac{dY}{dX} = -1.5 + \frac{2X}{16} (3.50) = 0$$
$$7X = 24$$

and $X = 3.43$ stations from the P.V.T. or $X = 100 + 57$.

3.12   1, false; 2, true; 3, true; 4, true; 5, false; 6, true; 7, false; 8, false; 9, true; 10, false.

3.13   Using test data,

$$d = \frac{V^2}{30(f - 0.05)} = 45 \text{ ft}$$

and

$$f = \frac{25^2}{30 \times 45} + \frac{1.5}{30} = 0.51$$

For accident car: $V^2 = 130 \times 30(0.51 - 0.05) + 5^2 = $ **42.7** mph **b**

3.14 $p = \frac{1}{12} = 0.08333$, $q = 1 - p = 0.91667$, $n = 10$.

   1. $p(1) = \dfrac{10!}{1! \, 9!} (0.08333)^1 (0.9167)^9$

          $= (10)(0.08333)(0.458) = 0.381$ **(38.1** per cent)   **1d**

   2. $p(0) = \dfrac{10!}{0! \, 10!} (0.08333)^0 (0.9167)^{10}$

          $= (1)(1)(0.419) = 0.419$ **(41.9** per cent)   **2a**

3.15   1, d; 2, f; 3, l; 4, a; 5, u; 6, b; 7, j; 8, s; 9, t; 10, h.

3.16   It may be inferred from (3) that a distribution reservoir is to be provided, rather than pumping directly from a small, clear well to the mains. Therefore filter capacity can be designed for normal rather than maximum flow.

   1. Daily consumption $= 14,000 \times 100 = 1.4$ mgd

Water filtered, including wash water $= 1.4/0.96 = 1.46$ mgd

   Required filter area $= (1.46/125) \times 43,560 = 509$ sq ft

   For a plant as small as this, the filters can be operated **on 1 eight-hour shift**

   Area $= 3 \times 509 \times (8/7.5) = 1,628$ sq ft

   Provide 3 units 20×27.5 ft in plan $= 1,650$ sq ft

   If one unit is out of service for extensive repairs the plant can be operated on a double shift to provide the required flow. An alternative would be to install 4 units.

   2. Alum $= 4.5$ grains per gal

       1 grain per gal $= 142.5$ lb per million gal

       $4.5 \times 142.5 \times 1.46 = 936$ lb of alum per day

   3. National Board of Fire Underwriters

$$Q = 1,020 \sqrt{P} \, (1 - 0.01 \sqrt{P})$$

when P = population in thousands

$$Q = 1{,}020 \sqrt{14} \, (1 - 0.01 \sqrt{14})$$
$$= 1{,}020 \times 3.74(1 - 0.037)$$
$$= 1{,}020 \times 3.74 \times 0.963 = 3674 \text{ gal per min}$$

National Board of Fire Underwriters recommends providing for a 10-hr fire in towns exceeding 2,500 in population. Therefore required storage = $3{,}674 \times 60 \times 10 = 2{,}204{,}000$

Peak of daily flow may occur during fire. Therefore domestic storage = 2.25×1.46 mgd = 3.29 million gal. Total storage, fire plus domestic use = $2.20 + 3.29 = 5.49$ million gal.

3.17 *a.* Detention period 30 to 60 sec. Sixty seconds recommended by most states.

b. Maximum rate of flow 0.5 to 1 ft per sec. The higher figure is generally used.

c, d, e. $wd = Q/V$, $wL = A$

For medium-size grit particles the recommended rate of flow is 50,000 gal per day per sq ft of surface area. For 1 mgd the required surface area would be 20 sq ft. Assume $w = 1$ ft (d). Then

$$L = 20 \text{ ft } (c) \text{ and } d = \frac{Q}{wV} = \frac{1.547}{1 \times 1} = 1.547 \text{ ft since 1 mgd} =$$

1.547 cfs. To provide grit storage capacity (g) make depth = 2 ft (e).

f. Use three units in parallel to take care of variations in flow.

g. Grit will vary in amount from 0.5 to 10 cu ft per million gal. Increasing depth to 2 ft will provide storage for $1 \times 20 \times 0.453$ = 9+ cu ft

h, i. Clean daily, manually, for so small a plant.

3.18 *a.* For raw sewage containing 250 ppm suspended solids 1.5 hr sedimentation should effect a reduction of ± 45 per cent or ± 138 ppm in the effluent (a).

b. $V = 1.5 \times 500{,}000/24 = \mathbf{31{,}250}$ **gal**

c. $V = 31{,}250/7.48 = 4{,}180$ cu ft

Assume $\pm 1{,}000$ gal per sq ft per day loading. Then $wl = 500{,}000/1{,}000 = 500$ sq ft. Take $l = 4w$, then $4w^2 = 500$, $w^2 = 125$, $w = 11.2$ using width = 12 ft; length = 48 ft; surface area = 576 sq ft; depth = $V/A = 4{,}180/576 = 7+$.

Use 8 ft.

$$w = \mathbf{12} \text{ ft}, \ L = \mathbf{48} \text{ ft}, \ D = \mathbf{8} \text{ ft}$$

d. On basis of 45 per cent removal we have 112 ppm settled out. This equals $\dfrac{112 \times 8.33}{2} = 468$ lb per day.

468 lb solids + 8,900 lb water = 9,368 lb

9,368 lb at 63 lb per cu ft = 148+ cu ft, say 150 cu ft

3.19 *a.* 500 ft maximum spacing between manholes. 300 ft preferable. On small sewers spacing is controlled by maximum convenient length of cleaning rods. On lines large enough for a man to enter, a greater spacing might be used, but is undesirable because of difficulty of removing a possible victim of noxious gases.

b. Twenty-five years minimum for sewers, ten years for treatment works. Sewers are buried in trenches. Additional units are readily constructed at a treatment plant and improvements in treatment methods may increase the efficiency of new units. Treatment-plant area should be adequate for long-term expansion.

c. 50 to 300 gal per capita per day average depending on character of district sewered. 130 gal is a reasonable value in most localities. Peak flow may be double average flow. Factor of safety should be applied to take care of errors in population estimates, etc. $130 \times 4 = 520$ gal per capita per day, sewer flowing full.

d. Sanitary sewers—2 ft per sec, sewer flowing full. Storm drains—3 ft per sec, sewer flowing full.

e. Sanitary sewers—8 in. minimum to minimize clogging. Storm drains are seldom less than 12 in.

f. Rational method is best for small areas. $Q = ciA$ where Q is flow in ft per sec, A is drainage area in acres, c is relative imperviousness of the contributory surface, and i is the rate of precipitation in inches per hour. A time-intensity curve of appropriate frequency is selected and the rainfall rate for the concentration time is used. This concentration time consists of time of flow to an inlet plus time of flow in the sewer.

g. Catch basins should be used on unpaved streets and on paved streets when the slopes are so steep that the velocity of surface wash exceeds the velocity of flow in the sewer at points further down the valley.

h. Crowns should be continuous. If invert of 24 in. is at elevation 100.00, invert of 15 in. would be at elevation 100.75 and invert of 8 in. would be at elevation 101.33.

3.20 The rational method of storm-drain design employs the equation $Q = ciA$ in which Q is the runoff in ft per sec to be carried by the drain from an area of A acres; c is the runoff coefficient or expected ratio of runoff to rain fall; i is the intensity of precipitation in second-feet per acre. Since 1 in. of rain per hour falling on an area of an acre is substantially equal to 1 cu ft per sec, the intensity of precipitation may be expressed more conveniently in inches per hour.

The acreage is measured directly from a map of the area. The value of c is estimated from a consideration of the relative imperviousness and the slope of the surface.

The value of i must be taken from a time-intensity curve of precipitation for a storm of a certain frequency of occurrence or from a rainfall formula derived for such a storm. The frequency of storm for which the curve in formula is to be used will depend upon the value of the property to be served by the drain and the magnitude of probable flood damage resulting from inadequate capacity of the drain. The greater the probable flood damage, the more it is necessary to design for infrequent storms of great magnitude.

Theoretically, the drain should be designed for such a storm that the increased flood damage resulting from a storm of greater magnitude would be slightly less than the increased construction cost to carry the flow from the latter storm. Practically, much judgement is required in selecting the storm-frequency curve to be used; for, while construction costs of various-sized drains can be estimated closely, it is difficult to estimate flood damages resulting from inadequate capacity of drain.

When an appropriate rainfall curve or formula has been selected, it is necessary to determine the concentration time since the rate of precipitation decreases with an increase in the concentration time t. This concentration time is made up of the inlet time, that is the time required for water to flow over the surface from the upper limit of a drainage area to the inlet, and the time of flow in the drain to the point under consideration.

For example, if the area contributing to the upper inlet of a drain is 2 acres, the runoff coefficient is 0.8, the inlet time is 5 min, and the rate of rainfall corresponding to this inlet time is 6 in. per hr, the drain would be designed for a Q equal to $0.8 \times 6 \times 2 = 9.6$ ft per sec. By drawing the profile of the street from the first to the second inlet we can determine the necessary slope for the drain. Then, with Q and S known, we can use pipe-flow diagrams or nomographs to determine the diameter of pipe necessary and the velocity of flow in the drain.

Suppose that this velocity works out to be 3 ft per sec and that the distance from the first to the second inlet is 360 ft. Then the time of flow in the drain is 2 min. At the second inlet then, the quantity of flow would be the product of the total contributary area, the runoff coefficient and the intensity of precipitation corresponding to a concentration time of $5 + 2 = 7$ min. This general procedure is continued as the drain is designed from inlet to inlet to the point of outfall.

3.21 Obtain or prepare a topographic map of the subdivision showing street and lot lines. A scale of 1 in. = 100 ft is generally suitable. Contour interval will vary from 1 to 10 ft depending upon steepness of slopes.

Show on this map the existing sewer into which the new sewers are to discharge and locate the new sewers to flow by gravity wherever possible.

Obtain or run profile levels for all streets to be sewered.

Locate all known underground utilities or structures on both plan and profiles. Obtain data on soil characteristics and, particularly, depths to ground water or rock.

Obtain data on probable population to be served 25 years hence, expected sewage flow per capita, probable ground-water infiltration, and expected flow from any commercial or industrial districts within the area.

Obtain data on depths of cellars, particularly on the downhill side of sidehill streets so the sewers may be placed deep enough to drain such cellars, using normal slopes for house connections.

Starting at upper end of a lateral carry design to a junction, then start at upper end of another lateral entering this junction. Proceed in this manner until entire system has been designed. Profiles of 1 in. = 100 ft horizontal and 1 in. = 10 ft vertical, should be prepared as work is carried out on plan since breaks in grade on profile will control location of intermediate manholes.

Manholes should be located at all intersections, at every change in grade or alignment, and at intervals not exceeding 500 ft, preferably less. Manholes should be numbered in a logical sequence.

Computations for each line should be arranged in tabular form and should show streets, manhole numbers, population, domestic sewage and infiltration, distances between manholes, slopes, size and kind of pipes required, surface elevations at manholes to 0.1 ft, invert elevations of sewers at manholes to 0.01 ft, velocity of flow when full, and capacity when full.

Much of this information is shown on both plan and profiles, together with arrows to indicate direction of flow.

3.22 *a.* The separate sludge-digestion process comprises the collection of sewage solids in primary sedimentation tanks (and also in the final tanks of certain secondary treatment processes) and the passing of these solids, by pumping or by gravity flow, to one or more tanks where digestion of the sludge occurs. Digestors are generally provided with floating covers, devices for the collection and utilization of sludge gases, and arrangements for heating and controlling the reaction of the sludge to reduce the time required for digestion. Many plants use two-stage digestion, the primary stage of about a week taking place in one digestor and the sludge then being transferred to a second digestor for about three weeks.

b. Sludge may be dewatered on open or covered drying beds or on vacuum filters. Sand beds require considerable acreage and may be high in first cost, particularly when covered. Drying is slow but does not require skilled attendance.

Vacuum filters require little space and dry the sludge quickly. They are not low in either first cost or maintenance. Skilled attendance is necessary since conditioning of the sludge may be required to obtain a satisfactory filter cake.

3.23   *a,* 2; *b,* 2; *c,* 4; *d,* 4; *e,* 4; *f,* 2; *g,* 2; *h,* 1; *i,* 4; *j,* 2.

3.24   Recommended capacity of septic tank for 10 persons = 900 gal. Recommended dimensions 3.5 ft wide, 8.5 ft long, 4.5 ft liquid depth, 5.5 ft total depth. If a domestic garbage grinder is to be installed, the capacity should be increased by 50 per cent. Capacity shown above provides for 2 years' accumulation of sludge plus 50 gal per capita sewage flow per day.

For fairly porous soil allow 450 sq ft of absorption area in bottom of disposal trenches. For 18-in. width of trench, length will equal 450/1.5 = 300 ft.

Sewage from house will flow into septic tank shown in Fig. 3.24a. Clarified effluent will flow out to distribution box and thence to 300 ft of 4-in. open-joint tile drain, Fig. 3.24b. Top of tile should be about 15 in. beneath ground surface. Tile should be surrounded with gravel, coarse sand, or cinders for a distance of at least 6 in. on sides and bottom and at least 2 in. above top of pipe. Open joints should be covered by tar paper. A more effective septic tank than that shown in Fig. 3.24b would involve a small secondary chamber following the main septic tank. This would be provided with an automatic siphon to permit dosing the entire tile field periodically instead of permitting the effluent

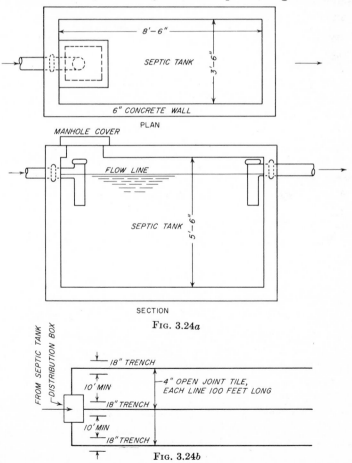

PLAN

SECTION

Fig. 3.24a

Fig. 3.24b

to trickle slowly through the first few joints as is likely to occur when no siphon is provided.

3.25 Malaria is transmitted by various mosquitoes of the *Anopheles* group. *Anopheles quadrimaculatus* is the most important vector of the disease in the eastern United States.

The mosquito must bite persons whose blood contains the malaria parasites. Certain stages in the life cycle of the parasite must then take place in the mosquito before the insect can transmit the disease by biting others. Obviously, if there are no cases in the area, there can be no transmission of the disease even though the density of *quadrimaculatus* infestation may be heavy.

Where there are cases in the area there are a number of control measures possible: (1) Infected persons should receive proper medical treatment to reduce or eliminate the parasites from their blood. (2) Infected persons should be protected from the bites of the insects. This is not difficult with proper screening since *quadrimaculatus* is predominantly a dusk and night biter. (3) The raising of cattle in rural areas reduces the incidence of the disease since the insect prefers cattle to man for a blood meal. (4) Standard mosquito-control measures aimed principally at the insect in the larval stage are effective in reducing mosquito density. (5) Blood tests of all persons in a heavily infected area should be carried out so that all infected persons can receive proper medical treatment. (6) DDT can be used as a residual spray in inhabited buildings so that adult mosquitoes are killed before they can transmit the disease. (7) Educational campaigns are effective in informing the public concerning the steps necessary to eliminate the disease or to reduce its incidence.

3.26   *a.*   1. Case rate = 2,891/2,000 = 1.446 per thousand

            2. Death rate = 1,304/2,000 = 0.652 per thousand

            3. Fatality rate = 1,304/2,891 × 100 = 45 per cent

b. Morbidity is the number of cases of a disease occurring during a year in terms of the population in hundred-thousands. While generally computed for the entire population, it may also be determined for certain age groups or classes.

3.27   *a.* While many diseases are transmitted by milk, tuberculosis and undulant fever are those most readily prevented by protecting the dairy herd. The tuberculin test is applied periodically, usually annually or semiannually, to detect tubercular cows in the herd. Such animals are destroyed. Undulant fever can be controlled by removal of infected animals from the herd and by pasteurization of the milk.

b. Measures to protect the milk line are designed to assure healthy cows, clean and healthy workmen, clean and dustless barns, prompt cooling of the milk in a separate milk house, proper utensils and the effective cleaning and sterilization thereof, and pasteurization of the milk.

c. The above steps will minimize or eliminate the transmission of typhoid, scarlet fever, diphtheria, undulant fever, and other diseases by milk.

d. Milk is an ideal medium for the development of many bacteria and bacterial growth is rapid in dirty milk. Pasteurization is not a sterilization process and hence a high-count raw milk still has considerable numbers of bacteria in it after pasteurization. The most effective measures to eliminate this danger are the provision of sanitary conditions in the dairy and the proper pasteurization and bottling of the milk.

e. Milk may be pasteurized at 143°F for 30 min or flash pasteurized at 160°F for 15 sec. Control of early flash pasteurizers was uncertain and the method was consequently but little used until recently. Modern flash pasteurizers provide excellent control and are as safe as the holder types. They require much less space and are consequently increasing in use.

3.28 *a.* Bubonic plague is transmitted to men by the bite of fleas from an infected rat.

b. Murine typhus fever (endemic) is transmitted by the bite of fleas or mites from an infected rat.

c. Malaria is carried by various mosquitoes of the *Anopheles* genus.

d. The common vector of dengue is the *Aedes aegypti* mosquito.

e. Smallpox is commonly communicated by discharges from the nose and throat of persons suffering from the disease. It may also be transmitted through the agency of material from pustules or skin lesions.

f. Undulant fever may be transmitted by unpasteurized milk, by contact with infected slaughtered animals, by contact with the dust of pens containing infected animals, and by contact with these animals when they abort since the disease causes infectious abortion in cows, sows, and goats.

g. Tularemia is usually communicated to man in skinning or plucking rabbits, quail, and other game. It may also be transmitted by certain biting flies.

h. The common vector of yellow fever is the *Aedes aegypti* mosquito. In jungle areas other mosquitoes are involved including *Sabethini* and *Haemagogus.*

i. Rocky Mountain spotted fever is transmitted from various small wild animals to man by ticks.

3.29 The steps to be taken and data to be collected in determining the source of the typhoid outbreak would include:

1. Prompt reporting to the local health department of all cases by the attending physicians.

2. The filling in of questionnaires relative to the sources of milk, water, shellfish, and other foods consumed by those persons affected by the disease.

3. The preparation of spot maps showing the areas within the community where there are cases of the disease.

4. The prompt bacteriological analysis of supplies of water, milk, shellfish, and other foods.

5. The check of restaurants and other establishments where food had been served or sold to victims of the disease for possible typhoid carriers.

6. A check of private well supplies and of cross connections between potable and industrial water supplies.

3.30 Two methods suitable for the disposal of garbage are sanitary fill and incineration.

Sanitary fill differs from open dumping in that the garbage is covered with about 2 ft of earth at the end of each working day. It is suitable where there are low areas, abandoned borrow pits, and other suitable land within or near the municipality and where suitable earth, preferably sandy soil, is available for covering. The method is generally cheaper than incineration and permits the eventual utilization of tracts of land for parks, housing developments, and other uses.

Incinerators will provide for the complete stabilization of organic materials and a great reduction in volume of the wastes collected. First costs and operating costs are high. They are particularly useful where extensive tracts of land for sanitary fill are not locally available.

3.31　There are several methods for determining the safe yield of a watershed from the storage which can be developed economically at a given reservoir site or, conversely, the storage which will be necessary to furnish a desired yield. One of the most satisfactory procedures is to use a mass curve of modified runoff.

Actual records of runoff, usually monthly, are obtained for a stream-gaging station on the stream relatively close to the site of the proposed dam. The longer the period of years for which records are available the greater the likelihood of the record including a period of extreme drought which will control the yield or the required storage.

If the records are given in cubic feet per second at the gage, they are converted to inches of depth on the watershed area above the gage and these figures are then used for the slightly different watershed area above the site of the proposed dam.

When the proposed reservoir will have a large surface area the actual runoff records can be modified downward to account for the expected increased evaporation losses from the reservoir. The amount of water to be released downstream for the use of riparian owners below the reservoir site is also deducted from the monthly values of actual runoff. The remainder will be water available for storage, for power or water supply use, and for waste over the spillway at times when the reservoir is full.

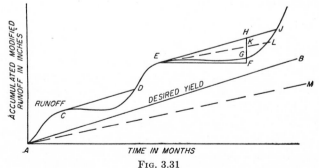

Fig. 3.31

The accumulated totals of modified runoff are then added month by month and plotted as a mass curve (Fig. 3.31). If, let us say, a yield of 100 mgd is desired, this value is expressed in inches depth per month on the watershed area and plotted as the sloping line AB. Lines CD and EJ are drawn parallel to AB from points C and E where the runoff starts to decline. It is obvious that the worst period of drought has occurred between E and F. During this period the total inflow is represented by the ordinate FG. The total required yield is represented by FH. Therefore water in the amount GH must be released from storage to supply the demand. This storage, shown here as inches depth on the watershed, can be converted into acre feet, billions of gallons, or other convenient units.

It may be uneconomical to develop this volume of storage at the site. If the storage which can be developed is GK we draw the dotted line EKL. The safe yield is then reduced to that represented by AM, drawn parallel to EJ.

3.32 *a.* The reduction of losses through leakage and waste in a water-distribution system is becoming ever more important as the cost of supplying water to a community increases year after year. Water pressures will also improve as waste is minimized. In some cities the actual flow of water in the system may be from $1\frac{1}{2}$ to 2 times the necessary consumption.

b. The complete metering of a city is most effective in reducing wastes. Leakage from broken mains may be detected by visual inspection. The Cole pitometer is widely used to detect leaks in mains. Audio devices to amplify the sound of escaping water are also used. Periodic waste-water surveys and careful maintainance of the distribution system may reduce the unaccounted-for water to perhaps 2 per cent of the delivered water.

c. When water shortages exist, educational campaigns have been quite effective in reducing actual consumption and minimizing the actual waste of water.

3.33 A rapid sand water filtration plant normally includes the following units or items: (1) screens for the removal of leaves and other large objects; (2) a chemical storage house for the storage of chlorine, ammonia, lime, alum, activated carbon, and soda ash; (3) a mixing device such as a flash, mixer, baffled mixing chamber, hydraulic jump, a low-lift pump suction to provide intimate mixing of activated carbon, lime, and alum, or other coagulant with the raw water; (5) one or more flocculators in which gentle stirring action permits satisfactory floc formation; (6) one or more sedimentation basins in which the coagulant can settle out, removing turbidity, color, and many bacteria; (7) one or more filter units as described in 3.27*b*; (8) a clear well for the storage of filtered water; (9) provisions for final chlorination and pH adjustment in the clear well; (10) wash-water pumps and a wash-water tank to permit backwashing the filters; (11) air compressors for air washing of filters in some cases; (12) a sewer connection for the dirty wash water; (13) suitable meters and gages to show the quantity of water being processed at any time; and (14) a well-equipped laboratory, adequately staffed, for analyses of the raw, finished, and delivered water. Low-lift and high-lift pumps and power-generation equipment will be required in some cases.

3.34   *a.* The zeolite softening process is essentially one of base-exchange. The zeolite, through which the water passes, consists of a chemical compound so loosely bound that it will readily exchange its sodium radical for the calcium and magnesium radicals in the water to be treated. The process is also reversible, depending upon the concentration of dissolved chemicals. Zeolites may be natural, such as glanconite or greensand, or synthetic, such as Permutit.

In passing the raw water through the zeolite the reactions are as follows:

$$NaZ \;+\; \frac{Ca}{Mg}(HCO_3)_2 \rightarrow \frac{Ca}{Mg}(Z) \;+\; ZNaHCO_3$$

Sodium zeolite	Calcium or magnesium bicarbonate hardness	Calcium or magnesium zeolite	Sodium bicarbonate

$$NaZ \;+\; \frac{Ca}{Mg}(SO_4) \rightarrow \frac{Ca}{Mg}(Z) \;+\; Na_2SO_4$$

Sodium zeolite	Sulfate hardness	Calcium or magnesium zeolite	Sodium sulfate

The end products of these reactions are soluble and do not produce hardness in the water.

When the softening capacity of the zeolite is exhausted, the raw-water flow is cut off and the zeolite is backwashed with a brine solution. The reaction of the salt with the calcium zeolite is as follows:

$$CaZ \ + NaCl \rightarrow \ NaZ \ + \ CaCl$$

Calcium zeolite		Sodium zeolite	Calcium chloride

After such backwashing, the restored zeolite can again be used to soften water.

   b. The effect of H_2S in raw water to be chlorinated is to increase the required dosage of chlorine to provide a disinfecting residual in the water since a part of the chlorine added will react with the hydrogen sulfide.

   c. Lime and soda ash are added to water to remove the bicarbonates and sulfates of calcium and magnesium which cause hardness. The chemical reactions with the bicarbonates and sulfates of calcium are as follows:

$$Ca(HCO_3)_2 + Ca(OH)_2 \rightarrow 2CaCO_3 + 2H_2O$$

Calcium bicarbonate	Lime	Calcium carbonate	Water

$$CaSO_4 \ + \ Na_2CO_3 \rightarrow \ CaCO_3 \ + \ Na_2SO_4$$

Calcium sulfate	Soda ash	Calcium carbonate	Sodium sulfate

The reactions with magnesium compounds are similar.

The addition of the lime and soda ash results in insoluble compounds which will settle out or in soluble compounds which do not cause hardness.

3.35   a. Chloramines are used in water treatment: (1) for disinfection; (2) for taste and odor control; (3) as algicides; (4) for weed control in reservoirs.

   b. Sodium hexametaphosphate (Calgon) is used to control the solubility equilibrium of calcium carbonate in water-softening plants; in the processing of industrial water supplies; and in the "threshold treatment" of water in an attempt to inhibit corrosion.

   c. Sodium bisulfite is a reducing agent. As such, it may be used in the deaeration of water to minimize its corrosiveness. It may also be used in dechlorination, to avoid chlorinous tastes after heavy applications of chlorine. Because careful chemical control is necessary, its use has not become widespread in water treatment.

d. Sulfur dioxide in aqueous solution has been used to clean rapid sand filters which have become coated with iron or manganese. This process is patented. Sulfur dioxide is also used in dechlorination.

e. Activated carbon is chiefly useful in water treatment to control tastes and odors.

f. Aluminum sulfate (alum) has a variety of uses in water treatment. The chief use is as a coagulant to remove color, turbidity, and bacteria in connection with rapid sand filtration.

g. Hydrated lime is sometimes used before filtration in an attempt to reduce the bacterial load on filters; it is frequently used after filtration to combat the corrosive quality of the water resulting from the presence of oxygen and carbon dioxide and also to adjust the pH value; it is widely used in water softening.

h. Sodium chloride is most widely used in water treatment in the regeneration of zeolite water-softening tanks.

i. Sodium carbonate is added to waters of low alkalinity to permit the use of alum as a coagulant; it is used in cleaning filters; and it is widely used with lime in water softening.

3.36   *a.* (1) *A single-stage centrifugal pump* is one having a single impeller. It is not, in general, economical to use when the operating head exceeds 300 ft (*b*).

(2) In a *centrifugal* pump a rotating impeller imparts velocity to the water entering from the suction line and forces it to flow under pressure into the discharge line.

(3) For higher heads (*c*) a *multistage* pump is used. Here two or more impellers are connected in series, the discharge from the first impeller passing to the suction of the second, etc.

(4) Both the reciprocating and the rotary pumps are *positive-displacement* pumps.

(5) *Reciprocating pumps* are provided with a piston or plunger moved by reciprocating action in a cylinder provided with the necessary inlet and outlet valves. The movement of the piston alternately draws water into the cylinder from the suction line and forces it out under pressure into the discharge line.

(6) In the single-acting reciprocating pump the flow is intermittent and unsteady. This condition is largely corrected by using a *double-acting plunger pump* in which there are two sets of inlet

and outlet valves so that continuous discharge of water is obtained.

(7) *A rotary pump* has cams or gears which rotate in a casing, thus forcing the incoming water around the casing and out the discharge line on each revolution.

3.37 *a. Bacillus coli* and other bacteria of *coli-aerogenes* group are normally found in the intestinal tracts of man and animals. The presence of these organisms in water supplies is indicative of sewage pollution. Although not in themselves pathogenic, their presence indicates possible pollution of the water by pathogenic intestinal bacteria such as *B. typhosus*, and hence the need for purification of the supply.

The *E. coli* index for a sample of water is the reciprocal of the smallest quantity of the sample, in cubic centimeters, which gives a positive test for coliform organisms. The determination of the "most probable number" is in greater use today for determining the numbers of *E. coli* present in a sample.

b. Anaerobic bacteria thrive in the presence of suitable nutrients and in the absence of free oxygen.

Aerobic bacteria require free oxygen for their development.

c. Hardness in water is commonly expressed in terms of parts per million of $CaCO_3$ even though the total hardness may be made up of the carbonates and sulfates of calcium and magnesium. A carbonate hardness of 100 ppm is indicative of a water supply about on the border between slightly and moderately hard. Such a supply would not, as a rule, be softened for municipal use.

3.38 The factors which control the accuracy of computation of the lengths of the sides of a triangulation network include: (1) precision of measurement of base line; (2) precision of measurement of angles; (3) size of "distance angles" involved in the computations; (4) number of side and angle equations provided by the geometrical figures used in the network; (5) number of stations occupied by the transit party; and (6) frequency with which check base lines are measured.

The effects of items (1) and (2) are obvious. Since the lengths of triangle sides are computed by the law of sines, the angles used in the computations (distance angles) should be of such size that their sines are changing slowly so that errors of angular measurement will have a minimum effect upon the computations. Thus, the closer the angles used approach 90°, rather than 0° or 180°, the less effect errors in angular measurement will have on the computed lengths of network sides. The effect of the size of the angles is indicated by

$$\delta A^2 + \delta A \delta B + \delta B^2$$

in the strength of figure equation. δA equals the tabular difference for 1 sec. in the log sine of angle A in the sixth decimal place. δB is the corresponding value for log sine of angle B.

Items (4) and (5) reflect the number of geometrical checks on the computed lengths permitted by the network. They are included in the $(D - C)/D$ term in the strength of figure equation. D represents the number of new directions observed. C is the number of geometrical conditions (side and angle equations) in the network figure.

If a triangulation network is extended indefinitely, its strength gradually becomes less. Specifications accordingly provide for the measurement of a check base periodically in extending a network.

3.39 Watershed control should include the right to approve or disapprove sewage-disposal facilities in the resort area around the lakes; constant vigilance with respect to records of the quality of the effluent from the sewage-treatment plants; provision for the application of copper sulfate for algae control in the lakes; and the establishment of many sampling stations so that the source of any unusual pollution or contamination may be quickly determined and steps taken to adjust the situation. A chlorinator at the outlet of the lowest of the lakes may reduce the load on the plant.

A boom should protect the influent canal from the entrance of large floating objects. Mechanically cleaned screens should be provided to remove leaves and other objects and to protect the pumps. Gates are desirable to control the water intake.

While a flash mixer may be used for mixing various chemicals with the raw water, satisfactory mixing should be provided for

by applying the chemicals to the suction line of the low-lift
pumps. Therefore, the low-lift pumphouse should have, in addition to space for pumps and motors, adequate space for chemical
storage, for solution-feed or dry-feed machines, and for measuring
devices so that the quantity of chemicals added can be proportionate to the flow. At this point, we may prechlorinate, apply
activated carbon, add alum or other coagulant, and apply soda
ash when the alkalinity of the raw water is low. Rail or truck
delivery of chemicals should be provided for. A storage yard for
chlorine containers is necessary. Provision should be made for the
fork lifting of pallets of bagged chemicals and transportation,
via elevator, to overhead storage from which the chemicals can
be fed via hoppers by gravity to the proportionate feeders.

The discharge from the low-lift pumps will pass to a flocculation tank and thence to a sedimentation basin. This basin may be
cleaned periodically or continuously, depending upon the volume
of sludge which may be anticipated. The sludge may be returned
to the river below the plant, lagooned, or dried and used as fill,
as may be dictated by local conditions.

The settled effluent will pass to rapid sand filters of the usual
type and then, through rate controllers, to the clear well. Pumps
to raise filtered water to the wash-water tank shall be provided.
Customary provisions for backwashing the filters shall be provided. The backwash water shall pass to the sewer or to the river
below the plant. Provision may be made for surface and air wash
as well. In the latter case, air-compression equipment must be
installed in the control house.

Postchlorination should be provided for in the clear well.
Lime will probably be necessary for pH adjustment and for CO_2
removal, although aeration is sometimes used for the latter purpose. Necessary facilities for chemical storage must be provided
near the clear well.

The high-lift pumps will raise the water from the clear well to
the distribution reservoir. Both low-lift and high-lift pumps will
normally be powered by electric motors. Standby power units,
either gasoline or diesel engines, must be provided in the event
of primary power failure.

Within the plant buildings space must be provided for an

administrative office, a chemical and bacteriological laboratory, heating equipment, change room, lavatories, garage space for maintenance trucks, storage space not only for chemicals but for maintenance supplies as well, fuel storage, motor and pump rooms, etc.

3.40 The inflow hydrograph will have the form shown in Fig. 3.40a. Determine the values of inflow in acre-feet for each time interval. Accumulate these and plot the accumulated inflow

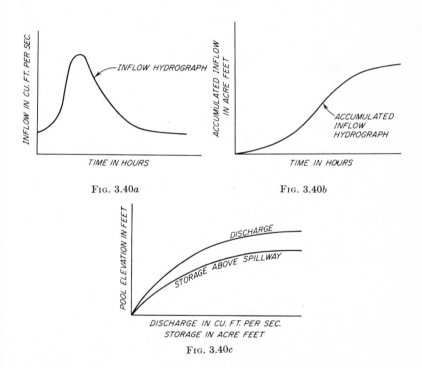

FIG. 3.40a   FIG. 3.40b

FIG. 3.40c

in acre-feet as shown in Fig. 3.40b. On a separate sheet, plot the pool elevation-discharge curve and the pool elevation-storage curve as shown in Fig. 3.40c. The first is determined by assuming various pool elevations and computing the corresponding discharge from the equation $Q = CbH^{1.5}$. The latter is determined from a contour map of the area flooded at various elevations of flow above the spillway.

Now assume a storage increment in acre-feet small enough so that the discharge curve during the period of accumulation of this storage is practically a straight line. This amount of storage (from Fig. 3.40c) will give a certain spillway discharge at the end of the period (from the same figure). Take the mean discharge for the period and convert it to acre-feet for a time period of

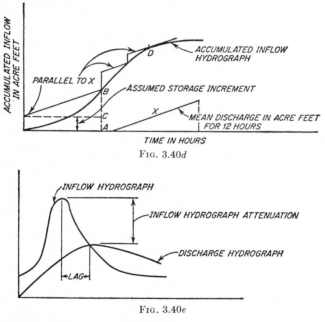

Fig. 3.40d

Fig. 3.40e

say 12 hr. Plot this at the right of Fig. 3.40b (as shown in Fig. 3.40d) and, from the assumed inflow storage in acre-feet, draw a parallel line until it intersects the curve of accumulated inflow. At this point (Fig. 3.40d) the total inflow is represented by AB, the total discharge is shown by CB, and the volume stored above the spillway AC is equal to the assumed value.

Continue adding storage increments, of greater and then lesser values, and follow the same procedure until the discharge line becomes tangent to the curve of accumulated inflow at D. At this point the inflow and outflow will be equal and the flood will have reached its maximum height above the spillway. The process may be continued, using negative increments of storage, to determine the form of the discharge hydrograph. As shown in Fig. 3.40e this may be compared with the inflow hydrograph to indicate the time lag in cresting and the attenuation, or reduction in the crest height, resulting from reservoir storage.

3.41 a, false; b, false; c, true; d, true; e, false; f, true; g, false; h, false; i, true; j, false; k, true; l, true; m, true; n, false; o, true; p, false; q, true; r, false; s, true; t, false; u, true; v, false; w, false; x, false; y, true.

3.42 a. Precipitation in northern New Jersey results from cyclonic storms of continental origin during the fall, winter, and spring months. Much of the summer precipitation results from local convective air movements accompanied by thunderstorms. In the late summer and early fall, hurricanes originating in the equatorial regions of the Atlantic may bring copious rainfall to this area. The net result is to produce reasonably uniform precipitation throughout the months of the year, although dry and wet periods are observed depending upon the effectiveness of each of these factors. Some increase in orographic precipitation is observed in the northern highlands, but this may be counterbalanced by the proximity of the coastal regions to hurricane paths.

b. The Thiessen method weights each station in proportion to the part of the watershed closest to it and is useful when there are few stations on or near the watershed and when conditions in the area are reasonably homogeneous. When many stations are located on or near the watershed, isohyetal lines may be drawn on the map and the mean precipitation determined from these. The mean value of all Weather Bureau records in the neighborhood is the least satisfactory.

c. Precipitation in the San Francisco area occurs chiefly during the winter months when the prevailing winds from the Pacific are cooled to below the dew point in crossing the cooler land areas.

d. In the Upper Mississippi Valley, cold fronts from Canada provide relatively dry air during the winter months and hence precipitation is light. During the summer, warm moisture-laden air from the Gulf moves up the valley. Because of high temperatures, its moisture is not dropped in the lower valley. With increased altitude and latitude, it is cooled. If cooled below the dew point, precipitation will occur during the crop-growing season. When the summer temperatures in this region are above normal, precipitation will not occur and droughts and dust storms have their inevitable effect.

3.43 *a.* 1. Biochemical reactions are accelerated at high temperatures so that oxidation and reduction are likely to occur in the sewer system rather than being delayed until the changes can occur under controlled conditions at the disposal plant. Also, the corrosion of iron, steel, and concrete is accelerated at high temperatures.

2. The appearance of the stream will suffer. Spawning beds for fish may be destroyed. Reaeration will be retarded in the stream.

3. The water will become turbid and unsightly. Sludge deposits will form which may interfere with bottom life or with navigation. Odors are likely to develop. Fish may be killed by clogging of their gills.

4. The water will become turbid. Light penetration will be reduced, thus reducing photosynthesis. Fine organic solids will increase the B.O.D. loading and cause a reduction in dissolved oxygen. Fish may be killed if the waste is toxic or if it injures the gills.

5. Fire and explosion hazards may result. The waste will interfere with many phases of treatment-plant operation.

b. 1. Heat-recovery systems may be shown to be of economic value to the company discharging the wastes. Otherwise the wastes may be cooled in holding basins or diluted with cold water.

2. Sawdust may be burned with waste lumber to provide steam power. It may be removed by fine screens or settling basins.

3. Lagoons may be used. Vacuum filters may be used for drying the sludge. The dried sludge may be burned, with lime recovered for reuse.

4. Usually chemical or biological treatment is required. Coagulation followed by flocculation and sedimentation will remove much of the solid matter. Treatment of the effluent in trickling filters, contact aerators, or activated-sludge tanks will result in further removal.

5. Oil and grease traps should be placed in waste lines. Additives may break up emulsions chemically. Coagulants and filter aids may be used.

3.44 *a.* Physical processes: dispersion, sedimentation, aeration, action of sunlight.

Chemical processes: oxidation, reduction, disinfection, coagulation.

Biological processes: aerobic and anaerobic bacteria oxidize and reduce the organic compounds; algae, fungi, protozoa, rotifera, crustacea, and larger forms of plant and animal life also play a part.

b. Hygienic considerations: contamination of water supplies, waters used for bathing, shellfish, ice; pollution which affects the public health in any way.

Economic considerations involve depreciation of the value of abutting property, damage to fish life, industrial water supplies, livestock which are watered from the polluted stream, river and harbor improvements and navigation, impairment of recreational facilities.

Legal considerations involve the right of other riparian owners to receive the water in a reasonable condition and to make reasonable use of it.

Aesthetic considerations are involved in the appearance and odor of the receiving stream.

Whenever any of the above is seriously affected by the disposal of raw sewage by dilution, either primary or primary and secondary treatment must be provided before dilution, the extent of the treatment depending upon the pollutional load and the ability of the receiving water to handle this load satisfactorily.

3.45 *a.* Racks, coarse and fine screens, grit chambers, preliminary and secondary plain sedimentation basins, grease traps, skimming tanks, chemical precipitation tanks.

b. Intermittent sand filters, standard and high-rate trickling filters, contact aerators, the activated-sludge process and modifications thereof.

c. Chlorination.

d. Separate sludge-digestion tanks (preferably heated), Imhoff tanks, septic tanks (for use in rural areas).

e. Sludge-drying beds (either covered or uncovered), sludge elutriation and thickening, vacuum filters, centrifuges, filter presses, kiln dryers, incinerators. Disposal at sea, burial, lagooning. Use of dried digested sludge as fill or as a soil conditioner.

f. Dilution, lagooning, irrigation.

3.46 a. 2.

b. Ferric chloride, a coagulant used to precipitate and condition sludge.

c. The sewer is being cleaned.

d. To kill tree roots which have entered the sewer through cracks or poorly constructed joints.

e. That a test was being made for the presence or absence of adequate oxygen in the sewer after having first made certain that no explosive gases were present.

f. Two feet per second.

g. Upgrade.

h. Manholes should be placed at every change in alignment, grade, size of pipe, and at intervals not to exceed 500 ft.

i. The velocity in a circular sewer is a maximum at slightly above 0.8 depth, since it varies directly with the hydraulic radius. At greater depths, the wetted perimeter increases more rapidly than does the area, and hence the hydraulic radius decreases. Thus the maximum discharge, the product of velocity and area, is obtained at about 0.93 depth, when both velocity and area are somewhat below their maximum values but when their product is a maximum.

j. In the septic tank, sedimentation and sludge digestion take place in the same chamber. Rising masses of sludge, buoyed up by entrapped gas, interfere with the sedimentation process. The required detention period is therefore much longer than when the sedimentation and digestion processes are separated. Septic tanks are therefore uneconomical for municipal use. They are useful in rural disposal systems where there are no provisions for skilled operating personnel.

3.47   1. $\dfrac{5 \times 10^6}{4} \times \dfrac{6}{24} \times \dfrac{1.25}{7.48} = 52{,}500$ cu ft

The BOD of the entering raw sewage at aeration tanks is equal to $(5 \times 8.33 \times 300)0.70 = 8{,}750$ lb per day. Using four tanks the volume required is

$$\dfrac{8{,}750/4}{38/1{,}000} = \textbf{55{,}750} \text{ cu ft} \qquad \textbf{1d}$$

Area (using depth of say 15 ft) $= \dfrac{55{,}750}{15} = 3{,}750$ sq ft

Use 30 by 125 ft tanks.

2. Aeration capacity $= \dfrac{1.5 \times 5 \times 10^6}{24 \times 60} = \textbf{5{,}200}$ cfm $\qquad \textbf{2c}$

3. Diffuser capacity $= \dfrac{8{,}750 \times 1.5 \times 1{,}000}{4(24 \times 60)} = \textbf{2{,}280}$ cfm $\quad \textbf{3a}$

3.48   Use Streeter-Phelps equation; 400 mgd $\times 1.55 = 620$ cfs

Weighted BOD (5 day at 20°C) $= \dfrac{620 \times 150 + 2{,}000 \times 0.05}{2{,}620}$

$$= 35.9 \text{ ppm}$$

$$\text{BOD} = LA(1 - 10^{-k_1 t})$$

or $\qquad 35.9 = LA[1 - 10^{-(0.5)(5)}]$

Solving LA at 20°C = 52.5 ppm.

$$LA_{T°C} = LA \text{ at } 20°C(0.02T°C + 0.6)$$

hence LA at 15°C = $52.5(0.02 \times 15 + 0.6) = 47.25$ ppm

At 15°C; dissolved oxygen (D.O.) saturation concentration

$$= 10.15 \text{ ppm}$$

Weighted D.O. $= \dfrac{2{,}000 \times 10.15 + 0 \times 620}{2{,}620} = 7.75$ ppm.

Initial deficit = $10.15 - 7.75 = 2.40$ ppm.

Time in stretch $= \dfrac{80}{20} = 4$ days flow downstream

$$\text{Deficit (at 4 days flow)} = \frac{k_1 A}{k_2 - k_1} (10^{-k_1 t} - 10^{-k_2 t}) + \text{D.O.}(10^{-k_2 t})$$

$$\text{or} \quad \text{Deficit} = \left\{ \frac{0.08(47.25)}{0.28 - 0.08} [10^{(-0.08)(4)} - 10^{(-0.28)(4)}] \right.$$

$$\left. + 2.4[10^{(-0.28)(4)}] \right\} = \textbf{7.80 ppm} \qquad \textbf{b}$$

3.49 a, 1; b, 1; c, 2; d, 1; e, 1; f, 4; g, 4; h, 2; i, 2; j, 2.

3.50 Total hardness due to Ca^{++}, Mg^{++}, Sr^{++} is:

Cation	Equivalent weight	Hardness (as mg per liter $CaCO_3$)
Ca^{++}	20.0	$20(50/20) = 50.0$
Mg^{++}	12.2	$15(50/12.2) = 61.5$
Sr^{++}	43.8	$5(50/43.8) = 5.8$
		117.3

Alkalinity is:

Anion	Equivalent weight	Hardness (as mg per liter $CaCO_3$)
HCO_3	61.0	$5(50/61) = 4.1$
CO_3	30.0	$10(50/30) = 16.7$
		20.8

Carbonate hardness = **20.8** mg per liter as $CaCO_3$ (20 + 30 = 50). Non-carbonate hardness = 117.3 − 20.8 = **96.5** mg per liter as $CaCO_3$.

3.51 *a. Atterberg limits* comprise the liquid, plastic, and shrinkage limits. The liquid limit is the water contents at which the soil has such a small shearing strength that it flows to close a preformed groove of standard width when jarred in a specified manner. The plastic limit is the water content at which a soil begins to crumble when rolled into threads of a specified diameter. The shrinkage limit is the water content that is just sufficient to fill the pores when the soil is at its minimum volume attained by drying. These limits furnish a sound basis for the identification and classification of soils and predict qualitatively their general behavior when they are correlated with soil tests of a more specific nature. The information furnished by these tests may be misleading, since, for the most part, they are performed on remolded soil.

For determining the liquid limit, a special standard-cup device is used. All the dimensions of the device are fixed and the jarring is accomplished by turning the crank at the rate of 2 revolutions per second. The moisture content when the groove is closed by 25 blows of the cup is defined as the liquid limit.

To find the plastic limit, a thread of soil is rolled out on a glass plate to a diameter of $\frac{1}{8}$ in. When the moisture content has been reduced to the point where crumbling starts, the moisture content is measured.

The shrinkage limit is found by measuring the volume (by displacement of mercury) of a soil pat which has been dried in a desiccator. The volume is then used to determine volume of the voids remaining in the soil pat.

b. Permeability test. This test is used to determine the coefficient of permeability in Darcy's equation, $q = kiA$, for a given soil. In the constant-head permeameter, the net head of water on the soil sample of a known length and area is kept constant, and the rate of flow is measured. The permeameter tube or container holds the sample in place by means of screens or fitted porous stones. Before running the test it is necessary to rid the system of any entrapped air.

c. The direct-shear test is used to measure the cohesion and angle of internal friction of a given soil under somewhat arbitrary conditions of loading rates, seepage conditions, and internal stresses.

The test is performed on a soil sample which is either an undisturbed sample or is representative of the true soil condition which is being investigated. The equipment for the test consists principally of the following: (1) a split rectangular box for holding the specimen constructed so that the upper half of the box can move horizontally with respect to the lower half; (2) a jacking arrangement by which a given vertical pressure can be applied to the specimen; (3) a means of pulling the upper half of the box with respect to the lower half; (4) a means of measuring the shearing resistance of the soil to the horizontal shearing displacement; and (5) a means of measuring the change of thickness of the sample during the test. At the start of a particular run the vertical pressure is applied and maintained constant. As shearing displacement is applied the shearing resistance is measured by a proving ring dial and the vertical displacement by means of an Ames dial.

In cases where seepage takes place under actual conditions, the test is performed with porous stones or plates top and bottom of the specimen and the load applied slowly or quickly depending on whether or not consolidation takes place.

d. Consolidation test. In this test the time rate of consolidation for an undisturbed soil sample is measured. From these measurements the coefficient of consolidation, Cv, in $T = Cvt/H^2$ can be found.

A carefully prepared undisturbed soil sample is cut to fit into the consolidometer. The consolidometer is equipped with porous stones to allow seepage of the pore water to take place. A constant vertical pressure is applied to the specimen during which time its change in thickness is observed at various intervals. Several such runs are made with different soil pressures in order ·to include the pressures to be encountered in the actual soil.

3.52 *a.* Pour cinders into the water along the outside face of sheet piling. The flow of water will carry the cinders into the interlocks in the sheet piling and greatly reduce these leaks. The sump pumps will take care of the remainder of the leakage.

b. Using a mix of 1 to 1 sand and cement, "grout" the river bottom for about 10 ft around the outside of the sheet piling. The mix is forced through a pipe by air pressure and the pipe is "air jetted" down through the ooze of the river bottom. Bags filled with sand and bags containing cinders should be stacked at convenient points around the cofferdam to be available if and when needed.

c. A pile should be completely driven before stopping for lunch, particularly when driving in a clay soil. If stoppage is permitted, redriving to completion may be very difficult because of the additional shearing resistance that the soil has acquired during the lunch interval (frictional take-up on the pile).

d. Fairly heavy boulders if the velocity of the stream is apt to be high.

e. Prefabricating timber that is to be pressure-creosoted allows every part of the completed member to be protected. The exposed timber in holes bored subsequent to creosoting could be treated by plugging one side of the hole and applying creosote from a pressure gun at the other side of the hole.

3.53 *a.* A relatively shallow trench dug in hardpan in dry weather would not require sheeting. In the case of a deep trench which might stand exposed for some time before backfilling, some bracing would be required because of the possibility of rain and water entering the trench with the subsequent loss of cohesion in the soil.

b. Paint the steel with a coat of oil. Lead paints under the heat of the torch produce a gaseous vapor that may be poisonous to the welder.

c. Oiling the forms is beneficial to concrete since it prevents drying out of the surface and renders the concrete more impervious.

d. Gunite is a rich cement mortar which is applied by spraying under high air pressure.

e. The major advantage of a-c over d-c welding is in the reduction of arc blow. Alternating current usually increases welding speeds and produces a higher quality weld. Where single-phase current is available the a-c current will be more economical.

3.54

Weight of container plus dry soil	582.04 grams
Weight of container	194.40 grams
Weight of dry soil	387.64 grams
Weight of dish plus dry soil	175.50 grams
Weight of dish	118.77 grams
Weight of dry soil	56.73 grams
Weight of bottle plus water plus soil	732.22 grams
Weight of bottle plus water	687.39 grams
Submerged weight of soil	35.83 grams
Weight of dry soil	56.73 grams
Weight of submerged soil	35.83 grams
Weight of displaced water	20.90 grams

$G_s = 56.73/20.90 = 2.72$

Initial height of specimen, $H_i = 1.235$

$$+ (0.4200 - 0.4010)$$
$$= 1.254 \text{ in.}$$

Cross-sectional area of specimen $= 0.7854 \times 4.23^2$
$$= 14.1 \text{ sq in.}$$

Volume of solids $= \dfrac{387.6 \times 2.2 \times 1,728}{1,000 \times 2.72 \times 62.4} = 8.69 \text{ cu in.}$

Height of solids $= H_s = 8.69/14.1 = 0.615 \text{ in.}$
Height of voids $= H_v = 1.254 - 0.615 = 0.639 \text{ in.}$
Initial voids ratio $= 0.639/0.615 = 1.04$

Pressure, kg per cm²	ΔH	$\Delta e = \Delta H / H_s$	Voids ratio e
0			1.040
¼	0.0069	−0.011	1.029
½	0.0044	−0.008	1.021
1	0.0077	−0.012	1.009
2	0.0182	−0.030	0.979
4	0.0401	−0.065	0.914
8	0.0513	−0.0845	0.830

(See Fig. 3.54.)

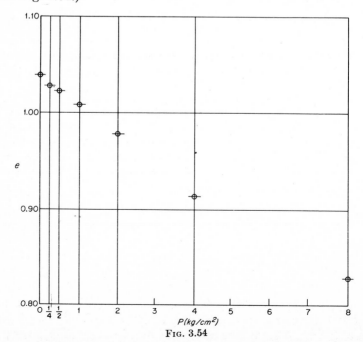

Fig. 3.54

3.55   For horizontally layered soil, use the Westergaard theory, and refer to D. W. Taylor, "Fundamentals of Soil Mechanics," Wiley, 1948, chart on p. 260.

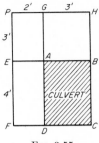

FIG. 3.55

Rectangle $PHCF$ (Fig. 3.55)

$\qquad M = \frac{7}{9} = 0.777$

$\qquad N = \frac{5}{9} = 0.555$

$\qquad K = 0.075$ (from chart)

Rectangle $PHBE$

$\qquad M = \frac{5}{9} = 0.555$

$\qquad N = \frac{3}{9} = 0.333$

$\qquad K = 0.043$

Rectangle $PGDF$

$\qquad M = \frac{7}{9} = 0.777$

$\qquad N = \frac{2}{9} = 0.222$

$\qquad K = 0.036$

Rectangle $PGAE$

$\qquad M = \frac{3}{9} = 0.333$

$\qquad N = \frac{2}{9} = 0.222$

$\qquad K = 0.021$

$$K_{ABCD} = K_{PHCF} - K_{PHBE} - K_{PGDF} + K_{PGAE}$$
$$= 0.075 - 0.043 - 0.036 + 0.021 = 0.017$$

Load on culvert = 7,000 lb $\times$ 0.017 = 119 lb

3.56   *a.* The foundations may consist of *two unreinforced-concrete pedestals* running down to bedrock. The top of the pedestal should have an area approximately twice to three times as great as that of the column cross section. The compressive stress in the pedestal should be somewhat less than $0.25f'_c$. The pedestal should be approximately twice as deep as the amount of flare or taper at each side of the column.

Dowels, whose area equals the area of the main steel in the column, should run down into the pedestal a distance sufficient to develop their full strength in bond. These dowels should not be hooked. A mat of rods should be used across the top of the pedestal. Tie rods should be used around the outside perimeter. Vertical rods should be used parallel to the four faces of the pedestal. The purpose of this system of reinforcement is to prevent the upper corners of the pedestal from cracking loose.

Investigate punching shear around the perimeter of the column, using the full depth of the pedestal and an allowable stress equal to $0.05f_c'$. The shear force should be computed as that on the bottom of the pedestal outside the perimeter of the column.

Bending stress need not be investigated since the pedestal sits on a roughened rock surface and cannot elongate laterally. Hence no tensile stress may develop.

    b. The foundations may consist of *two separate reinforced-concrete footings*. The footings may be placed directly under the column, or a pedestal may be used above the footing proper. Sometimes the use of a pedestal may so reduce the necessary size of footing as to make the cost of the combination less than the cost of a footing alone.

The necessary bearing area of the spread footing, which is preferably made square, is determined by considering the soil's allowable bearing pressure and the weight of the footing, the column, and the live load minus the soil pressure which originally existed at that depth before construction.

Steel to be used in the bottom of the footing (3 in. cover) may be computed on the basis of moment on a section through the face of the column or pedestal, multiplying the computed moment by 0.85 and using $M = A_s f_s j d$.

Investigation for punching shear seems unnecessary since failures due to this cause in reinforced-concrete spread footings are virtually unknown.

Diagonal tension (shear) is computed at a section which is a distance d away from the face of the column or the pedestal toward the edge of the footing, where d is the effective depth of the footing. The shear V on this section is taken as the total pressure on the bottom of the footing acting between the critical section, the outside face of the footing, and two 45° lines drawn out from the center of the column. $v = V/bjd$, where b is the width of the critical section between 45° lines. v should not exceed $0.03f_c'$, assuming that the main reinforcement is hooked as it should be.

Bond should be checked, using the shear as the total pressure acting on an area bounded by the outside of the footing, the face of the column, or pedestal, and 45° lines from the column center. Bond stress should not exceed $0.05f_c'$.

c. A *combined footing* may be used. It should be so proportioned that the center of gravity of the loads and the center of gravity of the footing area coincide. This can be accomplished by making the footing rectangular and extending it beyond the heavier of the two columns for a greater distance than beyond the lighter or by making the footing trapezoidal in plan with the wider portion near the heavier column. The footing is designed in its length as an overhanging beam with two supports (the columns). Appropriate shear and moment diagrams are drawn and reinforcing is placed according to the principles of reinforced-concrete beam design.

The short direction is designed as is an ordinary spread footing, taking critical sections as described above in answer *b.*

In the long direction, shears for computation of diagonal tension are taken from the shear diagram.

Bottoms of all footings should be below the expected depth of frost penetration.

The variable depth to bedrock beneath the two columns may cause unequal settlement. Thus there would be a tendency for the combined footing to tilt. This should be considered in choosing among the three alternate possibilities.

3.57   *a. Wet unit weight*

Volume of sample $= V_t = 298.1/62.4 = 4.78$ cu ft
$\gamma_w = 587.4/4.78 = 122.8$ lb per cu ft

b. Dry unit weight

Dry weight of sample $= 103.6 \times 587.4/112.4 = 540.5$ lb
$\gamma_d = 540.5/4.78 = 113.2$ lb per cu ft

c. Moisture content

$$w = \frac{(112.4 - 103.6)}{103.6} \times 100 = 8.49 \text{ per cent}$$

d. Voids ratio

$v_s = 540.5/(2.67 \times 62.4) = 3.25$ cu ft
Volume of voids $= v_v = v_t - v_s = 4.78 - 3.25 = 1.53$ cu ft
$e = v_v/v_s = 1.53/3.25 = 0.471$

e. *Porosity*

$$n = (v_v/v_t) \times 100 = 1.53 \times 100/4.78 = 32.0 \text{ per cent}$$

f. *Saturation*

Weight of water in original sample $= 587.4 \times 8.8/112.4 = 46 \text{ lb}$
$s = (46 \times 100)/(62.4 \times 1.53) = 48.2 \text{ per cent}$

g. *Relative density*

$$D_R = \frac{(1/\gamma_{min}) - (1/\gamma_d)}{(1/\gamma_{min}) - (1/\gamma_{max})} \times 100 = \frac{(1/103.9) - (1/113.2)}{(1/103.9) - (1/117.4)} \times 100$$
$$= \frac{0.00963 - 0.00883}{0.00963 - 0.00852} \times 100 = 72 \text{ per cent}$$

3.58 *a.* Interbedded fine silts and coarse sands with possibly gravels at depth. High ground water possible. General absence of coarse gravels but plenty of clean sands and fine gravels. Bridges will require deep and expensive foundations.

b. Subsoils often wet and plastic. Beach ridges will be sandy and gravelly in texture. Transition may require special care due to differential settlement. Adequate drainage will be a major problem. Material is difficult to handle when wet. High volume changes will occur as the moisture content changes.

c. The low areas probably will have poor internal drainage. Better on the higher slopes. Soil probably clay with high organic content. Excavation will be easy. Borrow is probably available. Frost action can be expected. Limestone honeycombing possible. Till will reflect presence of limestone base.

d. Generally quartz sand in dunes. Excavation is easy but excessively rough on equipment. Probably dunes will have to be stabilized.

e. Excavation will be easy. Excellent source of borrow. Highly stable material and makes good road base and foundation material.

3.59 Assume $\phi = 35°$. From design charts obtain $D/h = 0.71$. Then, depth of sheet pile penetration $= 0.71 \times 15 = 10.7$ ft. Also

$$\frac{M_{max}}{M'} = 1.58 \qquad \text{or} \qquad M_{max} = 1.58M'$$

$$p_a = K_a \gamma$$

where $K_a = \dfrac{1 - \sin \phi}{1 + \sin \phi} = 0.27$

Using $\gamma = 125$ lb per cu ft, $p_a = (0.27)125 = 34.$

$$M_{max} = 1.58 \times 5 \times 0.5 \times 15^2 \times 34 = 30{,}217.5 \text{ ft-lb}$$

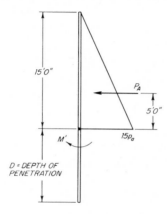

F<small>IG</small>. 3.59

This is a specialty problem. If a "USS Steel Sheet Piling Design Manual" is available or the book "Foundation Design" by Wayne C. Teng (Prentice-Hall) is available, the reader will find the problem worked out in detail.

3.60 *a.* A balanced design is one in which both the concrete and the steel are so proportioned as to work to their full working stresses when the member carries its full allowable load.

b. The transformed section is one in which the flexure steel is conceived to be replaced by a large area of imaginary concrete which can take tension. This gives a homogeneous section of concrete to which ordinary beam analyses may be applied.

c. 1. In a new design the neutral axis is located by the allowable working stresses in the steel and the concrete.

2. In an investigation of an existing beam the neutral axis is located at the centroid of the existing transformed area.

d. The fiber stress in a beam is assumed to be proportional to the distance of the fiber from the neutral axis.

e. Fireproofing in columns should consist of at least 1½ in. of concrete or a thickness of concrete equal to 1½ times the maximum size of the coarse aggregate. Where the concrete is to be exposed to weather or be in contact with the ground, the fireproofing shall be not less than ⅝ in. in diameter and 1½ in. for bars ⅝ in. or less in diameter.

f. In a T beam the shear is assumed to be carried by an area whose width is equal to the width of the stem and whose depth is equal to the depth of the T beam.

g. The function of the stirrups is to prevent diagonal tension cracks.

h. Stirrups are not needed in slabs because the depth of the slab is so proportioned that the intensity of shearing stress is kept below the critical limit which indicates that diagonal tension would require stirrups.

i. Diagonal tension is computed by means of a formula for the shearing stress, but the failure to be guarded against is one of diagonal tension and not shear. Concrete is rarely used in a manner which would subject it to shearing failure and thus shear is only used as a measure of tensile stress.

j. Three inches of concrete should intervene between the reinforcement and the ground surface of the footing.

3.61

List 1	List 2
1	*b*
2	*e*
3	*a*
4	*c*
5	*d*
6	*f*

3.62 *a.* A sole plate is a steel plate attached to the lower flange of the girder below the end stiffeners which distributes the girder reaction to the masonry bearing plate. It shall have a thickness of not less than ¾ in. and not less than the thickness of the flange angles plus ⅛ in. Preferably it shall not be longer than 18 in.

    b. 1. Stiffeners shall be placed at all points of end bearing and bearing of concentrated loads. According to the American Railway Engineering Association "if the depth of the web between the flanges or side plates of a plate girder exceeds 60 times its thickness, it shall be stiffened by pairs of angles riveted to the web."

2. The clear distance between stiffeners shall not exceed 72 in.

3. The clear distance between stiffeners shall not exceed

$$\frac{10,500t}{\sqrt{S}}$$

where t = thickness of web in inches

S = unit shearing stress, gross section, in web at point considered

Other specifications have similar requirements.

c. Stiffener angles at points of concentrated load shall have milled ends.

d. One-eighth of the web may be included in the net area of the flange and one-sixth of the web may be used for the gross area.

e. Anchor bolts shall be not less than $1\frac{1}{4}$ in. in diameter. Anchor bolts not carrying uplift shall extend 12 in. into the masonry. Those carrying uplift shall be designed to engage a mass of masonry the weight of which is $1\frac{1}{2}$ times the uplift.

f. Combined stresses include dead-load stresses, live-load stresses, impact stresses, wind stresses, snow stresses, ice stresses, etc., in such a combination as to give both maximum negative and maximum positive effect.

g. Alternate stresses are those used to design a member which goes through a reversal of stress. The alternates would be the maximum positive and the maximum negative stresses.

h. A gusset plate is a plate used to connect the various members of a truss which comes together at a joint. It is made large enough to include the necessary number of connection rivets.

i. A filler plate is used at truss joints to compensate for any difference in the depth or thickness of the members being joined so that connection may be made to a flat gusset plate.

j. Where the open sides of truss members are joined by latticing, stay plates are used as near the end of the member as possible and at intermediate points where the lacing is interrupted. The purpose of the stay plates is to lend a little additional rigidity to the member.

3.63 *a.* Fills may be compacted by: (1) jetting, which involves forcing water into the fill under pressure using nozzles on high-pressure hoses and relying on the downward pressure of the water to align the grains in a dense condition; (2) tamping, which may be accomplished using hand tampers or pneumatically or gasoline-driven vibratory tampers and compacting in layers; and (3) rolling in layers using a sheep's-foot roller and a number of passes for each layer. Jetting has the advantage that it can be used for relatively thick layers and can be accomplished rather quickly. It has the disadvantage of introducing large quantities of water into the embankment which greatly decreases the embankment's strength and stability. Hand tamping when properly executed is very effective, but it has the disadvantage of being expensive and lengthy. Rolling using a sheep's-foot roller is probably the most effective method for getting adequate compaction on jobs of fairly large extent. It has the disadvantage that its effectiveness is not quite as great with noncohesive soils as with cohesive ones.

b. Shrinkage refers to the fact that a cubic foot of soil in its natural state will not yield a cubic foot as artificially compacted in the embankment due to greater relative density in the embankment and incidental losses as the material is transported from one place to the next. The amount of shrinkage is determined by the comparative relative densities from the natural state to the embankment and the amount of material lost in transit. A shrinkage factor must be used in balancing cuts and fills or it will be found that there is not sufficient material to complete the embankment. In order to compensate for the shrinkage all fill quantities are multiplied by an amount equal to $1/(1 - S)$ where S = shrinkage factor expressed as a ratio.

c. The temporary pavement should be a bituminous macadam. Lay a foundation of 6 to 8 in. of crushed stone ranging 4 to 1 in. in size. Roll using a 10-ton three-wheel roller moving longitudinally with subsequent passes from the edges of the pavement toward the center. Then sweep in choke stone and continue to roll wetting down the surface as you go. The wearing course should consist of broken stone with a range of sizes from $2\frac{1}{2}$ to 1 in. The depth of the wearing course is about 4 in. This course is rolled just as the base course was rolled. High-penetration asphalt is then applied at the rate of 2 gal per sq yd. Choke stone is then spread immediately on the surface and broomed and rolled in.

Then a seal coat of asphalt is applied at the rate of ¾ to 1 gal per sq yd. Then sand or cover stone is applied and rolled into the surface. All excess is swept off and the surface is ready for use. The strength and durability of this pavement depends partly on the subgrade and partly on the waterbound macadam base course used. This type of pavement was recommended because it is relatively inexpensive, does not require a curing period before it can be used, and is flexible so that it can withstand a certain amount of settlement in the newly compacted approach embankment.

3.64 Referring to Fig. 3.64, a base *AB* is taped along one bank, the quadrilateral *ABCD* is triangulated and tied into the bridge center line *EF*. *CD* is measured as a check base and the control net is adjusted by least squares.

With the design stationing of *E*, *F*, and the center lines of piers 1, 2, 3, 4, and 5 known, we may set range flags on shore at *G*, *H*, *J*, *K*, *L*, *M*, *N*, and *O*, drive batter frames upstream from each pier as shown at pier 2, and compute the necessary angles to be laid off from the bases on either shore to set *P* and *Q* on these batter frames on the center line of pier 2. Then range flags can be set at *R*, *S*, *T*, and *U*, to permit driving the sheet piles

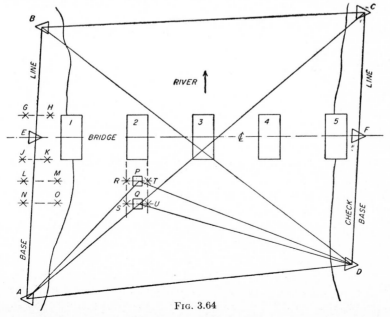

Fɪɢ. 3.64

for the long side of the pier-2 cofferdam. A similar procedure is applied to all other piers.

With the cofferdams in place the system of triangulation is applied to the center of each cofferdam in turn to permit the establishment of points controlling the placing of forms, stone facing, and arch hinges.

3.65 Such a traverse should be executed using taping bucks or stools since stakes cannot be driven. A minimum of three bucks is required. Tape measurements, each of substantially the full length of the tape where possible, should be made from the end monument to a mark on the first buck, and then carried successively from buck to buck. A tension balance and thermometers should be provided. Where a long tape (200 ft or 50 m) is used, a center support, adjusted to grade by eye, may be necessary.

A transit party consisting of an instrument man and one helper is required to align the bucks as they are successively moved from the rear to the head of the line. A levelman and rodman must establish the differences in elevation between each pair of bucks to permit computation of the inclination corrections. The taping party will normally consist of front and rear chainmen, front and rear stretchermen, and a recorder who may also hold the center support when needed. Necessary corrections include standardization, temperature, slope, and sag.

ELECTRICAL ENGINEERING

4.01 *References:* "I.E.S. Lighting Handbook," 1947.
Krachenbuehl, "Electric Illumination," Wiley, 1951.
Sharp, "Introduction to Lighting," Prentice-Hall, 1951.

For parking lots, recommended illumination is 1 to 2 ft-c.
For a downtown area assume 2 ft-c. A broad-beam (30° or more)
general-service type of floodlight is indicated for this area of
$200 \times 300 = 60,000$ sq ft. Select a 500-watt floodlight unit
having initial beam lumens = 3,400. Assume a depreciation
factor = 0.7.

$$\text{No. of units} = \frac{\text{ft-c} \times \text{area}}{0.7 \times \text{initial beam lumens}} = \frac{2 \times 60,000}{0.7 \times 3,400} = 50$$

Refer to table 8-8, "I.E.S. Lighting Handbook," 1947, p. 8–27.
For a 25-ft mounting height = D, assume $Z = 40$ ft for a
40° beam. The coverage of one unit = 2,300 sq ft. $60,000/50 =$
1,200 sq ft minimum coverage required, so the 50 (500-watt)
units will give ample uniformity. The 50 units could be mounted
five to a pole on 10 poles spaced 60 ft along the two long sides
of the lot. A more efficient but less diversified arrangement
could be made by using 1,000-watt general-service lamps in
18- to 24-in. broad-beam reflectors which would give 8,900 initial
beam lumens.

$$\text{No. of 1,000-watt units} = \frac{2 \times 60,000}{0.7 \times 8,900} = 20$$

Area to be covered by one unit = 60,000/20 = 3,000 sq ft.
Refer to table 8-8, "I.E.S. Lighting Handbook," p. 8–27.

For D = 25 and Z = 50 ft, a 35° beam would cover 3,270 sq ft.
The 1,000-watt units could be mounted in groups of two on 10 poles spaced 60 ft along each long side of the lot.

4.02 *Reference:* "I.E.S. Lighting Handbook," 1947, pp. 8–38 to 8–40.

The formula for this point-by-point calculation is

$$E_h = (I/h^2) \cos^3 \theta_h$$

where E_h = the horizontal illumination in foot-candles at a point p in a plane h ft below a point-source of intensity I candlepower in the direction of p, and θ_h = the angle between a vertical line at p and the line connecting p and the source.

a. At the center, the contribution of one lamp is

$$E_{h_1} = (800/10^2) \cos^3 21.8° \quad \text{(where tan } \theta_h = \frac{4}{10} \text{ and } \theta_h = 21.8°\text{)}$$
$$= 6.4 \text{ ft-c}$$

The contribution of the other lamp is the same, so at the center of the table (both ways) E_h = 2 × 6.4 = 12.8 ft-c.

b. At the center of either end of the table, from the nearer lamp, E_h = 6.4 ft-c. (as in *a*), but from the farther lamp, E_h = $(800/10^2) \cos^3 50.2°$ = 2.1 ft-c. The resultant E_h = 6.4 + 2.1 = 8.5 ft-c.

c. At any corner, E_h = 8 $\cos^3 26.5°$ + 8 $\cos^3 51.2°$ = 5.6 + 2.0 = 7.6 ft-c.

4.03 *Reference:* "I.E.S. Lighting Handbook," p. 6–10.

Take efficiency of a 750-watt PS Mazda lamp as 20 lumens per watt. Assume the same luminaire efficiency and depreciation factor for incandescent and fluorescent lamps:

The power ratio of fluorescent to incandescent for 20 ft-c = $\frac{20}{42}$. If the illumination is raised to 30 ft-c, $\frac{30}{20}$ more lumens will be required. The ration of power for a change would be $\frac{20}{42} \times \frac{30}{20}$ = 0.715, and the power-cost saving would be (1 − 0.715)100 or 28.5 per cent.

4.04 *Reference:* "I.E.S. Lighting Handbook," 1947.

Recommended illumination level = 50 ft-c (Table A-1, p. A-2).

A general diffusing or semi-indirect type of luminaire, as shown in Table 8-2, pp. 8–9 to 8–11, is indicated. The room index, Table 8-3, $=F$. Assume ceiling and wall reflectances = 50 per cent. From Table 8-2, pp. 8–9 to 8–10, the coefficient of utilization for semi-indirect luminaires is about 0.30. Assume the maintenance factor = 0.60. The total initial light flux required is: $F = 50(20 \times 30)/0.30 \times 0.60 = 167,000$ lumens. The maximum spacing of luminaires is 12 ft, and if six luminaires are used, they could be arranged in two rows of three per row with 10-ft spacing. The lumens per luminaire would be $167,000/6 = 28,000$. If incandescent lamps are used, one 1,500-watt, two 750-watt, or three 500-watt lamps per luminaire would be required (Table 6-3, p. 6–10). If fluorescent lamps are used, seven 100-watt lamps per luminaire would be required (Table 6-8, p. 6–35). A more uniform installation with lower brightness contrast would be to use more luminaires, say 12, spacing them in three rows, 7 ft between rows, and four in each row spaced on 7.5-ft centers. The lumens per luminaire would then be $167,000/12 = 14,000$. One 750-watt incandescent lamp per luminaire would be sufficient, or four 100-watt fluorescent lamps, soft white.

Wiring. Reference: "The National Electrical Code," 1949.

If two rows of three luminaires per row are used, a separate two-wire branch circuit should be used for each row. Assuming 120-volt lamps, the current per circuit would be $3 \times 1,500/120 = 37.5$ amp. Two #6 AWG for each circuit with 50-amp protection would be adequate. If three rows of four luminaires per row are used, separate branch circuits for each row should be used. The current per circuit would be $4 \times 750/120 = 25$ amp. Two #10 AWG for each circuit with 30-amp protection would be adequate.

4.05 *Reference:* "I.E.S. Lighting Handbook," 1947.

a. Table A-1. For rough bench and machine work, 20 ft-c illumination level is recommended. For medium bench and machine tool work, 30 ft-c is recommended.

b. Page 10–105. Current practice is to use direct fluorescent-lamp luminaires for general industrial lighting and to provide uniform illumination level throughout every work area.

c. Assume 30 ft-c illumination on 30-in. work plane. Assume ceiling and wall reflectances = 50 per cent, and luminaires mounted 2 ft below ceiling. The mounting height above floor =

$18 - 2 = 16$ ft. From Table 8-3, p. 8-13 the room index = D. Select a direct fluorescent-lamp luminaire, as on Table 8-2, p. 8–6. The coefficient of utilization is 0.58 for the two 100-watt types. Maintenance factor, assumed poor, = 0.45. The total lumens = $30(40 \times 250)/0.58 \times 0.45 = 1,150,000$. From Table 6-8, p. 6–35, the 100-watt white lamp has an initial output of 4,300 lumens. The number of two-lamp luminaires required is $1,150,000/2 \times 4,300 = 134$. These luminaires are approximately 5 ft long, and for uniform illumination the spacing should not be greater than the mounting height of 16 ft. Four rows of two-lamp luminaires, with 34 luminaires lengthwise per row (about 7-ft 4-in. centers) would be a suitable arrangement.

d. Four-wire three-phase branch circuits for each row with the luminaires balanced between the three phases and neutral would be a suitable circuit arrangement.

e. The walls and ceiling should be painted a light color. In part c., 50 per cent reflectance was assumed. If higher reflectance than this could be maintained, fewer lamps would have been required for the same level of illumination and brightness contrast would have have been lower.

4.06 No. "The National Electrical Code," Vol. V, 1949, Table 4, p. 326, specifies that a 1-in. conduit be used for two or three #6 wires.

4.07 *a.* $P_{12} = \dfrac{1}{R^2 + X^2} (R|\bar{V}_1|^2 - R|\bar{V}_1| \, |\bar{V}_2|\cos \delta$

$$+ X|\bar{V}_1| \, |\bar{V}_2|\sin \delta)$$

$$= \frac{1}{5^2 + 40^2} [5(345)^2 - 5(345)(360)\cos 10°$$

$$+ 40(345)(360)\sin 10°]$$

$$= 520.8 \text{ Mw}$$

$Q_{12} = \dfrac{1}{R^2 + X^2} (X|V_1|^2 - X|\bar{V}_1| \, |\bar{V}_2|\cos \delta$

$$- R|\bar{V}_1| \, |\bar{V}_2|\sin \delta)$$

$$= \frac{1}{5^2 + 40^2} [40(345)^2 - 40(345)(360)\cos 10°$$

$$- 5(345)(360)\sin 10°]$$

$$= -147.3 \text{ Mvar}$$

$$P_{21} = \frac{1}{5^2 + 40^2} [5(360)^2 - 5(124,200)\cos 10°$$
$$- 40(124,200)\sin 10°]$$
$$= -508.5 \text{ Mw}$$
$$Q_{21} = \frac{1}{1,625} [40(360)^2 - 40(124,200)\cos 10°$$
$$+ 5(124,200)\sin 10°]$$
$$= \textbf{245.7 Mvar}$$

b. $P_{\text{loss}} = P_{12} + P_{21} = 520.8 - 508.5 = \textbf{12.3}$ Mw
   $Q_{\text{loss}} = Q_{12} + Q_{21} = -147.3 + 245.7 = \textbf{98.4}$ Mvar

4.08   Rated current, flowing before the fault, is

$$I_L = P_L/E_L = 10,000/250 = 40 \text{ amp}$$

The generated emf under these conditions is

$$E_g = E_L + I_0R_a = 250 + 40 \times 0.50 = 270 \text{ volts}$$

On short-circuit, it is assumed that E_g, the inductance L, and the armature resistance R_a remain constant. The circuit resistance becomes $R = R_a + R_{sc} = 0.5 + 0.2 = 0.7$ ohm. The equivalent circuit from

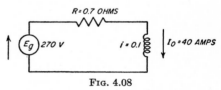

FIG. 4.08

$t = 0$, when the fault occurs, until the breaker opens is 0.02 sec.
   The current I_0 is the initial current flowing in the inductive circuit at $t = 0$, and the current $I = E_g/R = 270/0.7 = 386$ amp is the current which will flow eventually if the breaker never operates. The transient equation for this case is

$$i = I - (I - I_0)e^{-Rt/L} \text{ for } 0 \leq t \leq 0.02$$

The maximum current withstood by the armature will occur at $t = 0.02$ sec:

$$I_{\text{max}} = 386 - (386 - 40)e^{-0.7 \times \frac{0.02}{0.1}} = 386 - 346 \times 0.869$$
$$= 386 - 301 = 85 \text{ amp}$$

4.09 At the instant the field is disconnected from the line, the circuit will be as shown, where R_d is the field-discharge resistance, and by Ohm's law, $R_f = 120/8.4 = 14.3$ ohms, is the field resistance. Because of the field inductance, L_f, the field current cannot change instantaneously; hence it remains at 8.4 amp.

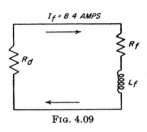

FIG. 4.09

The voltage across the two resistances will be equal to the induced voltage across L_f, and this value must be kept to one-third of the standard high-potential test voltage of 1,000 plus twice the voltage rating of the machine. We then have the voltage

equation: $I_f(R_d + R_f) = \dfrac{(1{,}000 + 2 \times 120)}{3} = 413.3$ whence,

$R_d = \dfrac{(413.3 - I_fR_f)}{I_f} = \dfrac{(413.3 - 120)}{8.4} = 34.9$ ohms required. It

will be noted that the values of the field inductance and, for that matter, field resistance, were not required for a solution.

4.10 From the first measurement we have (1) $3C_g = 0.5$ where C_g = the capacitance of each conductor to the sheath. From the second measurement we have (2) $2C_m + C_g = 0.6$ where C_m = the capacitance between any two conductors. From (1), $C_g = 0.5/3 = 0.167$ μf. Substituting (1) in (2), $C_m = (0.6 - 0.167)/2 = 0.217$ μf. We thus have a balanced Y load of three 0.167-μf capacitors and a balanced Δ load of three 0.217-μf capacitors. By a ΔY impedance transformation, the balanced Δ is equivalent to a balanced Y having capacitors of $0.217 \times 3 = 0.651$ μf each.

The charging current is thus due to an equivalent balanced Y load each leg of which consists of a capacitor of $0.167 + 0.651 = 0.818$ μf.

Then the charging current per phase is

$$I_L = I_\phi = E_\phi B_\phi = E_\phi(2\pi fC) \Rightarrow \frac{25{,}000 \times 377 \times 0.818}{\sqrt{3} \times 10^6}$$

$$= 4.45 \text{ amp}$$

4.11   In the new installation we tolerate total losses in the amount:

$$(0.8)(0.5/100)(500) = 2,000 \text{ watts}$$

New core losses: the voltage is the same but the frequency is only $50/60 = 0.833$ per unit.

$$P_{ec} = K(f\phi)^2 = KE^2 = 625 \text{ watts (same as given)}$$
$$P_{\text{hyst.}} = K_1 fB^{1.6} = K_1 \frac{(fB)^{1.6}}{f^{0.6}} = \frac{K_1 E^{1.6}}{f^{0.6}} = \frac{K_2}{f^{0.6}}$$
$$= 625 \text{ watts} \left(\frac{1}{0.833^{0.6}}\right) = 700 \text{ watts}$$

Permitted new copper loss:

$$2,000 - (700 + 625) = 675 \text{ watts}$$

The old copper losses were related to the current as follows: $1,250 = I^2R$. In the new installation, $675 = I^2R$. Thus,

$$I^2 = \frac{675}{1,250} = 0.735 \text{ per unit}$$
$$\text{kva} = (0.735)(500) = 370 \text{ kva}$$

New rating: 50 Hz, 1,000/500 volts, 370 kva.

4.12   *a.* $N_{600} = \dfrac{600 \times 200}{2,400} = 50 \text{ turns}$

$$N_{240} = \frac{240 \times 200}{2.400} = 20 \text{ turns}$$

b and *c.* $I_p = \dfrac{200,000}{2,400} = 83.3 \text{ amp}$

d. $I_{600} = \dfrac{100,000}{600} = 166 \text{ amp}$

$$I_{240} = \frac{100,000}{240} = 416 \text{ amp}$$

e. $I_p = A\underline{/\alpha°} + B\underline{/\beta°} = \dfrac{600 \times 166}{2,400} \underline{/-45.6°}$

$$+ \frac{416 \times 240}{2,400} \underline{/0°} = 29.5 - j29 + 41.6$$

$$I_p = 76.6 \text{ amp}$$

4.13 $I_R = 15,000/2,300 = 6.52$ amp

$$Z_H = \frac{V}{I} = \frac{65}{6.52} = 9.96 \text{ ohms}$$

$$R_H = \frac{P}{I^2} = \frac{350}{(6.52)^2} = 8.22 \text{ ohms}$$

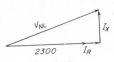

FIG. 4.13

$$X_H = \sqrt{Z_H^2 - R_H^2} = 5.65 \text{ ohms}$$

     a. $Ix = 6.52 \times 5.65 = 36.8$ volts

$IR = 6.52 \times 8.22 = 53.6$ volts

$V_{NL} = \sqrt{(2,354)^2 + (37)^2} = 2,354$ volts

$$\text{Regulation} = \frac{(V_{NL} - V_{FL})}{V_{FL}} \times 100 = \frac{(2,354 - 2,300)}{2,300}$$
$$\times 100 = 2.34 \text{ per cent}$$

     b. Maximum efficiency occurs when $I^2R = CL$

$I^2R = I^2 \times 8.22 = 245$ watts

$I = 5.45$ amp

Power out (pf = 1.0) = $2,300 \times 5.45 = 12.5$ kw (neglecting regulation)

$$\text{Maximum efficiency} = \frac{\text{power out}}{\text{power in}} = \frac{12,500}{12,500 + 245 + 245}$$
$$= \frac{12,500}{12,990} = 0.962 = 96.2 \text{ per cent}$$

4.14 Assuming rated voltage out:

$Z_{\text{rated}} = 2,200/45.5 = 48.5$ ohms

$Z_{A_{po}} = 72/2,200 = 0.0326$

$R_{A_{po}} = 1,000/100,000 = 0.01$

$\cos \theta_A = 0.01/0.0326 = 0.307$, $\theta_A = 72.1°$

$Z_{B_{po}} = 75/2,200 = 0.0341$

$R_{B_{po}} = 1,100/100,000 = 0.011$

$\cos \theta_B = 0.011/0.0341 = 0.322$, $\theta_B = 71.1°$

$\dot{V} = \dot{E}_1 + \dot{I}_1 Z_1 = \dot{E}_2 + \dot{I}_2 Z_2$

$\dot{E}_1 = \dot{E}_2$ and $\dot{I}_1 + \dot{I}_2 = \dot{I}_T = 2\underline{/-37°}$

$E_1 + \dot{I}_1(0.0326\underline{/72.1°}) = 1$

$E_1 + \dot{I}_2(0.0341\underline{/71.1°}) = 1$

$$\dot{I}_1 = \frac{0.0341\underline{/71.1°}}{0.0326\underline{/72.1°}} \dot{I}_2 = (1.04\underline{/-1°})\dot{I}_2$$

$$(1.04)\underline{/-1°}\dot{I}_2 + \dot{I}_2 = 2\underline{/-37°}$$

$$\dot{I}_2(1.04 - j0.017 + 1) = (1.6 - j1.2)$$

$$\dot{I}_2 = \frac{2\underline{/-37°}}{2.04\underline{/-0.5°}} = 0.98\underline{/-36.5°}$$

$$\therefore |\dot{I}_2| = 0.98PO = 98 \text{ per cent rated}$$

$$\dot{I}_1 = (1.04\underline{/-1°})\dot{I}_2 = (1.04\underline{/-1°})(0.98\underline{/-36.5°})$$
$$= 1.02\underline{/-37.5°}$$

$$\therefore |\dot{I}_1| = 1.02PO = 102 \text{ per cent rated}$$

4.15 Since unit 2 has a smaller per unit Z on its own base, it will be the limiting transformer. Let unit 2 then pass only 30 Mva.

$$\text{Mva}_1 = \text{Mva}_2 \left(\frac{Z_2}{Z_1}\right)$$

where $Z_2 = j0.07(^{45}\!\!/_{30}) = j0.105$ per unit on 45-Mva base.

$$\text{Mva}_1 = 30 \left(\frac{0.105}{0.08}\right) = 39.4 \text{ Mva}$$

$$\text{Total Mva} = 30 + 39.4 = \textbf{69.4 Mva}$$

4.16 Although the power could be transmitted at 11,000 volts, it would be preferable to increase the transmission voltage to about 33 kv. Transformers at each end would then be required. Assuming the use of Y-Y banks, the line voltages and turn ratios are shown in the following diagram, with each bank rated at 3,000 kva.

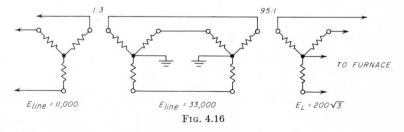

$E_{line} = 11,000 \qquad E_{line} = 33,000 \qquad E_L = 200\sqrt{3}$

Fɪɢ. 4.16

At 33,000 volts each wire will have

$$I_{\text{line}} = \frac{VA}{\sqrt{3} \times E_{\text{line}}} = \frac{3,000 \times 10^3}{\sqrt{3} \times 33 \times 10^3} = 52.5 \text{ amp}$$

The allowable loss for each of the three wires is

$$P_{\text{loss/wire}} = P_{\text{phase}} \times 0.05 = \frac{3,000 \times 10^3 \times 0.8}{3} \times 0.05$$

$$= 40,000 \text{ watts}$$

Hence the maximum allowable resistance per wire per mile (where "d" is the length of the transmission line in miles) is

$$R_{\text{wire mile}} = \frac{P_{\text{loss/wire}}}{I_L{}^2 \times d} = \frac{40,000}{(52.5)^2 \times 10} = 1.45 \text{ ohms per mile}$$

This requirement may be met with selection of #4 AWG standard annealed copper wire having 1.34 ohms per mile at 25°C. On wooden poles with 250-ft span, a minimum spacing of 5 ft between wires should be allowed. The resistance drop in this line is $IR = 52.5 \times 14.5 = 760$ volts. Tables for this size wire and spacing give 0.80 ohm reactance per mile per wire at 60 cps. Hence, the reactive drop for 10 miles is $IX = 52.5 \times 8.0 = 420$ volts. Therefore, the total drop (neglecting charging current) is

$$IZ = \sqrt{(IR)^2 + (IX)^2} = \sqrt{(760)^2 + (420)^2} = 870 \text{ volts}$$

This is only 4.6 per cent of $33,000/\sqrt{3}$ volts and is completely negligible for a furnace application. Regulation of generator and transformer have not been included.

4.17 Lower transmission voltages result in larger wire size or in high resistance and low efficiency. On the other hand, higher transmission voltages require more expensive tower and protective equipment, but the larger spacing required results in higher inductive reactance. Furthermore, for economical pole spacing, there is a minimum size wire for sufficient mechanical strength. In this problem construction costs are neglected.

$$I_{\text{line}} = \frac{P_{\text{delivered}}}{(\text{pf} \times \sqrt{3} \times E_{\text{line}})} = 25,000 \times \frac{10^3}{(0.85 \times \sqrt{3} \times E_{\text{line}})}$$

The allowable voltage drop and power loss in each wire are respectively

$$IZ_{\text{per wire}} = 0.08 E_{\text{phase}} = 0.08 \times E_{\text{line}}/\sqrt{3}$$
$$P_{\text{loss per wire}} = 0.06 P_{\text{phase}} = 0.06 \times 25,000 \times 10^3/3 = 500 \times 10^3$$
$$\text{watts}$$

Based on allowable power loss only, the allowable resistance of wire per mile is, where $d = 25$ miles,

$$R_{\text{wire mile}} = \frac{P_{\text{loss per wire}}}{d \times I_{\text{line}}^2} = \frac{500 \times 10^3}{25 \times I_{\text{line}}^2}$$

For the three available transmission voltages, the values are tabulated below per wire:

kv	I_{line} (amp)	P_{loss} (kw)	R_{mile} (ohms)	Wire size, AWG
12.5	1,360	500	0.0108	
66.0	256	500	0.303	4-0
115.0	143	500	0.970	2

The last column gives the nearest standard copper-wire size, taken from wire tables at 25°C, having a resistance a trifle lower than the allowable ohms per mile. On this basis, the 12.5 kv is immediately eliminated as requiring impracticable size (large) wire.

In determining IZ drop, as this line is short, neglect capacitance (charging current). The inductive reactance depends on spacing, and minimum spacing for safety and minimal corona loss depend on voltage. Consider the data in the following table (at sea level):

Kv	Wire size, AWG	$R_{(\text{ohms per mile})}$	Minimum allowance spacing (ft)	$X_{(\text{ohms per mile})}$
66	4-0	0.264	9	0.770
115	2	0.840	60	
115	2-0	0.420	20	0.895

None of these choices satisfies the regulation requirement. The 60-ft spacing (approximated) is too great even to consider the reactance. Choosing a larger wire size, e.g., #2-0 as shown in the third line of the table, gives $Z_{\text{wire mile}} = \sqrt{R^2 + X^2} = 0.990$ ohm and the drop per wire is then $(IZ)_{\text{drop per wire}} = I_{\text{line}} \times Z_{\text{wire mile}} \times d = 143 \times 0.990 \times 25 = 3{,}550$ volts which satisfies with ample margin both the loss and regulation requirements.

4.18

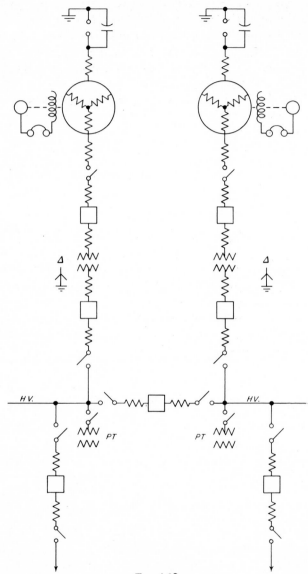

FIG. 4.18

4.19 If all reactive kva are supplied by the synchronous condenser, 5,000 kw, including the 300-kw loss in the condenser, may be added.

a. This leaves 4,700 kw for the new motors. At 80 per cent pf, this means 4,700/0.8 = **5,875** kva in motors.

b. For a pf of 0.8, the reactive factor is 0.6, and 0.6 × 5,875 = 3,525 kvars in the new motors. Total kvars to be supplied by the condenser are 8,667 + 3,525 = **12,192** kva = condenser rating.

c.

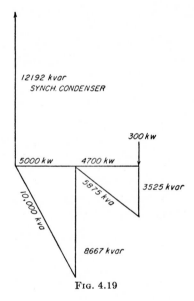

Fɪɢ. 4.19

4.20

(1)	(2)	(3)	(4)	(5)	(6)	(7)	(8)	(9)
	Load		Effi-ciency	Input	Factors		Input	
Rating	hp	kw		kw	Power	reactance	kva	kvar
(Given)	(Given)	$0.746 \times (2)$	(Given)	$(3)/(4)$	(Given) $\cos\theta$	$\sin\theta$	$(5)/(6)$	$(7)\times(8)$
1–50 hp	30	22.4	0.86	26.05	0.70	0.715	37.20	26.60
1–100 hp	75	56.0	0.89	63.00	0.80	0.60	78.75	47.20
2–15 hp	15	11.2	0.92	12.20	0.85	0.527	14.35	7.56
1–300 hp	310	231.5	0.92	252.00	0.85	0.527	296.00	156.00
1–33 kw	—	33.0	1.00	33.00	1.00	0	33.00	0
							459.35	−237.36
1–500 hp	{ (×0.746)	373	1.000	373.00	0.80	0.60	466.0	+280.00
	{ 300	224	0.925	242.00				
Total				628.25				42.64

This combination can be adjusted to unity pf where kw = kva. Otherwise

$$\text{kva} = \sqrt{(\text{kw})^2 + (\text{unbalanced kvar})^2}$$

The above figures indicate almost unity power factor so that kva in this case will be very close to 629.5 kva.

4.21

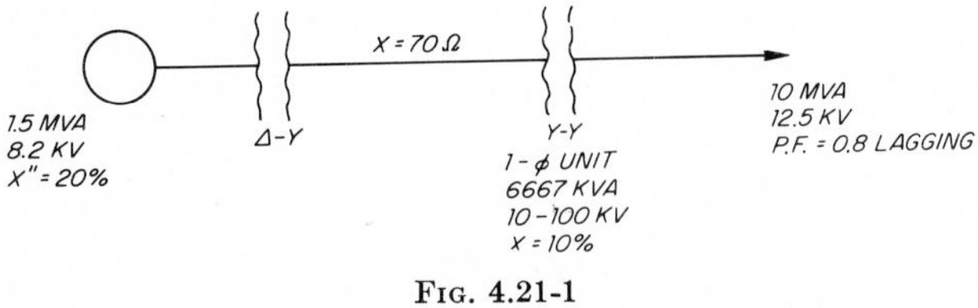

1.5 MVA
8.2 KV
X″ = 20%

Δ-Y

X = 70 Ω

Y-Y
1 - φ UNIT
6667 KVA
10 - 100 KV
X = 10%

10 MVA
12.5 KV
P.F. = 0.8 LAGGING

Fig. 4.21-1

Base chosen: 10 Mva, 12.5 kv in the load circuit.

$$Z_{\text{base}}(\text{for H.T. line}) = \frac{125^2}{10}$$

$$X_{\text{line}} = \frac{70}{(125^2/10)} = 0.0448 \text{ per unit}$$

$$X_{\text{transf.}} = 0.1 \left(\frac{1}{3 \times 6.667}\right)\left(\frac{\sqrt{3} \times 10}{12.5}\right)^2 = 0.0096$$

$$X_{\text{gen.}} = 0.2 \left(\frac{10}{15}\right)\left(\frac{8.5}{7.22}\right)^2 = 0.1845$$

where $\bar{V}_{\text{base}(G)} = \left(\frac{125}{\sqrt{3}}\right)\left(\frac{1}{10}\right) = 7.22 \text{ kv.}$

$$I_L = \frac{1}{0.8 + j0.6} = 0.8 - j0.6 = 1 \,\underline{/-36.8°}$$

$$\bar{V}_t = 1.0 + (0.8 - j0.6)(j0.0448 + 2Xj0.0096)$$
$$= 1.0384 + j0.0512 = 1.04 \,\underline{/2.8°} \text{ per unit}$$

Generator line-to-line voltage $= 1.04 \times 7.22 \text{ kv per unit}$
$$= \mathbf{7.79} \text{ kv} \qquad \mathbf{d}$$

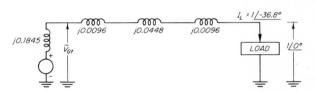

Fig. 4.21-2

4.22 The propagation constant per loop mile for the transmission line with uniformly distributed parameters is

$$\gamma = \alpha + j\beta = \sqrt{(r + j1\omega)(g + jc\omega)}$$

At a frequency of 796 cps, $\omega = 2\pi f = 5,000$

$$\gamma = \sqrt{(10.44 + j18.3)(0.300 + j41.9)(10^{-6})}$$
$$= \sqrt{21.1/60.2° \times 42.0 \times 10^{-6}/90°}$$
$$= 0.0298\underline{/75.1°} = 0.00766 + j0.0287$$

whence

$$\alpha = 0.00766 \text{ neper per loop mile (attenuation)}$$
$$\beta = 0.0287 \text{ radian per loop mile (phase shift)}$$

a. The wavelength is $\lambda = 2\pi/\beta = 6.28/0.0287 = 219$ miles.

b. The phase velocity is given by $v = \omega/\beta = 5{,}000/0.0287 = 174{,}000$ miles per sec or $(174{,}000/186{,}000) \times 100 = 93.5$ per cent of the speed of light.

c. The attenuation, if the line is properly matched at the load is $\alpha = 0.00766$ neper per loop mile. There are 8.68 db in one neper, hence the attenuation is

$$0.00766 \times 8.68 = 0.0665 \text{ db per loop mile}$$

d. For distortionless transmission it is necessary that

$$1g = rc \text{ or } 1 = \frac{rc}{g} = \frac{(10.44 \times 0.00838 \times 10^{-6})}{(0.300 \times 10^{-6})}$$
$$= 0.290 \text{ henry per loop mile}$$

As the line has but 0.00366 henry per loop mile, it is necessary to use additional loading of $0.290 - 0.00366 = 0.286$ henry per loop mile.

4.23 To match properly the generator to the line, the latter must have a characteristic impedance $Z_0 = 100 + j0$ ohms, equal to the generator impedance. Since the stub is to be connected in parallel with the line, it is convenient to work with admittances.

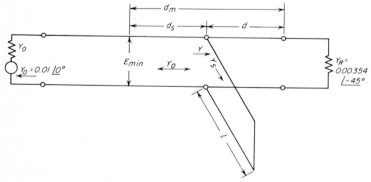

Fig. 4.23

In this case there is a conductive characteristic admittance

$$Y_0 = \frac{l}{Z_0} = G_0 = 0.01 \text{ mho}$$

The stub must be connected to the line at a distance d, preferably as near the load as possible, where the line admittance looking toward the load, $Y = G + jB$, has a real or conductive component $G = G_0$. To determine this location, compute first the receiving-end reflection factor

$$N_R = \frac{[(Z_R/Z_0) - 1]}{[(Z_R/Z_0) + 1]} = \frac{(2 + j2 - 1)}{(2 + j2 + 1)}$$
$$= 0.62\underline{/30°}$$

The standing-wave ratio of the line prior to stubbing is

$$\frac{(1 + N_R)}{(1 - N_R)} = \frac{(1 + 0.62)}{(1 - 0.62)} = 4.27$$

Next compute the location, d_m, at which the standing wave of voltage is at its minimum, $d_m = \dfrac{(\varphi + \pi)\lambda}{4\pi}$ where

φ = angle, in radians, of $N_R = 0.525$

λ = wavelength, in meters, $= \dfrac{(3 \times 10^8)}{f} = 0.5$

Consequently, $d_m = (0.525 + \pi) \times \dfrac{0.5}{4\pi} = 0.136$ meter. Now the stub must be connected a distance d_s either side of the point of voltage minimum, where

$$d_s = \frac{(\lambda \times \cos^{-1} N_R)}{4\pi} = 0.5 \times \frac{(51.6/57.3)}{4\pi} = 0.0358 \text{ meter}$$

As it is desirable to place the stub as close to the load as possible, locate it at $d = d_m - d_s = 0.136 - 0.036 = 0.1$ meter. As mentioned above, a point has been located at which the admittance of the line looking toward the load is $Y = G_0 + jB$. The lossless stub has an admittance $Y_s = 0 + jB_s$. By adjusting the length of the stub so that $B_s = -B$, then the admittance of the line and stub in parallel, Y', will be $Y' = Y + Y_s = G_0 + jB + 0 - jB = G_0$, and the line is matched, as desired, to the left of the stub.

The length of the lossless short-circuited stub required is given in terms of the standing-wave ratio of the line without the stub as

$$1 = \frac{\lambda}{2\pi} \times \tan^{-1} \frac{\sqrt{\rho}}{(\rho - 1)} = \frac{0.5}{6.28} \times \tan^{-1} \frac{\sqrt{4.27}}{(4.27 - 1)}$$

$$= 0.045 \text{ meter}$$

It should be noted that answers to questions of this type are obtained with much greater facility by means of a universal transmission line diagram, such as the Smith chart.

4.24 a.

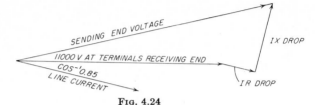

FIG. 4.24

b. Line current $= \dfrac{\text{watts}}{E \cos \theta} = \dfrac{1{,}200 \times 10^3}{11{,}000 \times 0.85} = \mathbf{128.4}$ **amp**

7.5 per cent of 1,200 kw = 90 kw

$I^2R = (128.4)^2 R \le 90{,}000$ watts

$$R \le \frac{90{,}000}{(128.4)^2} = 5.45 \text{ ohms or } 0.2725 \text{ ohms}$$

per mile of wire

#4-0 solid copper wire has 0.264 ohms per mile and radius = 0.23 in.

c. Resistance per wire $= 0.264 \times 10 = \mathbf{2.64}$ **ohms**

d. Inductance per mile $= 0.7411 \times \log_{10} (18/0.23 = 78.2) + 0.080 = 0.7411 \times 1.8932 + 0.080 = 1.485$ millihenrys per mile

Reactance per wire $= 2 \times \pi \times 50 \times 1.485 \times 10^{-3} \times 10 = \mathbf{4.66}$ **ohms**

e. Sending-end voltage, where $0.527 = \sin (\cos^{-1} 0.85)$,

$$= 2 \sqrt{(5{,}500 \times 0.85 + 128.4 \times 2.64)^2 + (5{,}500 \times 0.527 + 128.4 \times 4.66)^2}$$

$$= 2 \sqrt{(4{,}670 + 339)^2 + (2{,}900 + 596)^2} = \mathbf{12{,}220} \text{ volts}$$

f. Regulation, assuming no load, the receiving-end voltage becomes equal to the sending-end voltage, is

$$\frac{(12{,}220 - 11{,}000)}{11{,}000} = \frac{1{,}220}{11{,}000} = \mathbf{11.1} \text{ per cent}$$

g. Line loss $= I^2R = (128.4)^2 \times 2.64 \times 2 = 87{,}000$ watts

$$\text{Efficiency} = \frac{1{,}200{,}000}{(1{,}200{,}000 + 87{,}000)} \times 100 = \textbf{93.2 per cent}$$

4.25 $Z = zl = (0.35 + j0.8)100$

$$= 100 \sqrt{0.1275 + 0.64} \underline{/\tan^{-1}8\%_{35}} = 87.3\underline{/66.4°}$$

$$Y \frac{\tanh \theta}{\theta} = \frac{Is}{Es} = \frac{j50}{100{,}000} = 5 \times 10^{-4}\underline{/90°} \text{ (from tests with}$$

100,000 volts)

Assuming, for the moment, that $\dfrac{\tanh \theta}{\theta} = 1\underline{/0°}$ and $Y =$

$5 \times 10^{-4}\underline{/90°}$

$$ZY = (87.3\underline{/66.4°})(5 \times 10^{-4}\underline{/90°}) = 436.5 \times 10^{-4}\underline{/156.4°}$$
$$= 0.04365\underline{/156.4°}$$

From chart in Woodruff, "Principles of Electric Power Transmission," Chap. 5,

$$\frac{\tanh \theta}{\theta} = 1.003\underline{/-0.1°} \text{ and } Y = \frac{5 \times 10^{-4}\underline{/90°}}{1.003\underline{/-0.1°}}$$

$$Y = 5 \times 10^{-4}\underline{/90.1°}$$

The correction is negligible except in the angle.
Then for $ZY = 0.04365\underline{/156.5°}$

$$\cosh \sqrt{ZY} = \cosh \theta = 0.98\underline{/0.05°}$$

and
$$\frac{\sinh \theta}{\theta} = 0.993\underline{/0.018°}$$

$$\text{Voltage to neutral} = \frac{132{,}000}{\sqrt{3}} = 76.3 \text{ kv}$$

$$\text{Line current} = \frac{50{,}000}{3 \times 76.3 \times 0.80} = 273 \text{ amp}$$
$$= 273(0.8 - j0.6) = (218.4 - j164) \textbf{ amp}$$

From Woodruff (as above),

$$E_S = E_R \cosh \theta + I_R Z \frac{\sinh \theta}{\theta}$$

$$= 76.3 \times 0.98\underline{/0.05°} + \frac{(273\underline{/-36.9°})(87.3\underline{/66.4°})(0.993\underline{/0.018°})}{1{,}000}$$

$$= 74.8\underline{/0.05°} + 23.68\underline{/29.68°}$$
$$\cos 29.68° = 0.8688$$
$$\sin 29.68° = 0.4952$$
$$E_S = 74.8 + j0 + 23.68(0.8688 + j0.4952)$$
$$= 74.8 + 20.55 + j11.7 = (95.35 + j11.7) \text{ kv}$$
$$I_S = I_R \cosh \theta + E_R Y \frac{\sinh \theta}{\theta} = (273\underline{/-36.8°})(0.98\underline{/0.05°})$$
$$+ (76,300\underline{/0°})(5 \times 10^{-4}\underline{/90.1°})(0.993\underline{/0.018°})$$
$$= 267.5\underline{/-36.75°} + 37.85\underline{/90.12°}$$
$$\cos 36.75 = 0.5983$$
$$\sin 36.75 = 0.8012$$
$$I_S = 267.5(0.8012 - j0.5983) + j3785 = 214 - j160 + j37.85$$
$$= (214 - j122.15) \text{ amp}$$

Sending-end power
$$= \text{real part of } (95.35 + j11.7) \times (214 + j122.15)$$
$$= 20,400 - 1,430 = 18,970 \text{ kw per phase}$$

$$\text{Efficiency} = \frac{\text{receiving-end power}}{\text{sending-end power}} = \frac{50,000}{3 \times 18,970} \times 100$$
$$= \frac{50}{56.9} \times 100 = 87.9 \text{ per cent}$$

Check by finding approximate value of line current by assuming half of the charging current supplied at each end.

Charging current at 76.3 kv to neutral is
$$I_c = \frac{76.3}{100} \times 50 = 38.15 \text{ amp}$$

With half at each end, current at receiving end is
$$I_R + \frac{I_c}{2} = \text{line current} + \frac{38.5}{2} = 218.4 - j164 + j19.1$$
$$= 218.4 - j144.9 = 262.5\underline{/-33.5°}$$
$$I^2R \text{ per line} = (262.5)^2(0.35 \times 100) = 2,890 \text{ kw}$$

$$\text{Efficiency} = \frac{50,000}{(50,000 + 3 \times 2,890)} \times 100 = \mathbf{87.4 \text{ per cent}}$$

4.26 One phase of the circuit is

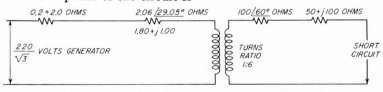

Fıg. 4.26a

The equivalent circuit referred to generator voltage is

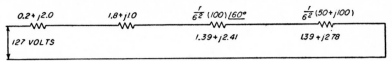

Fɪɢ. 4.26*b*

Simplified, this becomes

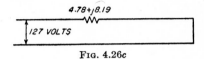

Fɪɢ. 4.26*c*

The current at the generator is

$$I_G = \frac{127}{4.78 + j8.19} = \frac{127}{9.48\underline{/59.7°}} = 13.4\underline{/-59.7°} \text{ amp}$$

The current at the short circuit is $\frac{1}{6}$ of this or

$$I_{\text{short circuit}} = \frac{1}{6} \times 13.4 = 2.63 \text{ amp}$$

4.27   From blocked test

$$R_{\text{eq}} \text{ per phase} = \frac{P_B}{3I^2} = \frac{40,500}{3 \times (170)^2} = 0.466 \text{ ohm}$$

$$Z_{\text{equiv}} = \frac{V}{I} = \frac{440}{\sqrt{3} \times 170} = 1.5 \text{ ohm}$$

$$X_{\text{equiv}} = \sqrt{Z_{\text{equiv}}^2 - R_{\text{equiv}}^2} = \sqrt{2.25 - 0.22} = 1.42 \text{ ohm}$$

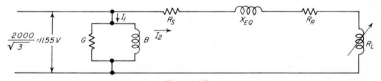

Fɪɢ. 4.27

$$R_S + R_R = 0.466 \text{ ohm}$$

Using d-c ratios for split

$$R_{S_{\text{dc}}} = 0.165 \text{ ohm}$$
$$R_{R_{\text{dc}}} = 0.0127 \times n^2 = 0.0127 \times 4^2 = 0.203 \text{ ohm (referred to stator)}$$

$$R_s = \frac{0.165}{(0.165 + 0.203)} \times 0.466 = 0.21 \text{ ohm}$$

$$R_R = \frac{0.203}{(0.165 + 0.203)} \times 0.466 = 0.256 \text{ ohm}$$

$$R_L = R_R \left(\frac{1 - S}{S}\right) = 0.256 \left(\frac{1 - 0.015}{0.015}\right) = 16.8 \text{ ohm}$$

From no-load test (assuming F and W inherent in R_L)

$$G = \frac{(10,100 - 2,000)}{3 \times (1,155)^2} = 0.00202 \text{ mho}$$

$$Y = \frac{I}{V} = \frac{15.3}{1155} = 0.0132 \text{ mho}$$

$$\text{pf} = \frac{p}{\sqrt{3}\,VI} = \frac{10,100}{\sqrt{3} \times 2,000 \times 15.3} = 0.191$$

$$B = Y \sin \theta = 0.0132 \times 0.9815 = 0.013 \text{ mho}$$

$$I_2 = \frac{V}{Z} = \frac{1,155}{0.21 + 0.256 + 16.8 + j1.42} = \frac{1,155}{17.27 + j1.42}$$

$$= \frac{1,155}{17.4/4.7^\circ} = 66.4\underline{/-4.7^\circ} \text{ amp} = 66 - j5.5$$

$$I_1 = V(G - jB) = 1,155(0.00202 - j0.013)$$
$$= 2.33 - j15 = 15.2\underline{/-81.2^\circ}$$

a. $I_{\text{stator}} = I_1 + I_2 = 68.3 - j20.5 = \mathbf{71.3\underline{/-16.7^\circ}} \textbf{ amp}$

$I_{\text{rotor}} = nI_2 = 4 \times 66.4 = \mathbf{265.6} \textbf{ amp}$

b. $P_{\text{out}} = (3I_2^2 R_L - F \text{ and } W)$

$$= [3 \times (66.4)^2 \times 16.8 - 2,000] = 220 \text{ kw}$$

hp $= 220/0.746 = \mathbf{295}$ **hp**

c. $N = \dfrac{120f}{p}(1 - s) = 1,000 \times 0.985 = \mathbf{985}$ **rpm**

d. Torque developed $= 3I_2^2(R_L + R_R) = 3 \times (66.4)^2(17.1)$

$$= 226,000 \text{ sync watts} = \frac{226,000}{105} = \mathbf{2,150} \textbf{ newton meters}$$

$$\text{Net torque} = \frac{(226,000 - 2,000)}{105} = \mathbf{2,130} \textbf{ newton meters}$$

e. pf $= \cos \underline{/I_{\text{stator}}} = \cos 16.7^\circ = \mathbf{0.958}$

f. Efficiency $= \dfrac{P_0}{P_I} \times 100 = \dfrac{220,000 \times 100}{\sqrt{3} \times 2,000 \times 71.3 \times 0.958}$

$\qquad\qquad\qquad\qquad\qquad\qquad\qquad\qquad = \mathbf{93.0}$ per cent

4.28 Present excitation $= 480 \times 11 = 5,280$ amp-turns per pole

Output current $= 35,000/125 = 280$ amp
Series field turns per pole $= 5,280/280 = 18.85$
Therefore, wind 19.5 turns per pole

4.29 *a.* At unity pf:

Output $= 2,500 \times 1$ kw

Current $= \dfrac{2,500}{(\sqrt{3} \times 6.6 \times 1)} = 219$ amp

Armature copper loss $= \dfrac{[3 \times (219)^2 \times 0.072]}{1,000} = 10.4$ kw

Field loss $= \dfrac{[(200)^2 \times 0.43]}{1,000} \qquad\qquad = 17.2$ kw

Friction loss $\qquad\qquad\qquad\qquad\qquad = 35.0$ kw
Core loss $\qquad\qquad\qquad\qquad\qquad\quad = 47.5$ kw
Total loss $\qquad\qquad\qquad\qquad\qquad\; = 110.1$ kw

Efficiency $= \dfrac{\text{output}}{(\text{output} + \text{losses})} = \dfrac{(2,500 \times 100)}{2,610.1}$

$\qquad\qquad\qquad\qquad\qquad\qquad\qquad = 95.8$ per cent

b. At 0.8 pf:
Output $= 2,500 \times 0.8 = 2,000$ kw

Current $= \dfrac{2,000}{(\sqrt{3} \times 6.6 \times 0.8)} = 219$ amp

Armature copper loss $= \dfrac{[3 \times (219)^2 \times 0.072]}{1,000} = 10.4$ kw

Field loss $= \dfrac{[(240)^2 \times 0.43]}{1,000} \qquad\qquad = 24.8$ kw

Friction loss $\qquad\qquad\qquad\qquad\qquad = 35.0$ kw
Core loss $\qquad\qquad\qquad\qquad\qquad\quad = 47.5$ kw
Total loss $\qquad\qquad\qquad\qquad\qquad\; = 117.7$ kw

Efficiency $= \dfrac{(2,000 \times 100)}{2,117.7} = 94.4$ per cent

4.30 *a.* Reads average. $V_{\text{av}} = \dfrac{(3V \times 1 \text{ msec})}{3 \text{ msec}} = 1$ volt

 b. Reads rms. $V_{\text{rms}} = \sqrt{\dfrac{(3)^2(1)}{3}} = \sqrt{\dfrac{9}{3}} = \sqrt{3} = $ **1.732** volts

 c. Peak above average $= 3V - 1V = 2$ volts
      rms reading $= 2/\sqrt{2} = $ **1.414** volts

d. Leads reversed.

$$\text{Average} = \frac{3 \times 2}{3} = 2 \text{ volts}$$

$$\text{Peak above average} = 3 - 2 = 1 \text{ volt}$$
$$\text{rms reading} = (1/\sqrt{2}) = \mathbf{0.707} \text{ volts}$$

See Fig. 4.30 with question.

4.31 *a.* Current at start varies directly as the applied voltage. For rated current at start, $V_{\text{start}} = 550/4.25 = 130$ volts.

 b. Starting torque varies as the square of the voltage.   At 440 volts, $T_{\text{start}} = [(440)^2/(550)^2] \times 160$ per cent $= 102$ per cent of rated torque, and $I_{\text{start}} = (440/550) \times 425$ per cent $= 340$ per cent of rated current.

4.32 If 85 per cent efficiency is assumed, the rated motor current will be

$$I_m = \frac{(746 \times \text{hp})}{(E \times \text{efficiency})} = \frac{746 \times 5}{120 \times 0.85} = 36.6 \text{ amp}$$

Rated field current is, by Ohm's law

$$I_f = \frac{E}{R_f} = \frac{120}{50} = 2.4 \text{ amp}$$

Whence rated armature current is

$$I_a = I_m - I_f = 36.6 - 2.4 = 34.2 \text{ amp}$$

The starter is to limit the armature current to 200 per cent of rated value, or 68.4 amp. If the armature circuit resistance is R_a and the maximum starter resistance is R_s, then by Ohm's law

$$68.4 = \frac{E}{(R_a + R_s)}$$

Whence the starter resistance must be

$$R_s = \frac{E}{68.4} - R_a = \frac{120}{68.4} - 0.8 = 1.755 - 0.800 = 0.955 \text{ ohm}$$

4.33 $I_{28.8} = 100$ amp

$$= \frac{28,800}{(\sqrt{3} \times 208 \times 0.8)}$$

Using $V_{A\text{ground}}$ as reference

$I_B = 100\underline{/120°} - 36.7°$
$\quad = 100\underline{/83.3°}$ amp
$I_A = 60\underline{/0°} + 100\underline{/-36.7°}$
$\quad = 60 + 80 - j60$
$\quad = 152\underline{/-23.2°}$ amp
$I_C = 60\underline{/-120°} +$
$\quad\quad 100\underline{/-120°} - 36.7°$
$\quad = 152\underline{/-143.2°}$ amp
$I_{\text{ground}} = 60\underline{/0°} + 60\underline{/-120°}$
$\quad = 60\underline{/-60°}$ amp

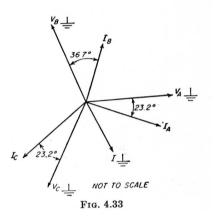

Fɪɢ. 4.33

4.34   Gain with feedback is $A_f = 100$. Gain without feedback is

$$A = (10^2 \times 10^2) \pm 2,000 = 10^4 \pm 2,000 \text{ approx.}$$

Since $A_f = \dfrac{A}{1 - A\beta}$, from any textbook on electronics,

$$100 = \frac{10^4}{1 - A\beta}$$

or $1 - A\beta = 100$

Also $\dfrac{dA_f}{100} = \dfrac{1}{1 - A\beta} \dfrac{dA}{A}$, from any electronics textbook

or $\dfrac{dA_f}{100} = \dfrac{1}{100} \times \dfrac{2,000}{10^4} = 20 \times 10^{-4}$

$$= \textbf{0.2} \text{ per cent maximum variation in gain.} \quad \textbf{b}$$

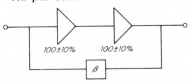

Fɪɢ. 4.34

4.35 Motor input $= 62.5 + 30.5 = 93$ kw

$$\text{Tangent of p-f angle} = \sqrt{3} \times \frac{(62.5 - 30.5)}{(62.5 + 30.5)}$$
$$= \sqrt{3} \times {}^{32}\!/_{93} = 0.596 = \tan 30.8°$$

Power factor $= \cos 30.8° = 0.859$
Power input $= \sqrt{3} \times EI \cos \theta$
$\qquad\qquad = \sqrt{3} \times 600 \times I \times 0.859 = 93,000$ watts
$I = 104.3$ amp in each line.
Power output $= 0.90 \times 93,000/746 = 112.2$ hp

4.36 Using complex notation
$\text{kva} = \sqrt{3} \times \text{kv} \times I \cos \theta + j \sqrt{3} \times \text{kv} \times I \sin \theta$
Total load kva $= 4,000 + j0$
Alternator No. 1
$\text{kva}_1 = \sqrt{3} \times 6.6 \times 200 \times 0.85 + j \sqrt{3} \times 6.6 \times 200$
$\qquad\qquad\qquad\qquad\qquad\qquad\qquad\qquad\qquad\qquad \times 0.527$

$\qquad = 1,943 + j1,205$
Alternator No. 2 must supply the difference
$\text{kva}_2 = \text{kva} - \text{kva}_1 = (4,000 - 1,943) + j(0 - 1,205)$
$\qquad = 2,057 - j1,205 = 2,380\underline{/-30.4°}$
Alternator No. 2 supplies 2,057 kw, at $\cos(-30.4°)$
$= 0.863$ pf lagging. $I_2 = 2,380/(\sqrt{3} \times 6.6) = 208$ amp

4.37 $a.$ $I = \dfrac{150/2}{100} = 0.75$ per unit

$\qquad E_{af} = \bar{V} + jX_d I$
$\qquad\qquad = 1.0 + (j1.1)(0.8 - j0.6)(0.75)$
$\qquad\qquad = 1.495 + j0.66$
$\qquad\qquad = 1.63 \underline{/23.85°}$ pu

$\qquad E_{af} = \dfrac{(1.63)(13.8 \text{ kv})}{\sqrt{3}} = 13$ kv per phase   **1a**

$b.$ $\delta = 23.85°$ **2c**

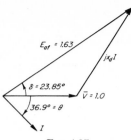

Fɪɢ. 4.37

4.38   *a.*   1. Capacitor. Capacitor changes the phasing between currents I_1 and I_2 in two windings spaced 90° from each other, treating a condition similar to a two-phase rotating field.

Fɪɢ. 4.38

     2. Repulsion. Wound rotor with commutator. Brushes are short-circuited creating an induced field which acts similarly to a series field.

     3. Shaded pole. Salient pole machine with short-circuited turn "*C*" as shown in Fig. 4.38. Flux through "*A*" is delayed because of Lenz's law causing a sweeping of flux across the pole face.

   b. Five hp for special applications.

   c. Rotate the brushes to opposite side of hard neutral.

   d. Reduced number of turns on field; compensating winding; laminated field structure; and increased armature size.

4.39   By definition, $db = 10 \log P_1/P_2$, and the power delivered to the coil based upon $P_2 = 0.005$ watts is

$$P_1 = 0.005 \text{ antilog } {}^{2}\!\%_{10} = 0.5 \text{ watts}$$

Inasmuch as this power is $P_1 = I^2R$, the rms current in the 10-ohm voice coil must be $I = \sqrt{P_1/R} = \sqrt{0.5/10} = 0.224$ amp. If mks units are desired, which are easier and require no conversion factors, the flux density of 10,000 lines psi is

$$B = 0.155 \text{ weber per sq meter}$$

and for the active length, which is the circumference of the voice coil

$$1 = -d = 3.14 \times 1 \times 0.0254 = 0.0799 \text{ meters}$$

Then the rms force exerted is simply $f = B1NI$

$$f = 0.155 \times 0.0799 \times 10 \times 0.224 = 0.0277 \text{ newtons}$$

(This force is equivalent to 1 ounce.) If the current is sinusoidal, the maximum thrust will be $\sqrt{2}$ times the rms value found above.

4.40 First find the Thevenin equivalent to the left of the base.

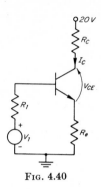

Fig. 4.40

$$V_1 = 20 \frac{R_b}{R_b + R_a} = 20 \frac{10}{100} = 2 \text{ volts}$$

$$R_1 = \frac{R_a R_b}{R_a + R_b} = \frac{(10 \text{ k}\Omega)(90 \text{ k}\Omega)}{100 \text{ k}\Omega} = 9 \text{ k}\Omega$$

From any textbook on electronics,

$$I_c = \frac{\beta(V_1 - V_{BE})}{R_1 + R_e(1 + \beta)} = \frac{40(2 - 0.7)}{9 \text{ k}\Omega + 1 \text{ k}\Omega(41)} = \mathbf{1.04 \text{ ma}} \qquad \mathbf{1d}$$

$$\begin{aligned} V_{CE} &= 20 - I_c(R_c + R_e) \\ &= 20 - I_c(R_c + R_e) \\ &= 20 - 1.04(5 + 1) = \mathbf{13.76 \text{ volts}} \qquad \mathbf{2a} \end{aligned}$$

4.41 The base current is

$$I_B = \frac{I_C}{\beta} = \frac{2 \text{ ma}}{10} = 200 \ \mu\text{a}$$

The emitter current is

$$I_E = I_C + I_B = 2 \text{ ma} + 0.2 \text{ ma} = 2.2 \text{ ma}$$

The voltage across R_c is

$$V_{R_c} = 20 - V_{CE} - I_E(2 \text{ k}\Omega) = 20 - 10 - 4.4 = 5.6 \text{ volts}$$

and $\quad R_c = \dfrac{V_{R_c}}{I_C} = \dfrac{5.6}{2 \times 10^{-3}} = \textbf{2.8 k}\Omega \quad$ **1c**

The voltage across R_a is

$$V_{R_a} = 20 - I_E(2 \text{ k}\Omega) - 0.7 = 20 - 4.4 - 0.7 = 14.9 \text{ volts}$$

The current through R_a is

$$I_{R_a} = I_B + \frac{4.4 + 0.7}{10 \text{ k}\Omega} = 0.2 \text{ ma} + 0.51 \text{ ma} = 0.71 \text{ ma}$$

$$R_a = \frac{V_{R_a}}{I_{R_a}} = \frac{14.9}{0.71 \text{ ma}} = \textbf{21.0 k}\Omega \quad \textbf{2b}$$

4.42   *a.* $\sqrt{2} \times 120 = 170$-volt   peak-to-neutral-peak   inverse

voltage per stack $= 2 \times \dfrac{\sqrt{3}}{2} \times 170 = 295$ volts

$$\frac{295}{20} \text{ volts} = 14.75 \text{ or } 15 \text{ disks}$$

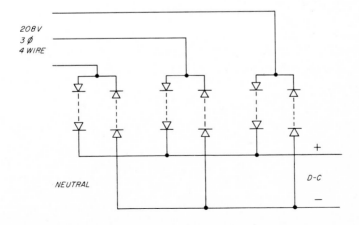

208V
3 ∅
4 WIRE

NEUTRAL

\+

D-C

\−

(a)

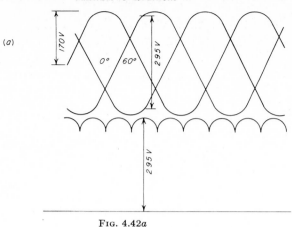

Fɪɢ. 4.42a

Six single stacks required, two stacks in series always. Current capacity, $3 \times 2 = 6$ amp.

Output voltage:

$$E_{av} = \frac{170}{\pi/6} \int_{0°}^{30°} [\cos \theta - \cos (120° + \theta)]\, d\theta$$

$$= \frac{170 \times 6}{\pi} \int_{0°}^{30°} (\cos \theta - \cos 120° \times \cos \theta + \sin 120° \times \sin \theta)\, d\theta$$

$$= 325 \int_{0°}^{30°} \left[(1 + 0.5) \cos \theta + \frac{\sqrt{3}}{2} \sin \theta \right] d\theta$$

$$= 325(1.5 \sin \theta - 0.866 \cos \theta) \Big]_{0°}^{30°}$$

$$= 325(1.5 \times 0.5 - 0 - 0.866 \times 0.866 + 0.866)$$

$$= 325 \times 0.866 = 282 \text{ volt no load}$$

or $0.85 \times 282 = 240$ volt full load

Output capacity $= 6 \times 240 = 1,440$ watts

b. Rectifier of part a is also the rectifier that gives the smallest voltage ripples.

c. The rectifier that takes the smallest number of disks and still draws a balanced load is one-half of the rectifier of part *a,* which uses voltages to neutral only once. Peak inverse voltage is as in part *a.*

Fig. 4.42c

Fifteen disks per stack are required but only three stacks current capacity $= 3 \times 2 = 6$ amp.

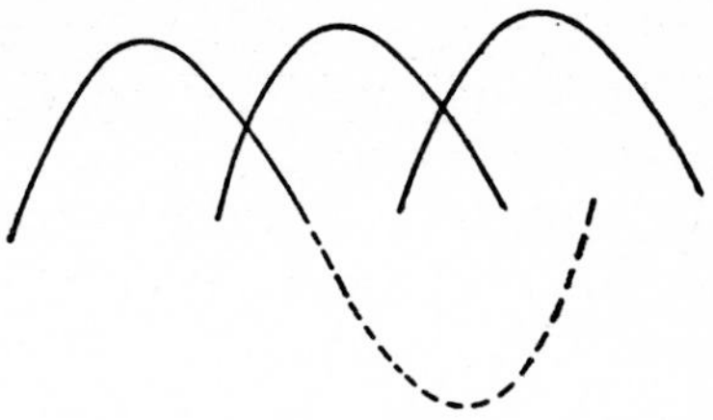

Output voltage

$$E_{av} = \frac{170}{\pi/3} \int_{0°}^{60°} \cos\theta \, d\theta = 162(\sin\theta) \Big]_{0°}^{60°}$$

$$= 162 \left(\frac{\sqrt{3}}{2} - 0 \right) = 140 \text{ volts no load}$$

$$= 0.85 \times 140 = 119 \text{ volts full load}$$

Voltage ripple in part $a = \dfrac{(295 - 255)}{2} = 20$ volts.

Voltage ripple in part $c = \dfrac{(170 - 85)}{2} = 42.5$ volts.

Voltage ripple in part c is $\dfrac{42.5}{20} = 2.13$ times voltage ripple in part *a.*

4.43   Power $= EI \cos\theta$

$550 = 220 \times 2.5 \cos\theta$

$\cos\theta = \dfrac{550}{220 \times 2.5} = 1$

$\theta = 0°$

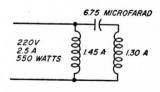

a. The current in the main winding, 1.45A, lags the line voltage. The condenser current, 1.3A, must lead the other current if their sum, 2.5A, is to be in phase with the line voltage.

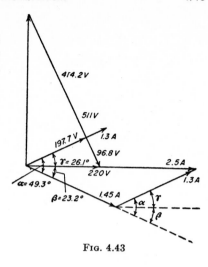

Fig. 4.43

By the cosine law

$$(2.5)^2 = (1.45)^2 + (1.3)^2 + 2 \times 1.45 \times 1.3 \cos \alpha$$
$$6.25 = 2.10 + 1.69 + 3.77 \cos \alpha$$
$$\cos \alpha = \frac{(6.25 - 3.79)}{3.77} = 0.652$$
$$\alpha = 49.3°$$
$$\sin \alpha = 0.758$$

By the sine law

$$\frac{2.5}{\sin (\pi - \alpha)} = \frac{1.3}{\sin \beta}$$
$$\sin \beta = (1.3/2.5) \times \sin \alpha$$
$$= 0.52 \times 0.758 = 0.394$$
$$\beta = 23.2°$$
$$\cos \beta = 0.918$$

Let $\gamma = \alpha - \beta = 49.3° - 23.2° = 26.1°$, $\cos \gamma = 0.898$

The components of the line voltage, 220, with respect to the condenser current, 1.3A, are $220 \cos \gamma = 220 \times 0.898 = 197.7$ volts.

$$220 \sin \gamma = 220 \times 0.440 = 96.8 \text{ volts}$$

The voltage drop across the condenser is

$$E_c = IX_c = \frac{1.3}{377 \times 6.75 \times 10^{-6}} = \frac{1,300}{2.545} = 511 \text{ volts}$$

Apparent impedance drop in condenser branch

$$220 = IZ = IR_W + j(IX_W - IX_C) = 197.7 - j96.8$$
$$IX_W - 511 = -96.8 \text{ volts and } IX_W = 511 - 96.8 = 414.2 \text{ volts}$$

Apparent $IZ_W = IR_W + jIX_W = 197.7 + j414.2$
Apparent $Z_W = 197.7/1.3 + j414.2/1.3 = 152 + j319$
$$= \sqrt{(152)^2 + (319)^2} = 353 \text{ ohms}$$

Main winding apparent $Z = 220/1.45 = 152$ ohms.

    b. Power to main winding $= 220 \times 1.45 \times \cos 23.2°$
$$= 293 \text{ watts}$$

    Power to auxiliary winding $= 197.7 \times 1.3 = 257$ watts.

4.44   *a.*

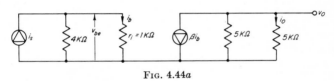

Fɪɢ. 4.44*a*

  b. The base current is

$$i_b = (\tfrac{4}{5})i_s$$

The output current is

$$i_0 = -\tfrac{1}{2}\beta i_b$$

Dividing these two equations we have

$$\frac{i_0}{i_s} = -\frac{0.5\beta i_b}{(\tfrac{5}{4})i_b} = -\frac{2}{5}\beta = -\frac{2}{5}40 = -\mathbf{16} \text{ current gain}$$

  c. $v_{be} = i_b r_i = 1 \text{ k}\Omega i_b$
     $v_0 = -\beta i_b(2.5 \text{ k}\Omega)$

$$\frac{v_0}{v_{be}} = -\frac{40i_b(2.5 \text{ k}\Omega)}{1 \text{ k}\Omega i_b} = -\frac{100 \text{ k}\Omega}{1 \text{ k}\Omega} = -\mathbf{100} \text{ voltage gain}$$

4.45   Using Laplace, we have

$$I(S) = \frac{V(S)}{Z(S)} = \frac{(V_1 + V_0)/S}{R + (1/CS)} = \frac{(V_1 + V_0)C}{RCS + 1}$$

$$I(S) = \frac{(V_1 + V_0)C}{RC} \frac{1}{S + (1/RC)} = \frac{V_1 + V_0}{R} \frac{V_1 + V_0}{S + (1/RC)}$$

Taking the inverse transform (from tables),

$$i(t) = \frac{V_1 + V_0}{R} \epsilon - \frac{t}{RC}$$

$$= \frac{200 + 150}{150} \epsilon - \frac{t}{2.25(10^{-3})} = 2.33^{-445t}$$

At $t = 0$, $i(t) = 2.33$. At $t = \infty$, $i(t) = 0$. At $t = 2.25$ msec,

$$i(t) = 2.33\epsilon^{-1} = \frac{2.33}{\epsilon} = \textbf{0.86 amp}$$

FIG. 4.45

4.46 No resistance values being furnished, assume a dissipationless parallel or antiresonant circuit for which the angular frequency of antiresonance is merely

$$\frac{1}{\sqrt{LC}} = 2\pi f_0$$

Hence the required capacitance is given by

$$C = \frac{1}{(2\pi f_0)^2 xL} = \frac{1}{(6.28 \times 60 \times 10^3)^2 \times 260 \times 10^{-6}}$$
$$= 0.0271 \times 10^{-6} \text{ f}$$

4.47 *a.*

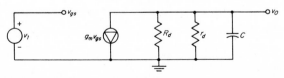

FIG. 4.47a

$b.$ $v_0 = -g_m v_{gs} \dfrac{R(1/jwc)}{R + (1/jwc)}$

where $\qquad R = \dfrac{R_d r_d}{R_d + r_d} \ 12 \text{ k}\Omega$

Solving, we have

$$\frac{v_0}{v_{gs}} = -g_m \frac{R}{1 + jwRc}$$

$c.$

$$\frac{v_0}{v_{gs}} = -g_m R \frac{1}{1 + j(w/w_1)}$$

where $\quad w_1 = \dfrac{1}{Rc} = \dfrac{1}{12 \times 10^3 \times 0.01 \times 10^{-6}}$

$$= 8.34 \times 10^3 \text{ radians per sec}$$

$$g_m R = 10^{-2} \times 12 \times 10^3 = 120$$

$$20 \log 120 = \mathbf{41.6} \text{ dB}$$

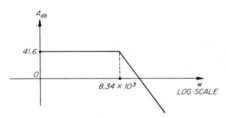

Fig. 4.47c

4.48 $a.$ $R = 12.6 + 50 = 62.6$ ohms. The general equation for the current is

$$i = I_0 \epsilon^{-(R/L)t}$$
$$e_{ab} = i(50) = 50 I_0 \epsilon^{-(R/L)t}$$
$$20 = 50(9.2)\epsilon^{(R/L)}$$

$$\epsilon^{(R/L)t} = 23, \ \frac{R}{L}t = 3.14$$

$$L = \frac{Rt}{3.14} = \frac{(62.6)(2)}{3.14} = \mathbf{40} \text{ henries}$$

$b.$ Induced voltage is

$$e_i = iR = I_0 R \epsilon^{-(R/L)t}$$

$$= \frac{(9.2)(62.6)}{23} = \textbf{25.1} \text{ volts}$$

c. Since $e_{ab} = 50 I_0 \epsilon^{-(R/L)t}$, max e_{ab} occurs at $t = 0$ or max $e_{ab} = 50(9.2) = \textbf{460}$ volts

d. $i = I_0 \epsilon^{-(R/L)t}$ or $0.01 I_0 = I_0 \epsilon^{-(R/L)t}$

$$\frac{R}{L} t = 4.6 \quad \text{or} \quad t = 4.6 \frac{L}{R} = \frac{(4.6)(40)}{62.6} = \textbf{2.94} \text{ sec}$$

4.49 $P_M = 3{,}600 \times R \times \dfrac{K_H}{S} = \dfrac{(3{,}600 \times 15 \times \frac{2}{3})}{50} = \textbf{720}$ watts

$P_{\text{system}} = P_M \times (CT \text{ ratio})(PT \text{ ratio}) = 720 \times 20 \times 20$
$= \textbf{288}$ kilowatts

4.50   Overhead ground wires are used to protect transmission lines against lightning by intercepting direct strokes and by providing multiple paths for stroke current. They also protect against induced strokes by increasing the capacitance between conductors and ground, thereby reducing the voltage from conductor to ground. Low ground resistance is important.

Alternative methods of affording a certain degree of protection to transmission lines are protector tubes, ground-fault neutralizers, and automatic reclosing.

Protective tubes simply provide gaps to ground the high-voltage surge on the line and some means to extinguish the power-follow arc.

Ground-fault neutralizers are reactance coils placed between neutral and ground and especially proportioned in relation to the charging current of the system so as to neutralize the ground-fault current.

Circuit breakers disconnect the lines from the power source when lightning or switching disturbances cause overvoltage or overcurrent. Automatic reclosing permits the lines to be kept in service except for the momentary interruption necessary to clear the disturbance.

Reference: Lewis, "Protection of Transmission Systems Against Lightning," Wiley, 1950.

4.51 When the bridge shown in Fig. 4.51 is balanced,

a. $$Z_A Z_C = Z_B Z_D \text{ or } Z_C = \frac{Z_B Z_D}{Z_A} = \frac{Y_A Z_B}{Y_D}$$

$$Y_A = \frac{1}{R_A} + j\omega C_A$$

$$Z_B = R_B + \frac{1}{j\omega C_B}$$

$$Y_D = j\omega C_D$$

Then $$Z_C = \frac{[(1/R_A) + j\omega C_A][R_B + (1/j\omega C_B)]}{j\omega C_D}$$

$$= \left(\frac{R_B}{R_A} + j\omega C_A R_B + \frac{1}{j\omega C_B R_A} + \frac{C_A}{C_B} \right) \frac{1}{j\omega C_D}$$

$$= \left(\frac{C_A R_B}{C_D} - \frac{1}{\omega^2 C_B C_D R_A} \right) + \frac{1}{j\omega C_D} \left(\frac{R_B}{R_A} + \frac{C_A}{C_B} \right)$$

The impedance Z_C to be connected between C and D should be a

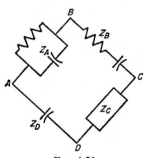

Fig. 4.51

resistor R_C in series with a capacitor C_C, where

$$R_C = \left(\frac{C_A R_B}{C_D} - \frac{1}{\omega^2 C_B C_D R_A} \right)$$

$$= \frac{0.053 \times 1{,}500}{0.265} - \frac{10^{12}}{25 \times 10^6 \times 0.053 \times 0.265 \times 10^3}$$

$$= 300 - 285 = 15 \text{ ohms}$$

$$C_C = \frac{C_D}{(R_B/R_A) + (C_A/C_B)} = \frac{0.265}{(1{,}500/1{,}000) + (0.053/0.53)}$$
$$= 0.166 \ \mu f$$

b. $$\frac{E_{BA}}{E_{AC}} = \frac{Z_A}{Z_A + Z_B} - \frac{Z_D}{Z_C + Z_D}$$

$$Y_A = \frac{1}{10^3} + j5 \times 10^3 \times 0.053 \times 10^{-6} = (1.0 + j0.265)10^{-3}$$

$$Z_A = \frac{1}{Y_A} = \frac{10^3}{1.0 + j0.265} = 965\underline{/-14.85°} = 936 - j247$$

$$Z_A + Z_B = 936 - j247 + 1{,}500 - j372 = 2{,}510\underline{/-14.6°}$$

$$\frac{Z_A}{Z_A + Z_B} = \frac{965\underline{/-14.85°}}{2{,}510\underline{/-14.6°}} = 0.385\underline{/-0.25°}$$

$$Z_D = \frac{10^6}{j5 \times 10^3 \times 0.27} = 740\underline{/-90°}$$

$$\frac{Z_D}{Z_C + Z_D} = \frac{740\underline{/-90°}}{15 + 10^6/(j5 \times 10^3 \times 0.166) - j740}$$
$$= \frac{740\underline{/-90°}}{15 - j(1{,}210 + 740)} = \frac{740\underline{/-90°}}{1{,}950\underline{/-90°}} = 0.379\underline{/0°}$$

$$\frac{E_{BA}}{E_{AC}} \doteq 0.385 - 0.379 = 0.006 = 0.6 \text{ per cent}$$

4.52 a. Over-all efficiency at full load equals

$$\frac{\text{Generator output}}{\text{Motor input}} \times 100 = \frac{6{,}000 \times 100}{220 \times 42} = 65 \text{ per cent}$$

b. Field current = $^{220}\!/_{110} = 2$ amp

$$\text{Approximate armature current} = \frac{\text{stray power}}{\text{voltage}} = \frac{500}{220}$$
$$= 2.27 \text{ amp}$$
$$\text{First estimate of line current at no load} = 2.27 + 2$$
$$= 4.27 \text{ amp}$$

For speed determination, assume constant flux, and

$$S = \frac{K(V - R_A I - V_{\text{brush}})}{\phi} = K_1 E_{\text{full load}}$$

$E_{\text{full load}} = 220 - (42 - 2)0.5 - 1 = 199$ volts
$K_1 = 1{,}480/199 = 7.45$ rpm per volt
$E_{\text{no load}} = 220 - 0.5 \times 2.27 - 1 = 218$ volts

$S_{\text{no load}} = 7.45 \times 218 = 1{,}620$ rpm

$\text{S.P.}_{\text{no load}} = \dfrac{1{,}620 \times 500}{1{,}480} = 546$ watts

Then the corrected armature current $I_{\text{s.p.}} = {}^{546}\!/_{220} = 2.48$ amp

$R_A I_A^2 = 0.5 \times (2.48)^2 = 3.08$ watts

$V_b I_A = 1 \times 2.48 = 2.48$ watts

Armature copper plus brush loss $= 3.08 + 2.48 = 5.56$ watts

$\Delta I_A = 5.56/220 = 0.025$ amp

Total armature current $= 2.48 + 0.025 = 2.5$ amp

Line current at no load $= 2.5 + 2 = 4.5$ amp

Power $= 4.5 \times 220 = 990$ watts

    c. Try #6 RW wire, good for 55 amp (N.E.C.)

               Ohms per 1,000 ft $= 0.4028$
               R (300 ft) $= 0.3 \times 0.4028 = 0.1208$ ohm
               Voltage drop $= 0.1208 \times 42 = 5.07$ volts

    Per cent drop $= 5.07 \times {}^{100}\!/_{220} = 2.30$ per cent. This is a permissible voltage drop in the line.

    Try #8 RH wire, good for 45 amp

               Ohms per 1,000 ft $= 0.64$
               Voltage drop $= 0.3 \times 0.64 \times 42 = 8.07$ volts
               Per cent drop $= 8.07 \times {}^{100}\!/_{220} = 3.67$ per cent
               Use #6 RW wire

4.53   The model is (neglecting R_a and R_b)

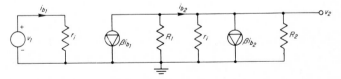

$$\text{F}_{\text{IG}}. \ 4.53. \quad \text{Model A.}$$

$$v_1 = i_{b_1}(r_i)$$

$$i_{b_2} = \beta i_{b_1}\left(\frac{R_1}{R_1 + r_i}\right)$$

$$v_2 = -\beta i_{b_2}(R_2)$$

$$v_2 = -\beta(\beta i_{b_1})\left(\frac{R_1}{R_1 + r_i}\right)R_2 = -\beta^2 i_{b_1}\left(\frac{R_1 R_2}{R_1 + r_i}\right)$$

$$v_2 = -\beta^2 \left(\frac{v_1}{r_i}\right)\left(\frac{R_1 R_2}{R_1 + r_i}\right)$$

Solving for the gain,

$$\frac{v_2}{v_1} = -\beta^2 \left(\frac{R_1 R_2}{r_i(R_1 + r_i)}\right)$$

4.54 The reactances are transformed to per unit values on a common kva base by the equation

$$Z_{\text{per unit}} = \frac{\text{kva}_{\text{base}} \times Z_{\text{ohms}}}{\text{kv}^2 \times 1,000}$$

Taking 65,000 kva as the base, the circuit becomes as shown in Fig. 4.54a.

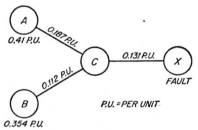

A
0.41 P.U.
0.187 P.U.
C
0.131 P.U.
X
FAULT
0.112 P.U.
B
0.354 P.U.
P.U. = PER UNIT

Fig. 4.54a

The network which represents this system is shown in Fig. 4.54b.

The equivalent reactance between the generator and the fault is

$$X = \frac{(0.41 + 0.187)(0.354 + 0.112)}{(0.41 + 0.187 + 0.354 + 0.112)} + 0.131 = 0.392 \text{ per unit.}$$

For a 65,000-kva base, the normal line current at the fault is

$$I = \frac{65,000}{13.2\sqrt{3}} = 2,850 \text{ amp.}$$

With a three-phase short circuit, the current in each line at the fault is $I = 2,850/0.392 = 7,270$ amp.

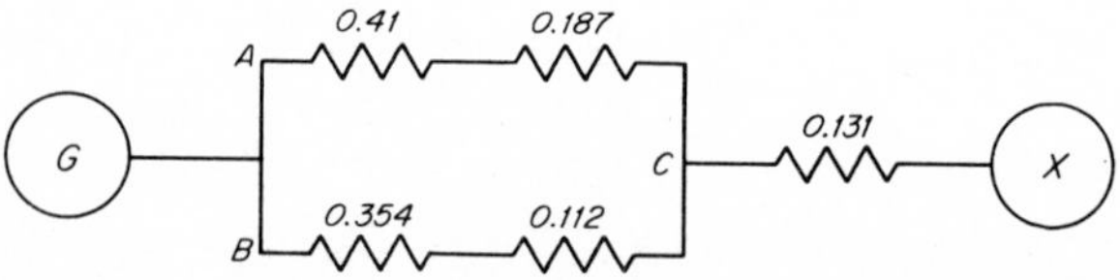

0.41   0.187
A
0.131
G   C   X
0.354   0.112
B

Fig. 4.54b

For generator A and transmission line AC, each line current is $I = \dfrac{7{,}270 \times (0.41 + 0.187)}{0.41 + 0.187 + 0.354 + 0.112} = 4{,}080$ amp.

For generator B and transmission line BC, each line current is $I = \dfrac{7{,}270 \times (0.354 + 0.112)}{0.41 + 0.187 + 0.354 + 0.112} = 3{,}180$ amp.

4.55

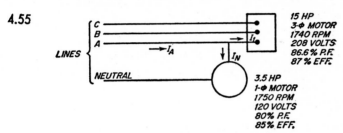

Fɪɢ. 4.55

From Fig. 4.55,
For the 15-hp motor:

$$I_L = \frac{\text{power out}}{\sqrt{3} \times E_L \times \cos \theta \times \text{eff.}}$$

$$= \frac{15 \times 746}{\sqrt{3} \times 208 \times 0.866 \times 0.87} = 41.3 \text{ amp, } 30° \text{ lag}$$

For the 3.5-hp motor:

$$I_L = \frac{\text{power out}}{E_L \times \cos \theta \times \text{eff.}} = \frac{3.5 \times 746}{115 \times 0.8 \times 85} = 33.4 \text{ amp, } 36.8° \text{ lag}$$

$$I_1 = 41.3\underline{/-30°} = 35.8 - j20.7$$
$$I_N = 33.4\underline{/-36.8°} = 26.7 - j20.0$$
$$I_A = I_1 + I_N = 62.5 - j40.7 = |74.6|$$

The line currents are:

$$I_A = 74.6 \text{ amp}$$
$$I_B = I_C = 41.3 \text{ amp}$$
$$I_N = 33.4 \text{ amp}$$

4.56 The complete system is shown in Fig. 4.56a. It is assumed that the Y-connected supply voltages are equal. Then the ground

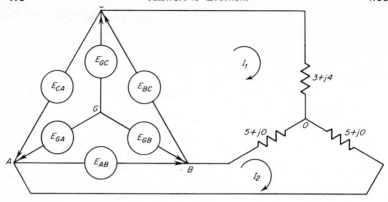

FIG. 4.56a

point G lies at the centroid of the equilateral vector triangle ABC, and since the line voltages are 1,000 volts, the phase voltages at the supply will be $1{,}000/1.732 = 577$ volts and $E_{GC} = 577\underline{/90°}$, $E_{GB} = 577\underline{/-30°}$, and $E_{GA} = 577\underline{/-150°}$.

Considering the load, there are two meshes shown, for which the self-impedances are $Z_{11} = 8 + j4$ and $Z_{22} = 10 + j0$. There is one neutral impedance $Z_{12} = 5 + j0$. There are two unknown mesh currents I_1 and I_2; so two Kirchhoff voltage rule equations may be written traversing any paths through the supply:

$$(1) \qquad\qquad E_{BC} = Z_{11}I_1 - Z_{12}I_2$$
$$(2) \qquad\qquad E_{AB} = -Z_{12}I_1 + Z_{22}I_2$$
$$(1) \quad -1{,}000 \cos 60° + j1{,}000 \sin 60° = (8 + j4)I_1 - (5 + j0)I_2$$
$$(2) \qquad\qquad +1{,}000 = -(5 + j0)I_1$$
$$+ (10 + j0)I_2$$

whence

$$(1) \qquad\qquad -500 + j866.7 = 8I_1 + j4I_1 - 5I_2$$
$$(2) \qquad\qquad +1{,}000 = -5I_1 + 10I_2$$

Solving the second equation for I_2 gives

$$I_2 = 100 + 0.5I_1$$

Putting this in the first equation gives

$$-500 + j866.7 = 8I_1 + j4I_1 - 500 - 2.5I_1$$
$$\text{and} \qquad (5.5 + j4)I_1 = j866.7$$
$$I_1 = 866.7\underline{/90°}\,/6.80\underline{/36°} = 127.5\underline{/54°}$$

The load phase voltage $= E_{oc} = I_1(3 + j4) = 127.5\underline{/54°} \times 5\underline{/53.2°} = 637.5\underline{/107.2°}$. From the subsidiary vector diagram,

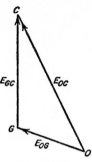

FIG. 4.56b

Fig. 4.56b, and by Kirchhoff's rule, $E_{oc} = E_{og} + E_{gc}$,

$$E_{og} = 637.5\underline{/107.2°} - 577\underline{/90°} = -188 + j618 - j577$$
$$= -188 + j41 = 192\underline{/167.7°} \text{ volts}$$

4.57   *a.* For maximum power to the load, allow the collector voltage to reach $2V_{cc}$.

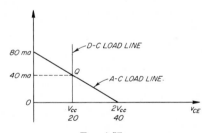

FIG. 4.57a

b. $R'_L = \dfrac{V_{cc}}{I_c} = \dfrac{20}{40 \text{ ma}} = 0.5 \times 10^3 = 500$ ohms resistance seen from primary.

$$R_L = \frac{R'_L}{(N_1/N_2)^2} = \frac{500}{25} = 20 \text{ ohms}$$

c. $P_L = E_{\text{rms}}I_{\text{rms}} = \dfrac{V_{cc}}{\sqrt{2}} \dfrac{I_c}{\sqrt{2}} = \dfrac{20 \times 40 \times 10^{-3}}{2} = 0.4$ watts

4.58 *a.* For a given transformer or magnetic circuit

$\phi_{max} = E_{rms}/4.44fN$, and since ϕ_{max} was equal in both tests,

$$\frac{E_{60}}{4.44(60)(N)} = \frac{E_{25}}{4.44(25)(N)}$$

$E_{25} = (2{,}200)(25)/60 = 917$ volts$_{rms}$

b. If the flux or flux density is maintained constant, the core-loss equation may be arranged as follows:

Core loss = hysteresis loss + eddy-current loss
$$CL = K_H f + K_E f^2$$

Dividing by f, the core loss per cycle equals

$$\frac{CL}{f} = K_H + K_E f$$

Substituting the two sets of data and solving the equations simultaneously for K_H and K_E gives

(1) $240/60 = 4 = K_H + 60K_E$
(2) $75/25 = 3 = K_H + 25K_E$
(1) $-$ (2) $1 = 35K_E$
                   $K_E = 0.0286$ and $K_H = 2.285$

Eddy-current loss $= K_E f^2 = (0.0286)(60)^2 = 103$ watts
Hysteresis loss $= K_H f = (2.285)(60) = 137$ watts

4.59 The equivalent current is as shown by Fig. 4.59.

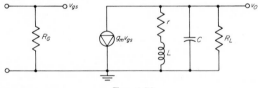

Fig. 4.59

Since the coil Q is high, the parallel resistance of the coil is

$$R_P = (Q_c)^2 r = \left(\frac{wL}{r}\right)^2 (10) = \left(\frac{10^6 \times 10^{-3}}{10}\right)^2 (10) = 100 \text{ k}\Omega$$

where $Q_c = $ coil Q.

a. $C = \dfrac{1}{w_0^2 L} = \dfrac{1}{(10^6)^2 10^{-3}} = 10^{-9}$ farads $= 0.001$ μf

$b.$ $Q_T = \dfrac{R_T}{w_0 L} = \dfrac{(R_L R_P)/(R_L + R_P)}{w_0 L} = \dfrac{50\text{ k}\Omega}{10^6 \times 10^{-3}} = 50$

$BW = \dfrac{w_0}{Q_T} = \dfrac{10^6}{50} = 0.02 \times 10^6 = 20R \text{ Hy}$

$c.$ At resonance the circuit is resistive and equal to

$R_T = 50$ ohms

$v_0 = -g_m v_{gs} R_T$

$G_{am} = \dfrac{v_0}{v_{gs}} = -g_m R_T = -5 \times 10^{-3} \times 50 \times 10^3 = -250$

4.60 The design formulas for constant-K filters are to be found in all communications handbooks. The low-pass T section has

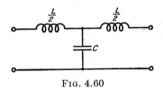

Fig. 4.60

the configuration shown in Fig. 4.60. If the resistive load R is made equal to the characteristic impedance Z_0, then

$L = R/\pi f_c = 600/\pi \times 1{,}000 = 0.191$ henry
$C = 1/\pi f_c R = 1/\pi \times 1{,}000 \times 600 = 0.53 \times 10^{-6}$ farad

Note that each coil used in Fig. 4.60 is to have inductance $L/2$.

ENGINEERING ECONOMICS AND BUSINESS RELATIONS

5.01 *a.* Standardization of equipment increases the productive ability of the average worker. The simplification which occurs permits the inexpert to become skilled more rapidly.

b. Certain materials are available only seasonally and substantial inventories of these must be maintained. The supply on hand of all raw materials must be enough to carry over a definite period of time. The comparison of the values of material and labor must indicate that proper economy of material use is being made. The kind and grade of material should be analyzed with the methods of manufacture, the use by the consumer, the expected life of the item and the price to be charged all being given proper weight.

c. A knowledge of this is prerequisite to establishing economy of operation. By recording the speed of operation, a base may thus be created for standardization of the materials, tools, equipment, and methods being used. The speed data gives some measure of the worker's ability and provides information for use in the training of new men.

d. Routing through the plant must be toward the end that maximum economy is attained. Not only the article itself but also its component parts must be so routed that the entire operation flows smoothly and interruptions due to breakdowns, overloaded departments, pileups, etc., are at a minimum. The functions of supply of stores, issue cards, labor timecards, and control forms must operate such that schedules may be maintained.

e. The work orders serve to alert the various departments of the work coming; thus, the engineering department may commence design work, the store clerks may anticipate stock requisitions, the production departments may look ahead and plans may be made towards providing any special equipment and facilities and adjusting the labor staff in accordance with anticipated volume of production.

5.02   *Job:* Specialization as applied to the job so subdivides the work that only one or a very few operations can be assigned to a worker. This both improves the quality and increases the quantity of output.

Individual: The end of improved quality and quantity of output is also achieved by assigning to each worker one or several manual or mental operations that he is particularly fitted to perform.

Machine: The simpler a machine or tool becomes, the less the attention and skill that is required for its operation or use.

Product: The quality is improved and the production cost reduced as the number of types and variety of sizes of a product are reduced.

5.03   *a. Craft union:* This is an association of employees possessing a common interest or skill or held together by employment in a single occupation or a group of closely related occupations, e.g., the Photo Engravers Union.

b. Industrial union: This is an association of all employees of an industry regardless of the type of work or the skill possessed by the individual, e.g., the United Automobile Workers.

c, and *d. Open shop* and *closed shop:* Normally these are phrases indicating that a plant has only nonunion members (open shop) or that it has only union members (closed shop) as employees. The answer, however, is more involved than this. A better distinction is to say that an open shop is one where the union is not recognized as such and is not therefore allowed to represent the workers in dealings with the employer, and that a closed shop is one where the union is the recognized bargaining agency.

e. Union shop: One where the union is the recognized bargaining agent. These may be open union shops, closed union shops with open unions, closed union shops with closed unions, or preferential union shops.

1. *Open union shop:* Both union members and nonunion members are employed.

2. *Closed union shop with open union:* Nonunion employees may be hired by the employer when union help is unavailable. However, these employees must join the union. The union cooperates by making membership easy to acquire.

3. *Closed union shop with closed unions:* The employer may hire new men only by going to the union. His right to discharge is also restricted.

4. *Preferential union shop:* As additional work becomes available, it is understood that union men are to receive initial preference. Nonunion men may be hired after the union file has been exhausted. In discharging during slack periods the nonunion men are to go first.

f. Industry-wide bargaining: Collective bargaining is conducted by the union with an association of employers. The agreements thus arrived at, therefore, may apply to an entire industry or major portion of it. The coal-mining industry and the clothing industry are good examples of this.

g. Down-time pay: A term used to describe pay provided to the employee for time lost on account of material supply shortages or machine breakdowns.

h. Maintenance of membership: This is a union security measure whereby a union employee pledges to remain a union member or an employee newly joining a union pledges to remain a member for the duration of the collective bargaining agreement if they have not resigned by a certain period.

i. Checkoff: Money which is remitted to the union is deducted from the employees' wages by the employer for union dues or assessments.

j. Seniority: A system by which length of service is the decisive element in deciding on promotions or layoffs. In some companies, periods of layoff or absence are deducted in computing length of service and in others it is not. Usually definite rules are set up regarding absences due to vacation periods, illness, leaves or governmental draft.

5.04 *a. Plant-wide seniority:* Length of service with the company regardless of the division or department is the yardstick employed.

b. *Departmental seniority:* Only length of service in a particular department is considered in computing seniority.

c. *Superseniority:* Agreements between the union and company may state that certain persons such as union shop stewards have seniority rights over and above those of any other employees in their departments.

Synthetic seniority: War veterans, even though not company employees before the war, may be given credit for their military service as years of company service.

d. *Probationary employees:* New employees who must serve a stated probation period varying from one month to even a year. It is understood that certain standards are to be met if their employment is to continue. Such employees have no seniority rights.

e. *Grievance:* This is a general term applied to an employee who feels that he is, or has, suffered an injustice. Such grievances may be those that affect only a few employees, e.g., a person feels that his work has been changed without pay adjustment or that he has been given work to do not within his duties. Another major category is that which affects all employees, e.g., a failure to agree between company and union.

f. *Arbitration of grievances:* This is the final step after a grievance has gone through the various channels of shop steward, department grievance committee, general company grievance committee, and possibly even the international union committee. Usually both parties agree in advance that the arbitration results shall be binding. The arbitration board may be a single person or a group of persons. Where several persons comprise the board, each party selects an equal number of members and these selectees choose an extra person to upset a possible tie vote.

g. *Call-in pay:* Where employees report for work and are sent home without working, a certain amount is paid to the employees. This varies from two to four hours pay. Similarly where an employee is called back to work after his regular day's work, if a minimum payment is required, it is termed call-back pay.

h. *Portal-to-portal pay:* A term used to describe payment of an employee for time spent on the property of an employer though not actually working at the job, e.g., time spent changing clothes to get ready for work or time spent traveling from the plant entrance to the place of work.

i. Featherbedding: A condition whereby payment is made for work not done. The term also applies to the case where more workers are used than are a reasonable requirement for efficient operation.

j. Boycotts: Primary boycott is one directed against an employer or company and consists of members of the union and their families withholding their patronage.

Secondary boycott is similar to the primary boycott but in addition attempts are made to induce others not directly concerned to withhold their patronage, e.g., the public may be asked not to purchase items offered or a contractor may be asked not to buy certain items.

5.05 Standard costs are planned predetermined costs which serve as a budget for setting standards of achievement and providing a check on attainment. They make for effective managerial control. Standards of costs (which must be revised from time to time) are set up for labor, material, and burden. For labor, operational studies are made and costs are the product of standard labor rates and standard labor times. Standard material costs are established for given periods from information provided by the purchasing department. Standard burden costs are the indirect expenses of the manufacturing departments charged against operations in the form of standard rates. Actual costs should be checked periodically against standard costs (usually by graphic charts) and variations investigated.

5.06 The law of diminishing returns, in very general terms, states that when the amount of a factor which is used in combination with comparatively fixed amounts of other factors is increased, the product will eventually fall after the increase reaches a certain variable amount. In the case of an office building, the rentable floor space is the profitable factor involved. As the height increases, the floor space will increase at a greater rate than the cost of adding more structure and providing more services. There will be a point, however, where increased height requires so much elevator space and service, so much lost traveling time, such large structural columns and members, and such increased utilities-installation costs and maintenance that the return per rentable square foot in spite of the additional floor space begins to fall off.

5.07 1. Provide ample aisles and stairs free of obstructions. Number of intersections of aisles kept at a minimum.

2. Ample room around all machines.

3. Sufficient and safe equipment-moving facilities; particularly hoisting and heavy-weight-lifting machinery.

4. Sufficient clear and well-marked exits to permit rapid escape in case of fire.

5. Design power system with minimum potentials and minimum exposure to human contact.

6. Provide for quick cutoff of electric power, steam, gas, and other utilities at individual machines, rooms, and buildings.

7. Provide quick ventilation facilities for removal of dust, fumes, and vapors.

8. Design adequate lighting for all areas.

9. Insure safe traffic rules.

10. Insist on proper housekeeping.

11. Keep visual aids and posters on safety effective, interesting, provoking, and new.

5.08 Preventive maintenance depends upon an adequate inspection program such that minor defects are remedied before they cause the need for major repairs and renewals may be made before failure of equipment.

Automobile Plant: All buildings, including foundations, walls, columns, girders, plaster cracks, etc., once a year. All roofs every six months. All floors depending on use from three months to two years. Paint, every six months, including not only cleanliness but also protection and light-reflection capacity. Drinking-water systems, toilets, and washbowls daily. Piping as follows: heat and low-pressure steam monthly; high-pressure steam weekly, oil and water every three months. All electric wiring every three months; all sprinklers, fire apparatus, and accessories every three months. Pressure tanks, ventilation, exhausters, and blowers weekly. Cranes, elevators, conveyors, lifting devices, hoists, crane runways, slings, ovens, furnaces, electric welders, grinders, and tools monthly. All fences, bridges, walks, driveways, fire escapes, manholes, and safety belts every six months.

5.09 The next best thing to a profit is to reduce the amount of loss. At half capacity, a loss of $111,000 − $96,000 (600 @ $160) = $15,000 will occur. Reduction of this loss is an obvious move.

$$\text{Cost of 1,200 items} = \$168,000$$

Less	600 @ 160	=	$ 96,000
	600 @ 100	=	60,000
			$156,000

$$\text{Net loss} = \$168,000 - \$156,000 = \$12,000$$

A loss of $12,000 instead of $15,000 is preferable and therefore the offer should be accepted.

5.10 This break-even type of problem is most easily solved by setting up an algebraic equation. In this case, let x = number of shoes to be manufactured. Then

$$3.00x - (0.90x + 0.80x + 0.40x) = 90,000$$
$$x = 100,000$$

Checking this:

$$100,000 @ \$3.00 = \$300,000$$
$$100,000 @ \$2.10 = \$210,000$$

Left to cover fixed charges: $ 90,000

5.11 Depreciation rate is the rate at which a physical asset lessens in value with the passage of time.

In the straight-line method, it is assumed that the depreciation is equal for each year of the asset's life. A formula for the depreciation per year would be

$$D = \frac{A - B}{n}$$

where D = depreciation per year
A = first cost
B = estimated salvage value at end
n = estimated life of asset in years

In the sinking-fund method of depreciation, one of a series of equal amounts is deposited into a sinking fund at the end of each year of the asset's life. This is usually compounded annually. This amount is equal to the estimated total depreciation times the sinking-fund factor corresponding to the estimated life and the interest-rate taken.

$D = (A - B)$sf where D, A, and B are as explained above and sf is a sinking-fund factor based on the well-known, equal-payment-series, sinking-fund-factor formula. The payments under the sinking-fund method becomes progressively greater. This method therefore is not as conservative as the straight-line method, particularly if there is any likelihood that an asset may be retired prior to the end of the estimated period of depreciation. The ICC, it would appear, is simply following the custom of imposing conservative practices on a public utility.

5.12 *a.* General partners are liable for all the firm's debts. In the case of joint liability, the individual is liable for his proportional part; e.g., if the firm is sued for $9,000 and there are three partners he is liable for $3,000. If the liability is joint and with several partners, he could become liable for the entire amount.

b. Limited partners have their liability limited to the original stake and are risking only this contribution.

c. Cumulative voting permits minority stockholders to elect members to the board of directors by concentrating their votes, e.g., ordinary voting permits each share of stock one vote each for directors. If five directors are to be elected, that would mean five votes per share or one for each director. Cumulative voting allows the five votes to be cast for one director.

5.13 *a.* In a partnership, two or more individuals associate themselves by contract in the common ownership, management, and control of a business enterprise. The partnership is not a separate legal entity but can only act in the names of its partners. Each partner may act for all and each generally bears unlimited liability. The death or withdrawal of one partner automatically dissolves the partnership. A corporation is a voluntary association of natural or legal persons, organized under and recognized by law as an artificial person separate from the persons who compose it, for the accomplishment of certain purposes specified in its charter. A corporation, unlike a partnership, can be formed only after authority has been granted by state (or Federal) government. The charter to do business nominally limits the corporation to certain lines of business. Ownership is vested in stockholders and shares are freely transferable. Liability is limited only to the amount paid for the stock shares. The enterprise has perpetuity. Capital is readily obtained through the sale of stock.

b. If an increase in the capital stock is made when actually no additional value has been paid either in cash or in money's worth, this is called "stock watering."

c. *Certified check:* A check drawn by a depositor against his account which the bank has marked "certified" indicating that sufficient funds have been set aside to satisfy it whenever presented. These funds are charged against the account of the person making the check. The bank has assumed an unconditional obligation to pay this amount.

d. *Corporation bond:* A written contract, under seal, whereby a corporation binds itself to pay a specific sum of money to the owner of the bond. Usually issued with coupons attached which, when severed, become promissory notes for the payments of installments of interest when the interest becomes due.

e. *Letter of credit:* A letter of request whereby one person, usually a merchant or banker, requests some other person to advance money or to give credit up to a certain amount to a third person named therein, charging it to the writer's account.

f. *Bill of lading:* A written acknowledgement by a carrier that he has received the goods for shipment. It is also a contract for transportation and a contract to make delivery to the person named or to his order.

g. *Promissory note:* A written promise to pay on demand, or at a fixed and determined future time, a certain sum of money to, or to the order of, a specified person or to bearer.

h. *Damages:* Liquidated damages means the sum agreed upon in advance by the parties as compensation for a breach. Unliquidated damages means sums such as a jury would award when the case is presented to them upon its own merits.

i. *Money and barter:* Money acts as a medium of exchange; as a common denominator of values; as an economic link between the present, past, and future. Barter is the process of exchange where no single medium of exchange is used. Where barter involves identical items (a rare situation) no problem is involved. Normally, however, barter involves different quantities of dissimilar items. Without some common denominator of value, as money, it becomes extremely difficult. Where the items are very different in value, as a diamond ring for a pound of steak, it is extremely tedious to trade for numerous commodities till the holder of the ring finally secures something approaching the value of the ring and still reasonable to the owner of the steak. When commodities are perishable, barter provides no link of present and future except promises to repay in the future for something received now.

5.14   The essential elements of a contract are:

1. *Competent parties:* Anyone can make a binding contract with these exceptions:

Minors (persons under 21)

Married women (permitted in most states by enabling statutes)

Lunatics (permitted in some states if made in good faith)

Drunken persons (permitted in some states if made in good faith)

2. *A lawful subject matter:* The subject contract may not violate a statute; may not be contrary to rules of common law; may not be forbidden by public policy.

3. *A proper consideration:* Consideration may be defined as the act or forbearance of one party which is given in exchange for the act or promise of the other. Examples would be a landlord giving up possession of premises in return for a promise to pay rent or a promise to buy goods at a fixed price when made in exchange for a promise to manufacture goods and sell them to the first party at that price. Love and affection are not regarded as a proper consideration in law.

4. *A genuine agreement or mutual assent:* Both parties must assent to the same thing in the same sense. The agreement must be voluntary and without force being used. In some cases, but not in all, the law requires that the agreement be in writing to be enforceable.

5.15   Contracts may be made illegal by the following ways:

1. By statute, e.g., gambling or betting contracts; usurious interest rates; Sunday contracts (in some states)

2. By being against public policy, e.g., contracts to bribe witnesses or jury members; contracts for votes in exchange for appointments after election; contracts agreeing never to marry

3. Certain mutual mistakes, as when both parties were mistaken about the item being sold at a sale

4. Fraud, where if the deceived party had known the true facts he would not have consented

5. Duress, when threats of actual physical violence against a person, his property, or someone related to him is used to force a person into a contract (the threat to sue, when there is ground for a lawsuit, is not duress)

6. Undue influence, where persuasion or capitalizing on another's trust and thereby taking unfair advantage of him can be proven

5.16 *a.* Contracts may be terminated before execution by
Mutual consent
Breach (damages may arise from this)
Death or disability in personal-service contracts
Subsequent change of law rendering subject matter illegal or impossible to perform
One party preventing the other from executing his part
Destruction of rare subject matter through no fault of the parties

b. A contract may be considered breached
When the contractor renounces his liability and refuses to perform
When by his acts he renders performance impossible
When he fails to perform what he agreed to do

5.17 Five rules of ethical behavior for a professional engineer:

1. No engineer should falsely or maliciously injure, directly or indirectly, the reputation, prospects, or business of another.

2. Every engineer should satisfy himself to the best of his ability that the enterprises with which he becomes identified are of legitimate character; and if not he should sever his connection as soon as practicable.

3. No engineer should attempt to supplant a fellow engineer after definite steps have been taken toward his engagement.

4. No engineer should attempt to conceal possible oversights or errors nor to shirk responsibility.

5. He should work towards a contract that is sound and fair to the owner, himself, and the contractor.

See also the next question.

5.18 It is not proper for a professional engineer to participate in competitive bidding against his colleagues to secure a professional engagement that is to go to the lowest bidder or to accept an engagement to review the work of another professional engineer without his knowledge. A professional man as such should be worthy of his hire. In committing his skill, education, and ability to a job it is presumed that he will give his best efforts and, therefore, to enter into competition with others implies that his abilities and fees are variable. Again, as a professional man of ethics, to review another's work without his knowledge is to question his ability and talents without giving him an opportunity to defend his work.

5.19 In these straight-line, depreciation-plus-average-interest problems, the most correct method of solution is to consider the first cost (minus salvage if any), divided by n (number of years involved) as capital recovery at 0 per cent capital recovery. Depreciation at 0 per cent $= \dfrac{(P - L)}{n}$. With a salvage value, the average interest $= (P - L) \dfrac{(i)}{(2)} \dfrac{(n + 1)}{(n)} + Li$

Reinforced bridge:

$$\text{Depreciation} = \frac{60{,}000 + 25{,}000 - 20{,}000}{5} \qquad = \$13{,}000$$

Maintenance and inspection $\qquad\qquad = \quad 1{,}500$

Average interest at 8 per cent; $(85{,}000 - 20{,}000) \times$

$$\frac{(0.08)}{(2)} \frac{(6)}{(5)} + 20{,}000(0.08) = 3120 + 1600 = \quad 4{,}720$$

Total annual cost $\qquad\qquad\qquad\qquad = \$19{,}220$

New bridge:

$$\text{Depreciation} = \frac{250{,}000 - 50{,}000}{30} \qquad = \$\ 6{,}666$$

Average interest at 8 per cent; $(250{,}000 - 50{,}000) \times$

$$\frac{(0.08)}{(2)} \frac{(31)}{(30)} + 50{,}000(0.08) = 8{,}266 + 4{,}000 = \$12{,}266$$

Total annual cost $\qquad\qquad\qquad\qquad = \$18{,}932$

Cheaper to replace.

5.20   *a.* Interest on 200,000 = 200,000(0.10) = $20,000
Interest on present worth of
    200,000 = 200,000(0.3855)(0.10) =   7,710
    Total annual cost   = $27,710

b. Interest on 300,000 = 300,000(0.10)   = $30,000

This being the total annual cost, plan *a.* is the more economical.

5.21   Furnishings type A:

First cost   = $10,000

Infinite renewals = $10,000 \dfrac{(0.07587)}{(0.08)}$ =   9,484

Annual maintenance $\dfrac{100}{0.08}$ =   1,250

Present worth of repairs $\dfrac{500(0.17046)}{0.08}$ =   1,065

Total capitalized cost   = $21,799

Furnishings type B:

First cost   = $15,000

Infinite renewals = $(15,000 - 3,000) \dfrac{(0.03683)}{(0.08)}$ =   5,524

Annual maintenance   =   0

Present worth of repairs $\dfrac{200(0.17046)}{0.08}$ =   426

                                                   = $20,950

5.22   The fundamental interest formula for this type of problem involving a systematic reduction of the principal of the debt by repayments of principal and interest uniformly is

$$ R = P \left[\frac{i(1 + i)^n}{(1 + i)^n - 1} \right] = P \frac{1}{\text{present worth factor}} $$

where R is the end of period payment
      P is the present sum of money
      i is the interest rate per period
      n is the number of periods

$$R = \frac{1,000}{3,993} = \$250.44$$

5.23   Haskold's valuation formula

$$P = R \left\{ \frac{1}{i' + i/[(1 + i)^n - 1]} \right\}$$

$$= \frac{R}{\text{return interest} + \text{sinking fund factor}}$$

is the most commonly used formula for finding present worth with two interest rates. Its greatest use seems to be in mine-valuation studies.

In this problem $R = 500,000$

$$i' = 0.10$$
$$i = 0.06$$
$$n = 20$$

$$P = 500,000 \left[\frac{1}{0.10 + 0.01842} \right] = \$3,931,306$$

This means that the $500,000 is actually divided into two parts. One part $393,130.60 is a 10 per cent return on $3,931,306. The other $106,869.40 invested each year at 6 per cent will yield $3,931,306 in 20 years, or $106,869.40 × (compound amount factor of 36.786) = $3,931,306. In other words, the syndicate will receive a 10 per cent return each year and in addition will recover the investment sum $3,931,306 at the end of 20 years.

5.24   Comparing by annual costs:

Method A:

Tunnel interest $= \$1,000,000 \times 0.08$	$=$	$\$\ 80,000$
Annual operation and upkeep	$=$	$25,000$
Total annual cost	$=$	$\$105,000$

Method B:

Ditch interest $= 350,000 \times 0.08$	$=$	$\$\ 28,000$
Annual operation and upkeep	$=$	$20,000$
Flume capital recovery at 5 per cent		
$= (200,000 - 30,000)(0.14903) + 30,000(0.08)$	$=$	$27,735$
Annual operation and upkeep	$=$	$40,000$
Total annual cost	$=$	$\$115,735$

Note that where annual cost of a structure with perpetual life is involved, interest on the investment is used, i.e., as the number of years approaches infinity the capital-recovery factor approaches the interest rate. Also that where salvage is involved, it is necessary to make the capital-recovery factor equal to the interest earned on the prospective salvage value plus the recovery with interest the difference between first cost and salvage.

5.25 The net return on the sale was $900,000 − $20,000 = $880,000. This problem becomes one of saying that $62,000 times the (present worth factor for 15 equal payments) plus $1,000,000 times the (P.W. factor for a single payment 15 years hence) should equal $880,000.

$$\text{Try 6 per cent; } P.W. = 62,000 \times 9,712 + 1,000,000 \times 0.4173$$
$$= \$1,019,444$$
$$\text{Try 7 per cent; } P.W. = 62,000 \times 9.108 + 1,000,000 \times 0.3624$$
$$= \$927,096$$
$$\text{Try 8 per cent; } P.W. = 62,000 \times 8.559 + 1,000,000 \times 0.3152$$
$$= \$845,858$$

By straight-line interpolation, the interest rate is found to be 7.42 per cent or, to the nearest 0.1 per cent, say 7.5 per cent.

5.26 Holders of simply participating, cumulative preferred stock are paid the stipulated rate of dividend, first for past unpaid dividends, then for current dividends, and after that the common stockholders are paid their dividend up to a rate equal to the preferred. Then both classes of stock participate at equal rates.

a. Earnings for 1958   = $78,500

    Balance preferred share, 1955 dividend =   2,500

                                         $76,000

    Preferred dividends 1956, 1957, 1958

        = (3 × $5 × 1,000)   =   15,000

                                           $61,000

    Common dividends 1958

        = ($5 × 10,000)   =   50,000

                                         $11,000

    Total shares ($1 × 11,000)   =   11,000

                                               0

Payment per common share = \$5 + \$1 = **\$6**

b. Earnings for 1959   = \$77,000

    \$5 × 1,000 (preferred shares)   =   5,000

                                    \$72,000

    \$5 × 10,000 (common shares)   =   50,000

                                      \$22,000

    \$2 × 11,000 (total shares)   =   22,000

                                          0

Payment per preferred share = \$5 + \$2 = **\$7**

c. **There are 11,000 shares of voting stock.**

d. **There are no redeemable shares.**

5.27 1. The normal ordering point is **1,800** units.   **1a**

2. The maximum storage space required is for 1,800 − 200 + 2,000 = **3,600** units.   **2c**

3. The normal maximum inventory is 1,800 − 30 × 50 + 2,000 = **2,300** units.   **3b**

4. The average inventory is equal to the minimum inventory limit plus one-half the standard order, or 1,800 − 30 × 50 + 2,000/2 = **1,300** units   **4d**

5. The annual inventory rate of turnover is the yearly total of units used divided by the average inventory, or 15,000/1,300 = **11.54.**   **5c**

5.28 Reference is made to Alford and Beatty, "Principles of Industrial Management," rev. ed., pp. 324–325, Ronald, 1951.

Data for Inventory Computation

Size of order	Unit pur- chase price	Unit buying expense	Average inventory in units	Value of average inventory
1,500	\$2.00	\$0.00500	1,050	\$2,100.00
2,000	1.85	0.00375	1,300	2,405.00
2,500	1.75	0.00300	1,550	2,712.50
3,000	1.70	0.00250	1,800	3,060.00

1. The value of the average inventory if lots of 2,000 units are purchased is **\$2,405.00.**   **1a**

2. The unit buying expense if a lot of 3,000 units is purchased is **$0.00250.**   **2d**

3. The cost to store 2,000 units for 1 year at the average inventory per unit charge is 2,000 × $0.30 = **$600.**   **3c**

4. The inventory carrying charge for an order of 2,500 units is $2,712.50 × 30 per cent = **$813.75.**   **4a**

5.29

Inventory Cost Computations

Units ordered	1,500	2,000	2,500	3,000
Purchase cost = 15,000 × unit purchase price	$30,000	$27,750	$26,250	$25,500
Buying expense = 15,000 × unit buying expense	75	56	45	38
Cost of storage = $0.30 × average inventory	315	390	465	540
Inventory carrying charge = 30 per cent × estimated value of average inventory	610	722	814	918
Total	$31,000	$28,918	$27,574	$26,996

The economic order size is **3,000** units.   **b**

5.30 1.

Date	Change	Accumulative valuation
June 1	500 units at $100	$ 50,000
June 12	+500 units at $125	112,500
June 25	−600 units	
	500 at $125	50,000
	100 at $100	40,000
July 6	+500 units at $110	95,000
July 15	−600 units	
	500 at $110	40,000
	100 at $100	30,000
Aug. 3	+500 units at $105	82,500
Aug. 19	−600 units	
	500 at $105	30,000
	100 at $100	20,000

The value of the inventory on Aug. 20 is **$20,000.**   **1d**

2.

Date	Change	Accumulative valuation
June 1	500 units at $100	$ 50,000
June 12	+500 units at $125	112,500
June 25	−600 units	
	500 at $100	62,500
	100 at $125	50,000
July 6	+500 units at $110	105,000
July 15	−600 units	
	400 at $125	55,000
	200 at $110	33,000
Aug. 3	+500 units at $105	85,500
Aug. 19	−600 units	
	300 at $110	52,500
	300 at $105	21,000

The value of the inventory on Aug. 20 is **$21,000.**   **2b**

3.

Date	Change	Weighted average evaluation
June 1	500 units at $100	$500 \times \$100$ $= \$ 50,000$
June 12	$+500$ units at $125	$1,000 \times \$112.5$ $= 112,500$
June 25	-600 units at $112.5	$400 \times \$112.5$ $= 45,000$
July 6	$+500$ units at $110	$900 \times \$111.111 = 100,000$
July 15	-600 units at $111.111	$300 \times \$111.111 = 33,333.33$

The value of the inventory on July 31 is **$33,333.33.**   **3a**

5.31 Reference is made to Alford and Beatty, "Principles of Industrial Management," rev. ed., pp. 390–392, Ronald, 1951, who in turn refer to formulas by J. W. Roe.

1.
$$N = \frac{I(A + B + C + 1/H) + LY}{S(1 + T)}$$

where N = number of pieces manufactured per year
$\quad\ I$ = cost of fixture
$\quad\ A$ = interest rate
$\quad\ B$ = fixed-charges rate
$\quad\ C$ = upkeep rate
$\quad\ H$ = estimated life of equipment
$\quad\ L$ = number of lots processed per year
$\quad\ S$ = estimated unit saving in direct labor cost
$\quad\ T$ = overhead saving due to direct labor saved
$\quad\ V$ = gross operating profit per year (less setups and fixed charges)
$\quad\ Y$ = estimated cost of each setup

$$N = \frac{\$1,000 \times (0.06 + 0.06 + 0.13 + \tfrac{1}{4}) + \$50}{(\$0.05)(1 + 0.40)}$$

$$= \frac{1,000 \times (0.25 + 0.25) + 50}{0.05 \times 1.4} = \frac{550}{0.07} = \mathbf{7,857} \text{ units}   \quad \mathbf{1b}$$

2. $H = \dfrac{I}{NS(1 + T) - LY - I(A + B + C)}$

    $= \dfrac{\$1,000}{9,200(\$0.05)(1.4) - \$50 - \$1,000(0.25)}$

    $= \dfrac{1,000}{644 - 50 - 250} = \dfrac{1,000}{344} = 2.91$ or **3** years   **2c**

3. $V = NS(1 + T) - LY - I(A + B + C + 1/H)$
      $= 10,000(\$0.05)(1 + 0.40) - \$50 - \$1,000(0.50)$
      $= \$700 - \$50 - \$500 = \mathbf{\$150}$ per year.   **3d**

4. $I = \dfrac{NS(1 + T) - LY}{(A + B + C + 1/H)}$

      $= \dfrac{12,000(\$0.05)(1 + 0.40) - 4 \times \$50}{0.50}$

  $= \dfrac{\$840 - \$200}{0.50} = \mathbf{\$1,280}$ **4c**

5.32 1. The total current liabilities to the tangible net worth $= \$10,000/\$29,000 = \mathbf{0.345}$.   **1a**

    2. The fixed assets to tangible net worth $= \$25,000/\$29,000 = \mathbf{0.861}$.   **2b**

    3. The net working capital is the difference between the total current assets and the total current liabilities, or $\$13,000 - \$10,000 = \$3,000$. The ratio of the net sales to net working capital $= \$30,000/\$3,000 = \mathbf{10}$.   **3c**

5.33 3,000 gal per year = 60 drums per year
Purchase price = 50 × \$3.30 = \$165 per drum

Size of order in drums	Price per drum	Estimated drum-buying expense	Average inventory in drums	Maximum inventory in drums	Estimated value of average inventory
6	\$165	\$3.500	7	10	\$1,155
8	165	2.625	8	12	1,320
10	165	2.100	9	14	1,485
12	165	1.750	10	16	1,650
15	165	1.400	11.5	19	1,898
20	165	1.050	14	24	2,310

	6	8	10	12	15	20
Buying expense = 60 × drum-buying expense	210.0	157.5	126.0	105.0	84.0	63.0
Cost of storage = 50 cents × maximum inventory	5.0	6.0	7.0	8.0	9.5	12.0
Inventory carrying charge = 12 per cent × estimated value of average inventory	138.6	158.4	178.2	198.0	227.7	277.2
Total variable expense	353.6	321.9	311.2	311.0	321.2	352.2

Order in 12-drum lots.

5.34 Since no data were given, cost of setting up the well-point system will be considered as part of the rental charge. Number of days April through September = 30 + 31 + 30 + 31 + 31 + 30 = 183 days = 26 weeks plus 1 day, including 3 legal holidays which fall on weekdays. Also, Apr. 1 and Sept. 30 fall on Sunday.

Wages one shift
   Weekdays (26 × 5) − 3   = 127 days @ $ 72 = $ 9,144.00
   Saturdays   = 26 days @ $108 =   2,808.00
   Sundays + 3 legal holidays = 30 days @ $144 =   4,320.00
                                                  _____
                                                   16,272.00
Wages on three shifts   =   3 × $16,272 = $48,816.00
   Payroll taxes and insurance @ 13 per cent ×
   $48,816   =   6,346.08
                                                  _____
Total wages $55,162.08
Rental charge = 6 × $1000   =   6,000.00
Fuel charge = 183 × $40   =   7,320.00
                                                  _____
Wages, rental, and fuel   = $68,482.08
Overhead and maintenance = 15 per cent ×
   $68,482.08   = 10,272.31
                                                  _____
                                                $78,754.39

Assume contractor borrowed $\dfrac{\$78,754}{0.96}$ at start of job

@ 8 per cent per year for 6 months, discount in
advance
Interest (apparent @ 4 per cent) = $82,035 −
$78,754   =   3,281.00
                                                  _____
                                                $82,035.39
Profit and (assume and/or) contingency @ 10 per
cent   =   8,203.54
                                                  _____
Minimum lump-sum bid   = **$90,238.93**

5.35 $2,000 = X ×$ capital-recovery factor (equal payment
         series)

$$= X \times \frac{0.04(1 + 0.04)^3}{(1 + 0.04)^3 - 1} = X \times 0.36035$$
$$X = \$5,000/0.36035 = \$13,875$$

$$\$13{,}875 + \$5{,}000 = Y \times [(1 + 0.04)^{17} - 1]/0.04$$
$$\$18{,}875 = Y \times 23.698$$
$$Y = \$18{,}875/23.698 = \mathbf{\$796.48} \; for \; \mathbf{17} \; equal \; yearly$$
$$payments.$$

5.36 Selling price   = $10,000.00

Commission @ 5 per cent	= $	500.00	
1965 charge	=	100.00	
1964 charge = $100 × 1.07	=	107.00	
1963 charge = $107 × 1.07	=	114.49	
1962 charge = $200 × (1.07)³	=	245.01	
		$1,066.50	1,066.50

Net return at time of selling   = $ 8,933.50

Amount invested	Number of years	Product
$5,000	6	30,000
100	5	500
100	4	400
($5,150)	6	30,900

Weighted amount invested for 6 years = $5,150
$$\$5{,}150 \,(1 + i)^6 = \$8{,}933.50$$
$$(1 + i)^6 = 8{,}933.50/5{,}150 = 1.73466$$

From compound-interest tables

$$(1.08)^6 = 1.587$$
$$(1.09)^6 = 1.772$$

$$(1.73466 - 1.587)/(1.772 - 1.587) = 0.14766/0.185 = 0.798$$
Rate of return = **8.8** *per cent*

5.37 From 6 per cent compound interest tables, use sinking-fund factors (SFF) starting with $n = 20$ and working up through $n = 13$, for uniform annual series payments.

End year	n	Value of mortgage × (SFF)	= Reduction in mortgage
0			= $100,000
	20	$100,000 × 0.02718	= − 2,718
1			$ 97,282
	19	$ 97,282 × 0.02962	= − 2,881
2			$ 94,401
	18	$ 94,401 × 0.03236	= − 3,054
3			$ 91,347
	17	$ 91,347 × 0.03544	= − 3,237
4			$ 88,110
	16	$ 88,110 × 0.03895	= − 3,432
5			$ 84,678
	15	$ 84,678 × 0.04296	= − 3,638
6			$ 81,040
	14	$ 81,040 × 0.04758	= − 3,856
7			$ 77,184
	13	$ 77,184 × 0.05296	= − 4,087
8			$ 73,097

Capital-recovery factor = 0.08718 for 20 equal payments of $8,718 for 6 per cent interest and principal reduction.

Capital-recovery factor = 0.08883, and 0.08883 × $73,097 = $6,493.20 for 30 equal payments for 8 per cent interest and principal reduction. Reduction of yearly payment = $8,718 − $6,493.20 = **$2,224.80**

5.38 The compound amount of a Fund A for a series of 20 $100 yearly payments at 5 per cent at the end of the twentieth year (from interest tables):

$$\text{Fund } A = \$100 \times [(1 + 0.05)^{20} - 1]/(0.05)$$
$$= \$100 \times 33.066 = \$3,306.6$$

but (Fund A − 0.10 × Fund A) = 0.90 × Fund A = $3,306.6 and Fund A = $3,306.6/0.90 = $3,674

Let ratio of actual fund needed to Fund $A = X$

$100X + 0.05(\$3,674X) = \$2,000$

$(100 + 183.7)X = 2,000$

$X = 2,000/283.7 = 7.05$

$\$3,674 \times 7.05 = \$25,902 = Maximum\ cost\ of\ sprinkler\ system$

Check:

Loss of 5 per cent interest per year on $\$25,902 = \$1,295$

$0.9 \times \$25,902 = \$23,312$

$(\$2,000 - \$1,295) \times 33.066 = \$705 \times 33.066 = \$23,312$

5.39 $\$500 = X \times$ capital-recovery factor (equal payment series)

$$= X \times \frac{[0.04\ (1 + 0.04)^8]}{[(1 + 0.04)^8 - 1]}$$

$$= X \times 0.14853$$

$$X = \$500/0.14853 = \$3,366.32$$

Present worth $= (\$500 + \$3,366.32)/(1 + 0.04)^{11}$

$$= \$3,866.32 \times 0.6496$$

$$= \$2,511.56 = single\ payment\ required$$

5.40

End year	Cost	Trade-in value	Net	Cost per year
1	$3,000 + 100	$1,800 - 200	$1,200	$1,200
2	3,100 + 200	1,600 - 200	1,500	750
3	3,300 + 300	1,400 - 200	1,900	633
4	3,600 + 400	1,200 - 200	2,400	600
5	4,000 + 500	1,000 - 200	3,000	600
6	4,500	800	3,700	617

Trade in car at end of 5 years

MECHANICAL ENGINEERING

6.01

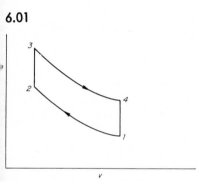

FIG. 6.01a   FIG. 6.01b

c_v = specific heat at constant volume = 0.1714 for air.

$$R_{(air)} = 53.3 \qquad p_2 = 155 \text{ psia}$$
$$t_1 = 100°\text{F} \qquad Q_k = 50 \text{ Btu}$$
$$p_1 = 13.75 \text{ psia} \qquad k = 1.4$$
$$V_1 = 1 \text{ cu ft}$$

a. $p_1 V_1{}^k = p_2 V_2{}^k$

$$(13.75 \times 144)(1)^k = (155 \times 144) V_2{}^k$$
$$V_2{}^{1.4} = \frac{(13.75)(1)}{155} = 0.0887$$
$$V_2 = 0.1772 \text{ cu ft}$$

$$\text{Compression ratio} = r_k = \frac{V_1}{V_2} = \frac{1}{0.1772} = \mathbf{5.642}$$

b. Per cent clearance $= \dfrac{V_2}{V_1 - V_2} = \dfrac{0.1772}{1 - 0.1772} = \mathbf{21.5}$ per cent

c. Q_A = area under $2 - 3$ on Ts diagram

$Q_A = wc_v(T_3 - T_2)$

T_3 = temperature after combustion

p_3 = pressure after combustion

$$w = \frac{p_1}{RT_1} = \frac{13.75 \times 144}{(53.3)(100 + 460)} = 0.0663$$

$$T_2 = \frac{p_2 V_2}{wR} = \frac{(155 \times 144)(0.1772)}{(0.0663)(53.3)}$$

or

$$T_2 = 1119°R$$

$Q_A = 50 = (0.0663)(0.1714)(T_3 - 1119)$ or $T_3 = \mathbf{5518°F}$

$$p_3 = \frac{wRT_3}{V_3} \qquad V_3 = V_2$$

$$= \frac{(0.0663)(53.3)(5518)}{(0.1772)} = 110{,}000 \text{ psf} = \mathbf{764.2} \text{ psia}$$

d. Net work per cycle = area enclosed on Ts diagram

$$W = Q_A - Q_R = Q_A - wc_v(T_4 - T_1)$$

To find T_4:

$p_3 V_3{}^k = p_4 V_4{}^k \qquad (764.2 \times 144)(0.1772)^{1.4}$

$\qquad\qquad\qquad\qquad = p_4(1)^{1.4}$ or $p_4 = 9{,}759$ psf

$p_4 V_4 = wRT_4 \qquad (9{,}759)(1) = (0.0663)(53.3)T_4 \qquad T_4 = 2762°R$

$W = 50 - (0.0663)(0.1714)(2762 - 560) = 24.98$ Btu

$\qquad\qquad\qquad\qquad\qquad = (24.93)(778) = \mathbf{19{,}432}$ ft-lb

e. Heat rejected = $Q_R = wc_v(T_4 - T_1) = (0.0663)(0.1714)$
$(2762 - 560) = \mathbf{25.02}$ Btu

f. m.e.p. = net work per $\dfrac{\text{cycle}}{\text{stroke}}$ length

$$= \frac{w}{V_1 - V_2} = \frac{19{,}432}{1 - 0.1772} = 23{,}617 \text{ psf}$$

$$= \frac{23{,}617}{144} = \mathbf{164} \text{ psi}$$

6.02 $q = 10$ Btu/(hr)(sq ft) of outside area of steel pipe. (*Reference:* Kreith, "Principles of Heat Transfer," 3d. ed., Intext, 1973.)

$$q = \frac{t_1 - t_0}{\dfrac{r_{oa}}{h_s r_{is}} + \dfrac{r_{oa}}{k_s} \ln\left(\dfrac{r_{os}}{r_{is}}\right) + \dfrac{r_{oa}}{k_a} \ln\left(\dfrac{r_{oa}}{r_{os}}\right) + \dfrac{1}{h_o}}$$

where r_{is} = inside radius of pipe
$\quad\quad r_{oa}$ = outside radius of asbestos

$$10 = \cfrac{300 - 70}{\cfrac{r_{oa}}{(20)(2.4485)} + \cfrac{r_{oa}}{(30)(12)} \ln\left(\cfrac{3.3125}{2.4485}\right) + \cfrac{r_{oa}}{(0.06)(12)} \ln\left(\cfrac{r_{oa}}{3.3125}\right) + \cfrac{1}{6}}$$

$$1 = \cfrac{23}{\cfrac{r_{oa}}{48.97} + \cfrac{r_{oa}}{360}(1.1977 - 0.8955) + \cfrac{r_{oa}}{0.720}(\ln r_{oa} - 1.1977) + \cfrac{1}{6}}$$

$$+ 0.0204 r_{oa} + 0.0008 r_{oa} + 1.389 r_{oa}(\ln r_{oa} - 1.1977)$$
$$+ 0.1666 = 23$$

$$1.389(\ln r_{oa}) - 1.642 = \frac{22.833}{r_{oa}}$$

By trial: $r_{oa} = 12$ in.

$$1.389(2.485) - 1.642 = 22.833/12$$
$$3.41 - 1.64 = 1.77 < 1.90$$

Try $r_{oa} = 3.31 + 9 = 12.31$ in.

$$1.389(2.510) - 1.64 = 22.833/12.31$$
$$3.49 - 1.64 = 1.85 = 1.85$$

Use **9**-in.-thick pipe covering ($r_{oa} = 12.31$ in.)　　**d**

6.03　*a.* Figure 6.03a is a line diagram of the apparatus showing the flow of steam and condensate in the system.

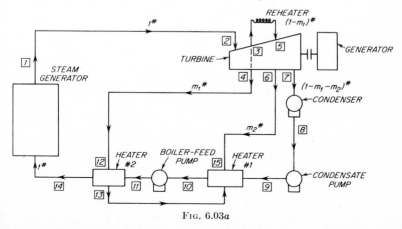

Fig. 6.03a

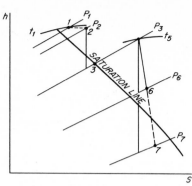

Fɪɢ. 6.03*b*

b. The *h-s* diagram is shown in Fig. 6.03*b*. All values taken from a Mollier chart or steam tables are as follows:

Point	p	t	h
1	700	744.1	1371.1
2	655	740.0	1371.1 (turbine)
3	109	334.1	1188.7
4	109	334.1	1188.7
5	109	700.0	1378.4
6	15	270.0	1177.6
7	1	101.7	973.0
8	1	101.7	69.7
9	15	101.7	69.7 (approx)
10	15	213.0	181.1
11	700	214.0	183.2
12	109	334.1	1188.7
13	109	334.1	305.0
14	700	324.1	296.8
15	15	270.0	1177.6

c. Temperature leaving boiler:

$p_1 = 700$ psia, $h_1 = 1371.1$ Btu per lb, so $t_1 = $ **744.1°F**

d. Temperature of feedwater leaving the last heater:

$$t_{14} = t_{13} - 10° = 334.1 - 10 = \textbf{324.1°F}$$

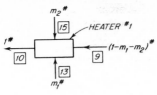

Fig. 6.03e

e. The heat balance around heater no. 1 is shown in Fig. 6.03e.

$$m_2 h_{15} + (1 - m_1 - m_2)h_9 + m_1 h_{13} = h_{10}$$
$$m_2(1177.6) + (1 - 0.1262 - m_2)(69.7) + (0.1262)(305) = 181.1$$
$$m_2 = 0.074 \text{ lb per lb throttle steam}$$

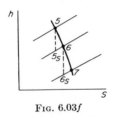

Fig. 6.03f

f. The h-s diagram is shown in Fig. 6.03f.

$$(h_5 - h_6)/(h_5 - h_{5_s}) = 0.80$$
$$h_6 = h_5 - 0.80(h_5 - h_{5_s})$$
$$= 1378.4 - 0.80(1378.4 - 1171) = 1212 \text{ Btu per lb}$$
$$(h_6 - h_7)/(h_6 - h_{6_s}) = 0.80$$
$$h_7 = h_6 - 0.80(h_6 - h_{6_s})$$
$$= 1212 - 0.80(1212 - 1026) = 1063 \text{ Btu per lb}$$

For $p_7 = 1$ psia and $h_7 = 1063$ Btu per lb (from the Mollier chart), there is 3.2 per cent moisture. The quality of the steam at exhaust = 96.8 per cent.

g. From the Mollier diagram using h_2 as h for the turbine entrance condition,

$$h_{2_s} = \text{enthalpy of the exhaust} = 900 \text{ Btu per lb}$$

6.04

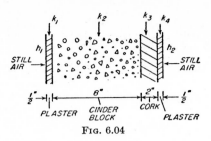

Fig. 6.04

Assume cement plaster: $k_1 = k_4 = 8.0$

$$Q = \frac{A(t_1 - t_2)}{1/h_1 + L_1/K_1 + L_2/K_2 + L_3/K_3 + L_4/K_4 + 1/h_2}$$

$$= \frac{1 \times 40}{1/1.65 + 0.5/8 + 1.61 + 2/0.30 + 0.5/8 + 1/1.65}$$

$$= \frac{40}{0.606 + 0.063 + 1.61 + 6.67 + 0.063 + 0.606}$$

$$= 40/9.62 = 4.16 \text{ Btu}/(\text{sq ft})(\text{hr})$$

where $t_1 - t_2 = 40°\text{F}$

A = area = 1 sq ft

L = thickness of material in inches

K = thermal conductivity [(Btu)(in.)/(hr)(sq ft)(°F)]

H = surface coefficient of heat transmission [Btu/(hr)(sq ft)(°F)]

(From Heating, Ventilating, and Air Conditioning Guide," ASHVE, 1950.)

For 8-in. cinder block, $L_2/K_2 = 1.61$

6.05 See preceding answer.

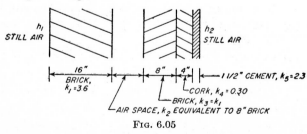

Fig. 6.05

$$Q = \frac{A(t_1 - t_2)}{1/h_1 + L_1/K_1 + L_2/K_2 + L_3/K_3 + L_4/K_4 + L_5/K_5 + 1/h_2}$$

$$= \frac{1(70 - 30)}{1/1.65 + 16/3.6 + 8/3.6 + 8/3.6 + 4/0.3}$$
$$+ 1.5/2.3 + 1/1.65$$

$$= 40/24.1 = 1.66 \text{ Btu per sq ft} \times \text{hr} = 39.8 \text{ Btu per sq ft}$$
$$\times \text{day}$$

6.06

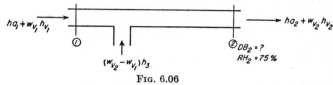

$ha_1 + w_{v_1} h_{v_1} \longrightarrow$

$\longrightarrow ha_2 + w_{v_2} h_{v_2}$

①   ② $DB_2 = ?$
$RH_2 = 75\%$

$(w_{v_2} - w_{v_1}) h_3$

FIG. 6.06

Where RH = relative humidity
    DB = dry-bulb temperature, °F
    ha = enthalpy of dry air, Btu per lb dry air
    hf = enthalpy of water spray = Btu per lb water
    w_v = lb vapor per lb dry air
    v = cu ft per lb dry air
    w_a = lb dry air per cu ft

Evaluating the entering energy to the leaving energy for 1 lb dry air,

$$h_{a_1} + w_{v_1} h_{v_1} + (w_{v_2} - w_{v_1}) h_3 = h_{a_2} + w_{v_2} h_{v_2} \qquad (1)$$

a. $h_{a_1} + w_{v_1} h_{v_1} = 23.7$ Btu per dry air*

w_{v_1} = 44 grains vapor per lb dry air* = $44/7{,}000$ = 0.0063 lb
    vapor per lb dry air.

$$\therefore 23.7 + (w_{v_2} - 0.0063)18.1 = h_{a_2} + w_{v_2} h_{v_2} \qquad (2)$$

A trial-and-error solution is applicable. Assume DB_2. Use chart to find $(h_{a_2} + w_{v_2} h_{v_2})$ and w_{v_2} based on this assumption. Apply equation (2) to check validity of assumption. As an aid in choosing a trial DB_2, note that $(w_{v_2} - 0.0063)18.1$ is very small compared to the other quantities in equation (2). $\therefore (h_{a_2} + w_{v_2} h_{v_2})$ is approximately equal to 23.7 and RH_2 = 75 per cent (given)

First trial: $DB_2 = 61°F$, RH_2 = 75 per cent, gives: $(h_{a_2} + w_{v_2} h_{v_2})$
    = 24 Btu per lb dry air* and $w_{v_2} = 60/7{,}000 = 0.0086$
    lb vapor per lb dry air*

Check: $23.7 + (0.0083 - 0.0063)(18.1) = 23.74$ or 24

Second trial: Assume $DB_2 = 60.5°F$
    $\therefore (h_{a_2} + w_{v_2} h_{v_2}) = 23.8$ Btu per lb dry air.*

$$w_{v_2} = 59/7{,}000 = 0.0084 \text{ lb vapor per lb dry air*}$$

Check: $23.7 + (0.0084 - 0.0063)(18.1) = 23.74 \text{ or } 23.8$

This is close enough. Therefore $v_1 = 13.48$ cu ft per lb dry air* and $w_{a_1} = 1/13.48 = 0.0743$ lb dry air per cu ft

Weight of incoming air $= (1{,}800)(0.0743) = 134$ lb per min

$\therefore$ Water required $= 134(w_{v_2} - w_{v_1}) = 134(0.0084 - 0.0063)$
$= \mathbf{0.282}$ lb per min

$$DB_2 = \mathbf{60.5°F} \text{ and } WB_2 = \mathbf{55.8°F}$$

b. Equation (2) becomes $23.7 + (w_{v_2} - 0.0063)1{,}150$
$$= h_{a_2} + w_{v_2}h_{o_2}$$

First trial: As in (*a*), assume $DB_2 = 71°F$

$\therefore (h_{a_2} + w_{v_2}h_{v_2}) = 30.4$ Btu per lb dry air*

$w_{v_2} = 85/7{,}000 = 0.01214$ lb steam per lb dry air*

Check: $23.7 + (0.0121 - 0.0063)1{,}150 = 30.4$

Steam required $= \left(\dfrac{\text{lb air}}{\text{min}}\right)\left(\dfrac{\text{lb steam}}{\text{lb air}}\right) = \text{lb per min}$

$134(w_{v_2} - w_{v_1}) = 134 \times 0.0058 = \mathbf{0.785}$ lb steam per min

$$DB_2 = \mathbf{71°F} \text{ and } WB_2 = \mathbf{65.5°F}$$

6.07 See previous answer for symbol notation.

Fɪɢ. 6.07. Energy diagram for a cooling tower where weight of dry air is 1 lb. $h_{v2} = h_{o2}$ since air leaves in a saturated condition; P_v = vapor pressure, psi; P_{sat} = saturated vapor pressure, psi.

$w_{v_B} = w_{v_A} - (w_{v_2} - w_{v_1}) = $ lb of departing water per lb dry air

$$\dfrac{\text{Dry air, lb}}{\text{Water, lb}} = \dfrac{1}{w_{v_A}} = \dfrac{(h_{f_A} - h_{f_B})}{0.24(t_2 - t_1) + w_{v_2}(h_{v_2} - h_{f_B}) - w_{v_1}(h_{v_1} - h_{f_B})}$$

$$= \dfrac{(93.91 - 48.0)}{0.24(115 - 85) + w_{v_2}(1{,}111.6 - 48) - w_{v_1}(1{,}098.8 - 48)}$$

* Using psychrometric chart.

$$\text{Eq. (1)} = \frac{45.9}{7.2 + w_{v_2}(1{,}063.6) - w_{v_1}(1{,}050.8)}$$

$$P_{v_1} = RH_1 \times P_{sat} = 0.47 \times 0.596 = 0.280 \text{ psi and } P_{v_2}$$
$$= 1.471 \text{ psi}$$

$$w_{v_1} = 0.622\left(\frac{P_{v_1}}{P_a - P_{v_1}}\right) = 0.622\left(\frac{0.280}{14.7 - 0.280}\right) = 0.0121 \text{ lb}$$
$$\text{vapor per lb dry air}$$

$$w_{v_2} = 0.622\left(\frac{1.471}{14.7 - 1.47}\right) = 0.0693 \text{ lb vapor per lb dry air}$$

a. $$\text{Eq. (1)} = \frac{45.9}{7.2 + 0.0693 \times 1063.6 - 0.0121 \times 1050.8}$$
$$= 0.674 \text{ lb dry air per lb water}$$

$$V = \frac{WRT}{P} = \frac{0.674 \times 53.3 \times (85 + 460)}{(14.7 - 0.28) \times 144} = 9.43 \text{ cu ft}$$
$$\text{dry air per lb water}$$
$$= 9.43 \text{ cu ft wet air per lb water (by Dalton's law)}$$

$$\frac{\text{Wet air, lb}}{\text{Water, lb}} = \frac{w_{v_1}}{w_{v_A}} + \frac{1}{w_A} = 0.0121 \times 0.674 + 0.674 = \textbf{0.682}$$

b. $$\frac{\text{Air, cfm}}{\text{Air, cu ft per lb water}} = \frac{2{,}000}{9.43} = \textbf{212} \text{ lb water per min}$$

6.08   Outside air = 7.5 cfm per person (not smoking)

a. Total outside air = 7.5 × 1,800 = 13,500 cfm

b. Sensible heat loss per person = 225 Btu per hr
Sensible heat loss of audience
$$= 225 \times 1{,}800 = 405{,}000 \text{ Btu per hr}$$
Solar heat loss $$= 120{,}000 \text{ Btu per hr}$$
Total heat loss = Q_s $$= 525{,}000 \text{ Btu per hr}$$

$$Q_s = W_s(0.24)(t_i - t_e)$$

where t_i = temperature in auditorium (°F)
   t_e = temperature of conditioned air
   W_s = conditioned air per hour, pounds

$$525{,}000 = W_s \times 0.24(78 - 65)$$
$$W_s = 168{,}000 \text{ lb air per hr}$$

From psychrometric chart, air at 65°F and at an estimated relative humidity = 70 per cent shows specific volume = 13.4 cu ft per lb dry air

$$\text{Air circulated} = \frac{\text{Air, lb per hr}}{60 \text{ min}} \times \frac{\text{cu ft}}{\text{lb dry air}} = \frac{168,000}{60} \times 13.4$$

$$= 37,400 \text{ cfm}$$

c. From "Heating, Ventilating, and Air Conditioning Guide,"
Chap. 6, Fig. 7, curve D, ASHVE, 1950.

165 Btu per hr = latent heat loss per person

Latent heat loss of audience = $165 \times 1,800 = 296,000$

Btu per hr

Moisture per person = 1,100 grains

Moisture from audience, grains = $1,100 \times 1,800$

$$= 1,980,000 \text{ grains per hr}$$

$$\frac{\text{Moisture per hr}}{\text{Dry air circulated per hr, lb}} = \frac{1,980,000}{168,000} = 11.8 \text{ grains}$$

moisture per lb dry air

Moisture content of inside air at 78°F to 67°F = 82 grains

per lb dry air (from chart)

Moisture content of entering air = $82.0 - 11.8 = 70.2$

grains per lb dry air

Air with 70.2 grains moisture per lb dry air and 65°F

has a wet-bulb temperature = **60.2°F**

Specific volume = 13.43 cu ft per lb dry air is sufficiently

close to 13.40 cu ft per lb dry air used in part b.

6.09 $Q = mc_p \, \Delta t$ or $m = Q/(c_p \, \Delta t)$,

where Q = Btu per hr

$c_p = 0.24$

w = weight standard air

$= 0.075$ lb per cu ft

$$\text{Cu ft per min} = \frac{Q}{c_p \, \Delta t} \times \frac{1}{60} \times \frac{1}{w}$$

$$= \frac{53,100}{0.24(75 - 61) \times 60 \times 0.075}$$

$$= 3,510 \text{ (through fan)}$$

The air-conditioner load is based only on the 1,200 cu ft per min
of outside air.

$$Q = (h_2 - h_1)(1,200/13.5)$$
$$= (37.75 - 27.9)(88.9) = 876 \text{ Btu per min}$$

where $v = 13.5$ cu ft per lb, and h_2 and h_1 come from the psychrometric chart, Fig. 6.09 .

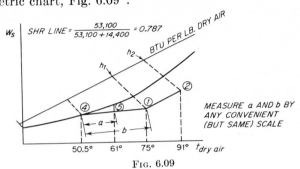

FIG. 6.09

One ton refrigeration = 200 Btu per min.

Tonnage = $^{876}\!\!/_{200}$ = 4.38 tons to bring the outside air up to inside conditions.

Tonnage total = 4.38 + (53,100 + 14,400)/(60 × 200)
                = 4.38 + 5.62 = 10.0 tons

b. Assume the bypass factor = 0 (necessary by virtue of data). Referring to both the numbered points on the layout diagram (Fig. 6.09 of the question) and the correspondingly numbered points on the psychrometric chart, point (4) is on the saturation line at $t_{d.a.} = 50.5°F$. With G = lb air per hr, w_s = grains water per lb dry air, and $h_{fg} = 1060$ Btu per lb,

$$Q_L = G(w_{s_1} - w_{s_5})h_{fg}/7,000$$
$$14,400 = (3,510 \times 60)(0.075)(\Delta w_s)(^{1060}\!\!/_{7000})$$
$$\Delta w_s = 6.02 \text{ grains per lb dry air}$$
$$w_{s_1} = 63 \text{ grains per lb dry air}$$
$$w_{s_5} = 63 - 6.02 = 56.98 \text{ or } 57 \text{ grains per lb dry air}$$

This checks since it falls on the SHR line. Point (5) must lie on the SHR line joining points (1) and (4) on the chart, namely, at $t_d = 61°F$. From the chart, the values for $\dfrac{\text{points } (5) - (4)}{\text{points } (1) - (4)} \times$

$100 = \dfrac{a}{b}$ (measured by any convenient scale on the chart) ×

100 = per cent of point (1) air bypassed, or (1.37/3.18) × 100 = 43 per cent air bypassed. Recirculated air = 3,510 cu ft per min

3,510 × 0.43 = **1,509** cu ft per min. **Air bypassed.**

6.10   *a.*   Sensible heat load $= Q_s = W \times C_p(t_a - t_b)$

$$W = \frac{Q_s}{c_p(t_a - t_b)} = \frac{450{,}000}{(0.24)(75 - 65)} = \textbf{187,500} \text{ lb air per hr}$$

b. $\dfrac{\text{Moisture, grains}}{\text{Dry air, lb}} = \dfrac{\text{grains moisture per hr}}{\text{lb dry air per hr}}$

$$= \frac{1{,}200{,}000}{187{,}500} = 6.40 \text{ grains per lb dry air to be removed}$$

$$\frac{\text{Moisture, grains}}{\text{Supply air, lb}} = 78 - 6.40 = 71.6 \text{ grains per lb air}$$

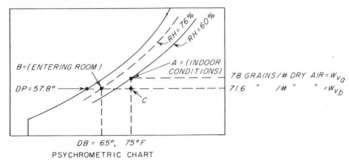

PSYCHROMETRIC CHART

Fɪɢ. 6.10

From Fig. 6.10 at $BD = 65°$F at 71.6 grains per lb air and dew point $= \textbf{57.8}°$F and $RH = \textbf{76}$ per cent

c. $\dfrac{1{,}200{,}000}{7{,}000}$ 171.2 lb moisture per hr

Heat load picked up in auditorium

$$= \left(\frac{\text{Pound moisture}}{\text{Hour}}\right)\left(\frac{\text{Btu}}{\text{lb moisture}}\right) = \left(\frac{\text{lb moisture}}{\text{hr}}\right)h_{f_g}$$

$$= 171.2 \times 1{,}051 = \textbf{180,000} \text{ Btu per hr}$$

6.11 h_f at 1 psia $= 69.70$ Btu per lb

$w_{cw} = $ lb circulating water per hr

Heat given up by steam $=$ heat picked up by water

$(1 \times 10^6)(1090 - 69.70) = w_{cw}(95 - 85)$

$w_{cw} = (1 \times 10^6)(1020.3)/10 = 102 \times 10^6$ lb per hr

$= (102 \times 10^6)/(3600 \times 62.4) = 454$ cu ft per sec

Flow area $= Q/V = {}^{454}\!/_6 = 75.8$ sq ft

Inside tube area $= \pi(0.875^2)/(4 \times 144) = 0.00417$ sq ft

Number of tubes per pass $= 75.8/0.00417 = 18{,}200$ tubes

$q = vA_0\,\Delta t_m$, where $\Delta t_m = $ log mean temperature difference

$$= \frac{(\theta_{\max} - \theta_{\min})}{\log_e (\theta_{\max}/\theta_{\min})}$$

$$= \frac{16.74 - 6.74}{\log_e (16.74/6.74)} = \frac{10}{0.912} = 11°F$$

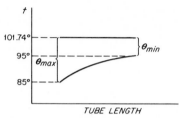

FIG. 6.11

where Fig. 6.11 shows $\theta_{\max} = 101.74 - 85 = 16.74°F$
and $\theta_{\min} = 101.74 - 95 = 6.74°F$
$(1 \times 10^6)(1090 - 69.70) = 480 \times A_0 \times 11$
$A_0 = (1 \times 10^6 \times 1020.3)/(480 \times 11) = 193,000$ sq ft
$= \pi \times d_0 \times L \times$ (tubes/pass) $\times$ (2 passes), where
$$d_0 = \tfrac{1}{12} \text{ ft}$$
$L = (193,000 \times 12)/(3.14 \times 18,200 \times 2) = 20.3$ ft

6.12   The equation of combustion is

$$C_{12}H_{26} + 18.5O_2 + 69.56N_2 + \underbrace{18.5O_2 + 69.56N_2}_{\text{Excess air}} \rightarrow$$

$$12CO_2 + 13H_2O + 69.56N_2 + \underbrace{18.5O_2 + 69.56N_2}_{\text{Excess air}}$$

The heat available to the heating system is the change in enthalpy of the products of combustion from 600°F (1060°R) to 75°F (535°R). Since it is possible that some of the H_2O vapor will condense at 75°F, giving off latent heat, this should be investigated. Assuming the products to be at 14.7 psia in the heating system, the partial pressure of the vapor will be

$$pp_{H_2O} = \frac{\text{No. of moles of } H_2O}{\text{No. of moles of products}} \times 14.7 \text{ psia} = \frac{13}{182.62} \times 14.7$$
$$= 1.048 \text{ psi or } 2.13 \text{ in. Hg}$$

The saturation pressure at 75°F = 0.875 in. Hg, and so H_2O will partially condense.

The number of moles of H_2O vapor at 75°F is

$$\frac{0.875}{29.92} \times 182.62 = 5.35 \text{ moles}$$

The number of moles that will condense $= 13 - 5.35 =$ 7.65 moles.

The saturation temperature of steam corresponding to 2.13 in. of Hg is 103°F (563°R).

The ΔH of the H_2O from 1060°R to 535°R will be considered as: ΔH (5.35 moles from 1060 to 535°R) $+ \Delta H$ (7.65 moles from 1060 to 563°R) $+ \Delta H$ (7.65 moles condensing from 563 to 535°R)

h (Btu per lb-mole)

	CO_2	N_2	O_2
1060°R	9,350	8,129	7,544
535°R	−4,013	−3,716	−3,103
Δh	5,337	4,413	4,441

ΔH (Btu per lb-mole of product) $= 12 \times 5,337(CO_2) +$ 139.12 $\times$ 4,413 $+$ 18.5 $\times$ 4,441 $+ \Delta H(H_2O)$

$\Delta H(H_2O) = 5.35(8.595 - 4,242) + 7.65(8,595 - 4,474)$ $+ 7.65(1,098 - 43)18$

$\Delta H = 64,000 + 615,000 + 82,200 + 23,300 + 31,500$ $+145,000$

$= 961,000$ (Btu per mole of fuel)

Higher heating value $= 20,550$ (Btu per lb) $\times$ 170.2 (lb per lb-mole) $= 3,500,000$ (Btu per lb-mole)

per cent utilization $= \dfrac{961,000}{3,500,000} \times 100 = 27.5$ per cent

6.13 For $P_1 = 1$ psig $= 15.75$ psia at 12 per cent moisture, $h_1 = 1,037$ Btu per lb

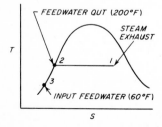
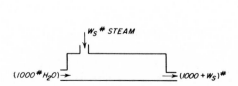

Fig. 6.13

Neglecting the effect of subcooling, which is nil in this case,

For $t_2 = 200°F$, $h_2 = 168$ Btu per lb

For $t_3 = 60°F$, $h_3 = 28.1$ Btu per lb

Heat absorbed by water = heat given up by steam.

$$1,000(h_2 - h_3) = w_s \text{ lb } (h_1 - h_2)$$
$$1,000(168 - 28) = w_s \text{ lb } (1,037 - 168)$$
$$w_s = 1,000 \times {}^{140}\!/\!_{869} = \textbf{161 lb}$$

6.14 c_{P_A} = specific heat of air (constant pressure)

        c_{P_G} = specific heat of flue gas (constant pressure)

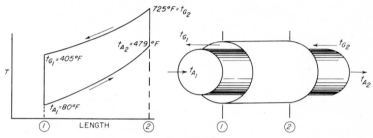

Fig. 6.14

Δt_m = log mean temperature difference = $\dfrac{\Delta t_1 - \Delta t_2}{\log_e (\Delta t_1/\Delta t_2)}$

$= \dfrac{(t_{G_1} - t_{A_1}) - (t_{G_2} - t_{A_2})}{\log_e (t_{G_1} - t_{A_1})/(t_{G_2} - t_{A_2})} = \dfrac{(325 - 246)}{\log_e {}^{325}\!/\!_{246}} = \dfrac{79}{\log_e 1.321}$

$= \dfrac{79}{0.278} = \textbf{284°F}$

$Q = W_A c_{P_A}(t_{A_2} - t_{A_1}) = 357,000 \times 0.24 \times 399$

$= 34.2 \times 10^6$ Btu per hr

$Q = W_G c_{P_G}(t_{G_2} - t_{G_1})$ or $W_G = \dfrac{Q}{c_{P_G}(t_{G_2} - t_{G_1})}$

$W_G = \dfrac{34.2 \times 10^6}{0.24 \times 320} = 447,000$ lb of flue gas

Note: The specific heat of the flue gas is equal to the weighted average of the specific heats of its constituents. A typical flue-gas analysis is

$$
\begin{aligned}
CO_2 &= 0.15 \times 0.202 = 0.0303\\
N_2 &= 0.80 \times 0.248 = 0.1983\\
O_2 &= \underline{0.05} \times 0.219 = \underline{0.0110}\\
&\quad\; 1.00 \qquad\qquad\quad 0.2396 \text{ or } 0.24
\end{aligned}
$$

6.15   Let V_1 = steam velocity leaving nozzle (ft per sec)

      V_{1_r} = relative velocity entering blade (ft per sec)

      V_2 = absolute exit velocity (ft per sec)

      V_{2_r} = relative velocity leaving blade (ft per sec)

      Δh = available energy (Btu per lb)

      K = velocity coefficient

      V_b = blade velocity (ft per sec)

      β = blade-entrance angle (degrees)

      F = force per lb steam per sec

      w = lb steam per sec

a. $V_1 = 223.8 \, k \, \sqrt{\Delta h} = 223.8 \times 0.95 \times \sqrt{100}$

$$= 2{,}120 \text{ ft per sec}$$

b, c, d, and e.

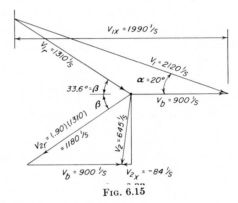

Fig. 6.15

f. $F = \dfrac{V_{1_x} - V_{2_x}}{g} = \dfrac{1{,}990 - (-84)}{32.2} = 64.4 \dfrac{\text{lb}}{\text{(lb steam per sec)}}$

g. Blade hp $= \dfrac{W(V_{1_x} - V_{2_x})V_b}{g \times 550} = \dfrac{1{,}800}{3{,}600} \times \dfrac{64.4 \times 900}{550}$

$$= 52.6 \text{ hp}$$

h. Shaft hp = blade hp − friction loss hp − windage loss hp

$$= 52.6 - 6 - 2 = 44.6 \text{ hp}$$

i. Losses converted into enthalpy of steam

     (1) Nozzle loss $= (1 - k^2) \, 100$

           $= [1 - (0.95)^2] \, 100$ $= 9.75 \text{ Btu per lb}$

     (2) Blade friction loss $= \dfrac{V_{1_r}{}^2 - V_{2_r}{}^2}{2gJ}$

$$= \frac{(1,310)^2 - (1,180)^2}{64.4 \times 778} \qquad = 6.52 \text{ Btu per lb}$$

(3) Residual velocity loss $= \dfrac{V_2{}^2}{2gJ}$

$$= \frac{(645)^2}{64.4 \times 778} \qquad = 8.30 \text{ Btu per lb}$$

(4) Friction between steam and wheel

$$= \frac{\text{hp} \times 550}{wJ} = \frac{6 \times 550 \times 2}{778} \qquad = 8.48 \text{ Btu per lb}$$

$$\text{total} = 33.05 \text{ Btu per lb}$$

Enthalpy of exhaust steam $= 1{,}200 - 100 + 33 = 1133$ Btu per lb

6.16　Load factor $=$ (average power)/(peak power)

Efficiency of steam-generating unit $= 80$ per cent

Heating value of Pittsburgh bituminous coal $= HV = 13{,}400$ Btu per lb

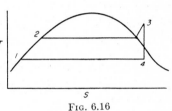

Fig. 6.16

Average steam rate $= 11$ lb steam per kwhr

Average power $=$ (load factor)(peak power per turbine)

$$= 0.75 \times 10{,}000 = 7{,}500\text{-kw output per turbine}$$

$$\text{Efficiency} = \frac{w_n(h_3 - h_1)}{w_f(HV)}$$

Obtain h_3 and h_1 from tables and charts

a. $w_n =$ lb steam supplied per hr $=$ (steam rate)(kw output)

$$= 11 \times 7{,}500 = 82{,}500 \text{ lb steam per hr}$$

$w_f =$ weight of fuel fired

$$= \frac{82{,}500(1{,}318 - 69.1)}{0.80 \times 13{,}400} = 9{,}620 \text{ lb/hr/turbine}$$

$$\text{Tons coal per day for four units} = \frac{4 \times 9{,}620 \times 24}{2{,}000} = \mathbf{461}$$

b. Assume 10°F rise in cooling water and no subcooling.

$Q_R =$ (heat rejected by four turbines) $= 4w_n(h_4 - h_1)$

$$= w_w(\Delta T) = 4 \times 82{,}500(921 - 69.1) = 281{,}000{,}000$$
Btu per hr

$$w_w = Q_R/\Delta T = 281{,}000{,}000/10 = 28{,}100{,}000 \text{ lb per hr}$$

$$\text{Gallons water per min} = \frac{28{,}100{,}000 \times 7.48}{60 \times 62.4} = 56{,}200$$

6.17 R_f = reheat factor
q_r = stage reheat, Btu per lb
n_{st} = stage efficiency
v_b = blade velocity, ft per sec
v_1 = steam entrance velocity,
ft per sec

At 180 psia and 600°F,

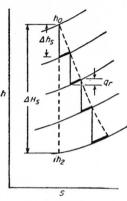

$$h_0 = 1323.5 \text{ Btu per lb}$$
$$ih_2 = 1111.0 \text{ Btu per lb}$$
$$\overline{\Delta H_s = \quad 212.5 \text{ Btu per lb}}$$

$$\rho = \frac{V_b}{V_1} = \frac{\cos \alpha}{2} \text{ for maximum efficiency}$$

Fig. 6.17

Let nozzle angle = $\alpha = 20°$.

$$\rho = \frac{\cos 20°}{2} = \frac{0.940}{2} = 0.470$$

However, considering disk friction and fanning, ρ for maximum efficiency will be less, or say $\rho = 0.420$

$$V_1 = \frac{V_b}{\rho} = \frac{625}{0.420} = 1{,}490 \text{ ft per sec}$$

Assume nozzle velocity coefficient = $K_n = 0.95$

$$\text{Ideal entrance velocity} = iV_1 = \frac{V_1}{k_n} = \frac{1{,}490}{0.95} = 1{,}565 \text{ ft per sec}$$

$$\Delta h_s = \frac{iV_1{}^2}{2g778} = \frac{(1{,}565)^2}{64.4 \times 778} = 48.9 \text{ Btu per lb}$$

Find number of stages

$$Z = \frac{\Delta H_s \times R_f}{\Delta h_s} = \frac{212.5 \times 1.06}{48.9} = 4.6$$

Select turbine of four stages.

Find Δhs:

$$\Delta hs = \frac{\Delta H_s \times R_f}{Z} = \frac{212.5 \times 1.06}{4} = 56.3 \text{ Btu per lb, first trial}$$

$$q_r = \Delta h_s(1.00 - \eta_{st}) = 56.3(1.00 - 0.75) = 14.1 \text{ Btu per lb}$$

Stage	Nozzle or wheel	Enthalpy	Pressure
1	N	1,323.5 56.3	180
	W	1,267.2 14.1	107
2	N	1,281.3 56.3	
	W	1,225.0 14.1	59
3	N	1,239.1 56.3	
	W	1,182.8 14.1	31
4	N	1,196.9 56.3	
	W	1,140.6	15

Exhaust pressure should be 16 psia, which for this trial would give an enthalpy = 1146.0 Btu per lb. Another trial value of Δh_s will be necessary. An estimate of second trial Δh_s may be made using the following approach:

$$\Delta h_s = \text{first trial } \Delta h_s - \frac{\text{error in enthalpy}}{Z\eta_{st}}$$

$$\Delta h_s = 56.3 - \frac{(1{,}146 - 1{,}140.6)}{4(0.75)} = 54.5 \text{ Btu per lb}$$

$$q_r = 54.5(1.00 - 0.75) = 13.6 \text{ Btu per lb}$$

Assuming that the second trial yields sufficiently accurate results, the calculations may be continued as follows:

Total internal work per lb steam $= E_i = Z(\Delta h_s - q_r)$
$$= 4(54.5 - 13.6) = 164 \text{ Btu per lb}$$

Actual engine efficiency $= \dfrac{E_i}{\Delta H_s} = \dfrac{164}{212.5} \times 100 = 77.2 \text{ per cent}$

Ideal steam rate $= \dfrac{3{,}413}{212.5} = 16 \text{ lb per kwhr}$

Assuming radiation and mechanical losses as 2 per cent of the internal work of the turbine,

$$\frac{\text{Ideal steam rate}}{\text{Actual steam rate}} = 0.98 \times \eta_i = 0.98 \times 0.772 = 0.756$$

Actual steam rate $= \dfrac{16}{0.756} = 21.2 \text{ lb per kwhr}$

6.18 $P_1 = 20 \text{ psia}$
$P_2 = 190 \text{ psia}$
$t_3 = 80°\text{F}$
$S_1 = S_2 = 1.3700$
$h_1 = 606.2 \text{ Btu per lb}$
$h_2 = 753 \text{ Btu per lb}$
$h_3 = h_4 = 132 \text{ Btu per lb}$

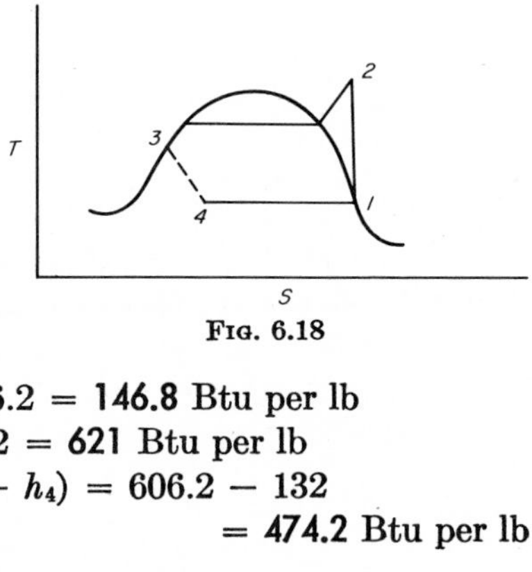

FIG. 6.18

a. $W = h_2 - h_1 = 753 - 606.2 = 146.8 \text{ Btu per lb}$
b. $Q_R = h_2 - h_3 = 753 - 132 = 621 \text{ Btu per lb}$
c. $Q_A = \text{refrigeration} = (h_1 - h_4) = 606.2 - 132$
$$= 474.2 \text{ Btu per lb}$$

d. $\text{cop} = \dfrac{QA}{W} = \dfrac{474.2}{146.8} = 3.23$

e. $\text{Circulation} = \dfrac{200 \text{ Btu per min}}{474.2 \text{ Btu per lb}} = 0.422 \text{ lb per min per ton of refrigeration}$

6.19 The operation of this stoker-fired plant during the second year was essentially carried out with a reduced amount of excess air. This is apparent from figures given: with 0 per cent CO and 10 per cent combustible in the ash pit there was 6.25 per cent

O_2 left in the flue gas during the first year's operation. In the second year's operation, besides the 0.1 per cent CO and the increased combustible (16 per cent $-$ 10 per cent $=$ 6 per cent) in the ash pit, both of which would require O_2 to burn to CO_2, as in the first year, there was only 3.9 per cent O_2 left in the flue gas. This shows that much less air was supplied in the second year, causing more combustion losses due to incomplete combustion, i.e., CO in the flue gas and unburned carbon in the ash pit. The increase in per cent CO_2 occurred because less air passed into the flue gas during the second year. The per cent N_2 should not change much; therefore in the flue-gas analysis the per cent O_2 decreased and the per cent CO_2 increased.

The efficiency $=$ heat transferred through boiler and tubes per pound of coal fired, divided by the heating value per pound of coal fired.

During the second year, because of reduced air supply and therefore less efficient combustion, less heat was liberated from each pound of coal. The temperature in the furnace, however, was higher because there was less excess air to absorb the heat that was released. The higher temperature in the furnace meant a greater temperature difference above the steam in the boiler and tubes. This temperature difference forced a greater rate of heat transfer to the steam. That is, although each pound of coal released less heat than in the first year, a greater portion of this heat was transferred to the steam because of the higher temperature difference. This compensated for the smaller amount of heat released from each pound of coal so that the numerator in the efficiency equation above remained unchanged, thus generating the same weight of steam from the same weight of coal each year. Since the same type of coal was used both years, the denominator did not change. Therefore the efficiency did not change.

The increased 2 per cent cost per thousand pounds of steam can be explained by (a) the faster deterioration and consequent replacement of furnace refractory and boiler tubes, due to operation at higher temperatures, (b) the increased slagging of boiler tubes at higher temperatures, requiring more frequent soot-blower operation, and (c) the greater refuse-handling cost, because of increased combustible in the ash pit.

6.20
$$\frac{W_{bo}}{Wf} = 12.77$$

$$\text{Boiler efficiency} = \frac{W_{bo}(h_2 - h_1)}{Wf \times q_H}$$

From chart, h_2(200 psia 87°F superheat) = 1251.5 Btu per lb
$$h_1(93°F) = 61 \text{ Btu per lb}$$

$$q_H = \frac{12.77(1{,}251.5 - 61)}{0.828} = \textbf{18,400} \text{ Btu per lb}$$

6.21 Over-all boiler efficiency $= \dfrac{W_{bo}(h_2 - h_1)}{Wfq_H}$

where W_{bo} = lb steam delivered by boiler per hr
h_2 = enthalpy of delivered steam
h_1 = enthalpy of water entering boiler
W_f = lb fuel fired per hr
q_H = higher heating value of fuel per lb

$$W_{bo} = \frac{885{,}000}{4} = 221{,}250 \text{ lb steam per hr}$$

h_2(at 400 psi, 700°F) = 1362.7 Btu per lb, $h_1(280)$ = 249.1

$$Wf = \frac{221{,}250(1362.7 - 249.1)}{0.825 \times 13{,}850} = 21{,}500 \text{ lb per hr}$$

$$= 21{,}500/2{,}000 = \textbf{10.75} \text{ short tons per hr}$$

6.22 *a.* A boiler hp is equal to the power necessary to evaporate 34.5 lb water per hr from saturated water at 212°F to saturated steam at 212°F.

$$1 - \text{boiler hp} = \textbf{33,475} \text{ Btu per hr}$$

b. 1 − mechanical hp = 33,000 ft-lb per min,   1 Btu = 778 ft-lb

$$1 - \text{mechanical hp} = 33{,}000/778 = \textbf{42.4} \text{ Btu per min}$$

$$\therefore 1 - \text{boiler hp} = \frac{33{,}475/60}{42.4} = \textbf{13.15} \text{ mechanical hp}$$

c. The factor of evaporation is the ratio that changes the actual evaporation rate (lb per hr) to the rate that would result if the boiler were operated from and at 212°F.

Equivalent evaporation = (actual evaporation)(factor of evaporation) = 4,000 × 1.06 = **4,240** lb per hr

$$\text{Boiler hp} = \frac{(\text{evaporation})(\text{latent heat @ }212°F)}{33{,}475}$$

$$= \frac{4{,}240 \times 970}{33{,}475} = 123 \text{ boiler hp}$$

6.23 $\text{Horsepower input} = \dfrac{0.0001573 \times \text{cfm} \times \text{static pressure}}{\text{static efficiency}}$

$$\text{Horsepower input} = \frac{0.0001573 \times (60{,}000/60) \times 2}{0.40}$$

$$= 0.786 \text{ hp}$$

Use 1 hp motor.

6.24 Velocity of air $= Q/A = \dfrac{9{,}300 \ (\text{cfm})}{(\pi/4) \times (3)^2 \ (\text{sq ft})} = 1{,}317$ ft per

min or 21.95 ft per sec

$$V = \sqrt{2g\frac{Dh_v}{12d}}$$

and

$$h_v = \frac{12dV^2}{2gD}$$

where $D = 62.1$ lb per cu ft H_2O at gage temperature

h_v = velocity pressure (in. H_2O)

d = density of air flowing

V = ft per sec

h_t = total pressure difference created by fan

h_s = static pressure

$$PV = WRT \quad \text{and} \quad d = \frac{W}{V}$$

$$\therefore d = \frac{P}{RT} = \frac{144 \times 14.1}{53.3(460 + 83)} = 0.0703 \text{ lb per cu ft}$$

where $P = \left(\dfrac{28.75}{29.9}\right)(14.7) = 14.1$ psi

$$h_v = \frac{12 \times 0.0703 \times (21.95)^2}{2 \times 32.2 \times 62.1} = 0.102 \text{ in. water}$$

$$h_t = h_s + h_v = 0.85 + 0.102 = 0.952 \text{ in. water}$$

$$\text{Air hp} = \frac{Qh_t D}{12 \times 33{,}000} = \frac{9{,}300 \times 0.952 \times 62.1}{12 \times 33{,}000} = 1.39 \text{ hp}$$

$$\text{Fan mechanical efficiency} = \frac{\text{air hp}}{\text{input}} = \frac{1.39}{3.55} \times 100$$

$$= \textbf{39.2 per cent}$$

$$\text{Static efficiency} = \text{mechanical efficiency} \left(\frac{h_s}{h_t}\right)$$

$$= 39.2 \left(\frac{0.850}{0.952}\right) = \textbf{35 per cent}$$

6.25 $Q = A \times V = (2 \times 3) \times 20 = 120$ cfm

For standard air conditions

$\quad V = 0.33$ fps

$$= 66.76 \sqrt{h_v} \text{ and } h_v = \left(\frac{0.33}{66.67}\right)^2 = 0.000025 \text{ in. water}$$

$h_t = h_s + h_v = 3.00$ in. water

$$\text{Air hp} = \frac{Qh_t}{6,355} = \frac{120 \times 3.00}{6,355} = \textbf{0.0566 hp}$$

6.26 $\text{Horsepower input} = \dfrac{Qh_t}{6,355 \times \text{eff}} = \dfrac{(60,000/60)(2)}{6,355 \times 0.40}$

$$= \textbf{0.786 hp}$$

Use 1-hp motor.

6.27 $N = $ standard commercial ton

$\quad \text{cop} = $ coefficient of performance

$\quad$ 1-ton capacity $= 12,000$ Btu per **hr**

$\quad$ 20-ton capacity $= 240,000$ Btu per **hr**

a. $Q = W \times C_p(t_2 - t_1)$

$$W = \frac{240,000}{0.24(90 - 70)} = \textbf{50,000 lb per hr}$$

b. Assuming cop $= 4$

$$\text{hp} = \frac{(4.71)N}{\text{cop}} = \frac{4.71 \times 20}{4} = \textbf{23.5 hp}$$

6.28   Refer to ammonia tables or charts for property values.

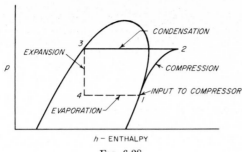

h – ENTHALPY

FIG. 6.28

a. Volume efficiency = $\dfrac{\text{volume ammonia delivered per min}}{\text{piston displacement (rpm)}}$

Volume ammonia delivered per min to compressor = 2 × (volume efficiency) × (piston displacement) (rpm)

$$= 2 \times 0.82 \left\{ (1) \left[\frac{\pi}{4} \times (1)^2 \right] \right\} (150) = 193 \text{ cfm}$$

Ammonia at 31.16 psi (from tables)—saturated, $v_1 = 8.92$ cu ft per lb

Weight to be circulated = $\dfrac{193}{8.92}$ = 21.60 lb per min

Refrigeration effect = $h_1 - h_3 = 612 - 143 = 469$ Btu per lb

Capacity = $\dfrac{\text{lb per (min)(Btu per lb)}}{200} = \dfrac{(21.6)(469)}{200}$

= **50.5** standard commercial tons

b. $\dfrac{\text{Btu extracted}}{\text{Pounds ice at 16°F}} = C_{p_w}(t_2 - 32) + \text{heat of fusion} +$

$c_{p_I}(32 - 16) = 1(80 - 32) + 144 + 0.5(32 - 16) = 48$

$+ 144 + 8 = 200$ Btu per lb ice = 400,000 Btu per ton ice

$\dfrac{\text{Ice, tons}}{24 \text{ hr}} = \dfrac{\text{capacity} \times 288,000 \text{ Btu per 24 hr}}{\text{Btu extracted per ton ice}} = \dfrac{50.5 \times 288,000}{400,000}$

= **36.4** tons ice

c. Ideal horsepower = $\dfrac{778W(h_2 - h_1)}{33,000}$ (for isentropic expan-

$$\text{sion}) = \frac{778 \times 21.6(724 - 612)}{33,000} = 57.0 \text{ hp}$$

$$\text{Brake horsepower} = \frac{\text{ideal hp}}{\text{over-all efficiency}} = \frac{57.0}{0.80} = 71.2 \text{ hp}$$

6.29 P_D = diametral pitch

N = number of teeth

D = pitch diameter

T_V = train value

w = rpm

$$P_D = \frac{N}{D}, \quad N_A = P_D D_A = 12 \times 3 = 36 \text{ teeth}$$

$$N_C = N_A + 2N_B = 36 + 2 \times 18 = 72 \text{ teeth}$$

$$\frac{W_A - W_D}{W_C - W_D} = T_V = \frac{N_C}{N_A} = \frac{72}{36} = -2$$

Case I: Gear A clutched in, gear C fixed.

$$\frac{W_A - W_{D(\mathrm{I})}}{0 - W_{D(\mathrm{I})}} = -2, \quad W_A - W_{D(\mathrm{I})} = 2W_{D(\mathrm{I})}, \quad W_{D(\mathrm{I})} = \frac{W_A}{3} \quad (1)$$

Case II: Gear A fixed, gear C clutched in.

$$\frac{0 - W_{D(\mathrm{II})}}{W_C - W_{D(\mathrm{II})}} = -2, \quad W_{D(\mathrm{II})} = 2W_C - 2W_{D(\mathrm{II})}, \quad W_{D(\mathrm{II})} = \frac{2W_C}{3} \quad (2)$$

(For same input speed, $W_A = W_C$)

$$\therefore \; \frac{W_{D(\mathrm{II})}}{W_{D(\mathrm{I})}} = \frac{\frac{2}{3}W_C}{\frac{1}{3}W_A} = \frac{2W_C}{1W_C} = 2$$

6.30 *a.* Displacement volume = stroke $\times$ area

$$= 6.00 \times \frac{\pi}{4} \times (5.375)^2 = 136 \text{ cu in.}$$

Initial volume = displacement volume + clearance
volume = $136 + 23.4 = 159.4$ cu in.

$$\text{Compression ratio} = \frac{\text{initial volume}}{\text{clear volume}} = \frac{159.4}{23.4} = 6.82 \text{ to } 1$$

b. Brake hp $= \dfrac{Tn}{63,000} = \dfrac{(12 \times 2,100) \times 2,500}{63,000} = 1,000 \text{ hp}$

$$b_{\text{mep}} = \frac{33,000 \times \text{bhp}}{LAN} = \frac{33,000 \times 1,000}{0.5[(\pi/4)(5.375)^2] \times 17,500} = 166 \text{ psig}$$

where T = torque (in-lb)

n = rpm

L = stroke in feet

A = piston area (sq in.)

$$N = \text{power cycles per min} = \frac{14 \times 2,500}{2}$$

c. Brake hp per lb $= 1,000/1,450 = 0.689$ hp per lb

6.31   a. Mean torque $= \dfrac{\text{maximum torque} + \text{minimum torque}}{2}$

$$T_m = \frac{20,000 + (-6,000)}{2} = 7,000 \text{ ft-lb}$$

Variable torque $= \dfrac{\text{maximum } T - \text{minimum } T}{2}$

$$T_v = \frac{20,000 - (-6,000)}{2} = 13,000 \text{ ft-lb}$$

Mean component of stress $= S_m = \dfrac{16 T_m}{\pi D^3} = \dfrac{16 \times 7,000}{\pi D^3}$

$$= \frac{35,600 \text{ psi}}{D^3}$$

Variable component of stress $= S_v = \dfrac{16 T_r}{\pi D^3} = \dfrac{16 \times 13,000}{\pi D^3}$

$$= \frac{66,200 \text{ psi}}{D^3}$$

S_n(reversed torsion) $= S_{t_{\text{ult}}} \times \dfrac{0.50}{2} = \dfrac{97,000}{4} = 24,250$ psi

$S_{s_{yp}} = 0.55 \, S_{t_{yp}} = 0.55 \times 58,000 = 31,900$ psi

$$\frac{1}{\text{Factor of safety}} = \frac{S_v}{S_n} + \frac{S_m}{S_y}$$

$$\frac{1}{2} = \frac{66,200}{24,250 \, D^3} + \frac{35,600}{31,900 \, D^3} = \frac{2.73}{D^3} + \frac{1.12}{D^3} = \frac{3.85}{D^3}$$

$$D^3 = 7.70, \ D = 1.975 \text{ in.}$$

$$b. \quad \frac{S_{s_{yp}}}{\text{Factor of safety}} = \frac{16T}{\pi D^3} = \frac{31{,}900}{2} = \frac{16 \times 20{,}000}{\pi D^3}$$

$$D^3 = 6.39, \quad D = 1.855 \text{ in.}$$

c. Working stresses which have been determined from the ultimate or yield-point values of a material with a factor of safety will give safe and reliable results only for static loading. Failures in machine parts subjected to variable loading occur at stresses considerably below yield point. Because of this "fatigue" phenomenon, the diameter calculated on the basis of a static torque loading in part *b* would not stand up under the variable loading acting on the shaft in part *a*.

6.32   For Corliss engine diagram factor = 0.85.

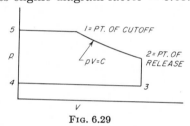

FIG. 6.29

Conventional diagram $\dfrac{V_1}{V_2} = \dfrac{3}{8} = \dfrac{1}{r_e}$, where $r_e = 2.67$

Head end piston area $= \dfrac{\pi}{4}(28)^2 = 615.0$ sq in.

Piston rod area $\quad = \dfrac{\pi}{4}(5)^2 \quad = \underline{19.6 \text{ sq in.}}$

Crank end piston area $\quad\quad\quad = 595.4$ sq in.

a. p_{m_c} = mean efficiency pressure (conventional)

$$= P_1\left(\frac{1 + \log_e r_e}{r_e}\right) - P_3 = 150\left(\frac{1 + \log_e 2.67}{2.67}\right) - P_3$$

$$= 56.3(1 + 0.981) - 15 = 96.8 \text{ psia}$$

P_{m_a} = mean effective pressure (actual)
  = (diagram factor) $\times P_{m_c}$
  = $0.85 \times 96.8 = 82.3$ psia

b. ihp head end $= \dfrac{P_{m_a}LA_HN}{33{,}000} = \dfrac{82.3 \times (^{54}\!\!/_{12}) \times 615 \times 60}{33{,}000}$

$$= 413 \text{ hp}$$

$$i\text{hp crank end} = \frac{P_{m_a}LA_cN}{33,000}$$

$$= \frac{82.3 \times (^{54}\!/_{12}) \times 595.4 \times 60}{33,000}$$

$$= \textbf{401 hp}$$

$$\text{Indicated hp} = \overline{814}$$

Assuming 90° mechanical efficiency, hp delivered at crankshaft = (mechanical efficiency)($i$hp) = 0.90×814

$$= \textbf{733 hp}$$

c. Horsepower delivered to mill

$$= (\text{brake hp})(0.90)(0.90)(0.85) = \textbf{505 hp}$$

6.33   Assuming uniform wear which is proportional to product of velocity and pressure

$$T = \pi\mu P_{(\max)}r_i(r_0{}^2 - r_i{}^2)$$

with one side effective and twice with two sides effective
where T = torque(in-lb)

μ = coefficient of friction

$P_{(\max)}$ = maximum pressure

r_i = inside radius

r_0 = outside radius

    = $1.25\, r_i$

N = total axial pressure

$\therefore T = 2\pi \times 0.3 \times 12 \times r_i{}^3[(1.25)^2 - (1)^2] = 6{,}516$

$$r_i{}^3 = \frac{6{,}516}{22.7(1.562 - 1{,}000)} = 512$$

$r_i = \textbf{8.00 in.}$ and $r_0 = \textbf{10.00 in.}$

$$N = \frac{T}{\mu(r_0 + r_i)} = \frac{6{,}516}{0.3(10 + 8)} = \textbf{1,205 lb}$$

6.34   Indicated work per revolution $= \dfrac{(\text{output hp}) \times 33{,}000}{N \times \text{efficiency}}$

$$= \frac{20 \times 33{,}000}{1{,}800 \times 0.80} = 458 \text{ ft-lb per revolution}$$

Let N = mean rpm

N_2 = maximum rpm

N_1 = minimum rpm

ω = radians per second

c_f = coefficient of fluctuation

J = mass moment of inertia = lb-sec^2 per ft

$\bar{r}$ = radius to concentrated mass (in.)

$\omega = \dfrac{2\pi N}{60}$

Excess energy delivered to flywheel = change in kinetic energy of fly wheel = ΔkE = $(0.20)(458)$

$$\Delta kE = 91.6 \text{ ft-lb} = \frac{1}{2} J(\omega_2{}^2 - \omega_1{}^2)$$

$$= \frac{1}{2} \frac{W}{g} \left(\frac{\bar{r}}{12}\right)^2 \left[\left(\frac{2\pi N_2}{60}\right)^2 - \left(\frac{2\pi N_1}{60}\right)^2\right]$$

$$= \frac{W}{2g} \left(\frac{\bar{r}\pi}{360}\right)^2 (N_2{}^2 - N_1{}^2)$$

since $N = \dfrac{N_2 + N_1}{2}$ and $c_f = \dfrac{N_2 - N_1}{N}$,

$$\begin{aligned} N_2 + N_1 &= 2N \\ N_2 - N_1 &= c_f N \\ \hline N_2{}^2 - N_1{}^2 &= 2c_f N^2 \end{aligned}$$

$$\Delta kE = 91.6 = \frac{W}{2g}\left(\frac{\bar{r}\pi}{360}\right)^2 2c_f N^2 = \frac{50}{64.4}\left(\frac{7\pi}{360}\right)^2 2c_f(1,800)^2$$

$c_f = 0.00488$

$$\begin{aligned} N_2 + N_1 = 2N &= 3,600 \text{ rpm} \\ N_2 - N_1 = c_f N &= \quad\quad 8.8 \text{ rpm} = \text{total speed variation} \\ \hline 2N_2 &= 3,608.8 \text{ rpm} \\ N_2 &= 1,804.4 \text{ rpm} \\ N_1 &= 1,795.6 \text{ rpm} \end{aligned}$$

6.35 W = indicated work

t = hours

T = torque (in-lb)

Q_A = heat added

e = thermal efficiency

w_f = fuel consumption (lb per hr)

L = stroke (ft)

q_L = lower heating value (Btu per lb)

A = area (sq in.)
P_m = mean effective pressure (psi)
N = explosions per min

Indicated power = $P_m LAN$ = $110 \times 1 \times \left(\dfrac{121\pi}{4}\right) \times \left(\dfrac{300}{2}\right)$

$$= 1{,}570{,}000 \frac{\text{ft-lb}}{\text{min}} = \frac{1{,}570{,}000}{33{,}000} = 47.6 \text{ hp}$$

W per hr = $\dfrac{1{,}570{,}000 \times 60}{778}$ = 121,000 Btu per hr

$Q_A = q_L w_f = 18{,}000 \times 24 = 433{,}000$ Btu per hr

$a.$ $e = \dfrac{W \text{ per hr}}{Q_A} = \dfrac{121{,}000}{433{,}000} \times 100 = 28$ per cent

$b.$ Mechanical efficiency = brake hp/indicated hp

Brake hp = $\dfrac{T \times \text{rpm}}{63{,}000} = \dfrac{12 \times 700 \times 300}{63{,}000} = 40$ hp

Mechanical efficiency = $\dfrac{40 \times 100}{47.6} = 84.2$ per cent

$c.$ Specific gravity of fuel = 0.75
7.5 gal per cu ft
$\dfrac{62.4 \times 0.75}{7.5} = 6.24$ lb per gal,

and $\dfrac{24 \text{ lb per hr}}{6.24 \text{ lb per gal}} = 3.84$ gal per hour

Fuel cost per hr = 3.84×10 cents = \$0.384

Fuel cost per hp per hr = $\dfrac{\$0.384}{40} = \$0.0096$

6.36 Dimensions are shown in Fig. 6.60 of the question.

Stress due to centrifugal force = $\dfrac{w}{g} \omega^2 r^2$

where w = specific weight
ω = radians per sec
r = mean radius of thin cylinder

$\Delta S_s'$(due to centrifugal force) = $\dfrac{w_s}{g} \left(\dfrac{6.28N}{60}\right)^2 \left(\dfrac{d + h_s}{2}\right)^2$

$$= \frac{490}{1,728 \times 386}$$
$$\times (0.0523N)^2(6 + 0.125)^1$$

$$= \frac{1}{1,360}(0.320N)^2 = \frac{N^2}{(115.2)^2}$$

$$\Delta S_c' \text{(due to centrifugal force)} = \frac{w_c}{g}\left(\frac{6.28N}{60}\right)^2\left(\frac{d - h_c}{2}\right)^2$$

$$= \frac{558}{1,728 \times 386}$$
$$\times (0.0523N)^2(6 - 0.063)^2$$

$$= \frac{1}{1,248}(0.310N)^2 = \frac{N^2}{(114.0)^2}$$

$\Delta S_s''$ (due to decrease in contact pressure) $= -Xd/2h_s$
$\Delta S_c''$ (due to decrease in contact pressure) $= +Xd/2h_c$

where X = change in contact pressure

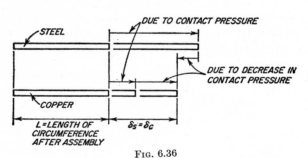

FIG. 6.36

Figure 6.36 with this answer shows the increments of change of length due to the forces acting.

Total change in stress in the steel $= \Delta S_s = \Delta S_s' + \Delta S_s''$
Total change in stress in the copper $= \Delta S_c = \Delta S_c' + \Delta S_c''$

$$\delta_c = L\epsilon_c = L\,\Delta S_c/E_c = 2,000L/E_c$$
$$\epsilon_c = \Delta S_c/E_c = 2,000/E_c$$
$$\delta_s = L\epsilon_s = L\,\Delta S_s/E_s = \delta_c = 2,000L/E_c$$
$$\epsilon_s = \Delta S_s/E_s = 2,000/E_c$$

$$(1) \qquad \frac{2,000}{E_c} = \frac{\Delta S_c}{E_c} = \frac{1}{E_c}\left[\left(\frac{N}{114}\right)^2 + \frac{Xd}{2h_c}\right]$$

$$(1a) = h_c(1) \qquad 2,000 h_c = \left[\left(\frac{N}{114}\right)^2 h_c + \frac{Xd}{2}\right]$$

$$(2) \qquad \frac{2,000}{E_c} = \frac{\Delta S_s}{E_s} = \frac{1}{E_s}\left[\left(\frac{N}{115.2}\right)^2 - \frac{Xd}{2h_s}\right]$$

$$(2a) = h_s(2) \quad 2,000\left(\frac{E_s}{E_c}\right)h_s = \left[\left(\frac{N}{115.2}\right)^2 h_s - \frac{Xd}{2}\right]$$

$$(1a) + (2a)\ 2,000\,[h_c + (15\!\!/\!\!8)h_s] = 2,000(0.0625 + 0.2345)$$

$$= \left(\frac{N}{114}\right)^2\left[\left(\frac{1}{1.02}\right) \times 0.125 + 0.0625\right]$$

$$N = 1,140\,\sqrt{(20 \times 0.297)/(0.185)} = 1,140\,\sqrt{32.1}$$
$$N = 1,140 \times 5.67 = \mathbf{6,450\ rpm}$$

6.37 $A_A = 2 \times 0.785 \times 1^2 = 1.570$ sq in.
$A_B = 0.785(2.50^2 - 1.50^2) = 3.14$ sq in. $= 2A_A$

After the screw has been lowered, *subsequent* deformations of tube B and rods A are equal, and $\delta_A = \delta_B$.

For static equilibrium

$$\Delta P_A + \Delta P_B = 0 \text{ and } \Delta P_A = -\Delta P_B$$
$$\Delta S_A A_A = -\Delta S_B A_B \text{ and } \Delta S_A = -2\Delta S_B$$
$$\Delta S_B = \text{final stress} - \text{initial stress (after screw has been lowered)}$$
$$= [(-10,000) - (-5,000)] = -5,000 \text{ psi}$$

$$\Delta S_A L_A/E_A + L_A\alpha_A\Delta t = \Delta S_B L_B/E_B + L_B\alpha_B\Delta t$$

where the first term on each side of the equation represents constraint, and the second term free temperature expansion. Substituting $-2\Delta S_B$ for ΔS_A in the last equation, and by combining terms,

$$\Delta t = \frac{\Delta S_B(L_B/E_B + 2L_A/E_A)}{L_A\alpha_A - L_B\alpha_B}$$

$$= \frac{-5{,}000[(23/14.5) + (2 \times 30/30)](1/10^6)}{[(30 \times 6.5) - (23 \times 10.5)](1/10^6)}$$

$$= \frac{-5{,}000(1.582 + 2)}{195 - 241.5} = \frac{-5{,}000 \times 3.582}{-46.5} = +384°F$$

Final temperature $= 70 + 384 = $ **454°F**

6.38 The torque-time relationship for load and input is shown in Fig. 6.38.

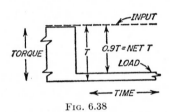

FIG. 6.38

Let $W = $ flywheel weight $= 3$ tons $= 6{,}000$ lb
$k = $ radius of gyration $= 5$ ft
$\alpha = $ angular acceleration
$T = $ torque input, ft-lb
$n_1 = $ initial rpm
$n_2 = $ final rpm
$\Delta\theta = $ revolutions after reduction of load $= 5$ rev
$T_{\text{net}} = J\alpha = 0.9T$
$\alpha = 0.9T/J$
$J = Wk^2/g = 6{,}000 \times 5^2/32.2 = 4{,}670$ ft-lb sec^2
$T = 63{,}000 \times \text{hp}/n = 63{,}000 \times 80/150 = 33{,}600$ in.-lb
$T = 2{,}800$ ft-lb
$\alpha = 0.9 \times 2{,}800/4{,}670 = 0.539$ radian per sec^2
$\alpha = 309$ rpm^2

For constant angular acceleration

$$n_2{}^2 - n_1{}^2 = 2\alpha\,\Delta\theta$$
$$n_2 = \sqrt{n_1{}^2 + 2\alpha\,\Delta\theta} = \sqrt{150^2 + 2 \times 309 \times 5}$$
$$= 160 \text{ rpm}$$

Change in speed $= n_2 - n_1 = 160 - 150 = $ **10 rpm**

6.39 T_1 = torque in
T_2 = torque out
n_1 = rpm in
n_2 = rpm out, and n_1/n_2 = **4.11**
r = radius of wheel (ft)
F = tractive force
v = velocity (fpm)

a. hp out = efficiency × hp in

$$T_2 n_2 = \text{efficiency} \times T_1 \times n_1$$

$$T_2 = \frac{n_1}{n_2} \times T_1 \times \text{efficiency} = 4.11 \times 230 \times 0.97 = \textbf{915 ft-lb}$$

$$F = \frac{T_2}{r} = \frac{915}{(15/12)} = \frac{915}{1.25} = \textbf{732 lb at both rear wheels}$$

b. hp $= \dfrac{F_v}{33,000} = \dfrac{732 \times (40 \times 5,280/60)}{33,000} = \dfrac{732 \times 3,520}{33,000}$
$$= \textbf{78.1 hp}$$

6.40   ihp = indicated horsepower per cylinder
bhp = brake horsepower
P_m = mean efficiency pressure (psi)
n = no. of cylinders
L = stroke (ft)
A = piston area (psi)
D = diameter (in.)
N = explosions per min

$$\frac{12 \times L(\text{ft})}{D(\text{in.})} = 1.4 \text{ and } D^2 = 73.5L^2$$

$$A = \frac{\pi D^2}{4} = 57.6L^2$$

$$\text{ihp} = \frac{\text{bhp}}{(\text{mechanical efficiency}) \times n} = \frac{30}{0.85 \times 6} = 5.88$$

$$\text{ihp} = \frac{P_m LAN}{33,000} = 5.88 = \frac{90 \times L \times 57.6L^2 \times \left(\dfrac{1,800}{2}\right)}{33,000}$$

$$L^3 = 0.0416, \; L = \textbf{0.346 ft or 4.15 in.}$$

$$D = \frac{4.15}{1.4} = \textbf{2.97 in.}$$

6.41   Let E_{ss} = shearing modulus of steel

E_{sa} = shearing modulus of aluminum

I_{ps} = polar moment of inertia of steel rod

I_{pa} = polar moment of inertia of aluminum tube

L = length of shaft = 5 ft = 60 in.

T_s = torque carried by steel rod

T_a = torque carried by aluminum tube

The angle of twist and the dimensions are shown in Fig. 6.41a,

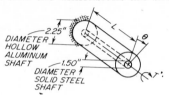

DIAMETER
HOLLOW
ALUMINUM
SHAFT
2.25"
1.50"
DIAMETER
SOLID STEEL
SHAFT

Fig. 6.41a

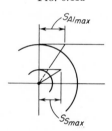

S_{Almax}

S_{Smax}

Fig. 6.41b

and the shear-stress variation is shown in Fig. 6.41b. The two basic equations are:

(1) $\theta = \theta_a = \theta_s$

(2) $T = T_a + T_s$

$$I_{ps} = \frac{\pi}{32}(1.5)^4 = 0.497 \text{ in.}^4$$

$$I_{pa} = \frac{\pi}{32}[(2.25)^4 - (1.5)^4] = 2.01 \text{ in.}^4$$

a. $\theta_a = T_a L / I_{pa} E_{sa}$ and $\theta_s = T_s L / I_{ps} E_{ss}$

$T = \theta_a I_{pa} E_{sa}/L + \theta_s I_{ps} E_{ss}/L = \theta (I_{pa} E_{sa} + I_{ps} E_{ss})/L$

$\theta = TL/(I_{pa} E_{sa} + I_{ps} E_{ss})$

$$= \frac{(15,000)(60)}{(2.01)(3.5 \times 10^6) + (0.497)(12 \times 10^6)}$$

$$= 0.0693 \text{ radian}$$

b. $S_{a(\text{max})} = T_a\left(\dfrac{(2.25)/2}{I_{pa}}\right) = \left(\dfrac{\theta_a I_{pa} E_{sa}}{L}\right)\left(\dfrac{(2.25)/2}{I_{pa}}\right)$

$$= \dfrac{\theta E_{sa}(2.25)}{2L}$$

$$= \dfrac{(0.0693)(3.5 \times 10^6)(2.25)}{(2)(60)} = 4{,}540 \text{ psi}$$

$S_{s(\text{max})} = \theta E_{ss}(1.5)/2L$
$= (0.0693)(12 \times 10^6)(1.5)/120 = 10{,}400 \text{ psi}$

6.42 N = rpm
     T = number of teeth
     x = number of gear pairs
     F = force, lb
     v = head velocity, fpm

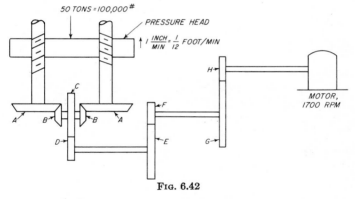

50 TONS = 100,000 #

PRESSURE HEAD

$1\ \dfrac{INCH}{MIN} = \dfrac{1}{12}$ FOOT/MIN

MOTOR,
1700 RPM

Fig. 6.42

½ in. travel of head per 1 revolution of A
1 in. per min travel of head = 2 rpm of A
N_A = 2 rpm,   and N_H = 1,700 rpm,   and reduction per train
    = ⅙

Train valve = $T_V = \dfrac{N_A}{N_H} = \dfrac{2}{1{,}700} = \dfrac{1}{850} = \left(\dfrac{1}{y}\right)^x$

when x = 4, as in sketch, y = 5.4 < 6 as required

$T_V = \dfrac{T_H}{T_G} \times \dfrac{T_F}{T_E} \times \dfrac{T_D}{T_C} \times \dfrac{T_B}{T_A}$

and assume y for spur-gear pairs to be 5, 5, and 6, and T multiples of 12

a. $T_V = \dfrac{1}{850} = \dfrac{12}{60} \times \dfrac{12}{60} \times \dfrac{12}{72} \times \dfrac{T_B}{T_A}$

$\dfrac{T_B}{T_A} = \dfrac{150}{850} = \dfrac{15}{85} = \dfrac{1}{5.67}$

Make T_B and T_A bevel gears, with teeth 15 and 85.

b. $\text{hp} = \dfrac{F_v}{\text{efficiency} \times 33{,}000} = \dfrac{100{,}000 \times \frac{1}{12}}{0.40 \times 33{,}000} = 0.632 \text{ hp}$

Use a ¾-hp motor.

6.43 V = pitch line velocity
       F_t = transmitted load
       P_D = diametral pitch
       n = number of teeth
       b = face width
       I = increment load
       S_n = endurance limit
       Y = form factor
       F_w = wear load
       N = rpm

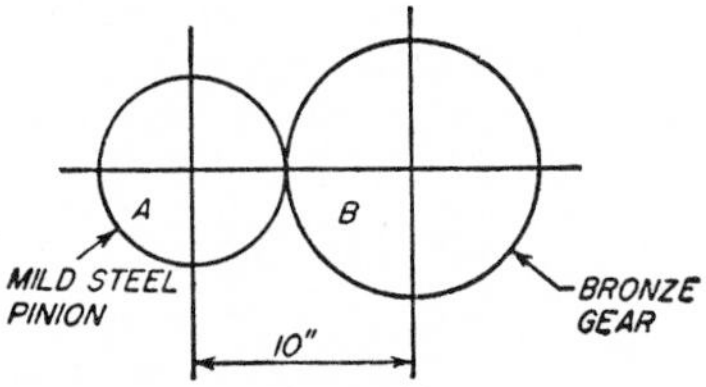

MILD STEEL PINION   BRONZE GEAR   10″

Fig. 6.43

$\dfrac{R_A}{R_B} = \dfrac{\omega_B}{\omega_A} = \frac{2}{3}$ (given)

$R_A = \frac{2}{3}R_B$

$R_A + R_B = 10$ in. (given)

$\frac{2}{3}R_B + \frac{3}{3}R_B = \frac{5}{3}R_B = 10$

$R_B = 6$ in. and $R_A = 4$ in.

$D_B = 12$ in. and $D_A = 8$ in.

$V = 2\pi N_A R_A = 2\pi \times 150 \times 4 = 3{,}770$ in. per min
                                = 314 ft per min

$\text{hp} = \dfrac{F_t V}{33{,}000}, \quad F_t = \dfrac{33{,}000 \times 8}{314} = 841 \text{ lb}$

All references to tables and pages are in Faires, "Design of Machine Elements," Macmillan, 1941.

e = maximum permissible error = 0.0045 (p. 216)
Use first-class commercial gears ($P_D > \frac{1}{2}$) (p. 216)
Assume $P_D = 6$ and expected error = 0.002 in. (p. 216)

Recommended proportions for cut teeth:

$$\frac{9.5}{P_D} < b < \frac{12.5}{P_D} \text{ or } 1.58 < b < 2.08$$

Use $b = 2$ in.

Find dynamic load F_D

For $14\frac{1}{2}°$ full depth, $k = 0.107e$

$$C = \text{deformation factor} = \frac{kEgEp}{(Eg + Ep)}$$

$$= \frac{0.107 \times 0.002 \times 15 \times 10^6 \times 30 \times 10^6}{(15 + 30)10^6} = 2{,}140$$

$$F_D = \frac{0.05V(bC + F_t)}{0.05V + (bC + F_t)^{0.5}} + F_t = I + F_t$$

$$(bC + F_t) = 2 \times 2{,}140 + 841 = 5{,}121$$

$$I = 910 \text{ (p. 214) and 920 (from the equation)}$$

$$F_D = 910 + 841 = \mathbf{1{,}751} \text{ lb}$$

Design is based on gear having smallest product of $S_n Y$, where

$$F_s = \frac{S_n Y_b}{P_D}$$

For gear (phosphor bronze):

$$T = P_D \times D = 6 \times 12 = 72 \text{ teeth}, \quad \text{and } Y = 0.360 \text{ (p. 204)}$$

$$S_n Y = 24{,}000 \times 0.360 = \mathbf{8{,}640}$$

For pinion (SAE 1035):

$$T = P_D D = 6 \times 8 = 48 \text{ teeth}, \quad \text{and } Y = 0.345 \text{ (p. 204)}$$

$$S_n Y = \frac{S_{ultimate}}{2} \times Y = \frac{87{,}000}{2} \times 0.345 = \mathbf{15{,}000}$$

$\therefore$ Bronze gear is weaker

$$F_s = \frac{24{,}000 \times 2 \times 0.360}{6} = 2{,}880 \text{ lb}$$

$$\text{Margin of safety} = \frac{F_s}{F_d} - 1 = \frac{2{,}880}{1{,}751} - 1 = 0.643$$

$\therefore$ Suitable for speed reducers, pumps, mining machinery, etc.
Check for wear:

$$Q = \frac{2D_B}{(D_A + D_B)} = \frac{2 \times 12}{(8 + 12)} = 1.2$$

Use SAE 1035 steel, tempering temperature = 600°F
Brinell No. 240 (Tables XXX and XXXIII)
By interpolation, $K = 126$

$$F_W = D_A b K Q = 8 \times 2 \times 126 \times 1.2 = 2{,}420 \text{ lb}$$
$$F_W > F_D, \quad 2{,}420 \text{ lb} > 1{,}751 \text{ lb}$$

$\therefore$ ($b = 2$ in.) and ($P_D = 6$) is suitable
b could be decreased somewhat to obtain a more economical design.

Recalculate for $b = 1\frac{7}{8}$ in. or $b = 1\frac{3}{4}$ in.

6.44 Assume:
1. The wrench moves in horizontal plane.
2. No energy is absorbed by the bolted connection.
3. The design stress = 40,000 psi.
4. No frictional losses.

Let Δ = max deflection
$\bar{F}$ = force causing Δ
W = 5 lb
v = 12 fps = 144 in. per sec
L = 18 in.
d = wrench diameter
g = 386 in. per sec^2
$\bar{M}$ = moment caused by $\bar{F}$

Figure 6.44 shows the wrench at the moment the 5-lb force strikes the handle at point (1) and the deflection to point (2) of

FIG. 6.44

the handle due to the blow. The total energy at position (1) = the total energy at position (2), or

$$PE_1 + KE_1 = PE_2 + KE_2$$
$$0 + KE_1 = PE_2 + 0$$
$$KE_1 = PE_2 = Wv^2/2g = \bar{F}\Delta/2 = \bar{M}^2 L/6EI_{xx}$$
$$= \bar{S}_t^2 I_{xx} L/c^2 6E$$

$$= (\bar{S}_t{}^2 L/6E)(\pi d^2/16)$$

where $\Delta = \bar{F}L^3/3EI$, $\bar{F} = \bar{M}/L$, $I = \pi d^4/64$, $c = d/2$, and $\bar{S}_t = \bar{M}c/I_{zz}$, or $\bar{M} = (\bar{S}_t I_{zz})/(c)$.

$$d = \frac{4v}{\bar{S}_t}\sqrt{\frac{3WE}{\pi gL}} = \frac{(4)(144)}{(40,000)}\sqrt{\frac{(3)(5)(30 \times 10^6)}{\pi(386)(18)}} = 2.07 \text{ in.}$$

Use 2-in.-diameter wrench.

6.45   *a.* $v_1' = 500$ cu ft per min, and $c = 3$ per cent. Values are schematically shown in Fig. 6.45.

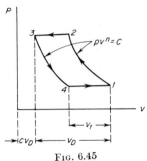

FIG. 6.45

Volumetric efficiency $= 1 + c - c(p_2/p_1)^{1/n}$
$= 1 + 0.03 - 0.03 \times (105/14.7)^{1/1.34}$
$= 1.03 - 0.03 \times (7.15)^{0.745}$
$= 1.03 - 0.03 \times 4.32 = 0.90 = 90 \text{ per cent}$
$= (\text{capacity of compressor})/\text{displacement}$
Displacement $= v_D = 500/0.90 = \mathbf{555} \text{ cu ft per min}$

b. $p_1 = 23.8$ in. Hg $= 23.8 \times 0.491 = 11.7$ psia, with $t_1 = 70°F$; since $v_2/v_1 = r_k$ is constant, but $p_1 v_1{}^n = p_2 v_2{}^n$ and $v_2/v_1 = (p_1/p_2)^{1/n} = $ constant, it means that the volumetric efficiency must be constant along with n and v_D.

Capacity in cubic feet per minute free air (measured at 23.8 in. Hg at 70°F) **would be the same.**
Since $r_k = c$, and $n = c$,

$$r_k = (p_2/p_1)^{1/n} = (p_2'/p_1')^{1/n}$$
$105/14.7 = p_2'/11.7$ and $p_2' = 11.7 \times 105/14.7 = \mathbf{83.5} \text{ psia}$

c. **Increase the rpm.**

6.46   *a.* One gallon = 8.34 lb water.

Work done by pump = $8.34 \times 500 \times 90 = 375{,}000$ ft-lb per min

With an efficiency of 60 per cent,

   Input to pump = $375{,}000/0.60 = 625{,}000$ ft-lb per min
or   Motor output = $625{,}000/33{,}000 = $ **18.93** hp

   b. Let N = rpm, H = head, and Q = capacity in gallons per minute.

$$\frac{N_1{}^2}{N_2{}^2} = \frac{H_1}{H_2}$$
$$N_2 = \left[\left(\frac{H_2}{H_1}\right)N_1{}^2\right]^{\frac{1}{2}} = [(^{120}\!/_{90})(100^2)]^{\frac{1}{2}} = \textbf{116 rpm}$$
$$Q_1/Q_2 = N_1/N_2$$
$$Q_2 = Q_1 \times N_2/N_1 = 500 \times {}^{116}\!/_{100} = \textbf{580 gal per min}$$

   c. Let P = power in horsepower.

$$P_1/P_2 = N_1{}^3/N_2{}^3$$
$$P_2 = P_1 \times (N_2{}^3/N_1{}^3) = 18.93 \times (116^3/100^3) = \textbf{29.6 hp}$$

A 25-hp motor would **not be adequate.**

6.47 $F = 50$ lb (in horizontal plane and at right angles to bar)

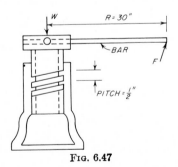

Fɪɢ. 6.47

$$\frac{\text{Work (out)}}{\text{Revolutions}}$$
$$= \text{efficiency} \times \frac{\text{Work (in)}}{\text{Revolutions}}$$
$$W \times \text{pitch} = \text{efficiency} \times F \times 2\pi R$$
$$W \times \tfrac{1}{2} = 0.80 \times 50 \times 2\pi \times 30$$
$$W = \textbf{15,100 lb}$$

6.48 $R = 12 + \tfrac{1}{2} \times \tfrac{7}{8} = 12.438$ in. = 1.0365 ft
Since lead of worm = two pitch distances, 40 revolutions of worm
= one revolutions of drum

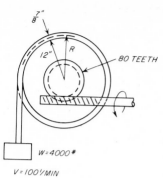

FIG. 6.48

$$\text{hp in} = \frac{\text{hp out}}{\text{efficiency}}$$

$$\text{hp in} = \frac{W \times V}{\text{efficiency} \times 33,000} = \frac{4,000 \times 100}{0.60 \times 33,000} = \textbf{20.2 hp}$$

$$\text{rpm drum} = \frac{V}{2\pi R} = \frac{100}{6.28 \times 1.0365} = 15.33 \text{ rpm}$$

$$\text{rpm worm} = 40 \times 15.33 = 613 \text{ rpm}$$

$$T_{(\text{in.})} = \frac{63,000 \times \text{hp}}{\text{rpm worm}} = \frac{63,000 \times 20.2}{613} = 2,075 \text{ in-lb}$$

$$T_{(\text{ft})} = \textbf{173} \text{ ft-lb}$$

6.49　$P = pA = 205 \times \dfrac{\pi}{4} \times (1.25)^2 = 252$ lb

N = effective coils = total coils $- 2 = 9.5 - 2 = \textbf{7.5}$ coils

D = diameter of coil

d = diameter of wire, and y = deflection (in.)

$$C = \frac{D}{d} = \frac{(4.5 - 0.5)}{0.5} = 8$$

$$y = \frac{8\,PC^3N}{E_s d} = \frac{8 \times 252 \times 8^3 \times 7.5}{11.5 \times 10^6 \times 0.5} = 1.345 \text{ in.}$$

Initial compressed length = free length $- y$

$$= 8.00 \text{ in.} - 1.345 \text{ in.} = \textbf{6.66} \text{ in.}$$

Solid length $= dN + 2d = 0.5 \times 7.5 + 2 \times 0.5 = 4.75$ in.
Deflection from free length to solid length $= y_c = 8.00 - 4.75$
$= 3.25$ in.

$$P_c = \frac{y_c}{y} \times P = \frac{3.25 \text{ in.}}{1.345 \text{ in.}} \times 252 \text{ lb} = 609 \text{ lb}$$
$$k = 1.18$$

(See curves of stress factors vs. spring index, Fig. 322, Faires, "Design of Machine Elements," Macmillan, 1941.)

$$S_s = \frac{k \times 2.55P_c D}{d^3} = 1.18 \times 2.55 \times 609 \times 4 \times 8 = 58{,}800 \text{ psi}$$

6.50

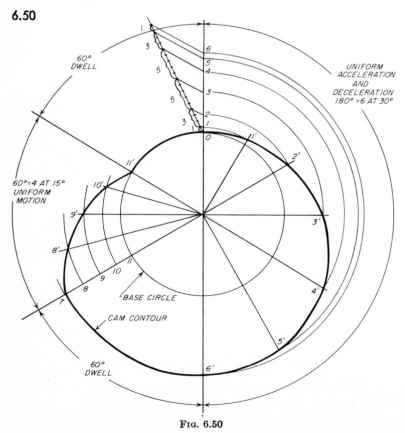

FIG. 6.50

6.51 N = number of teeth, w = angular velocity. Gear 1 is input gear, gear 4 is output gear.

$$\frac{w_4}{w_1} = \frac{1,131}{2,000} = \frac{N_1 N_3}{N_2 N_4}$$

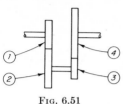

Fɪɢ. 6.51

Factoring, we have

$$1,131 = 3 \times 377 = 3 \times 13 \times 29 = 39 \times 29$$

Let $N_1 = $ **39** and $N_3 = 29$ **1a**

$$N_1 + N_2 = N_3 + N_4$$
$$39 + N_2 = 29 + N_4$$
or $$N_4 = N_2 + 10$$

Also

$$\frac{1,131}{2,000} = \frac{N_1 N_3}{N_2 N_4} = \frac{39 \times 29}{N_2 N_4} \quad \text{or} \quad N_2 N_4 = 2,000$$

or $$N_2(N_2 + 10) = 2,000$$
$$N_2{}^2 + 10N_2 - 2,000 = 0$$
$$N_2 = -5 \pm \sqrt{25 + 2,000} = -5 + 45 = 40$$
and $$N_4 = N_2 + 10 = 40 + 10 = \mathbf{50} \text{ teeth} \quad \mathbf{2c}$$

6.52 Let the smaller gear be 1 and the larger gear be 2. Let

$$\Sigma = \text{angle between shafts}$$
$$\psi = \text{helix angle}$$
$$p_n = \text{normal circular pitch}$$
$$N = \text{number of teeth}$$
$$w = \text{angular velocity}$$
$$c = \text{center-to-center distance}$$
$$\Sigma = \psi_1 + \psi_2$$
$$90° = 35° + \psi_2 \quad \text{and} \quad \psi_2 = 55°$$

$$\text{Speed ratio} = \frac{w_2}{w_1} = \frac{1}{2} = \frac{N_1}{N_2} = \frac{24}{N_2} \quad \text{or} \quad N_2 = 48 \text{ teeth}$$

$$c = \frac{p_n}{2\pi}\left(\frac{N_1}{\cos\psi_1} + \frac{N_2}{\cos\psi_2}\right) = \frac{0.785}{2\pi}\left(\frac{24}{\cos 35°} + \frac{48}{\cos 55°}\right)$$
$$= \textbf{14.125 in.} \quad \textbf{b}$$

where $\cos 35° = 0.819$ and $\cos 55° = 0.573$.

6.53 Call points of contact on bodies 1 and 2, B and P, respectively.

    a. Find angular velocity of body 1:

$$\mathbf{V}_P = \mathbf{V}_B + \mathbf{V}_{P/B}$$
$$V_P = w_2(\overline{O_2P}) = (10)(2) = 20 \text{ in. per sec}$$

Solve graphically or by geometry.

$$V_B = 0.5V_P\sqrt{2} = 14.142 \text{ in. per sec}$$
$$V_{P/B} = 0.5V_P + 0.866V_P = 1.366V_P = 27.32 \text{ in. per sec}$$
$$w_1 = \frac{V_B}{\overline{O_1B}} = \frac{0.5(V_P)\sqrt{2}}{1\sqrt{2}}$$
$$= 10 \text{ radians per sec} \quad \textbf{1d; Clockwise} \quad \textbf{2a}$$

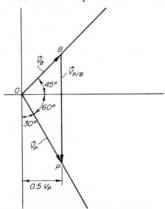

FIG. 6.53a. Velocity polygon.

    b. Find angular acceleration of body 1:

$$\mathbf{A}_P = \mathbf{A}_B + \mathbf{A}_{P/B} + 2\mathbf{w}_1\mathbf{V}_{P/B} = \mathbf{A}_B{}^N + \mathbf{A}_B{}^T$$
$$+ \mathbf{A}_{P/B}^N + \mathbf{A}_{P/B}^T + 2\mathbf{w}_1\mathbf{V}_{P/B}$$

$$A_P = w_2{}^2(\overline{O_2P}) = (10)^2(2) = 200.0 \text{ in. per sec}^{-2}$$
$$A_B{}^N = w_1{}^2(\overline{O_1B}) = (10)^2\sqrt{2} = 141.4 \text{ in. per sec}^2$$
$$A_{P/B}^N = \frac{V_{P/B}^2}{R_1} = \frac{\overline{27.32^2}}{(1)} = 746.4 \text{ in. per sec}^2$$
$$2w_1V_{P/B} = (2)(10)(27.32) = 546.4 \text{ in. per sec}^{-2}$$

By geometry $A_B{}^T = 669.2$ in. per sec^{-2}

$$\alpha_1 = \frac{A_B{}^T}{(O_1B)} = \frac{669.2}{\sqrt{2}}$$
$$= \textbf{473.2} \text{ radians per sec}^{-2} \qquad \textbf{3c ; Counterclockwise} \qquad \textbf{4b}$$

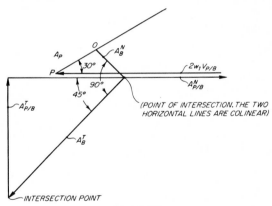

INTERSECTION POINT

Fig. 6.53b

6.54 Break unbalance effects into horizontal and vertical components. Find vertical components:

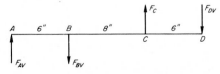

Fig. 6.54a

At $B: F_B = 4 \qquad F_{BV} = F_B \sin 30°$
$$= 2.000 \qquad F_{BH} = F_B \cos 30° = 3.464$$
$$\Sigma M_A = 0 \qquad 20F_{DV} + 6F_{BV} - 14F_C = 0$$
$$\text{or} \qquad 20F_{DV} + (6)(2.00) - (14)(2.00) = 0$$

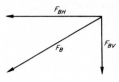

FIG. 6.54b

therefore $F_{DV} = 0.800$ oz-in.

$\Sigma M_D = 0 \qquad 20F_{AV} + (6)(2) - (14)(2) = 0$

$$F_{AV} = \mathbf{0.800} \text{ oz-in.}$$

Find horizontal components:

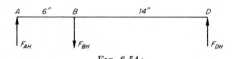

FIG. 6.54c

$\Sigma M_A = 0 \qquad 20F_{DH} - 6F_{BH} = 0$

$\qquad$ or $\qquad 20F_{DH} - (6)(3.464) = 0 \qquad F_{DH} = \mathbf{1.039} \text{ oz-in.}$

$\Sigma M_D = 0 \qquad 20F_{AH} - 14F_{BH} = 0$

$$\text{or} \qquad F_{AH} = \frac{(14)(3.464)}{20} = \mathbf{2.425} \text{ oz-in.}$$

At left end:

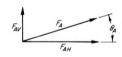

FIG. 6.54d

$F_A = \sqrt{F_{AV}^2 + F_{AH}^2} = \sqrt{(0.800)^2 + (2.425)^2} = \mathbf{2.554} \text{ oz-in.}$

$\theta_A = \tan^{-1}(0.800/2.425)$

$\theta_A = \mathbf{18.3°}$

At right end:

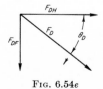

FIG. 6.54e

$$F_D = \sqrt{F_{DV}{}^2 + F_{DH}{}^2} = \sqrt{(0.800)^2 + (1.039)^2} = \mathbf{1.311} \text{ oz-in.}$$

$$\theta_D = \tan^{-1}\left(\frac{0.800}{1.039}\right)$$

$$\theta_D = \mathbf{37.6°}$$

6.55

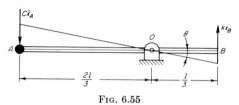

Fig. 6.55

$$J_0 = m\left(\frac{2l}{3}\right)^2, \ x_B = \frac{l}{3}, \ \dot{x}_A = \frac{2}{3}l\dot{\theta}$$

$$\Sigma T_0 = J_0\ddot{\theta}$$

$$-(Kx_B)\left(\frac{l}{3}\right) - C\dot{x}_A\left(\frac{2l}{3}\right) = J_0\ddot{\theta}$$

$$-\frac{Kl^2}{9}\theta - \frac{4Cl^2}{9}\dot{\theta} = \frac{4ml^2}{9}\ddot{\theta}$$

$$\ddot{\theta} + \frac{C}{m}\dot{\theta} + \frac{K}{4m}\theta = 0$$

Damped natural frequency $= w_{nd} = \sqrt{\dfrac{K}{4m} - \left(\dfrac{c}{2m}\right)^2}$

for critical damping $w_{nd} = 0$. Therefore,

$$\left(\frac{C_c}{2m}\right)^2 = \frac{K}{4m}$$

$$C_c = \sqrt{Km}$$

6.56 $C_1 = S_y(4 - e)b = (35,000)(4 - e)6 = 210,000(4 - e)$

$$C_2 = \frac{S_y}{2}eb = \frac{35,000}{2}(e)6 = 105,000e$$

$$M = 240,000 \times 12 = 2,880,000 \text{ lb-in.}$$

$$T_1 = C_1 \text{ and } T_2 = C_2$$

External moment is balanced by summation of internal moments:

$$M = 2\left(2 + \frac{e}{2}\right)C_1 + 2\left(\frac{2}{3}e\right)C_2 = (4 - e)C_1 + 1.33C_2e$$

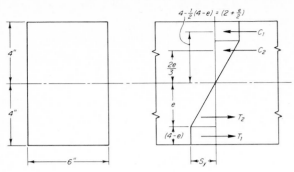

FIG. 6.56

or $2,880,000 = (4 + e)(210,000)(4 - e) + 1.333(105,000)e$

or $\qquad 70e^2 = 480 \qquad e^2 = 6.857 \qquad e = 2.619$ in.

Distance from surface $= 4 - e = 4 - 2.619 = $ **1.381** in.

6.57   *a.* The discharge $Q = \dfrac{\text{gal per min}}{448} = \dfrac{4,000}{448} = 8.93$ cfs

$$h_f = f \frac{L}{D} \frac{V^2}{2g} \tag{1}$$

$$Q = AV = \frac{\pi}{4} D^2 V \qquad \text{or} \qquad V^2 = \frac{16Q^2}{\pi^2 D^4} \tag{2}$$

where f = friction factor. If h_f = head loss, ft, then substituting equation (2) into (1), we have

$$h_f = f \frac{L}{D} \frac{Q^2}{2g(D^2\pi/4)^2}$$
$$D^5 = \frac{8LQ^2}{h_f g \pi^2} f = \frac{(8)(10,000)(8.93)^2}{(75)(32.2)\pi^2} f = 267.0f \tag{3}$$

The Reynolds number Re is

$$\text{Re} = \frac{VD}{\nu} = \frac{4Q}{\pi \nu} \frac{1}{D} = \frac{(4)(8.93)}{\pi(0.0001)D} = \frac{113,800}{D} \tag{4}$$

where ν = kinematic viscosity, sq ft per sec.

From Moody diagram, roughness height $\epsilon = 0.00015$ ft. Solve by trial and error. For first trial, let $f = 0.02$, get $D = 1.393$ from Eq. (3), and then Re $= 81,400$ from Eq. (4) and $\epsilon/D = 0.00011$. Using Re and ϵ/D in Moody diagram gives $f = 0.019$. Repeat above procedure using $f = 0.019$ resulting in the following values: $D = 1.382$, Re $= 82,300$, and $f = 0.019$. The trial value of f checks out. Therefore $D = (1.382)(12) =$ **16.6** in.   **1b**

    $b.$

$$\frac{V_1{}^2}{2g} + \frac{p_1}{\gamma} + z_1 = \frac{V_2{}^2}{2g} + \frac{p_2}{\gamma} + z_1 + h_f \qquad V_1 = V_2$$

$$\frac{(40)(144)}{(0.85)(62.4)} + 200 = \frac{p_2(144)}{(0.85)(62.4)} + 50 + 75$$

$$p_2 = \textbf{67.5} \text{ psi} \qquad \textbf{2d}$$

6.58   Principles of dynamic similitude will be applied.

$$\text{Inertia force} = \text{Ma} = \frac{\rho L^3 L}{T^2}$$

where $\rho =$ density. Since $V = L/T$,

$$\text{Ma} = \rho L^2 V^2$$

$$\text{Viscous force} = \tau A = \mu \left(\frac{dV}{dy}\right) L^2$$

where $\mu =$ viscosity
      $\tau =$ shear stress
      $V =$ velocity

$$\text{Viscous force} = \mu \left(\frac{V}{L}\right) L^2 = \mu VL$$

$$\frac{\text{Inertia force}}{\text{Viscous force}} = \frac{\rho L^2 V^2}{\mu VL} = \frac{\rho LV}{\mu}$$

$\dfrac{\mu}{\rho} = \nu =$ kinematic viscosity, sq ft per sec. The Reynolds number Re $= LV/\nu$.

$$\text{Re(model)} = \text{Re(prototype)}$$

$$\frac{L_m V_m}{\nu_m} = \frac{L_p V_p}{\nu_p}$$

$\nu_m = 1.57 \times 10^{-4}$ sq ft per sec, $\nu_p = 1.05 \times 10^{-5}$ sq ft per sec; from temperature vs. kinematic viscosity curve.

$$L_m = L_p \left(\frac{V_p}{V_m}\right)\left(\frac{\nu_m}{\nu_p}\right) = \left(\frac{6}{12}\right)\left(\frac{60}{200}\right)\left(\frac{1.57 \times 10^{-4}}{1.05 \times 10^{-5}}\right)$$

$$L_m = \textbf{2.24} \text{ ft}$$

6.59   Index of curvature $= c_1 = \dfrac{2R}{h} = \dfrac{(2)(5)}{4} = 2.500$

Area of cross section $= A = 4$ sq in.

Eccentricity $= e = \dfrac{hc_1}{2(3c_1{}^2 - 0.8)}$

$$= \dfrac{(4)(2.5)}{2[(3)(2.5)^2 - (0.8)]} = 0.2785$$

$$h_1 = \dfrac{h}{2} - e = \dfrac{4}{2} - 0.2785 = 1.7214$$

$$h_2 = \dfrac{h}{2} + e = \dfrac{4}{2} + 0.2785 = 2.2785$$

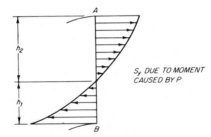

Fig. 6.59a

S_y due to moment caused by P:

At B: $S_B = \dfrac{PRh_1}{Aea} = \dfrac{(1,000)(5)(1.7214)}{(4)(0.2785)(3)}$

$$= 2,575.4\text{-psi compression}$$

At A: $S_A = \dfrac{PRh_2}{Aec} = \dfrac{(1,000)(5)(2.2785)}{(4)(0.2785)(7)} = 1,461.0\text{-psi tension}$

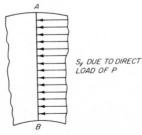

Fig. 6.59b

S_y due to direct load of P:

$$S_A = S_B = \frac{P}{A} = \frac{1,000}{4} = 250\text{-psi compression}$$

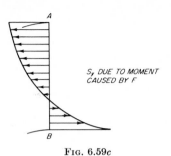

S_y DUE TO MOMENT
CAUSED BY F

Fɪɢ. 6.59c

S_y due to moment caused by F:

At B: $S_B = \dfrac{FRh_1}{Aea} = \dfrac{F(5)(1.7214)}{(4)(0.2785)(3)} = 2.5754F$ tension

At A: $S_A = \dfrac{FRh_2}{Aec} = \dfrac{F(5)(2.2785)}{(4)(0.2785)(7)} = 1.4609F$ compression

Total magnitude stress at A = total magnitude stress at B
    $1461.0 - 250.0 - 1.4609F = -(-2575.4 - 250 + 2.5754F)$
or $F = \mathbf{1{,}448}$ lb

6.60   For E6010, Tension yield point = S_{ty} = 50,000 psi
            Shear yield point = S_{sy} = 25,000 psi
            Allowable shear stress = $S_{sA} = \dfrac{S_{sy}}{2}$ = 12,500 psi

Point B most critically stressed.

    Left throat area = $6 \times 2 \times \frac{1}{4} \times 0.707 = 2.121$ sq in.
    Right throat area = $2 \times \frac{3}{8} \times 4 \times 0.707 = 2.121$ sq in.
        Total area = 4.242 sq in.

Find centroid taking moments about left end.

$$\bar{x} = \frac{\Sigma Ax}{\Sigma A} = \frac{(2.121)(3) + (2.121)(12)}{4.242} = 7.50 \text{ in.}$$

Find area moment of inertia about centroid.

$$J = \Sigma A \left(\frac{l^2}{12} + a^2 \right) = 2.121 \left(\frac{6^2}{12} + 4.5^2 \right)$$
$$+ 2.121 \left(\frac{4^2}{12} + 4.5^2 \right) = 95.1 \text{ in.}^4$$

where l = length of weld and a = distance from centroid to weld center.

$$S_{sP} = \frac{P}{\Sigma A} = \frac{P}{4.242} = 0.236P; \quad S_{sM} = \frac{Tr}{J}$$
$$= \frac{(5P)(7.5)}{95.1} = 0.394 \, P$$

$$S_{sT} = \sqrt{S_{sP}{}^2 + S_{sM}{}^2}$$
$$12{,}500 = \sqrt{(0.236P)^2 + (0.394P)^2} = 0.460P$$
$$P = 27{,}200 \text{ lb}$$

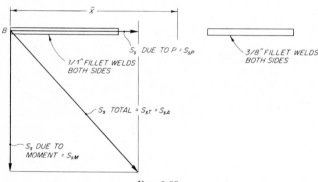

Fig. 6.60

STRUCTURAL ENGINEERING

7.01 Assume punched hole = $1\frac{3}{16}$ in. diameter
Area net section 1-1
$$= \frac{1}{4} \times (5 - 1\frac{3}{16}) = 1.04 \text{ sq in.}$$
Area net section 2-2
$$= \frac{1}{4} \times (5 - 2 \times 1\frac{3}{16}) = 0.84 \text{ sq in.}$$
Hanger load from gross section
$$= 5 \times \frac{1}{4} \times 16,000 = 20,000 \text{ lb}$$
Hanger load from net section 1-1 = $1.04 \times 16,000 = 16,650$ lb
Hanger load from net section 2-2 = $\frac{4}{3} \times 0.84 \times 16,000 =$
17,950 lb
Four rivets in shear = $4 \times 0.442 \times 12,000 = 21,000$ lb
Four rivets in bearing = $4 \times \frac{1}{4} \times 0.75 \times 24,000 = 18,000$ lb
Joint is weakest in tension in the net section 1-1
$$\text{Efficiency of the joint} = \frac{\text{smallest allowable load}}{\text{allowable load in gross section}} =$$
$(16,650/20,000) \times 100 = 83.3$ per cent

7.02 Area gross section = $\frac{3}{8} \times 9 = 3.37$ sq in.
Area net section 1-1 (2 holes) = $\frac{3}{8} \times (9 - 2 \times \frac{3}{4})$
$= 2.81$ sq in.
Area net section 2-2 (4 holes) = $\frac{3}{8} \times (9 - 4 \times \frac{3}{4})$
$= 2.25$ sq in.
Plate tension from gross section = $3.37 \times 20,000 = 67,500$ lb

501

Plate tension from net section 1-1 = 2.81 × 20,000 = 56,200 lb

Plate tension from net section 2-2 = $\frac{8}{6}$ × 2.25 × 20,000 = 60,000 lb

Eight rivets in shear = 8 × 0.442 × 16,000 = 56,500 lb

Eight rivets in bearing = 8 × $\frac{3}{8}$ × $\frac{3}{4}$ × 32,000 = 72,000 lb

Maximum permissible tension in the plate is 56,200 lb

Efficiency of the joint = (56,200/67,500) × 100 = 83.3 per cent

7.03 p = three rivet diameters = 3 × $\frac{3}{4}$ = 2.25 in., $2p$ = 4.50 in.

Investigate on the basis of $4\frac{1}{2}$ in. length of main plate

Area gross plate = 4.5 × $\frac{1}{2}$ = 2.25 sq in.

Area net section 1-1 (1 hole) = $\frac{1}{2}$(4.5 − $^{13}\!/_{16}$) = 1.84 sq in.

Area net section 2-2 (2 holes) = $\frac{1}{2}$(4.5 − 2 × $^{13}\!/_{16}$) = 1.44 sq in.

Tension in plate from gross section = 2.25 × 16,000 = 36,000 lb

Tension in plate from net section 1-1 = 1.84 × 16,000 = 29,500 lb

Tension in plate from net section 2-2 = $\frac{3}{2}$ × 1.44 × 16,000 = 34,600 lb

Three rivets in shear = 6 × 0.442 × 10,000 = 26,500 lb

Three rivets in bearing = 3 × $\frac{1}{2}$ × $\frac{3}{4}$ × 20,000 = 22,500 lb

Allowable load per inch of seam = 22,500/4.5 = 5,000 lb

Working efficiency = (22,500/36,000) × 100 = 62.5 per cent

7.04   Tension main plate must carry = $\dfrac{\text{diameter tank in inches}}{2}$

× psi = (60/2) × 500 = 15,000 lb per in. = 180,000 lb per ft.

Minimum thickness of plate is probably $\dfrac{12\!/_{11} \times 15,000}{20,000}$ = 0.818 in., or say $^{13}\!/_{16}$ in.

Assuming $\frac{7}{8}$-in. rivets are used, the pitch for the inside row of rivets = 3 × $\frac{7}{8}$+ = $2\frac{5}{8}$+ = 3 in.

Shear 11 rivets = 22 × 0.601 × 15,000 = 198,500 lb

Bearing 11 rivets (double shear) = 11 × $\frac{7}{8}$ × $^{13}\!/_{16}$ × 30,000 = 235,000 lb

Tension in plate from gross section = 12 × $^{13}\!/_{16}$ × 20,000 = 195,000 lb

Tension in plate from section 1-1

$= \frac{11}{11}(12 - 0.94) \times \frac{13}{16} \times 20,000 = 178,000$ lb

Tension in plate from section 2-2

$= \frac{11}{10}(12 - 1.88) \times \frac{13}{16} \times 20,000 = 181,000$ lb

Tension in plate from section 3-3

$= \frac{11}{8}(12 - 3.76) \times \frac{13}{16} \times 20,000 = 184,000$ lb

Assume splice plate $= \frac{9}{16}$ in. thick

Tension in splice plate at section 4-4

$= \frac{11}{11}(12 - 3.76) \times 2 \times \frac{9}{16} \times 20,000 = 185,000$ lb

Bearing 11 rivets (single shear)

$= 11 \times \frac{7}{8} \times 2 \times \frac{9}{16} \times 24,000 = 260,000$ lb

Design is satisfactory: $p = 3$ in., $t_{main\ plate} = \frac{13}{16}$ in.,
$t_{splice\ plate} = \frac{9}{16}$ in.

Efficiency $= (178,000/195,000) \times 100 = 91.2$ per cent

7.05 Longitudinal tension $= \dfrac{\text{diameter tank in inches}}{4} \times \text{psi} =$

$\frac{60}{4} \times 500 = 7,500$ lb per in.

Shear five rivets $= 9 \times 0.601 \times 15000 = 81,100$ lb (four in double shear, one in single shear)

$81,100 \div 7,500 = 10.8$ in. Use pitch $= 5\frac{1}{4}$ in.

Actual tension per $10\frac{1}{2}$ in. $= 10.5 \times 7,500 = 78,750$ lb

Tension gross section $= 10.5 \times \frac{13}{16} \times 20,000 = 171,000$ lb

Tension in two cover plates at net section (inner row of rivets)
$= (10.5 - 1.87) \times t \times 20,000 = 78,750$ lb

$$t = \frac{78,750}{8.63 \times 20,000} = 0.457 \text{ in.}$$

This is split $\frac{5}{9} \times 0.457 = 0.25$ in. minimum (to inside plate) and $\frac{4}{9} \times 0.457 = 0.20$ in. minimum (to outside plate)

Use two splice plates $\frac{7}{16}$ in. thick to take care of bearing of five rivets $= 9 \times \frac{7}{16} \times \frac{7}{8} \times 24,000 = 82,700$ lb

Efficiency of splice $[(81,100/171,000) \times 100 = 47.4$ per cent] is low because thickness of main plate was dictated by hoop tension and not by longitudinal tension

7.06 Eleven rivets; eight in double shear, three in single shear, eight in bearing on $\frac{3}{4}$-in. plate, 3 on $\frac{1}{2}$-in. plate

Single shear one rivet $= 0.994 \times 8,800 = 8,750$ lb per rivet

Bearing one rivet on $\frac{3}{4}$-in. plate $= 1.125 \times \frac{3}{4} \times 19,000 = 16,060$ lb per rivet

Bearing one rivet on $\frac{1}{2}$-in. plate = $1.125 \times \frac{1}{2} \times 19,000$ = 10,700 lb per rivet

Allowable shear in 11 rivets = $19 \times 8,750$ = 166,250 lb

Allowable bearing in 11 rivets = $3 \times 10,700 + 8 \times 16,060$ = $32,100 + 128,480$ = 160,580 lb

Tension gross section main plate = $16.5 \times \frac{3}{4} \times 11,000$ = 136,000 lb

Tension in main plate from section 1-1
   = $(16.5 - 1.19) \times \frac{3}{4} \times 11,000$ = 126,200 lb

Tension in main plate from section 2-2
   = $\frac{19}{18}(16.5 - 2.38) \times \frac{3}{4} \times 11,000$ = 123,000 lb

Tension in main plate from section 3-3
   = $\frac{19}{16}(16.5 - 4.76) \times \frac{3}{4} \times 11,000$ = 115,000 lb

Tension in the inner cover plates section 4-4
   = $\frac{19}{11} \times \frac{1}{2}(16.5 - 4.76) \times 11,000$ = 111,530 lb (critical)

Efficiency of the splice = $(111,530/136,000) \times 100 = 82$ per cent

7.07   Tension in tank shell per inch = $(D/2) \times \mathrm{psi} = 7\frac{2}{2} \times 125$ = 4,500 lb. Assume rivet pitch = 3 in., and rivet diameter = $\frac{7}{8}$ in., hole = $\frac{15}{16}$ in.

$$\text{Minimum thickness main plate} = \frac{1}{3 - 0.94} \times \frac{3 \times 4,500}{16,000} = 0.41 \text{ in.}$$

Use $\frac{7}{16}$-in. plate, two rows of rivets.

Shear one rivet = $0.601 \times 12,000$ = 7,200 lb, two rivets = 14,400 lb

Bearing one rivet = $0.875 \times 0.437 \times 24,000$ = 9,200 lb

Tension gross section of plate = $\frac{7}{16} \times 3 \times 16,000$ = 21,000 lb

Tension section 1-1 of plate = $\frac{7}{16}(3 - 0.94) \times 16,000$ = 14,420 lb

Efficiency of splice = $(14,400/21,000) \times 100 = 68.5$ per cent

7.08   Slot the 3-in. leg of the angle with a $\frac{1}{2}$-in. slot just inside the 4-in. leg for a length of $6\frac{3}{4}$ in.

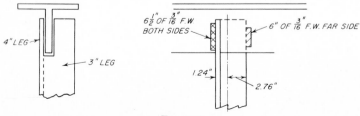

Fɪɢ. 7.08

Because of rounded edge of angle, fillet weld should not exceed $\frac{3}{16}$ in. in size (AISC)

Throat depth $= 0.707 \times \frac{3}{16} = 0.1325$ in.

Allowable load per inch of fillet weld $= 0.1325 \times 15,000 = 1,990$ lb per in.

Tensile strength $4 \times 3 \times \frac{1}{4}$ angle $= 1.69 \times 20,000 = 33,800$ lb

Number of inches of fillet weld needed $= 33,800/1,990 = 17.0$ in.

Take moments about back of 3-in. leg of angle

$$1.24 \times 33,800 = 4 \times L_R \times 1,990$$
$$L_R = \frac{1.24 \times 33,800}{4 \times 1,990} = 5.27 \text{ in.}$$

Use one length $= 5\frac{1}{2}$ in. $+ \frac{1}{2}$ in. $= 6$ in.

Take moments about edge of 4-in. leg of angle

$$2.76 \times 33,800 \times L_L \times 2 \times 4 \times 1,990$$
$$L_L = \frac{2.76 \times 33,800}{8 \times 1,990} = 5.86 \text{ in.}$$

Use two lengths $= 6$ in. $+ \frac{1}{2}$ in. $= 6\frac{1}{2}$ in.

7.09   Slot the stem of the T with a $\frac{3}{4}$-in. slot $10\frac{1}{2}$ in. long just beneath the flange.

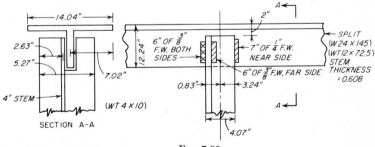

Fig. 7.09

(WT 4×10)

Area $= 2.94$ sq in.

Flange thickness $= 0.378$ in.

Stem thickness $= 0.248$ in.

Tensile strength T (WT 4×10) $= 2.94 \times 20,000 = 58,800$ lb

Take moments about back of flange (WT 4×10)

$$0.83 \times 58{,}800 = 4 \times 2{,}000 \times L_R$$

$$L_R = \frac{0.83 \times 58{,}800}{4 \times 2{,}000} = 6.10 \text{ in.}$$

Use $6\frac{1}{2}$ in. $+ \frac{1}{2}$ in $= 7$ in.

Take moments about edge of stem (WT 4×10)

$$3.24 \times 58{,}800 = 3 \times 4 \times 3{,}000 \times L_L$$

$$L_L = \frac{3.24 \times 58{,}800}{12 \times 3{,}000} = 5.30 \text{ in.}$$

Use three lengths of $5\frac{1}{2}$ in. $+ \frac{1}{2}$ in. or **6** in.

Check: $6.5 \times 2{,}000 + 3 \times 5.5 \times 3{,}000$

$\quad = 62{,}500 \text{ lb} > 58{,}800 \text{ lb}$

7.10   Let f = torsional force 1 in. from rivet group center of gravity

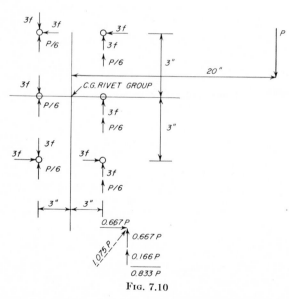

Fɪɢ. 7.10

Torsion M(about center of gravity of rivet group) $= 0 =$

$$+20P - 10 \times 3f \times 3, \quad \text{or } 3f = \frac{20p}{30} = 0.667P$$

Lower right-hand rivet = most stressed rivet

$$\sqrt{(0.667p)^2 + (0.833p)^2} = 1.075p$$
$$1.075p = 9{,}500 \text{ (given)}$$
$$\therefore p = 9{,}500/1.075 = \mathbf{8{,}850} \text{ lb}$$

A = area of bolt

$$A \times 16{,}000 = 9{,}500, \quad A = \frac{9{,}500}{16{,}000} = 0.593 \text{ sq in.}$$

Use $\frac{7}{8}$-in. diameter bolt $(A = 0.601)$
Thickness of plate for bearing = t_b

$$t_b \times \tfrac{7}{8} \times 32{,}000 = 9{,}500, \quad t_b = 9{,}500/28{,}000 = 0.34 \text{ in.}$$

Thickness of plate for tension (bending) = t_t

$$I_{\text{(net of plate)}} = t_t \times 9^3 \times \tfrac{1}{12} - 2 \times (1 \times t_t) \times 3^2 = 42.75 t_t$$

$$f = \frac{M_c}{I}, \quad 20{,}000 = \frac{17 \times 8{,}850 \times 4.5}{42.75 t_t}$$

$$t_t = \frac{17 \times 8{,}850 \times 4.5}{42.75 \times 20{,}000} = 0.79 \text{ in.}$$

Use $\frac{13}{16}$-in. plate

7.11 Let f be the force in a rivet 1 in. from the center of the rotation. Then torsion in the rivets of one plate (about center of gravity of rivet group)

$$
\begin{aligned}
10 \times 2.75 \times 2.75 f &= 76 f \\
4 \times 4 \times 4 f &= 64 f \\
4 \times 7 \times 7 f &= 196 f \\
\hline
& 336 f
\end{aligned}
$$

$$M = 0$$
$$= 2 \times 336 f - 14.75 \times 50{,}000$$
$$f = 1{,}094 \text{ lb}$$
$$7f = 7{,}660 \text{ lb}$$
$$2.75 f = 3{,}010 \text{ lb}$$

A 1-in. rivet in single shear and bearing on a $\frac{1}{2}$-in. plate may carry 11,780 lb.

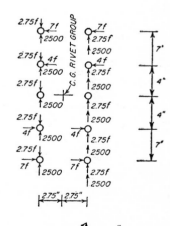

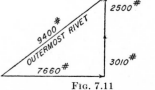

1d

Fig. 7.11

$I_{(net)}$ of plates $= \frac{1}{12} \times 1 \times 17^3 - 2 \times 1 \times 1 \times (4^2 + 7^2)$

$$f = \frac{M_c}{I} = \frac{12 \times 50,000 \times 8.5}{280} = 18,200 \text{ psi} \qquad \textbf{2c}$$

7.12 Let f = shearing value of the weld 1 in. from center of gravity of welds

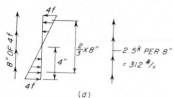

Fig. 7.12a. Shearing value of welds along one edge.

M of welds about center of gravity of welds =
$4 \times [4 \times 8 \times 4f + 4 \times (4f/2) \times (\frac{2}{3} \times 8)] = 512f + 171f = 683f$
$683f = 12 \times 10,000$, $f = 176$ lb per in., $4f = 704$ lb per in.

> Maximum weld force is 1,250 lb per in. Allowable for $\frac{1}{4}$-in. fillet weld = 2,000 lb per in., which is OK

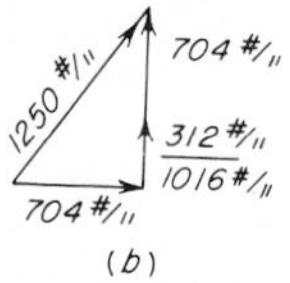

Fig. 7.12b

$I = \frac{1}{12} \times \frac{1}{2} \times 8 \times 8 \times 8 = 21.3$ in.^{4}

$f = \frac{M_c}{I} = \frac{8 \times 10,000 \times 4}{21.3} = 15,000$ psi, which is OK

A better use of welds is not to weld into corners.

If only 5 in. of weld is used on each of the four sides
$4[4 \times 5 \times 4f + 2.5 \times (2.5f/2) \times (\frac{2}{3} \times 5)] = 320f + 42f = 362f$
$362f = 12 \times 10,000$, $f = 331$, $2.5f = 824$, $4f = 1,324$ lb per in.
$$\frac{10,000}{4 \times 5} = 500 \text{ lb per in.}$$

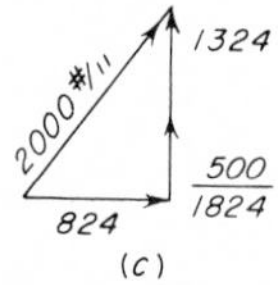

If four $\frac{5}{16}$-in. fillet welds are used, each 4 in. long, then

Fig. 7.12c

$4[4 \times 4 \times 4f + 2 \times (2f/2) \times (\frac{2}{3} \times 4)]$
$\qquad\qquad = 256f + 21f = 277f$
$277f = 12 \times 10,000$, $f = 433$, $2f = 866$, $4f = 1,733$ lb per in.
$$\frac{10,000}{4 \times 4} = 625 \text{ lb per in.}$$

2,500 lb per in. OK for $\frac{5}{16} =$ in. fillet weld.

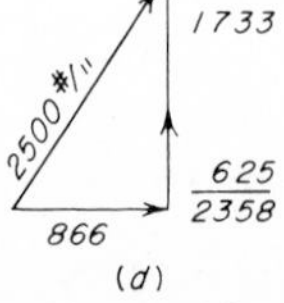

(d)
Fig. 7.12d

7.13 Gage from back of 4-in. leg to rivets in $3\frac{1}{2}$-in. leg = 2 in.

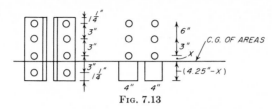

FIG. 7.13

Each $\frac{7}{8}$-in. rivet good for **9,020 lb in shear and bearing on** $\frac{3}{8}$-in. leg. Area of two $\frac{7}{8}$-in. rivets = 1.2 sq in. Assume that center of gravity of areas lies x distance below the second rivet from the bottom. Taking statical moments about the center-of-gravity line

$$\frac{8(4.25 - x)^2}{2} = 1.2[x + (x + 3) + (x + 6)] = 1.2(3x + 9)$$

$$(4.25 - x)^2 = \frac{1.2}{4}(3x + 9)$$

$$18.1 - 8.5x + x^2 = 0.9x + 2.7$$

$$(x^2 - 9.4x + 15.4) + 6.7 = 6.7$$

$$x - 4.7 = \pm 2.59, \quad x = +4.7 - 2.59 = 2.11 \text{ in.}$$

Let f be the intensity of tension or compression 1 in. from the center-of-gravity line

$$1.2(2.11^2 f + 5.11^2 f + 8.11^2 f) + \tfrac{8}{3} \times 2.14^3 f = 115.8f + 26.2f$$
$$= 142.0f = 8 \times 9,020 \times 2$$

= double shear in rivets through web of beam times the 2 in. of eccentricity.

$$f = 1,016 \text{ psi per in.}$$

Tension in top rivet = 0.601 × 1,016 × 8.11 = **4,950 lb**

7.14 Similar to 7.13, assume that the center of gravity lies x distance below the third rivet above the bottom. M about center-of-gravity line is

$$\frac{8(7.25 - x)^2}{2} = 1.2[x + (x + 3) + (x + 6) + (x + 9)$$
$$+ (x + 12) + (x + 15) + (x + 18) + (x + 21)]$$

$$(7.25 - x)^2 = \frac{1.2}{4}(8x + 84)$$
$$52.6 - 14.5x + x^2 = 2.4x + 25.2$$
$$(x^2 - 16.9x + 27.4) + 44 = 44$$
$$(x - 8.45)^2 = 44$$
$$x - 8.45 = \pm 6.63$$
$$x = +8.45 - 6.63 = 1.82 \text{ in.}$$

Let f be the intensity of tension or compression 1 in. from the center-of-gravity line.

$$
\begin{aligned}
1.2 \times 1.82^2 f &= 1.2f \times 3.3 \\
1.2 \times 4.82^2 f &= 1.2f \times 23.2 \\
1.2 \times 7.82^2 f &= 1.2f \times 61.2 \\
1.2 \times 10.82^2 f &= 1.2f \times 117.3 \\
1.2 \times 13.82^2 f &= 1.2f \times 191.0 \\
1.2 \times 16.82^2 f &= 1.2f \times 283.0 \\
1.2 \times 19.82^2 f &= 1.2f \times 393.0 \\
1.2 \times 22.82^2 f &= \underline{1.2f \times 522.0} \\
& \quad\; 1.2f \times 1596 = 1,915f
\end{aligned}
$$

$$
\begin{aligned}
8 \times (5.43)^3 \times f/3 &= 427f \\
20 \times 9,020 \times 2 &= \underline{} \\
& \quad 2,342f
\end{aligned}
$$

$$f = 154.0 \text{ lb/(sq in.)(in.)}$$

Tension in top rivets = $0.601 \times 154.0 \times 22.82 =$ **2,110 lb**

7.15 Side dimension of fillet weld must be ⅝ in. or less.

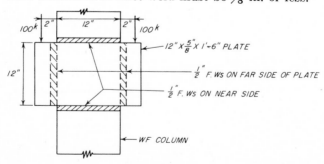

Fig. 7.15

Throat dimension of ½-in. fillet weld = $0.707 \times \frac{1}{2} = 0.354$ in.

½-in. fillet weld can carry $0.354 \times 13,000 = 4,600$ lb per in. in shear.

4 × 12 × 4,600 = 221,000 lb which is greater than 200,000 lb

7.16 Gross I of girder.

I of four angles about c.g of angles = 4 × 7.5 =		30 in.⁴
plus Ad^2 for angles		
= 4 × 5.86(21.25 − 1.03)²	=	9,600 in.⁴
I of two cover plates		
= 2 × 14 × ½ × 21.50²	=	6,470 in.⁴
I of web plate = 1/12 × 3/8 × 42³	=	2,320 in.⁴
	I =	18,420 in.⁴

$$f = \frac{Mc}{I} = 18,000 = \frac{\frac{1}{8} \times W \times 30 \times 30 \times 12 \times 21.75}{18,420}$$

W = 11,300 lb per ft = total uniform load per foot of girder

Center of gravity of flange is 1.05 in. below back of cover plate

Plate:	14 × ½ =	7.00 in.²	× ¼ in. =	1.75 in.³
Angles: 2 × 5.86 =	11.72	× 1.53	= 17.92
Flange:		= 18.72	(1.05)	= 19.67

$$sb = \frac{VQ}{I} = \frac{75,000 \times 18.72 \times 20.70}{18,420} = 1,580 \text{ lb per in.}$$

One ¾-in. rivet may carry 2 × 0.4418 × 15,000 = 13,250 lb in double shear and 11,300 lb in double bearing on ⅜-in. web maximum rivet pitch for one line of rivets = 11,300/1,580 = 7.15 in. Use 6 in. (Maximum usually allowed by specifications for this condition)

7.17 Try four angles 6 × 6 × ⅞ in. placed 36½ in. back to back of angles.

Gross I of girder:

I of four angles = 4 × 31.9	=	127 in.⁴
Ad^2 for angles = 4 × 9.73(18.25 − 1.82)²	=	10,508 in.⁴
I of web = 1/12 × 3/8 × 36³	=	1,458 in.⁴
I gross	=	12,093 in.⁴
I of holes = 2 × 1⅞ × 13/16(18.25 − 3.50)² =		750 in.⁴
I net	=	11,343 in.⁴

Weight four angles $= 4 \times 33$ lb $= 132$ lb per ft
Weight web plate $= {}^{490}\!/_{144} \times 36 \times \frac{3}{8} =$ $\underline{\quad 46}$ lb per ft
Weight girder $= 178$ lb per ft
Dead load moment
    $= \frac{1}{8} \times 178 \times 50 \times 50 \times 12$ $= \quad 666{,}000$ in-lb
Live load moment
    $= \frac{1}{4} \times 70{,}000 \times 50 \times 12$ $= \underline{10{,}500{,}000}$ in-lb
                               $M = 11{,}166{,}000$ in-lb

$$f = \frac{Mc}{I} = \frac{11{,}166{,}000 \times 18.25}{11{,}343} = 17{,}950 \text{ psi in bending}$$

$$V = 70{,}000 + 25 \times 178 = 74{,}450 \text{ lb}$$

$$sb = \frac{VQ}{I} = \frac{74{,}450 \times 19.46 \times 16.43}{11{,}343} = 2{,}085 \text{ lb per in.}$$

One $\frac{3}{4}$-in. rivet in bearing $= \frac{3}{4} \times \frac{3}{8} \times 40{,}000 = 11{,}250$ lb
Maximum pitch of $\frac{3}{4}$-in. rivets $= 11{,}250/2{,}085 = 5.40$ in.

7.18   Trial design by flange-area method.
     Let girder be $\frac{1}{15}$ by 60 ft or 4 ft in depth
     Shear at end $= 5{,}000 \times 30 = 150{,}000$ lb
     Web area needed at 13,500 psi $= 150{,}000/13{,}500 = 11.1$ sq in.

Use web plate $48 \times \frac{3}{8} = 18.0$ sq in.

Assume lateral support of girder sufficient to permit a bending stress of 18,000 psi in compression.
Moment $=$ flange area $\times 18{,}000 \times 45$ in. (couple arm)
        $= \frac{1}{8} \times 5{,}000 \times 60 \times 60 \times 12 = 27{,}000{,}000$
                                         in-lb

Flange area $= \dfrac{27{,}000{,}000}{18{,}000 \times 45}$ $= 33.33$ sq in.

$\frac{1}{8}$ area of web $=$ flange area $= \underline{\quad 2.25}$ sq in.
                                      31.08 sq in.
Area of two angles $8 \times 6 \times \frac{3}{4}$ $= \underline{19.88}$ sq in.
                                      11.20 sq in.
Area of one cover plate $18 \times \frac{5}{8} = $ 11.25 sq in.

Check design by gross-moment-of-inertia method.
Distance back to back of 8-in. legs of angles = $48\frac{1}{2}$ in.

I of two angles about horizontal center of gravity of angles	= 68 in.^{4}
Plus area of two angles × distance squared to N.A. of web = $19.88(48.50/2 - 1.56)^2$	= 10,250 in.^{4}
I of cover plate $= 18 \times \frac{5}{8} \times (48.50/2 + \frac{1}{2} \times \frac{5}{8})^2$	= 6,775 in.^{4}
I of top flange	= 17,093 in.^{4}
I of bottom flange	= 17,093 in.^{4}
I of web plate $= \frac{1}{12} \times \frac{3}{8} \times 48^3$	= 3,456 in.^{4}
I total	= 37,642 in.^{4}

$$f = \frac{Mc}{I} = \frac{27,000,000 \text{ in.-lb} \times 24.875 \text{ in.}}{37,642 \text{ in.}^4} = 17,850 \text{ psi, which is OK}$$

7.19 Center of gravity of T (above bottom of W)   18.16 in.

W: 28.22 in.^{2} × 9.08 in. = 256 in.^{3}	+ 0.40 in.
	18.56 in.
Channel: 9.90 in.^{2} × 17.77 in. = 176 in.^{3}	− 0.79 in.
38.12 in.^{2} (11.32 in.) = 432 in.^{3}	17.77 in.
	−11.32 in.
	6.45 in.

I of T (about center of gravity):

Channel: I about channel center of gravity =	8.2 in.^{4}
$A \times d^2 = 9.90 \times 6.45^2$ =	412.0 in.^{4}
W: I about W center of gravity =	1,674.7 in.^{4}
$A \times d^2 = 28.22 \times 2.24^2$ =	141.5 in.^{4}
I =	2,236.4 in.^{4}

Moment:

$$S = \frac{I}{c} \text{ (for top flange)}$$

$$= 2,236.4/7.24 = 309 \text{ in.}^3$$

$$M = f \times S = 17,050 \text{ psi} \times 309 \text{ in.}^3 = 5,260,000 \text{ in.-lb}$$

$$S = \frac{I}{c} \text{ (for bottom flange)}$$

$$= 2,236.4/11.32 = 197 \text{ in.}^3$$

$$M = f \times S = 20,000 \text{ psi} \times 197 \text{ in.}^3 = 3,940,000 \text{ in.-lb(use)}$$

$$\frac{P \times 30}{4} \times 12 + \frac{1}{8} \times 129.9 \times 30 \times 30 \times 12$$

$$= 3,940,000 \text{ lb}$$
$$90P = 3,940,000 - 175,500 = 3,764,500$$
$$P = \mathbf{41,900 \text{ lb}}$$

Shear:

Two $\frac{5}{8}$-in. rivets in single shear $= 2 \times 0.307 \text{ in.}^2 \times 15,000 \text{ psi}$
$$= 9,200 \text{ lb}$$

Horizontal shear per inch of beam (at end) may not

exceed $\dfrac{9,200 \text{ lb}}{6 \text{ in.}} = 1,534 \text{ lb per in.} = sb = \dfrac{VQ}{I}$

$V = \dfrac{sh \times I}{Q} = \dfrac{1,534 \times 2,236.4}{9.90 \times 6.45} = 53,600 \text{ lb} = \dfrac{P}{2} + 15 \text{ ft}$

$$\times 129.9 \text{ ft-lb}$$

$P = 2(53,600 - 1,950) = \mathbf{103,300 \text{ lb}} > \mathbf{41,900 \text{ lb}}$

Deflection:

$$= \frac{1}{360} \times 30 \times 12 = \frac{P \times (30 \times 12)^3}{48 \times 30,000,000 \times 2,236.4}$$

$$P = \frac{4 \times 1,000,000 \times 2,236.4}{30 \times 30 \times 12 \times 12} = \mathbf{69,000 \text{ lb}} > \mathbf{41,900 \text{ lb}}$$

Permissible $p = \mathbf{41,900 \text{ lb}}$

7.20   (Refer to answer 1.616.)

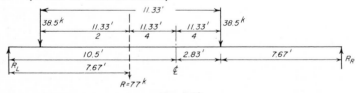

Fig. 7.20

Maximum live-load moment

$$= \frac{7.67 \text{ ft}}{21 \text{ ft}} \times 77 \text{ kips} \times 7.67 \text{ ft} \qquad\qquad = \mathbf{215 \text{ ft-kips}}$$

Dead-load moment at center line
$$= \frac{1}{8} \times 100.7 \text{ ft-lb} \times 21 \text{ ft} \times 21 \text{ ft} \qquad = \underline{\mathbf{5.5 \text{ ft-kips}}}$$

Total moment $\qquad\qquad\qquad\qquad\qquad\qquad\qquad = \mathbf{220.5 \text{ ft-kips}}$

Maximum end live-load shear

$$= 38.5 \text{ kips} + \frac{(21 \text{ ft} - 11.33 \text{ ft})}{21 \text{ ft}} \times 38.5 \text{ kips} = 56.2 \text{ kips}$$

Dead load shear = 10.5 ft × 100.7 lb   =   1.06 kips

Total shear   = 57.26 kips

Center of gravity of T (above bottom of W)   24.280 in.

W: 23.54 in.^{2} × 12 in.   = 282.5 in.^{3} − 0.70 in.

Channel: 6.03 in.^{2} × 23.58 in. = 142.0 in.^{3} 23.58 in.

           29.57 in.^{2} (14.35 in.)   = 424.5   −14.35 in.

                                            9.23 in.

I of T [about horizontal line through center of
gravity]   14.35 in.

                                          −12.00 in.

                                          2.35 in.

Channel: Ad^2 = 6.03 in.^{2} × 9.23 in. × 9.23 in. =   514 in.^{4}

         I about its center of gravity   =   4 in.^{4}

W: Ad^2 = 23.54 × 2.35 × 2.35   =   130 in.^{4}

      I about its center of gravity   = 2,230 in.^{4}

      I total   = 2,878 in.^{4}

$$S = \frac{I}{c} = \frac{2,878}{14.35} = 201 \text{ in.}^3 \text{ (section modulus for bottom}$$
fiber)

$$S = \frac{2,878}{9.93} = 290 \text{ in.}^3 \text{ (section modulus for top fiber)}$$

$$f(\text{bottom}) = \frac{Mc}{I} = \frac{220,500 \text{ ft-lb} \times 12 \text{ in. per ft} \times 14.35 \text{ in.}}{2,878 \text{ in.}^4}$$

$$= 13,170 \text{ psi}$$

$$f(\text{top}) = \frac{220,500 \times 12 \times 9.93}{2,878} = 9,130 \text{ psi}$$

$f(\text{allowable})$ for $\dfrac{L}{b} = 21 = 18,000$ psi in compression (also
tension)

Bridge crane runway girders are subjected to lateral loads = to
10 per cent of suspended load on crane plus an increase of vertical
live loads due to impact. Full data not given in question, but
fiber stresses will increase due to impact plus bending about

vertical axis. Above bending stresses do not appear to be unreasonable. Horizontal shear per running inch of beam at plane between channel and

$$W = sb = \frac{VQ}{I} = \frac{57,260 \text{ lb} \times 6.03 \text{ in.}^2 \times 9.23 \text{ in.}}{2,878 \text{ in.}^4}$$

$$= 1,108 \text{ lb per in.}$$

Single shear value of two $\frac{3}{4}$-in. rivets

$$= 2 \times 0.4418 \text{ in.}^2 \times 15,000 \text{ psi} = 13,250 \text{ lb}$$

$$\frac{13,250}{1,108} = 12 \text{ in.}$$

Use **6** in. maximum rivet pitch.

7.21 $R_L = R_R = \frac{3}{8}wL = \frac{3}{8} \times 10 \times 12 = 45$ kips
$R_C = 2 \times \frac{5}{8}wL = \frac{5}{4} \times 10 \times 12 = 150$ kips
M at R_C (no deflection of R_C) $= 45$ kips $\times$ 12 ft $-$ 10 kips per ft $\times$ 12 ft $\times$ 6 ft $= -180$ ft-kips

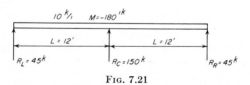

FIG. 7.21

1. Remove R_C and under 10 kips per ft load on 24 ft simple span Δ center line

$$= \frac{5}{384} \times \frac{240,000 \text{ lb} \times 24 \text{ ft} \times 24 \text{ ft} \times 24 \text{ ft} \times 1,728}{30,000,000 \text{ psi} \times 1,373 \text{ in.}^4}$$

$$= 1.81 \text{ in.}$$

2. R_C must push the load up 1.81 in. for no deflection or 1.31 in. for a $\frac{1}{2}$-in. deflection

$$R_C \text{ for a } \frac{1}{2}\text{-in. deflection} = \frac{1.31}{1.81} \times 150 \text{ kips} = 108.5 \text{ kips}$$

and $R_L = R_R = 120 - \dfrac{108.5}{2} = 120 - 54.25 = 65.75$ kips.

M at R_C ($\frac{1}{2}$-in. deflection of R_C)
$= 65.75$ kips $\times$ 12 ft $-$ 10 kips per ft $\times$ 12 ft $\times$ 6 ft
$= +69$ ft-kips

3. Change in bending moment at $R_C = 180 + 69 = 249$ ft-kips. This may be found by $(65.75 - 45) \times 12 = 249$ ft-kips.

Alternate: Change in bending moment at $R_C = \Delta$ bending

$$\text{moment}_C = \frac{\Delta_{R_C}}{1,000} \times \frac{2 \times 12}{4}$$

where Δ_{R_C} is found from the deflection at C

$$\Delta_C = \frac{1}{2} \text{ in.} = \frac{1}{48} \times \frac{\Delta_{R_C}(24)^3 \times 1728}{30,000,000 \times 1,373}$$

$$\frac{\Delta_{R_C}}{1,000} = 41.5 \text{ kips}$$

$$\Delta \text{ bending moment}_C = \frac{41.5}{2} \times 12 = \textbf{249.0 ft-kips}$$

7.22   Determine reactions:

$$M_{R_L} = 0 \qquad 6R_R - 4 \times 2,400 - 2 \times 4,800 = 0$$
$$R_R = 3,200 \text{ lb}$$
$$M_R = 0 \qquad 6R_L - 4 \times 4,800 - 2 \times 2,400 = 0$$
$$R_L = 4,000 \text{ lb}$$

Check reactions: $F_v = 0 \qquad 3,200 + 4,000 - 4,800$
$$- 2,400 = 0 \qquad \text{OK}$$

$$I_x = \frac{bh^3}{12} = \frac{1 \times 6^3}{12} = 18 \text{ in.}^4 \qquad Q = A'\bar{y} = (1 \times 1.5)2.25$$
$$= 3.375 \text{ in.}^3$$

$$S_{xy} = \frac{VQ}{I_x b} = \frac{4,000 \times 3.375}{18 \times 1} = 750 \text{ psi}$$
$$M_{a-a} = 1.5 \times 4,000 = 6,000 \text{ ft-lb} = 72,000 \text{ in.-lb}$$
$$S_x = \frac{Mc}{I_x} = \frac{72,000 \times 1.5}{18} = 6,000 \text{ psi}$$

Maximum normal stress:

$$S_1 = \frac{S_x}{2} + \sqrt{\left(\frac{S_x}{2}\right)^2 + S_{xy}^2} = \frac{6,000}{2} + \sqrt{\left(\frac{6,000}{2}\right)^2 + (750)^2}$$
$$= 6,090\text{-psi tension} \qquad \textbf{1d and 2a}$$

Minimum normal stress:

$$S_2 = \frac{S_x}{2} - \sqrt{\left(\frac{S_x}{2}\right)^2 + S_{xy}{}^2} = \frac{6,000}{2} - \sqrt{\left(\frac{6,000}{2}\right)^2 + (750)^2}$$

$$= -90\text{-psi compression} \qquad \textbf{3a} \text{ and } \textbf{4b}$$

Maximum shearing stress:

$$S_s = \frac{S_1 - S_2}{2} = \frac{6,090 - (-90)}{2} = 3,090 \text{ psi} \qquad \textbf{5c}$$

$$\tan 2\theta_n = -\frac{S_{xy}}{(S_x - S_y)/2} = -\frac{750}{6,000/2} = -0.25$$

$2\theta_n = 166°$ and $346°$ or $\theta_n = 83°$ and $173°$.

7.23 1. Beam *EF* carries a moment of 499.7 ft-kips and end shear = 20.7 kips. Use A36 structural steel.

$$\text{Section modulus required} = S \text{ in.}^3 = \frac{M \text{ in-lb}}{f \text{ psi}}$$

$$= \frac{499,700 \times 12}{22,000} = 273 \text{ in.}^3 \ (f = 0.6F_y = 22,000)$$

Use W24 × 110, $S = 274.4$ in.^{3} Use two connection angles 4 × 3½ × $\frac{5}{16}$ and four ¾-in. A307 bolts per angle.

2. Beam *CD* carries a moment of 251 ft-kips and end shear = 25.1 kips

$$S = \frac{251,000 \times 12}{22,000} = 137 \text{ in.}^3$$

Use W 24 × 68, $S = 153.1$ in.^{3}
Use same end connections as for 1, above.

3. Beam *AB* carries a moment of 931 ft-kips and end shear = 39.4 kips

$$S = \frac{931,000 \times 12}{22,000} = 508 \text{ in.}^3$$

Use W 36 × 160, $S = 541.0$ in.^{3} Use two connection angles 4 × 3½ × $\frac{5}{16}$ and six ¾-in. A307 bolts per angle.

7.24 (The following formulas for deflections of simple beams due to two concentrated loads at the third points are given in many texts and handbooks.)

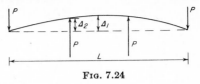

Fɪɢ. 7.24

First, neglect center line load of $4P$

$$\Delta_1(\text{at center line})\uparrow = \frac{P \times L/3}{24EI}\left(3L^2 - 4\frac{L^2}{9}\right)$$

$$= \frac{PL}{3 \times 9 \times 24EI}(27L^2 - 4L^2) = \frac{23PL^3}{648EI}$$

$$\Delta_2(\text{at third point})\uparrow = \frac{P \times L/3}{6EI}\left(3L \times \frac{L}{3} - 3\frac{L^2}{9} - \frac{L^2}{9}\right)$$

$$= \frac{4PL}{4 \times 3 \times 9 \times 6EI}(9L^2 - 3L^2 - L^2) = \frac{20PL^3}{648EI}$$

Deflection of center line point above third points $= \dfrac{3PL^3}{648EI}$

Second, neglect end loads of P

$$\Delta_3(\text{at center line})\downarrow = \frac{Fx^3}{48EI} = \frac{4P \times L^3/27}{48EI} = \frac{2PL^3}{648EI} = \text{deflection}$$

of center line point below third points

Resultant deflection at center line due to all P loads is $\dfrac{PL^3}{648EI}$ *upward.*

7.25 Solve by use of conjugate-beam method, requiring no handbook formulas.

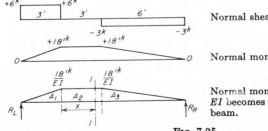

Normal shear diagram.

Normal moment diagram.

Normal moment diagram divided by EI becomes the loading for conjugate beam.

Fɪɢ. 7.25

R_L for conjugate beam, M about R_R

Area no.	Area	Arm (from R_R)	Area × arm = product

$$A_1 \; = \frac{18}{EI} \times 3 \times \tfrac{1}{2} = \frac{27}{EI} \times (9 + \tfrac{1}{3} \times 3) = \frac{270}{EI}$$

$$A_2 \; = \frac{18}{EI} \times 3 \qquad\;\; = \frac{54}{EI} \times (6 + \tfrac{1}{2} \times 3) = \frac{405}{EI}$$

$$A_3 \; = \frac{18}{EI} \times 6 \times \tfrac{1}{2} = \frac{54}{EI} \times (\tfrac{2}{3} \times 6) \qquad\; = \frac{216}{EI}$$

$$\Sigma \text{ areas} = \qquad\qquad\qquad \frac{135}{EI} \qquad\qquad \Sigma M = \frac{891}{EI}$$

$$\text{and } R_L = \frac{891}{12EI} = \frac{74.25}{EI} \left(\frac{\text{ft}^2\,\text{kips}}{EI} \right)$$

R_R for conjugate beam, M about R_L

Area	×	Arm	=	M

$$A_1 = \frac{27}{EI} \times (\tfrac{2}{3} \times 3) \qquad\quad = \frac{54}{EI}$$

$$A_2 = \frac{54}{EI} \times (3 + \tfrac{1}{2} \times 3) \; = \frac{243}{EI}$$

$$A_3 = \frac{54}{EI} \times (6 + \tfrac{1}{3} \times 6) \; = \frac{432}{EI}$$

$$\Sigma M = \frac{729}{EI} \text{ and } R_R = \frac{729}{12EI} = \frac{60.75}{EI}$$

$$R_L + R_R = \frac{135}{EI} = \frac{74.25}{EI} + \frac{60.75}{EI}$$

Find where shear for conjugate-beam loading passes through zero.

$$R_L - A_1 = \frac{74.25}{EI} - \frac{27}{EI} = \frac{47.25}{EI}$$

which is less than $\dfrac{54}{EI}$ or A_2

Therefore shear passes through zero x distance to right of A_1 area. So

$$\frac{18x}{EI} = \frac{47.25}{EI} \text{ and } x = \frac{47.25}{18} = 2.625 \text{ ft}$$

and deflection is maximum at section 1-1, 5.625 ft to right of R_L.
Δ max = moment of conjugate-beam loading about section 1-1.

$$= +R_L \times 5.625 - A_1 \times (1 + 2.625) - \frac{47.25}{EI} \times \frac{2.625}{2}$$

$$= \frac{74.25}{EI} \times 5.625 - \frac{27}{EI} \times 3.625 - \frac{47.25}{EI} \times 1.312$$

$$= (+417.65 - 97.87 - 62.01) \times \frac{1}{EI} = \frac{257.77}{EI} \left(\frac{\text{ft}^3 \text{ kips}}{EI}\right)$$

7.26   For simplicity, divide this problem into two parts and solve by the conjugate-beam method.

1. *Concentrated load*

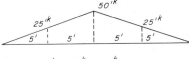

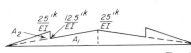

Max $M = \frac{1}{4} \times 10 \times 20 = 50$ ft-kips—normal moment diagram.

Conjugate beam loading—normal moment diagram divided by EI.

Fɪɢ. 7.26a

$$A_1 = \frac{1}{2} \times 10 \times \frac{25}{EI} = \frac{125}{EI}$$

$$A_2 = \frac{1}{2} \times 5 \times \frac{12.5}{EI} = \frac{31.25}{EI},$$

$$R_L = A_1 + A_2 = \frac{156.25}{EI}$$

$$M \text{ at center line} = \frac{156.25}{EI} \times 10 - \frac{125}{EI} \times \frac{10}{3} - \frac{31.25}{EI}\left(5 + \frac{5}{3}\right)$$

$$= (+1,562.5 - 416.7 - 208.3) \times \frac{1}{EI}$$

$$= \frac{937.5}{EI} \left(\frac{\text{ft}^3 \text{ kips}}{EI}\right)$$

2. Uniform load

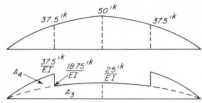

Max $M = \frac{1}{8} \times 1 \times 20 \times 20 = 50$ ft-kips—normal moment diagram is a parabola and M at quarter point = $\frac{3}{4}M$ at center line.

Conjugate beam loading — normal moment diagram divided by EI

Fig. 7. 26b

$A_3 = \dfrac{2}{3} \times \dfrac{25}{EI} \times 10 = \dfrac{166.67}{EI}$ (area under parabola from vertex outward = two-thirds area of enclosing rectangle)

$A_4 = \dfrac{1}{2} \times 5 \times \dfrac{18.75}{EI} = \dfrac{46.87}{EI}$ (it is close enough to treat area as a triangle)

$R_L = A_3 + A_4 = \dfrac{213.54}{EI},$ and M at center line

$= + \dfrac{213.54 \times 10}{EI} - \dfrac{46.875}{EI} \times \left(5 + \dfrac{5}{3}\right) - \dfrac{166.67}{EI} \times \dfrac{3}{8} \times 10$

M at center line $= (+2{,}135.4 - 312.5 - 625) \times \dfrac{1}{EI}$

$= \dfrac{1{,}197.9}{EI} \left(\dfrac{\text{ft}^3 \text{ kips}}{EI}\right)$

Total Δ at center line $= (937.5 + 1197.9) \times \dfrac{1}{EI}$

$= \dfrac{2{,}135.4}{EI} \left(\dfrac{\text{ft}^3 \text{ kips}}{EI}\right)$

(To understand the conversion of Δ to inches, assume I of beam $= 344$ in.[4])

Δ max $= \dfrac{2{,}135{,}400 \times 1728}{344 \times 30{,}000{,}000} = 0.357$ in. (use as 0.36 in. or 0.4 in.)

(The exact way in getting and dealing with A_4 would have shown in the fourth figure to the right of the decimal by about -1.)

7.27 Take statical moment from back of plate to find center of gravity.

Member	Area	$\times$	Arm	=	Product
Plate	4.88	$\times$	$\frac{3}{16}$	=	0.915
2 Channels	6.72	$\times$	$4\frac{3}{8}$		29.400
	11.60 in.^{2}	$\times$	(2.61 in.)	=	30.315 in.^{3}

Center of gravity lies 2.61 in. from back of plate and on axis of symmetry.

$I_{(1-1)}$

Plates $\frac{1}{12} \times 13 \times \frac{3}{8} \times \frac{3}{8} \times \frac{3}{8}$ =   0.06 in.^{4}

$+4.88 \times (2.61 - 0.19)^2$ =   28.80

2 channels 2×32.3 =   64.60

$+6.72 \times (4.37 - 2.61)^2$ =   20.82

I (total)   = 114.28 in.^{4}

$r^2_{(1-1)} = \dfrac{I}{A} = \dfrac{114.28}{11.60}$ =   9.85 in.^{2}

$\dfrac{L^2}{r^2} = \dfrac{(20 \times 12)^2}{9.85}$ = 5,840

Permissible fiber stress = $17,000 - 0.485 \times 5,840 = 14,170$ psi

Allowable load = $A_s f_s = 11.60 \times 14,170 =$ **164,300** lb

$I_{(2-2)}$

Plate $\frac{1}{12} \times \frac{3}{8} \times 13 \times 13 \times 13$ =   68.7 in.^{4}

2 channels 2×1.3 =   2.6

$+6.72 \times 4.58^2$ = 141.0

                                                                   212.3 in.^{4}

$r^2_{(2-2)} = \dfrac{212.3}{11.60}$ = 18.3 > 9.85

Use $r^2_{(1-1)}$ only.

7.28 15 channel 50, $A = 14.64$ in.^{2}, $x = 0.80$ in.,
$I_1 = 401.4$ in.^{4}, $I_2 = 11.2$ in.^{4}

Center-of-gravity distance from top of cover plate = 5.42 in.

Member		Area	×	Arm	=	Product
Plate 22 × $\frac{3}{4}$	=	16.50	×	$\frac{3}{8}$	=	6.18
2 Channels	=	29.28	×	8.25	=	242.00
		45.78 in.^{2}	×	(5.42 in.)	=	248.2 in.^{3}

$I_{(1\text{-}1)}$
Plate = $\frac{1}{12} \times 22 \times \frac{3}{4} \times \frac{3}{4} \times \frac{3}{4}$ =   0.8
$+16.50\,(5.42 - 0.37)^2$ =   420.2
2 channels = 2×401.4 =   802.8
$+29.28\,(8.25 - 5.42)^2$ =   234.2
$I_{(1\text{-}1)}$ = 1,458.0 in.^{4}
$r^2_{(1\text{-}1)} = \dfrac{I}{A} = \dfrac{1,458.0}{45.78}$ =   31.8 in.^{2}

$I_{(2\text{-}2)}$ (axis of symmetry)
Plate = $\frac{1}{12} \times \frac{3}{4} \times 22 \times 22 \times 22$ =   665.0
2 channels = 2×11.2 =   22.4
$+29.28 \times (6.8)^2$ = 1,358.0
$I_{(2\text{-}2)}$ = 2,045.4 in.^{4}
$r^2_{(2\text{-}2)} = \dfrac{2,045.4}{45.78}$ = 44.7 in.$^2 > 31.8$

Permissible fiber stress = $15,000 - \dfrac{(25 \times 12)^2}{4 \times 31.8} = 14,292$ psi

Allowable load = $45.78 \times 14,292 =$ **655,000** lb

7.29 A of four angles   =   9.92
        A of $8 \times \frac{3}{8}$ plate   =   3.00
                                      12.92 in.^{2}

        Plate = $\frac{1}{12} \times 8 \times \frac{3}{8} \times \frac{3}{8} \times \frac{3}{8}$ =   0.04
        Four angles = 4×4 = 16.00
        $+9.92\,(0.19 + 1.28)^2$ = 21.40
        $I_{(1\text{-}1)}$ = 37.44 in.^{4}
        $r^2_{(1\text{-}1)} = \dfrac{I}{A} = \dfrac{37.44}{12.92} = 2.90$ in.^{2}, $r_{(1\text{-}1)} = 1.70$ in.

Use $K = 1$, $\dfrac{Kl}{r} = \dfrac{1 \times 15 \times 12}{1.70} = 106 < 120 < C_c$

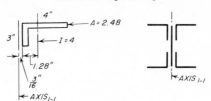

FIG. 7.29

$$C_c{}^2 = \frac{2\pi^2 E}{F_y} = \frac{2\pi^2 29,000,000}{36,000}, \; C_c = 126$$

$$\text{F.S.} = \frac{5}{3} + \frac{3}{8}\left(\frac{Kl/r}{C_c}\right) - \left(\frac{Kl/r}{2C_c}\right)^3 = 1.909$$

$$F_a = \left[1 - \frac{1}{2}\left(\frac{Kl/r}{C_c}\right)^2\right] \times F_y = 12,200 \text{ psi}$$

Allowable concentric load $= 12.92 \times 12,200 = \mathbf{157{,}600}$ lb

7.30 Find A, I, and r of section.

$$\begin{aligned}
A &= 84.37 + 2 \times 22.00 \times 2 = & 172.37 \text{ in.}^2 \\
I_{(1-1)}:\ \text{W } 14 \times 287 & & = 3{,}912.1 \text{ in.}^4 \\
\text{Two } 22 \times 2 \text{ in. plates} & & \\
2 \times \tfrac{1}{12} \times 22 \times 2 \times 2 \times 2 &= & 29.4 \\
2 \times 44 \times 9.40 \times 9.40 &= & 7{,}770.0 \\
& & \overline{11{,}711.5 \text{ in.}^4} \\
I_{(2-2)}:\ \text{W } 14 \times 287 & = & 1{,}466.5 \\
\text{Plates } 2 \times \tfrac{1}{12} \times 2 \times 22^3 & = & 3{,}550.0 \\
& & \overline{5{,}016.5 \text{ in.}^4}
\end{aligned}$$

$$r^2 = \frac{I}{A} = \frac{5{,}016.5}{172.37} = 29.1 \text{ in.}^2, \; r = 5.4 \text{ in.}$$

$$K = 1, \frac{Kl}{r} = 1 \times 20 \times 12/5.4 = 44.5 < 120 < C_c = 126$$

F.S. $= 1.797$, $F_a = 18{,}820$ psi, $F_b = 22{,}000$ psi, $F'_e = 75.44$ psi
(Table 2, Appendix A of AISC Specifications)

$$f_a = \frac{P}{A} = \frac{1{,}250{,}000}{172.37} = 7{,}250 \text{ psi}$$

$$f_b = \frac{Pec}{I} = \frac{1{,}250{,}000 \times 8.40 \times 10.40}{11{,}711.5} = 9{,}325 \text{ psi}$$

Assume $C_m = 1$. When $f_a/F_a \geq 0.15$, the AISC column formula is

$$\frac{f_a}{F_a} + \frac{C_m f_b}{[1 - (f_a/F_e')]F_b} \leq 1 \qquad \begin{array}{c} \text{AISC 1.6.1} \\ (1.6\text{-}1a) \end{array}$$

Check: $f_a/F_a = 7{,}250/18{,}820 = 0.385 > 0.15$.
The column formula becomes

$$0.385 + \frac{9{,}325}{[1 - (7{,}250/75{,}440)]22{,}000} = 0.385 + 0.469$$

$$= 0.854 < 1$$

Assumed section is OK.

7.31   Check the slenderness ratios about axes x and y.

$$\frac{K_x L}{r_x} = \frac{1.80(15)(12)}{6.42} = 50.47$$

$$\frac{K_y L}{r_y} = \frac{0.80(15)(12)}{4.01} = 35.91$$

The slenderness ratio about the x axis controls.

$$C_c = \sqrt{\frac{2\pi^2 E}{F_y}} = \sqrt{\frac{2(3.14)^2(29{,}000)}{36}} = 126.1$$

Find F_a. From above, $K_x L/r_x < C_c$, therefore,

$$F_a = \frac{\left[1 - \dfrac{(K_x L/r_x)^2}{2C_c^2}\right]F_y}{\dfrac{5}{3} + \dfrac{3(K_x L/r_x)}{8C_c} - \dfrac{(K_x L/r_x)^3}{8C_c^3}} \qquad \begin{array}{c} \text{AISC 1.5.1.3.1} \\ (1.5\text{-}1) \end{array}$$

$$\frac{(K_x L/r_x)^2}{2C_c^2} = \frac{(50.47)^2}{2(126.1)^2} = 0.08$$

$$\frac{3(K_x L/r_x)}{8C_c} = \frac{3(50.47)}{8(126.1)} = 0.15$$

$$\frac{(K_x L/r_x)^3}{8C_c^3} = \frac{(50.47)^3}{8(126.1)^3} = 0.01$$

$$F_a = \frac{(1.00 - 0.08)(36)}{1.66 + 0.15 - 0.01} = 18.30 \text{ ksi}$$

Unsupported length of compression flange $= \dfrac{76b_f}{\sqrt{F_y}} = \dfrac{76(15.6)}{\sqrt{36}} =$

197.60 in. or 16.47 ft $> 15'\text{-}0''$

Therefore, $F_b = 0.66F_y = 0.66(36) = 24$ ksi

$$f_a = \frac{P}{A} = \frac{500}{49.1} = 10.18 \text{ ksi}$$

$$f_{b_x} = \frac{M}{S_x} = \frac{220 \times 12}{267} = 9.89 \text{ ksi}$$

$$\frac{f_a}{F_a} = \frac{10.18}{18.30} = 0.56 > 0.15$$

Therefore, the following interaction formula has to be satisfied:

$$\frac{f_a}{F_a} + \frac{C_{mx}f_{bx}}{[1 - (f_a/F_{ex})]F_{bx}} \leq 1 \qquad \begin{array}{c} \text{AISC 1.6.1} \\ \text{(1.6-1a)} \end{array}$$

Since column is subject to sidesway, $C_m = 0.85$

$$F_e' = \frac{12\pi^2 E}{23\left(\dfrac{KL_b}{r_b}\right)^2} = \frac{12 \times 3.14^2 \times 29 \times 10^3}{23\left(\dfrac{1.80 \times 15 \times 12}{6.42}\right)^2}$$

$$= 58.56 \text{ ksi} \qquad \text{AISC 1.6.1}$$

$$C_{mx}f_{b_x} = 0.85 \times 9.89 = 8.42$$

$$\left(1 - \frac{f_a}{F_e'}\right)F_b = \left(1 - \frac{10.18}{58.56}\right)24 = 19.82$$

Substituting the above values in the interaction formula,

$$0.56 + \frac{8.42}{19.82} = 0.98 < 1.00$$

Therefore the column is **adequate.**

7.32 A ⅞-in. rivet in double shear and in double bearing on a ½-in. plate is good for 18,040 lb. A36 steel.
Four rivets required at each end.
Assume two angles 8 × 4 × 1 with 4-in. legs vertical and stitch-riveted.

Weight = 74.8 lb per ft
$\quad\quad A = 22.0$ in.²
$\quad\quad I = 23.3$ in.⁴
$\quad\quad x = 1.05$ in.
$\quad\quad r = 1.03$ in.

Rivet gage 4-in. leg = 2.50 in.

$$\frac{l}{r} = \frac{10 \times 12}{1.03} = 116.5 < 200, \text{ which is OK}$$

Distance of center of gravity for net section from back of 8-in. legs.

$$y \times A_{(net)} = y[A_{(gross)} - A_{(holes)}] = A_{(gross)} \times \text{arm} - A_{(holes)} \times \text{arm}$$
$$y(22.00 - 2.00) = 22.00 \times 1.05 - 2.00 \times 2.50$$
$$y = \frac{(23.10 - 5.00)}{20.00} = \frac{18.10}{20.00} = 0.905 \text{ in.}$$

$$
\begin{aligned}
I_{(net\ section)} &= I_{(gross\ section)} - I_{(hole)} - A_{(net)} \times d^2 \\
&= 23.30 - 2.00(2.50 - 1.05)^2 - 20.00(1.050 - 0.905)^2 \\
&= 23.30 - 2.00(1.45)^2 - 20.00(0.145)^2 \\
&= 23.30 - 4.20 - 0.42 = 18.68 \text{ in.}^4
\end{aligned}
$$

Tension stress at end of member caused by

1. Direct stress $= \dfrac{P}{A} = \dfrac{65,000}{20.00}$ $\qquad\qquad = 3,250$ psi

2. Eccentric bending $= \dfrac{Pec}{I}$

$$= \frac{65,000}{18.68}(2.50 - 0.905)(4.00 - 0.905)$$
$$= 3,480 \times 1.595 \times 3.095 \qquad\qquad = 17,200 \text{ psi}$$

3. Negative bending at end due to weight of

member $= \dfrac{Mc}{I} = \dfrac{1}{12}\, wl^2 \times \dfrac{c}{I}$

$$= \tfrac{1}{12} \times 74.8 \times 10 \times 10 \times 12 \times \frac{3.095}{18.68} \qquad = \underline{1,240 \text{ psi}}$$

$$21,690 \text{ psi}$$

7.33 Assume the weight of the member with rivets and lacing $=$ 112 lb per ft.

Moment due to dead weight $= \tfrac{1}{12} \times 112 \times 25 \times 25 \times 12 = 70,000$ in.-lb

Assume two 15-in. channels at 33.9 lb per ft with four holes out of web for $\tfrac{7}{8}$-in. rivets.

$$\frac{l}{r} = \frac{25 \times 12}{5.62} = 53.4,\ \text{which is OK for primary tension}$$
member

Area (gross) of 1 channel $\qquad = 9.90$ sq in.
Area of holes $= 4 \times 1 \times 0.400 = \underline{1.60}$ sq in.
Area (net) $\qquad\qquad\qquad = 8.30$ sq in.

$$f = \frac{P}{A_{(n)}} + \frac{Mc}{I_{(g)}} = \frac{350{,}000}{2 \times 8.30} + \frac{70{,}000 \times 7.5}{2 \times 312.5}$$

$$= 21{,}100 + 840 = 21{,}940 \text{ psi}$$

$\therefore$ Assumed section is OK

7.34   Try an ST 8 WF 22.5 lb per ft.

$A = 6.62$ in.^{2}

$d = 8.06$ in.

$I = 37.8$ in.^{4}

$r = 2.39$ in. and 1.52 in.

$y = 1.87$ in.

Negative moment due to weight

$= \frac{1}{12}wl^2 = \frac{1}{12} \times 22.5 \times 8 \times 8 \times 12 = $ 1,440 in-lb

Negative moment due to purlin $= \dfrac{PL}{8}$

$= \dfrac{4{,}000}{8} \times 8 \times 12$ $= $ 48,000 in-lb

M_T $= $ 49,400 in-lb

$\dfrac{Kl}{r} = \dfrac{1 \times 8 \times 12}{1.52} = 63.1$, F.S.

$= 1.838$, $C_m = 0.85$, $F'_e = 37{,}500$ psi

(Table 2, Appendix A, AISC Specifications)

$F_a = 17{,}130$ psi

$F_b = 22{,}000$ psi

$f_a = \dfrac{P}{A} = \dfrac{60{,}000}{6.62} = 9{,}070$ psi

$f_b = \dfrac{Mc}{I} = \dfrac{49{,}440}{37.8} \times (8.06 - 1.87) = 1{,}308 \times 6.19$

$= 8{,}090$ psi

Check: $f_a/F_a = 9{,}070/17{,}130 = 0.529 > 0.15$.

$$\frac{f_a}{F_a} + \frac{C_m f_b}{[1 - (f_a/F'_e)]F_b} = 0.529 + \frac{0.85 \times 8{,}090}{[1 - (9{,}070/37{,}500)]22{,}000}$$

$$= 0.529 + 0.413 = 0.942 < 1$$

Assumed section is OK

7.35   Choose two angles $3\frac{1}{2} \times 3\frac{1}{2} \times \frac{3}{8}$

$A = 4.96$ in.^{2}

$I = 5.8$ in.^{4}

$y = 1.01$ in.

$w = 17$ lb per ft

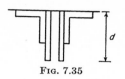

FIG. 7.35

Normal rivet gage for $3\frac{1}{2}$-in. leg is 2 in. from back of angle. Design plates to have center of gravity of plate and angle at rivet gage. Use $\frac{3}{8}$-in. plates.

For one angle and one plate: statical moment from back of L.

$$2.48 \times 1.01 + \frac{3}{8} \times d \times \frac{d}{2} = (\frac{3}{8}d + 2.48) \times 2$$

$$2.50 + 0.1875d^2 = 0.75d + 4.96$$

$$+d^2 = 4d + 13.12$$

$$d^2 - 4d + (4) = +13.12 + (4) = +17.12$$

$$(d - 2) = \pm \sqrt{17.12} = \pm 4.14 \text{ in.}$$

$$d = +4.14 + 2 = 6.14 \text{ in.}$$

Try two plates $6 \times \frac{3}{8}$ in. (weight = 16.0 lb)

$$A = 2 \times \frac{3}{8} \times 6 + 4.96 = 4.50 + 4.96 = 9.46 \text{ in.}^2$$

I (for two angles and two plates)

(weight = 16.0 + 17.0) = (33 lb per ft)

Two angles	=	5.80 in.^{4}
$+4.96 \times 0.99^2$	=	4.86
Two plates = $\frac{1}{12} \times \frac{3}{4} \times 6^3$	=	13.50
$+4.50 \times 1^2$	=	4.50
		28.60 in.^{4}

$$r^2 = \frac{I}{A} = \frac{28.60}{9.46} = 3.02 \text{ in.}^2, \text{ and } \frac{Kl}{r} = \frac{1 \times 8 \times 12}{1.74} = 55.4$$

$\dfrac{Kl}{r}$ in the other plane is 33.1, as the unbraced length is 4×12

$F_a = 17,860$ psi, $C_m = 0.85$, $F'_e = 48,670$ psi (Table 2, Appendix A, AISC Specifications)

$F_b = 22,000$ psi

$$f_a = \frac{P}{A} = \frac{75,000}{9.46} = 7,920 \text{ psi}$$

$$f_b = \frac{Mc}{I} = \frac{(48,000 + \frac{1}{12} \times 33 \times 8 \times 8 \times 12) \times 4}{28.60}$$

$$= \frac{(48,000 + 2,110)}{7.15} = 7,000 \text{ psi}$$

Check: $f_a/F_a = 7,920/17,860 = 0.443 > 0.15$

$$\frac{f_a}{F_a} + \frac{C_m f_b}{[1 - (f_a/F_e')]F_b} = 0.443 + \frac{0.85 \times 7,000}{[1 - (7,920/48,670)]22,000}$$
$$= 0.443 + 0.323 = 0.766 < 1$$

7.36   The vertical load is divided into two components:

Normal (to roof surface) $= \dfrac{2}{\sqrt{5}} \times 160 = 143 \text{ lb per ft} = w_n$

Parallel (to roof surface) $= \dfrac{1}{\sqrt{5}} \times 160 = 71.5 \text{ lb per ft} = \boldsymbol{w_p}$

Bending moment about the x-x axis
$= \tfrac{1}{8} w_{(n)} l^2 = \tfrac{1}{8}(143 + 90) \times 20 \times 20 \times 12 = 140,000 \text{ in-lb}$
Bending moment about the y-y axis
$= \tfrac{1}{10} w_p \left(\dfrac{l}{3}\right)^2 = \tfrac{1}{10} \times 71.5 \times \dfrac{20 \times 20 \times 12}{3 \times 3} = 3,820 \text{ in-lb}$

Assume 7-in. I beam 15.3 lb per ft $S_{x\text{-}x} = 10.4 \text{ in.}^3$
                                              $S_{y\text{-}y} = 1.5 \text{ in.}^3$

$$f = \frac{M_x}{S_x} + \frac{M_y}{S_y} = \frac{140,000}{10.4} + \frac{3,820}{1.5} = 13,460 + 2,540$$
$$= 16,000 \text{ psi}$$

Assumed section is OK.

7.37   Assume a beam weight of 140 lb per ft. Maximum moment induced in the beam is

$$\text{Maximum } M = \frac{5.14 \times 24^2}{8} = 370.08 \text{ ft-kips}$$

Assume allowable stress $F_b = 19.5 \text{ ksi}$.

$$\text{Required } S_x = \frac{370.08 \times 12}{19.5} = 227.74 \text{ in.}^3$$

Try a W 14×142, $S_x = 227 \text{ in.}^3$, $r_T = 4.33 \text{ in.}$, $b_f' = 15.5 \text{ in.}$
Compression flange, see AISC Specifications 1.5.1.4.1e

$$\frac{76b_f}{\sqrt{36}} = 12.67b_f = 12.67 \times 15.5 = \frac{196.39}{12} = 16.37 \text{ ft} < 24 \text{ ft}$$

Design the beam without lateral supports.

$$\frac{L}{r_T} = \frac{24 \times 12}{4.33} = 66.51$$

In this case $C_b = 1$, see AISC Specifications 1.5.1.4.6a

$$\sqrt{\frac{102 \times 10^3 \times 1}{36}} = 53.25 \leq \frac{L}{r_T} \leq \sqrt{\frac{510 \times 10^3 \times 1}{36}}$$
$$= 119 \qquad (1.5\text{-}6a)$$

From the above it can be seen that L/r_T is between the above calculated square roots, therefore the allowable stress F_b is

$$F_b = \left[\frac{2}{3} - \frac{(L/r_T)^2 F_y}{1{,}530 \times 10^3 C_b} \right] F_y \qquad (1.5\text{-}6a)$$
$$= \left[0.67 - \frac{(66.51)^2 \times 36}{1{,}530 \times 10^3 \times 1} \right] 36 = 20.37$$

Actual $f_b = \dfrac{M}{S_x} = \dfrac{370.08 \times 12}{227} = 19.56 < 20.37$. **W 14 × 142** is OK.   **1b**

7.38 Try a bolt of $1\frac{1}{2}$-in. diameter.

Diameter at the root of thread = 1.283 in.
Area at the root of thread   = 1.294 in.²

$$S = \frac{P}{A} = \frac{20{,}000}{1.294} = 15{,}500 \text{ psi}$$
$$S_t = \frac{Tc}{J} = \frac{6{,}000 \times d/2}{\pi d^4/32} = \frac{6{,}000 \times 16}{\pi d^3} = \frac{30{,}600}{(1.283)^3} = 14{,}600 \text{ psi}$$

Maximum normal stress $= \frac{1}{2}(S) + \sqrt{\left(\frac{S}{2}\right)^2 + (S_t)^2}$

$$= \frac{15{,}500}{2} + \sqrt{(7{,}750)^2 + (14{,}600)^2} = 7{,}750 + 16{,}500$$
$$= 24{,}250 \text{ psi}$$

$1\frac{1}{2}$-in. bolt is OK

7.39 $\dfrac{E_s}{E_w} = \dfrac{30{,}000{,}000}{1{,}500{,}000} = 20$

$A \times$ arm = product
$4 \times 6 = 24 \times 3$ = 72
$\frac{1}{2} \times (4 \times 20) = 40 \times 6\frac{1}{4} = 250$

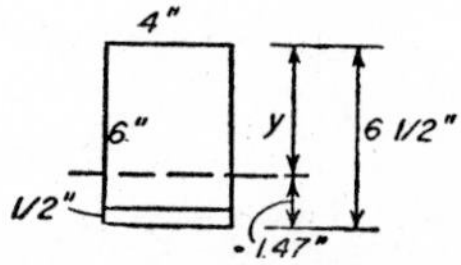

FIG. 7.39

$$64(5.03) = 322$$

$$\frac{f_t}{600} = \frac{1.47}{5.03}, \quad f_t = 600 \times \frac{1.47}{5.03} = 175.1 \text{ (equivalent wood)}$$

Stress in steel when wood is stressed to 600 psi
Compression $= 20 \times 175.1 = 3{,}502$ psi tension

7.40

$f_c' = 3{,}000$ psi,
$f_c = 1{,}350$ psi
$\dfrac{E_s}{E_c} = \dfrac{30{,}000{,}000}{3{,}000{,}000} = n = 10$
$I_w = 5110.3$ in.^{4},
$A = 47.04$ in.^{2}

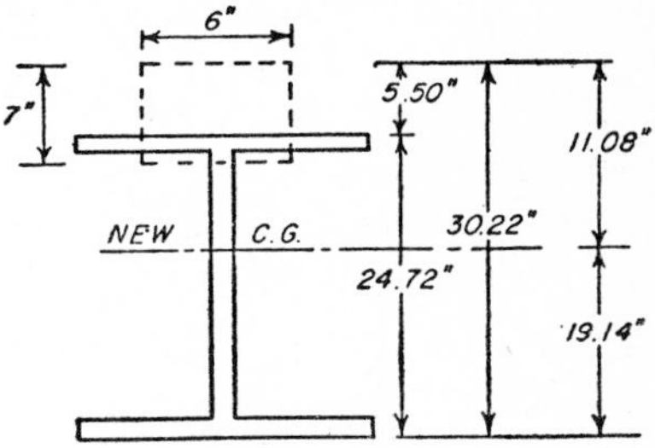

Fig. 7.40

Assume concrete slab to be
equivalent billet of steel $=$
$\dfrac{5 \times 12}{n} = \dfrac{60}{10} = 6$ in. wide. Find new center of gravity (of composite section) from bottom of W.

Member	Area	$\times$	Arm	$=$	Product
W	47.04	$\times$	12.36	$=$	582 in.^{3}
Steel block	42.00	$\times$	26.72	$=$	1,124 in.^{3}
	89.04	$\times$	(19.14 in.)	$=$	1,706 in.^{3}

I of composite section:

steel block $\frac{1}{12} \times 6 \times 7 \times 7 \times 7$ $\quad = \quad 171.5$ in.^{4}
$+42(26.72 - 19.14)^2 = 42 \times 7.58^2 \quad = 2{,}410.0$
W $\quad = 5{,}110.3$
$+47.04(19.14 - 12.36)^2$
$\qquad\qquad = 47.04 \times 6.78^2 = \underline{2{,}160.2}$
$\qquad\qquad\qquad\qquad\qquad\qquad 9{,}852$ in.^{4}

Weight of concrete per foot
$$= \frac{7 \times 60}{144} \times 150 = 438 \text{ lb per ft}$$

Weight of W per foot $\qquad\qquad = \underline{160 \text{ lb per ft}}$
Weight$_T$ $\qquad\qquad\qquad\qquad = 598$ lb per ft or say **600**

Dead moment = $\frac{1}{8} \times 600 \times 40 \times 40 \times 12 = 1,440,000$ in.-lb

Fiber stresses in W $= \frac{Mc}{I} = \frac{1,440,000 \times 12.36}{5,110} = 3,480$ psi

Fiber stress in the bottom of the steel beam due to concentrated load may be $20,000 - 3,480 = 16,520$ psi and the fiber stress in the top of the concrete may be $\frac{1}{10} \times \frac{11.08}{19.14} \times 16,520 = 956$ psi

Live load moment $= \frac{PL}{4} = \frac{P \times 40 \times 12}{4} = 120P$

$$f = \frac{Mc}{I} = 956 \times 10 = \frac{120P \times 11.08}{9,852}$$

$$P = \frac{9,852 \times 79.6}{11.08} = \mathbf{70,815}\ \text{lb}$$

7.41 $sb = \frac{VQ}{I} = \frac{33,000 \times 42 \times 7.58}{9,852} = 1,068$ lb per in.

Two $\frac{7}{8}$-in. rivets in single shear $= 2 \times 0.601 \times 15,000$
$= 18,040$ lb

End spacing of Z bars is $\frac{18,040}{1,068} = 16.9$ in.

Use **17**-in. spacing, maximum.

7.42 The total horizontal force at the wall $= (540 \times 18)/2 = 4,860$ lb per ft.

The maximum moment $= 4,860 \times \frac{18}{3} \times 12$

$= 350,000$ in.-lb   1971 ACI 8.6 (8-1)

The maximum $\rho = 0.75 \, (0.85$ 10.3.2

$\times 0.85 \times \frac{3.5}{50} \times \frac{87}{87 + 50}) = 0.024$ 7.13

Use a steel ratio of $\rho = 0.18\dfrac{f'_c}{f_y} = 0.18 \times \dfrac{3.5}{50} = 0.013$

The resisting moment of a section is

$$M_u = \phi\rho f_y bd^2 \left(1 - 0.59\rho\frac{f_y}{f'_c}\right) \qquad \text{1963 ACI (16-1)}$$

or

$$d^2 = \frac{350{,}000}{0.90 \times 0.013 \times 50{,}000 \times 12\left(1 - \dfrac{0.59 \times 0.013 \times 50{,}000}{3{,}500}\right)}$$

or $d = 7.48$ in.

Required d for shear:

$$v_c = 2\sqrt{f'_c} = 2\sqrt{3{,}500} = 118.32 \text{ psi} \qquad \text{1971 ACI } 11.4.1$$
$$d = \frac{4{,}860}{118.32 \times 12} = 3.42 \text{ in.}$$

Bending controls. Assume a $d = 7.5$ in. Then by 1971 ACI 11.0,

$A_s = \rho bd = 0.013 \times 12 \times 7.5 = 1.17$ in.^{2} per ft $= 1.2$ bars per ft

Use #9 at 10 in., ACI requires 3-in. cover. The diameter of a #9 bar is 1.13 in. Therefore, the minimum thickness of the wall is $h = 7.5 + 3 + (1.13/2) = 11.07$ in. Use $h = 11$ in. and $d_c = 3.57$ in.

Check distribution of steel:

$$z = 0.40 \times 50 \sqrt[3]{3.57\,\frac{2 \times 3.57 \times 12}{1.2}}$$

$$= 127 < 145 \text{ kips per in.} \qquad \text{OK}$$
$$\text{1971 ACI 10.6.3 and 10.6.4}$$
$$\text{(10-2)}$$

7.43 Deflection considerations require a minimum slab thickness h equal to

$$\min h = \frac{1}{20}\left(0.4 + \frac{f_y}{100,000}\right) \qquad \text{ACI 9.5.1}$$

$$= \frac{10 \times 12}{20}\left(0.4 + \frac{50,000}{100,000}\right) = 5.4 \text{ in.}$$

Assume a 6-in. slab.

$$\text{Slab weight} = \tfrac{6}{12}\,(0.150) = 0.075 \text{ kips per ft}$$
$$\text{Total load } w_u = 1.4D + 1.7L = 1.4(0.075) + 1.7(0.400)$$
$$= 0.785 \text{ kips per ft}$$

Maximum moment induced in slab:

$$\text{Maximum } M_u = \frac{w_u l^2}{8} = \frac{0.785(10)^2}{8} = 9.81 \text{ ft-kips}$$

To ensure that tensile yielding controls the strength of the section,

$$\text{Maximum } \rho = 0.75\rho_b \qquad \text{ACI 10.3.2}$$

$$= (0.75)(0.85)\beta_1\left(\frac{f_c'}{f_y}\right)\left(\frac{87,000}{87,000 + f_y}\right) \qquad \text{ACI 8.1}$$

$$= (0.75)(0.85)(0.85)\left(\frac{3}{50}\right)\left(\frac{87}{87 + 50}\right) = 0.021$$

Try #5 bars at $5\tfrac{3}{4}$ in., $A_s = 0.65$ sq in.

$$d = 6 - \left(0.75 + \frac{0.625}{2}\right) = 4.94 \text{ in.}$$

Supplied tensile reinforcement is

$$\rho = \frac{0.65}{12 \times 4.94} = 0.011 < 0.021 \qquad \text{OK}$$

To find the service moment of the slab, first find a.

$$a = \frac{A_s f_y}{0.85 f_c' b} = \frac{0.65 \times 50}{0.85 \times 3.00 \times 12}$$

$$= 1.062 \text{ in.} \qquad \begin{array}{l}\text{1971 ACI 10.2.7}\\\text{1963 ACI (16-1)}\end{array}$$

$$M_u = \phi A_s f_y \left(d - \frac{a}{2} \right) \qquad \text{ACI (16-1)}$$

$$= 0.90 \times 0.65 \times 50 \left(4.94 - \frac{1.062}{2} \right) \frac{1}{12}$$
$$= 10.75 \text{ ft-kips} \qquad \text{OK}$$

Check shear:

$$\text{Maximum } V_u = 0.785 \times \frac{10}{2} = 3.93 \text{ kips}$$

$$v_u = \frac{V_u}{\phi bd} = \frac{3930}{0.85 \times 12 \times 4.94}$$
$$= 77.88 \text{ psi} \qquad \text{1971 ACI (11-3)}$$

The maximum shear stress carried by the concrete is

$$v_c = 2 \sqrt{f'_c} = 2 \sqrt{3,000} = 109 \text{ psi} \qquad \text{ACI 11.4.1}$$

Since $v_u < v_c$, no stirrups are necessary (ACI 11.1.1).
Check distribution of tensile reinforcement (ACI 10.6.3).

$$d_c = \text{cover} + \tfrac{1}{2} \text{ bar diameter}$$
$$d_c = 0.75 + \tfrac{1}{2} \times 0.625 = 1.063$$
$$\text{No. of bars} = \frac{12}{5.75} = 2.087 \text{ per ft}$$
$$A = \frac{2d_c b}{\text{no. of bars}} = \frac{2 \times 1.063 \times 12}{2.087} = 12.22 \text{ in.}^2 \text{ per ft}$$
$$z = f_s \sqrt[3]{d_c A} = 0.60 \times 50 \sqrt[3]{1.063 \times 12.22}$$
$$= 70.50 < 175 \text{ kips per in.} \qquad \text{1971 ACI (10-2)}$$

7.44   The centroid of the tensile reinforcement is located at a distance of 3.74 in. from the bottom of the beam. Therefore,

$$d = 40 - 3.74 = 36.26 \text{ in.}$$

Check maximum tensile reinforcement requirements.

$$A_s = 2 \times 0.79 + 2 \times 1.00 + 4 \times 1.27 = 8.66 \text{ sq in.}$$
$$\text{Actual steel ratio} = \frac{A_s}{bd} = \frac{8.66}{18 \times 36.26} = 0.0133$$

Maximum allowed $\rho = 0.75\rho_b$ ACI 10.3.2

$$= 0.75(0.85)\beta_1\left(\frac{f_c'}{f_y}\right)\left(\frac{87{,}000}{87{,}000 + f_y}\right)$$

$$= 0.75 \times 0.85 \times 0.85\left(\frac{3}{40}\right)\left(\frac{87}{137}\right)$$

$$= 0.0278 \qquad \text{OK}$$

Find a.

$$a = \frac{A_s f_y}{0.85 f_c' b} = \frac{8.66 \times 40}{0.85 \times 3 \times 18} = 7.55 \text{ in.}$$

The usable moment is equal to

$$M_u = \phi A_s f_y\left(d - \frac{a}{2}\right) = 0.90 \times 8.66 \times 40\left(36.26 - \frac{7.55}{2}\right)\frac{1}{12}$$
$$= 843.96 \text{ ft-kips}$$

Dead load of beam $= 0.145 \times \dfrac{18 \times 40}{144} = 0.725$ kips per ft

Dead load moment $= \dfrac{0.725 \times 15^2}{8} = 20.39$ ft-kips

Live load moment $= \dfrac{P_u L}{4} = \dfrac{P_u(15)}{4}$

$\dfrac{P_u(15)}{4} = $ usable net $M_u = 843.96 - 20.39 \times 1.4$
$$= 815.41 \text{ ft-kips}$$

or $P_u = 217.44$ kips

or dividing by live load safety factor,

$$P = \frac{217.44}{1.7} = 127.91 \text{ kips}$$

If no stirrups are to be used, then

Maximum $v_u = 0.5v_c$ ACI 11.1.1(d)
$$v_c = 2\sqrt{f_c'} = 2\sqrt{3{,}000} = 109.6 \text{ psi} \qquad \text{ACI 11.4.1}$$

Maximum $V_u = \phi v_u bd = 0.85 \times 0.5 \times 109.6 \times 18$
$$\times 36.26\frac{1}{1{,}000} = 30.2 \text{ kips}$$

Dead load $V = \dfrac{0.725 \times 15}{2} = 5.4$ kips

Dead load $V_u = 5.4 \times 1.4 = 7.6$ kips

$P_u = 2(30.2 - 7.6)$

$\quad = 45.2$ kips.   Assuming there are no stirrups and dividing by live load safety factor,

$$P = \frac{45.2}{1.7} = 26.6 \text{ kips}$$

Check for deflection. According to the problem the allowable deflection is

$$\frac{L}{800} = \frac{15 \times 12}{800} = 0.225 \text{ in.}$$

or $\Delta = \dfrac{PL^3}{48EI} = 0.225$

$E = 57,000 \sqrt{f'_c} = 57,000 \sqrt{3,000}$

$\qquad\qquad\qquad = 3.12 \times 10^6 \text{ psi}$ ACI 8.3.1

$I = \left(\dfrac{M_{cr}}{M_a}\right)^3 (I_g) + \left[1 - \left(\dfrac{M_{cr}}{M_a}\right)^3\right] I_{cr}$ ACI 9.5.2.2
(9-4)

$f_r = 7.5 \sqrt{f'_c} = 7.5 \times 54.8 = 411 \text{ psi}$ ACI 9.5.2.2
(9-5)

$I_g = \dfrac{18 \times 40^3}{12} = 96,000 \text{ in.}^4$

$M_{cr} = \dfrac{411 \times 96,000}{20} = 1,972,800 \text{ in.-lb}$ ACI 9.5.2.2
(9-5)

To find I_{cr} find the moment of inertia of the transformed section about its centroidal axis.

$$n = \frac{29 \times 10^6}{3.12 \times 10^6} = 9$$

$$nA_s = 9 \times 8.66 = 77.94$$

The distance of the centroid from the top is

$$\bar{y} = \frac{(18 \times 7.55)3.78 + (77.94)36.26}{18 \times 7.55 + 77.94} = 15.62 \text{ in.}$$

Moments of inertia are

Compression area: $\dfrac{1}{12} \times 18 \times 7.55^3$

$$+ 18 \times 7.55 \left(15.62 - \dfrac{7.55}{2}\right)^2 = 19,712 \text{ in.}^4$$

Transformed area: $77.94(36.26 - 15.62)^2 = 33,203 \text{ in.}^4$

$$I_{cr} = 19,712 + 33,203 = 52,915 \text{ in.}^4$$

$$M_a = \dfrac{843.96}{0.90} = 937.73 \text{ ft-kips}$$

$$\left(\dfrac{M_{cr}}{M_a}\right)^3 = \left(\dfrac{1,972,800}{937.73 \times 12,000}\right)^3 = (0.175)^3 = 0.005$$

$$I = 0.005 \times 96,000 + (1 - 0.005)52,915 = 53,100 \text{ in.}^4$$

Therefore,

$$\dfrac{P(15 \times 12)^3}{48 \times 53,100 \times 3.12 \times 10^3} = 0.225$$

or

$$P = 306.5 \text{ kips}$$

From the above it can be seen that shear controls. The maximum live load $P = 26.6$ kips.

7.45 The total uniformly distributed ultimate load on the beam is

$$w_u = 1.4 \times 0.6 + 1.7 \times 1.2 = 2.88 \text{ kips per ft}$$

$$\text{Maximum } M = \dfrac{2.88 \times 20^2}{8} = 144.00 \text{ ft-kips}$$

Maximum steel ratio $\rho = 0.75\rho_b = 0.75$

$$\times 0.85\beta_1 \left(\dfrac{f_c'}{f_y}\right)\left(\dfrac{87,000}{87,000 + f_y}\right)$$

$$\rho = 0.75 \times 0.85$$

$$\times 0.85\left(\dfrac{4}{50}\right)\left(\dfrac{87}{87 + 50}\right) = 0.0275$$

Minimum ratio $\rho = \dfrac{200}{f_y} = \dfrac{200}{50,000} = 0.004$ ACI 10.5.1

Any steel ratio between the above limits is acceptable. The steel ratio that yields the most economical section is

$$\rho = 0.18 \frac{f_c}{f_y} = 0.18 \times \frac{4}{50} = 0.0144$$

To determine the values of b and d of the section, use the equation

$$bd^2 = \frac{M_u}{\phi f_y [1 - (0.59 f_y / f_c')]}$$

$$bd^2 = \frac{144.0 \times 12}{0.90 \times 0.0144 \times 50(1 - 0.59 \times 0.0144 \times (50/4)}$$

$$= 2{,}983 \text{ in.}^3$$

Assume $b = d/2$. This value yields an economical section. $d^3 = 5{,}966$ or $d = 18.14$ and $b = 9.07$ in.

Deflection considerations require that the minimum thickness h of the beam be

$$\text{Minimum } h = \frac{L}{16}\left(0.4 + \frac{f_y}{100{,}000}\right) \qquad \text{ACI 9.5.1}$$

$$\text{Minimum } h = \frac{20 \times 12}{16}\left(0.4 + \frac{50}{100}\right) = 13.5 \text{ in}$$

Use $b = 10$ in. and $d = 17.5$ in. Then $A_s = \rho bd = 0.0144 \times 10 \times 17.5 = 2.52$ sq in.

Use three #9 bars. Check to see if the width of the beam can accommodate the bars.

$$\text{Required } b = \left(2 \times 1.5 + 3 \times 1.128 + 2 + 2 \times \frac{1.128}{3}\right)$$

$$= 9.14 < 10 \qquad \text{OK}$$

The last term allows for possible stirrups.

Check for shear,

$$v_u = \frac{V_u}{\phi bd} = \frac{0.5 \times 2.88 \times 20 \times 1{,}000}{0.85 \times 10 \times 17.5} = 193.61 \text{ psi}$$

Shear stress at d distance from face of support is

$$v_u = \frac{(10 - 1.46) \times 193.61}{10} = 165.34 \text{ psi} \qquad \text{ACI 11.2.2}$$

Shear stress carried by concrete is

$$v_c = 2\sqrt{f_c'} = 2\sqrt{4{,}000} = 126.5 \text{ psi} \qquad \text{ACI 11.4.1}$$

Stirrups are required (ACI 11.1.1). Use #2 U bars placed vertically. Spacing required is

$$s = \frac{A_v f_y}{(v_u - v_c)b} \qquad \text{ACI 11.6.1}$$

$$s = \frac{2 \times 0.05 \times 50{,}000}{(165.34 - 126.4)10} = 12.84 \text{ in.} > \frac{d}{2} = 8.75$$

Place stirrups $8\frac{1}{2}$ in. on centers. See ACI 11.1.4(b). Place stirrups to a distance x from the center line of the beam. At this distance v_u becomes equal to $\frac{1}{2}v_c = 63.25$ psi. From the shear diagram,

$$x = \frac{(10 - 1.46)63.25}{165.34} = 3.27 \text{ ft}$$

Since f_y does not exceed 40,000 psi, it is not necessary to check the distribution of the tensile reinforcement (ACI 10.6.3).

7.46 Assume a clear span of $l_n = 18'\text{-}0''$.

$$\text{Negative moment at interior support } M = -\frac{w l_n{}^2}{10}$$

$$\text{Shear } V = 1.15\frac{w l_n}{2}$$

$$\text{Positive moment in end span } M = \frac{w l_n}{14} \qquad \text{ACI 8.4.2}$$

$$\text{Positive moment in center span } M = \frac{w l_n}{16}$$

$$w_u = 1.4 \times 0.6 + 1.7 \times 1.2 = 2.88 \text{ kips per ft}$$

Maximum moment in the slab $M_u = \dfrac{2.88 \times 18^2}{10} = 93.31$ ft-kips

Maximum allowable steel ratio $\rho = 0.75\rho_b$

$$= 0.75 \times 0.85\beta_1 \left(\frac{f_c'}{f_y}\right)\left(\frac{87}{87 + 50}\right)$$

$$\rho = 0.75 \times 0.85 \times 0.85 \left(\frac{4}{50}\right)\left(\frac{87}{87 + 50}\right) = 0.0275$$

$$\text{Minimum } \rho = \frac{200}{f_y} = \frac{200}{50,000} = 0.004$$

An economical steel ratio $\rho = 0.18\dfrac{f_c'}{f_y} = 0.18\dfrac{4}{50} = 0.0144$. Using

this ratio it is

$$bd^2 = \frac{M_u}{\phi f_y \left(1 - 0.59\rho\dfrac{f_y}{f_c'}\right)}$$

$$= \frac{93.31 \times 12}{0.90 \times 0.0144 \times 50(1 - 0.59 \times 0.0144 \times (50/4)}$$

$$= 1,934 \text{ in.}^3$$

Assume $b = 0.50d$, then $d^3 = 3,868$ or $d = 15.7$ in.
Deflection considerations allow

$$\text{Minimum } h = \frac{l}{16}\left(0.4 + \frac{f_y}{100,000}\right) = \frac{20 \times 12}{16}\left(0.4 + \frac{50,000}{100,000}\right)$$

$$= 13.5 \text{ in.} \qquad \text{ACI 9.5.1}$$

Use $b = 8$ in. and $d = 16$ in.

Then over interior supports

$$A_s = 0.0144 \times 8 \times 16 = 1.84 \text{ sq in., three #7 bars.}$$

Positive steel in center span

$$A_s = {}^{10}\!/_{16} \times 1.84 = 1.15 \text{ sq in., two #6 bars, one #5 bar.}$$

Positive steel in end span $= {}^{10}\!/_{14} \times 1.84$
$$= 1.31 \text{ sq in., three #6 bars.}$$

Check to see whether three #7 bars can fit in the beam.

Required $b = 2 \times 1.5 + 3 \times 0.875 + 2$
$$= 7.63 \text{ in.} < 8 \text{ in.} \qquad \text{OK}$$

Check for shear:

$$V_u = 1.15 \left(\frac{w_u l_n}{2} \right) = 1.15 \left(\frac{2.88 \times 18}{2} \right) = 29.81 \text{ ft-kpsi}$$

$$v_u = \frac{29,810}{0.85 \times 8 \times 16} = 273.99$$

$$v_c = 2 \sqrt{f_c'} = 2 \sqrt{4,000} = 126.5 \text{ psi}$$

Stirrups are required (see ACI 11.2.2). Use #3 U stirrups.

$$\text{Required spacing } s = \frac{A_v f_y}{(v_u - v_c)^b} \qquad \text{ACI 11.6.1}$$

Shear stress v_u at distance d from face of support (see ACI 11.2.2) is

$$v_u = \frac{9.0 - 1.33}{9} \times 273.99 = 233.50 \text{ psi}$$

$$s = \frac{2 \times 0.11 \times 50,000}{(233.50 - 126.5)8} = 12.85 \text{ in.} > 8 \text{ in.}$$

Place stirrups to a distance x from the center line of the beam equal to

$$x = \frac{(9.0 - 1.33)63.25}{233.50} = 2.08 \text{ ft}$$

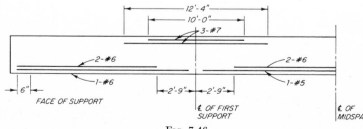

Fig. 7.46

7.47 By 1973 ACI specifications: $n = 9, f_s = 18,000$

$$8y \times \frac{y}{2} + 2 \times (18 - 1) \times 0.60 \times (y - 2)$$
$$= 2 \times 9 \times 1.27 \times (16 - y)$$

$4y^2 + 20.4 \times (y - 2) = 22.86 \times (16 - y)$

$y^2 + 5.1y - 10.2 = 91.44 - 5.71y$

$y^2 + 10.81y + (5.405)^2 = 101.64 + (5.405)^2 = 130.8$

$y = 11.42 - 5.40 = 6.02$ in.

Moment of inertia:

Compression area $= \frac{1}{3} \times 8 \times 6.02^3 = 580$
Compression steel $= 20.4 \times 4.02^2 = 330$
Tension steel $= 22.86 \times 9.98^2 = \underline{2,280}$
$ 3,190$ in.^{4}

If $f_s = 20,000$ psi, $f_c = \dfrac{6.02 \times 20,000}{9.98 \times 9} = 1,340$ psi

$f = \dfrac{Mc}{I}$ or $M = \dfrac{fI}{c} = \dfrac{1,340 \times 3,190}{6.02} = 710,066$ in-lb

This is $> 12 \times 50,000$ or $600,000$ in-lb. $\therefore$ Beam is OK to carry load.

7.48 Total ultimate moment $M_u = 1.4 \times 75 + 1.7 \times 120 = 309$ ft-kips
The maximum steel ratio for singly reinforced beam is

$$\rho_{\max} = 0.75\rho_b = 0.75 \times 0.85 \times 0.80 \left(\frac{5}{50}\right)\left(\frac{87}{87 + 50}\right)$$
$$= 0.0324$$

max $A_s = 0.0324 \times 12 \times 15.5 = 6.02$ sq in.

$\dfrac{a}{2} = \dfrac{A_s f_y}{1.70 f_c' b} = \dfrac{6.02 \times 50}{1.70 \times 5 \times 12} = 2.95$

$M_u = \phi A_s f_y \left(d - \dfrac{a}{2}\right)$

$ = 0.90 \times 6.02 \times 50(15.5 - 2.95)/12$
$ = 284$ ft-kips < 309 ft-kips

A doubly reinforced beam is needed and M_u above becomes M_{u_1} and $M_u = 309$ ft-kips.

$$M_{u_2} = M_u - M_{u_1} = 309 - 284 = 25 \text{ ft-kips}$$

$$A_s' = \frac{M_{u_2}}{\phi f_y(d - d')} = \frac{25 \times 12}{0.90 \times 50(15.5 - 2.5)} = 0.513 \text{ sq in.}$$

$$A_{s_1} = 6.02$$
$$\underline{A_{s_2} = 0.513}$$
$$A_s = 6.533$$

Use three #9 bars = 3.00 sq in. in upper row of tension steel

one #9 bar + two #10 bars = 3.54 sq in. in lower row of tension steel

$$\underline{\phantom{6.54 \text{ sq in}}}$$
6.54 sq in. total

Assume the compressive steel stress $f_s' = f_y$ at failure, but to ensure failure by tensile yielding use

$$A_s' = \frac{0.513}{0.75} = 0.69 \text{ sq in. min., or one #8 bar} = 0.79 \text{ sq in.}$$

Use $A_{s_1} = 6.54 - 0.79 = 5.75$ and $A_{s_2} = 0.79$ sq in.

$$\rho = \frac{6.54}{12 \times 15.5} = 0.0352$$

$$\rho' = \frac{0.79}{186} = 0.0042$$

$$(\rho - \rho') = \frac{5.75}{186} = 0.0310$$

$$(\rho - \rho')_{\max} = 0.75\rho_b = 0.0324$$

$$(\rho - \rho')_{\min} = 0.85 \times 0.80 \left(\frac{5}{50}\right)\left(\frac{2.5}{15.5}\right)\left(\frac{87}{37}\right) = 0.0258$$

$$(\rho - \rho')_{\text{used}} = 0.0310 < 0.0324 \text{ max} > 0.0258 \text{ min} \qquad \text{OK}$$

Recheck resisting moments.

$$M_{u_2} = \phi A_s' f_y(d - d') = \frac{0.90 \times 0.79 \times 50 \times 13}{12} = 38.4 \text{ ft-kips}$$

$$\frac{a}{2} = \frac{(A_s - A_s')f_y}{1.70 f_c' b} = \frac{5.75 \times 50}{1.70 \times 5 \times 12} = 2.82 \text{ in.}$$

$$M_{u_1} = \frac{0.90 \times 5.75 \times 50(15.5 - 2.82)}{12} = 274 \text{ ft-kips}$$

$M_u = 38.4 + 274 = 312.4$ ft-kips > 309 ft-kips   OK

Check distribution of tensile reinforcement.

$d_c = 1.50 + 1.27/2 = 2.13$ in.

$A = (1.50 + 2.13)12/3 = 14.5$

$z = 0.60 \times 50 \sqrt[3]{2.13 \times 14.5} = 94 < 145$ kips per in.   OK

7.49   The new total load is

$$U = 1.4 \times 0.6 + 1.7(1.2 + 0.6) = 3.9 \text{ kips per ft}$$

The new ultimate moment $M_u = \dfrac{3.90}{2.88} \times 144 = 195.00$ ft-kips

$$\text{Maximum } \rho = 0.75 \times 0.85 \times 0.85 \times \frac{4}{50} \times \frac{87}{137} = 0.0275$$

$$\text{Minimum } \rho = \frac{200}{50,000} = 0.004$$

Try maximum reinforcement to see if compressive reinforcement is required.

Maximum $A_s = 0.0275 \times 10 \times 17.5 = 4.81$ sq in.

Maximum usable moment the beam can carry is

$$M_u = \phi A_s f_y \left(d - \frac{a}{2} \right) = 0.90 \times 4.81 \times 50 \left(17.5 - \frac{a}{2} \right) \frac{1}{12}$$

$$= 18.04 \left(17.5 - \frac{a}{2} \right)$$

where $a = \dfrac{4.81 \times 50}{0.85 \times 4 \times 10} = 7.07$ in. Therefore,

$$M_u = 18.04 \left(17.5 - \frac{7.07}{2} \right) = 251.93 \text{ ft-kips}$$

Compressive reinforcement is not necessary. For new reinforcement try three #10 bars.

Required $b = 2(1.5) + 3(1.27) + 2(1.27) + 0.50$

$$= 9.85 < 10 \text{ in.} \quad \text{OK}$$

$A_s = 3.81$ sq in.

then $\rho = \dfrac{3.81}{10 \times 17.5} = 0.022 < 0.0275$ OK

$$a = \frac{3.81 \times 50}{0.85 \times 4 \times 10} = 5.60$$

$$M_u = 0.90 \times 3.81 \times 50 \left(17.50 - \frac{5.60}{2}\right) \frac{1}{12}$$
$$= 210.02 > 195.0 \qquad \text{OK}$$

Use three #10 bars. Check shear.

$$V_u = 0.5 \times 3,900 \times 20 = 39,000 \text{ lb}$$
$$v_u = \frac{39,000}{0.85 \times 10 \times 17.5} = 262.18 \text{ psi} \qquad \text{ACI (11-3)}$$

The concrete can carry

$$v_c = 2 \sqrt{4,000} = 126.5 \text{ psi} \qquad \text{ACI 11.4.1}$$

Stirrups are required (see ACI 11.1.1).

Use #3 U bars. The shear stress d distance from the face of the support is

$$v_u = \frac{10.0 - 1.46}{10.0} \times 262.18 = 223.90 \text{ psi}$$

Required spacing $s = \dfrac{2(0.11)50,000}{(223.90) - 126.5)10} = 11.31 > \dfrac{d}{2}$

$$= 8.75 \text{ in.}$$

Use #3 U stirrups $8\frac{1}{2}$ in. centers to a distance x from center line of beam

$$x = \frac{10.0 - 1.46}{223.90} 63.25 = 2.42 \text{ ft}$$

Check distribution of tensile reinforcement (see ACI 10.6.3).

d_c = cover + $\frac{1}{2}$ bar dia. + stirrup diameter
$\quad = 1.5 + \frac{1}{2} \times 1.27 + 0.38 = 2.51$ in.

$$A = \frac{2d_c b}{\text{no. of bars}} = \frac{2 \times 2.51 \times 10}{3} = 16.73$$

$z = 0.60 \times 50 \sqrt[3]{2.51 \times 16.73}$
$$= 104.5 < 145 \text{ kips per in.} \qquad \text{OK}$$

7.50　Determine the effective width of the beam:

$$\frac{L}{4} = \frac{20 \times 12}{4} = 60 \text{ in.}$$
$$16h_f + b_w = 16 \times 4 + 12 = 76 \text{ in.}$$

Beam spacing = 45 in.

The latter controls, therefore $b_f = 45$ in., $d = 24 - 2 = 22$ in.

Determine whether the beam has a real T-beam action. Assume $a = h_f = 4$ in. Then

$$A_s = \frac{M_u}{\phi f_y \left(d - \dfrac{a}{2} \right)} = \frac{985 \times 12}{0.90 \times 60(22 - 2)} = 10.94 \text{ sq in.}$$

$$a = \frac{10.94 \times 60}{0.85 \times 4 \times 45} = 4.29 \text{ in.}$$

Since a is larger than $h_f = 4$ in., the beam acts like a T beam. Tensile reinforcement due to compression in flanges is

$$A_{s_1} = \frac{0.85 f'_c (b_f - b_w) h_f}{f_y} = \frac{0.85 \times 4(45 - 12)4}{60} = 7.48 \text{ sq in.}$$

Corresponding moment is

$$M_1 = \phi A_{s_1} f_y \left(d - \frac{h_f}{2} \right) = 0.90 \times 7.48 \times 60(22 - 2)\frac{1}{12}$$
$$= 673.20 \text{ ft-kips}$$
$$M_2 = M - M_1 = 985.00 - 673.20 = 311.80 \text{ ft-kips}$$

For the rectangular part of the beam assume $\rho = 0.0138$ or $A_{s_2} = 3.65$ sq in.

$$a = \frac{3.65 \times 60}{0.85 \times 4 \times 12} = 5.37 \text{ in.}$$

$$M_u = 0.90 \times 3.65 \times 60 \left(22 - \frac{5.37}{2} \right) \frac{1}{12} = 317.25 \text{ ft-kips}$$

Total moment $M_u = 673.20 + 317.25$
$$= 990.45 > 985 \text{ ft-kips} \qquad \text{OK}$$

Total tensile reinforcement is

$$A_s = A_{s_1} + A_{s_2} = 7.48 + 3.65 = 11.13 \text{ sq in.}$$

Check to see if the beam fails by yielding of tensile reinforcement. Balanced reinforcement of the T beam is

$$\rho = \rho_b + \rho_f$$

where $\rho_b = 0.85(0.85)\left(\dfrac{4}{60}\right)\left(\dfrac{87}{87 + 60}\right) = 0.0285$

and $\rho_f = \dfrac{A_{s_1}}{b_w d} = 0.0283$

Maximum $\rho = 0.75(0.0285 + 0.0283) = 0.0426$

Actual $\rho = \dfrac{11.13}{12 \times 22} = 0.0421 < 0.0426$ OK

7.51 Weight per ft $= (2 \times 2 + \frac{1}{2} \times 4) \times 145 = 870$ lb per ft

$M_D = \frac{1}{8} \times 870 \times 40 \times 40 \times 12 = 2,080,000$ in-lb

$M_L = \quad 172,000 \times 12 \qquad\quad = 2,064,000$ in-lb

$M_I = \quad\;\; 52,000 \times 12 \qquad\quad = \underline{\;\;624,000}$ in-lb

$M_T \qquad\qquad\qquad\qquad\qquad = 4,768,000$ in-lb

Assuming steel and concrete is stressed to maximum allowable,

$k = 0.377$

$j = 0.874$

$kd = 0.377 \times 21.5$ in. $= 8.10$ in.

and 21.50 in. $- 8.10$ in. $= 13.40$ in.

$\left(\dfrac{8.10 - 6}{8.10}\right) \times 1,350 = 350$ psi

$350 \times 6 = 2,100 \times 3 \quad = \quad 6,300$

$\dfrac{1,000}{2} \times 6 = \dfrac{3,000 \times 2}{5,100} \dfrac{= \quad 6,000}{(2.41) = 12,300}$

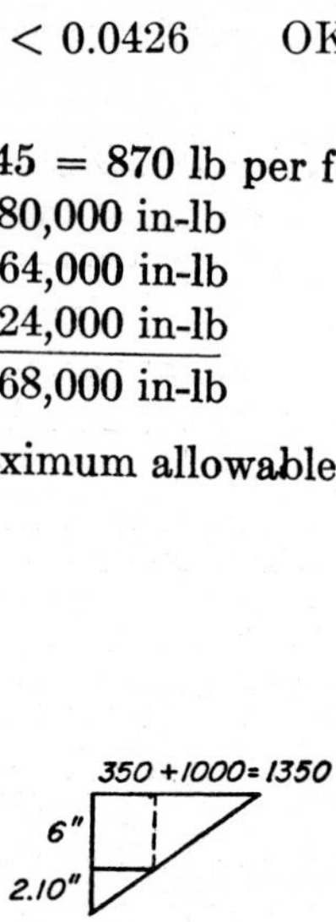

Fig. 7.51

Compressive force centers 2.41 in. from top of slab.

Use six #11 bars ($A_{s_1} = 9.36$ sq in.) in bottom row of steel.

Moment carried by #11 bars $= 20,000 \times 9.36 \times 19.09$

$= 3,580,000$ in-lb

Use second row of steel 3 in. above #11 bars.

$M = 4,768,000 - 3,580,000 = 1,188,000$

$= \dfrac{10.40}{13.40} \times 20,000 \times 16.09 A_{s_2}$

$A_{s_2} = \dfrac{1,188,000 \times 13.40}{16.09 \times 20,000 \times 10.40} = 4.75$ sq in.

Use six #8 bars. $A_{s_2} = 4.76$ sq in.

Center of gravity of $A_s = 3 \times 4.76/(4.76 + 9.36)$
$\qquad = 1.01$ in. above A_{s_1}

M beam can carry in concrete $= 72 \times 5,100 \times 18.08 =$
6,640,000 in-lb

and no A_s' is needed. $[b = 72 \text{ in.} < l/4 < (24 + 2 \times 8 \times 6)]$

$72 \times 6(y - 3) = 4.76 \times 9 \times (18.5 - y)$
$\qquad\qquad\qquad\qquad\qquad + 9.36 \times 9 \times (21.5 - y)$

$(y - 3) = 1.83 - 0.099y + 4.20 - 0.195y$

$1.294y = 9.03$, and $y = 6.98$ in.

$$
\begin{aligned}
I = \tfrac{1}{12} \times 72 \times 6 \times 6 \times 6 &= 1,296 \\
+ \; 72 \times 6 \times 3.98^2 &= 6,850 \\
+ \; 47.6 \times 11.52^2 &= 5,680 \\
+ \; 93.6 \times 14.52^2 &= \underline{17,820} \\
& \; 31,646 \text{ in.}^4
\end{aligned}
$$

$$
f_c = \frac{4,768,000}{31,646} \times 6.98 = 1,035 \text{ psi}
$$

$$
f_s = \frac{4,768,000}{31,646} \times 14.52 \times 9 = 19,700 \text{ psi}
$$

$$
v = \frac{V}{b'd} = \frac{40,000}{24 \times 21.5} = 77.5 \text{ psi} > 60 \text{ psi (no stirrups)}
$$

$v'bs < 2A_v f_v$, and $s = 21.5/2 = 10.75$ in.

Minimum stirrup steel $= 2A_v = 0.0015bs$

$A_v = 0.0015 \times 24 \times 10.75/2 = 0.194 < 0.20$ sq in.

Use #4 U stirrups.

$17.5 \times 24 \times 10.75 = 4,520 < 2 \times 0.20 \times 20,000 = 8,000$ lb

First stirrups 5 in. from ends, then on 10-in. centers

7.52 $f_c = 1,800$, $f_s = 20,000$

$n = 8$, $k = 0.419$, $j = 0.860$

$R = 324$

b for T beam $= l/4 = 16 \times 12/4$
$\qquad\qquad\qquad = 48$ in. $< b' + 2 \times (8 \times 4.5) < 96$ in.

Assume dead weight of T beam = 600 lb per ft

$+M = \frac{8}{10} \times (\frac{1}{8} \times 8,600 \times 16 \times 16 \times 12)$

$\qquad\qquad\qquad\qquad\qquad = 2,640,000$ in-lb

$-M = \frac{1}{4} \times (2,640,000) = 660,000$ in-lb

$M = Rbd^2 = 324 \times 48 \times d^2 = 2,640,000$ in-lb

$d^2 = 169$, $d = 13.0$, use 14 in.; $kd = 5.87$ in.

As neutral axis lies below slab, A_s' is required.

$d - kd = 14$ in. $- 5.87$ in. $= 8.13$ in.

$$jd = 14 \text{ in.} - 1.96 \text{ in.} = 12.04 \text{ in.}$$

(to bottom row of tension steel, approximate)

$A_s = \dfrac{M}{f_s jd} = \dfrac{2,640,000}{20,000 \times 12.04 \text{ in.}} = 10.96$ sq in. (if all A_s were in one row)

$b' = 24$ in. for six #11 bars in one row.

Use six #11 bars, $A_{s_1} = 9.36$ in bottom row.

$+M_1 = 9.36 \times 20,000 \times 12.04 = 2,255,000$ in-lb

$+M_2 = 2,640,000 - 2,255,000 = 385,000$ in-lb

Place A_{s_2} 3 in. above A_{s_1}

$M_2 = A_{s_2} \times \dfrac{5.13}{8.13} \times 20,000 \times 9.04 = 385,000$ in-lb

$A_{s_2} = \dfrac{385,000}{114,000} = 3.38$ sq in.

Use six #7 bars, $A_{s_2} = 3.60$ sq in.

Center of gravity of steel (below A_{s_2}

line) $= \dfrac{3 \times 9.36}{9.36 + 3.60} = 2.17$ in. or

0.83 in. above A_{s_1}

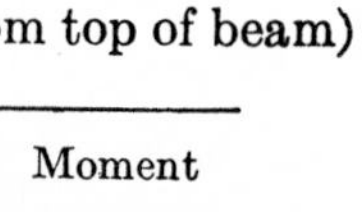

Fig. 7.52a

Center of gravity of compression forces (from top of beam)

	Force	Arm	Moment
$c_1 = 4.5 \times 420 =$	1,890 lb	$\times 2.25$	$= 4,253$
$c_2 = 4.5 \times 1,380/2 =$	3,105 lb	$\times 1.5$	$= 4,657$
	4,995 lb	(1.79)	$= 8,910$

New jd = 14.00 − 0.83 − 1.79 = 11.38 in. > 11.35 in.
$M_{(concrete)}$ = 4,995 × 48 × 11.38 = 2,730,000 in-lb
$\quad M_T$ = 2,640,000 in-lb. No A_s' is needed.

Check kd:

$$48 \times 4.5 \times (kd - 2.25) = 9.36 \times 8 \times (14 - kd)$$
$$+ 3.60 \times 8 \times (11 - kd)$$
kd = 5.80 < 5.87, but f_c and f_s are OK

At end of beam for negative moment

$$A_s \text{ (at top)} = \frac{660,000}{20,000 \times 0.860 \times 14} = 2.74 \text{ sq in.}$$

The six #7 bars bent up from A_{s_2} = 3.60 sq in.
$M_{(concrete stem)} = Rb'd^2 = 324 \times 24 \times 14 \times 14$
$\quad\quad\quad\quad\quad\quad\quad = 1,525,000 \text{ in-lb} > 660,000 \text{ in-lb}$
V [14 in. (= d) from end] = 8 × 8,600 × (96 − 14)/96
$\quad\quad\quad\quad\quad\quad\quad\quad\quad = 58,700 \text{ lb}$

$$u = \frac{V}{\Sigma_0 jd} = \frac{58,700}{6 \times 4.43 \times 11.38} = 194 < \frac{4.8\sqrt{4,000}}{1.41}$$
$$= 215 \text{ psi}$$

$$v \text{ (14 in. from end)} = \frac{58,700}{24 \times 14} = 175 > 60 \text{ and } < 190 \text{ psi}$$

Minimum stirrup area = $2A_v$ = 0.0015 bs
A_v = 0.0015 × 24 × $\frac{7}{2}$ = 0.127 or #4 U stirrup
Maximum v' for #4 U stirrup = $\dfrac{2 \times 0.20 \times 20,000}{24 \times 7}$ = 47 psi
$v'bs$ = (175 − 60) × 24 × 6 = 16,560 = $2A_v$ × 20,000
A_v = 0.413 or #6 U stirrups at end on 6-in. centers

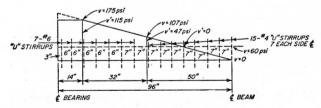

SHEAR DIAGRAM AND STIRRUP ARRANGEMENT
Fig. 7.52b

7.53 Total tensile reinforcement $A_s = 6 \times 1.00 = 6.00$ sq in. Check whether a extends in web.

$$a = \frac{A_s f_y}{0.85 b_f d} = \frac{6 \times 60}{0.85 \times 25 \times 24} = 4.24 > 4.00 \text{ in.}$$

A T-beam analysis is required. Check to see whether the beam will fail by yielding of tensile reinforcement. Tensile reinforcement due to compression in flanges is

$$A_{s_1} = \frac{0.85 f_c'(b_f - b_w)h_f}{f_y} = \frac{0.85 \times 4(25 - 10)4}{60} = 3.40 \text{ sq in.}$$

$A_{s_2} = 6.00 - 3.40 = 2.60$ sq in.

$$\rho_w = \frac{A_s}{b_w d} = \frac{6}{10 \times 24} = 0.025$$

$$\rho_f = \frac{A_{s_1}}{b_w d} = \frac{3.40}{10 \times 24} = 0.0142$$

$$\rho_b = (0.85)(0.85)\left(\frac{4}{60}\right)\left(\frac{87}{147}\right) = 0.0285$$

Maximum $\rho = 0.75(\rho_b + \rho_f) = 0.75(0.0285 + 0.0142)$
$$= 0.0427 > 0.025 \qquad \text{OK}$$

Find capacity.
Moment due to flanges:

$$\frac{M_1}{\phi} = A_{s_1} f_y \left(d - \frac{h_f}{2}\right) = 3.4 \times 60(24 - 2)\frac{1}{12} = 374.00 \text{ ft-kips}$$

Moment due to web:

$$a = \frac{2.6 \times 60}{0.85 \times 4 \times 10} = 4.59 \text{ in.}$$

$$\frac{M_2}{\phi} = A_{s_2} f_y \left(d - \frac{a}{2}\right) = 2.6 \times 60\left(24 - \frac{4.59}{2}\right)\frac{1}{12}$$
$$= 282.17 \text{ ft-kips}$$

Capacity of beam:

$$M = \phi(M_1 + M_2) = 0.90(374.00 + 282.17)$$
$$= 590.55 \text{ ft-kips} \qquad \textbf{d}$$

7.54 $P_u = 1.4 \times 266,000 + 1.7 \times 530,000$
$$= 372,400 + 901,000 = 1,273,400 \text{ lb}$$

Here compression controls. Use $e = 0.05\,D$ minimum

Try $D = 21$ in., $e = 1.05$ in.,

$$D_s = 21 - 2 \times \left(1.5 + 0.375 + \frac{1.693}{2}\right)$$

$$= 15.56 \text{ in.}$$

$A_{st} =$ eight #14S bars $= 8 \times 2.25 = 18.00$ sq in.

$A_g = 346$ sq in., $p_t = A_{st}/A_g = 18.00/346 = 5.2\% < 8\%$

Clear spacing of bars $= (3.14 \times 15.56/8) - 1.69 = 4.41$ in.

$$P_u = \phi\left(\frac{A_{st} \times f_y}{\dfrac{3e}{D_s} + 1} + \frac{A_g \times f'_c}{\dfrac{9.6 \times D \times e}{(0.8D + 0.67D_s)^2} + 1.18}\right)$$

$$= 0.75\left(\frac{18 \times 60,000}{\dfrac{3.15}{15.56} + 1} + \frac{346 \times 4,000}{\dfrac{9.6 \times 21 \times 1.05}{(1.68 + 10.4)^2} + 1.18}\right)$$

$$P_u = 0.75\left(\frac{1,080,000}{1.202} + \frac{1,384,000}{0.286 + 1.18}\right)$$

$$= 0.75 \times (896,000 + 944,000)$$

$$= 0.75 \times 1,840,000 = 1,380,000 > 1,273,400 \text{ lb}$$

Spiral steel:

$$\rho_s = 0.45\left(\frac{A_g}{A_c} - 1\right)\frac{f'_c}{f_y} = 0.45\left(\frac{21^2}{18^2} - 1\right)\frac{4,000}{60,000}$$

$$= \frac{0.45}{15}\,(1.360 - 1) = 0.0108$$

Try $\frac{3}{8}$-in. spiral, $a_s = 0.11$.

$$g = \frac{4a_s}{\rho_s D_c} = \frac{4 \times 0.11}{0.0108 \times 18} = 2.26 \text{ in.}$$

Use $g = 2.25 < 2.26 < 18/6 =$ or $<$3-in. clear maximum.

7.55 $P_u = 1,273,400$ lb as in answer to **7.54**

Here compression controls. Use $e = 0.05t$ minimum.

Try $t = 19$ in., $e = 0.95$ in., $D_s = 13.31$ in.

$A_{st} =$ six #14S + two #11 bars $= 6 \times 2.25 + 2 \times 1.56$

$$= 16.62 \text{ sq in.}$$

$A_g = 361$ sq in., $p_t = A_{st}/A_g = 16.62/361 = 4.61\% < 8\%$

Clear spacing of bars $= (3.14 \times 13.31/8) - 1.69 = 3.54$ in.

$$P_u = \phi\left(\frac{A_{st}f_y}{\dfrac{3e}{D_s} + 1} + \frac{A_sf_c'}{\dfrac{12te}{(t + 0.67D_s)^2} + 1.18}\right)$$

$$= 0.75\left(\frac{16.62 \times 60,000}{\dfrac{3 \times 0.95}{13.31} + 1} + \frac{361 \times 40,000}{\dfrac{12 \times 19 \times 0.95}{(19 + 8.91)^2} + 1.18}\right)$$

$$= 0.75 \times (821,000 + 991,000)$$

$$= 0.75 \times 1,812,000 = 1,359,000 > 1,273,400 \text{ lb}$$

For spiral design:

$$\rho_s = 0.45\left(\frac{19 \times 19}{3.14 \times 8.3 \times 8.3} - 1\right)\frac{4,000}{60,000} = \frac{0.45}{20}(1.67 - 1)$$

$$= 0.0201$$

Try ½ in. spiral, $a_s = 0.20$ sq in.

$$g = \frac{4a_s}{\rho_s D_c} = \frac{4 \times 0.20}{0.0201 \times 16.62} = 2.40$$

Use $g = 2.25 < 2.40 < 1\%$ < 3 in. clear maximum.

7.56 Total axial load and moment are

$$P_u = 1.4 \times 210 + 1.7 \times 140 = 532.00 \text{ kips}$$

$$\frac{P_u}{\phi} = \frac{532.00}{0.7} = 760.00 \text{ kips}$$

$$M_u = 1.4 \times 63 + 1.7 \times 42 = 159.60 \text{ ft-kips}$$

$$\frac{M_u}{\phi} = \frac{159.60}{0.7} = 228.00 \text{ ft-kips}$$

Actual eccentricity is

$$e = \frac{228 \times 12}{760.00} = 3.60 \text{ in.}$$

Check to see if there is a slenderness effect.

$$r = 0.3 \times 20 = 6.00 \qquad \text{ACI 10.11.2}$$

Since the column is braced against sidesway, $k = 1$ (ACI 10.11.3).

$$\frac{kl_u}{r} = \frac{1.0 \times 8.0 \times 12}{6} = 16.00$$

$$34 - 12\left(\frac{M_1}{M_2}\right) = 34 - 12\left(\frac{-1}{2.5}\right) = 38.8$$

Therefore,

$$\frac{kl_u}{r} < 38.8 \qquad \text{ACI 10.11.4}$$

No slenderness effect. For longitudinal reinforcement assume six #9 bars. Effective depth $d = 20.00 - 2.50 = 17.50$ in.

Find the balanced eccentricity of the section.

$$\frac{P_{u_b}}{\phi} = 0.85\beta_1 f'_c bd\left(\frac{87}{87 + f_y}\right) = 0.85 \times 0.85 \times 4 \times 14$$

$$\times 17.5\left(\frac{87}{87 + 60}\right) = 419.05 \text{ kips}$$

$$a_b = \beta_1 d\left(\frac{87}{147}\right) = 0.85 \times 17.5 \times \frac{87}{147} = 8.80 \text{ in.}$$

$$\frac{P_{u_b}e'_b}{\phi} = 0.85 f'_c a_b b\left(d - \frac{a_b}{2}\right) + A'_s f'_y(d - d')$$

$$419.05 e'_b = 0.85 \times 4 \times 8.80 \times 14\left(17.5 - \frac{8.8}{2}\right)$$

$$+ 3.0 \times 60(17.5 - 2.5)$$

$$e'_b = 19.5 \text{ in.}$$

then $e_b = 19.5 - 7.5 = 12.0$ in. $= e$. Therefore compression controls. Proceed by successive approximations. Assume $a = 14.5$, then $c = \dfrac{14.5}{0.85} = 17.06$.

The stress in the tension steel is

$$f'_s = 87\left(\frac{d - c}{c}\right) = 87\left(\frac{17.5 - 17.06}{17.06}\right) = 2.24 \text{ ksi}$$

$$e' = 3.60 + 7.50 = 11.10 \text{ in.}$$

$$\frac{P_u e'}{\phi} = 0.85 f'_c ab \left(d - \frac{a}{2} \right) + A'_s f'_y (d - d')$$

$$\frac{P_u 11.10}{\phi} = 0.85 \times 4 \times 14.5 \times 14 \left(17.5 - \frac{14.5}{2} \right)$$
$$+ 3 \times 60(17.5 - 2.5)$$

$$\frac{P_u}{\phi} = 880.59 \text{ kips}$$

Find a using the equation

$$P_u = 0.85 f'_c ab + A'_s f'_y - A_s f_s$$
$$880.59 = 0.85 \times 4 \times a \times 14 + 3.0 \times 60 - 3.0 \times 2.24$$

or $\qquad a = 14.8$

close enough to the assumed value.

For $a = 14.8$, $\dfrac{P_u}{\phi} = 880.59$ kips, $\dfrac{M_u}{\phi} = 880.59 \times \dfrac{3.60}{12}$

$$= 264.18 \text{ ft-kips}$$

Usable load $P_u = 0.70 \times 880.59 = 616.41$ kips > 532 kips
Usable moment $M_u = 0.70 \times 264.18 = 184.93$ kips > 160 kips
Use #3 ties (see ACI 7.12.3). Spacing of ties

$$16 \times 1.13 = 18.08 \text{ in.}$$
$$48 \times 0.38 = 18.24 \text{ in.}$$
$$\text{Least dimension} = 14 \text{ in.}$$

Use #3 ties 14 in. on centers.

7.57 Assume $d' = 2.5$ in. Then $d = 22 - 2.5 = 19.5$ in. Determine e_b:

$$P_u = 0.85 f'_c \beta_1 \left(\frac{87}{87 + f_y} \right) = 0.85 \times 4 \times 14 \times 19.5$$
$$\times 0.85 \left(\frac{87}{147} \right) = 466.94 \text{ kips}$$

$$a_b = \frac{466.94}{0.85 \times 4 \times 14} = 9.81 \text{ in.}$$

$$\frac{P_{u_b} e_b'}{\phi} = 0.85 f_c' a_b b \left(d - \frac{a}{2} \right) + A_s' f_y' (d - d')$$

$$\frac{466.94 e_b'}{0.70} = 0.85 \times 4 \times 9.81 \times 14 \left(19.5 - \frac{9.81}{2} \right)$$
$$+ 2 \times 1.27 \times 60(19.5 - 2.5)$$

$$e_b' = 14.10 \text{ in.}$$
$$e_b = 14.10 - 8.55 = 5.55 < 16 \text{ in.}$$

Therefore tension controls.

$$\rho = \frac{A_s}{bd} = \frac{2 \times 1.27}{14 \times 19.5} = 0.0093$$

$$e' = e + 8.5 = 16 + 8.5 = 24.5$$

$$m = \frac{f_y}{0.85 f_c'} = \frac{60}{0.85 \times 4} = 17.65$$

$$\left(\frac{e'}{d} - 1 \right)^2 = \left(\frac{24.5}{19.5} - 1 \right)^2 = (0.256)^2 = 0.066$$

$$2\rho m \left(1 - \frac{d'}{d} \right) = 2 \times 0.0093 \times 17.65 \left(1 - \frac{2.5}{19.5} \right) = 0.286$$

$$\frac{P_u}{\phi} = 0.85 f_c' bd \left[- \left(\frac{e'}{d} - 1 \right) \right.$$
$$\left. + \sqrt{\left(\frac{e'}{d} - 1 \right)^2 + 2\rho m \left(1 - \frac{d'}{d} \right)} \right]$$

$$\frac{P_u}{\phi} = 0.85 \times 4 \times 14 \times 19.5 \, [- 0.256$$
$$+ \sqrt{0.066 + 0.286}]$$
$$= 312.8 \text{ kips}$$

Usable load $P = 0.7 \times 312.8 = 218.97$ kips   **b**

7.58   Ultimate loads:
$$\frac{P_u}{\phi} = \frac{1}{0.7} (1.4 \times 350 + 1.7 \times 153) = 1071.43 \text{ kips}$$

At the top:
$$\frac{M_2}{\phi} = \frac{1}{0.7} (1.4 \times 50 + 1.7 \times 50) = 221.43 \text{ ft-kips}$$

At the bottom:

$$\frac{M_1}{\phi} = \frac{1}{0.7} \ (1.4 \times 20 + 1.7 \times 15) = 76.43 \ \text{ft-kips}$$

Slenderness effect is

$$r = 0.3 \times 24 = 7.2$$

Since the column is not braced against sidesway, $k > 1$ (ACI 10.11.3). Since it bends in double curvature, $k = 2$.

$$\frac{kl_u}{r} = \frac{2 \times 17 \times 12}{7.2} = 56.66 > 22$$

the slenderness effect has to be considered (ACI 10.11.4).

$$I_g = \frac{1}{12} \times 24 \times 24^3 = 27{,}648 \ \text{in.}^4$$

Distance of centroid of reinforcement to centroidal axis of section is

$$\frac{24}{2} - 2.14 = 9.86 \ \text{in.}$$

where $2.14 = d'$.

Total $A_s = 8 \times 1.27 = 10.16$ sq in.
Then $I_{se} = 10.16(9.86)^2 = 987.75$ in.^{4}
        $E_s = 29 \times 10^6$ psi
        $E_c = 57{,}000 \ \sqrt{f'_c} = 4.03 \times 10^6$ psi   ACI 8.3.1

Top moment M_2 governs.

$$\beta_d = \frac{50 \times 1.4}{50(1.4 + 1.7)} = \frac{70}{155} = 0.45$$

$$EI = \frac{(E_c I_g / 5) + (E_s I_{se})}{1 + \beta_d}$$

$$= \frac{4.03 \times 2.76 \times 10^{10} \times 0.2 + 29 \times 9.88 \times 10^8}{1.45}$$

ACI 10.11.5
(10-7)

or $EI = 3.5 \times 10^{10}$

Then

$$P_{cr} = \frac{\pi^2 EI}{(kl_u)^2} = \frac{9.86 \times 3.5 \times 10^{10}}{(2 \times 17 \times 12)^2 \times 10^3} = 2{,}073.00$$

ACI 10.11.5
(10-6)

Column is not braced against sidesway, therefore $c_m = 1$.

$$\delta = \frac{c_m}{1 - (P_u/\phi P_c)} = \frac{1}{1 - (1,071.43/2,073.00)}$$
$$= 2.08 \qquad \text{ACI 10.11.5}$$
$$\text{(10-5)}$$

Magnified moment $M_c = \delta M_2 = 2.08 \times 221.43 = 460$ ft-kips

Design for an axial load of $1,071.43/\phi$ kips and a moment of $460/\phi$ ft-kips. For these values

$$e = \frac{460 \times 12}{1,071.43} = 5.16 \text{ in.}$$

Find e_b:

$$a_b = \frac{1,071.43}{0.85 \times 5 \times 24} = 10.5 \text{ in.}$$

$$1,071.43 e_b' = 0.85 \times 5 \times 10.5 \times 24 \left(21.86 - \frac{10.5}{2} \right)$$
$$+ 5.08 \times 60(21.86 - 2.14)$$

$$e_b' = 22.21 \text{ in.}$$
$$e_b = 22.21 - 9.86 = 12.35 \text{ in.}$$

$e < e_b$, therefore compression controls.

Proceed using successive approximations. Assume $a = 16.00$ in.

$$c = \frac{16.00}{0.80} = 20.00$$

$$f_s = 87 \left(\frac{21.86 - 20.00}{20.00} \right) = 8.1 \text{ ksi}$$

$$e' = 5.16 + 9.86 = 15.02 \text{ in.}$$

$$15.02 \frac{P_u}{\phi} = 0.85 \times 5 \times 16.00 \times 24 \left(21.86 - \frac{16.00}{2} \right)$$
$$+ 5.08 \times 60(21.86 - 2.14)$$

$$\frac{P_u}{\phi} = 1,905 \text{ kips}$$

$$1,905 = 0.85 \times 5 \times 24a + 5.08(60 - 8.1)$$
or $\qquad\qquad a = 16.1 \text{ in.}$

close enough to the assumed value.

For $a = 16.1$, $P_u/\phi = 1,905 > 1,071.43$. Column is adequate.

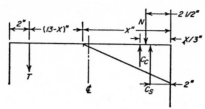

FIG. 7.59

Take $\Sigma M_N = 0$

Force	Area	Intensity	Arm	Factor (to simplify = moment products)	
$+C_c$	$24x$	$f_c/2$	$(+)\left(\dfrac{x}{3} - 2.5\right)$	$x/4f_c$	$= +x^2(x - 7.5)$
$+C_s$	$(18 - 1)3$	$\dfrac{(x - 2)}{x} f_c$	$(-)(0.5)$	$x/4f_c$	$= -6.375(x - 2)$
$-T$	9×3	$\dfrac{(13 - x)}{x} f_c$	$(+)(10.5)$	$x/4f_c$	$= -70.875(13 - x)$

$$\Sigma M_N = 0 = +x^2(x - 7.5) - 6.375(x - 2) + 70.875(x - 13)$$
$$x^3 - 7.5x^2 + 64.5x - 908.56 = 0$$

Solve for x by trial.

$$x = 10.00, \ +1,000 - 750 + 645 - 908.56 = -13.56$$
$$x = 10.06, \ +1,018.1 - 759 + 647.7 - 908.6 = -1.8$$
$$x = 10.07, \ +1,021.1 - 760.5 + 649.5 - 908.6 = +1.5$$

Use $x = 10.06$ in. $> \dfrac{15}{4}$ in. $= 3.75$ in. min

$$+C_c = 24 \times 10.06 \times f_c/2 \quad = +120.8f_c$$
$$+C_s = 17 \times 3 \times \frac{8.06}{10.06} \times f_c = +\ \underline{40.9f_c}$$
$$+161.7f_c$$

$$-T = \ 9 \times 3 \times \frac{2.94}{10.06} \times f_c = -\ \underline{7.9f_c}$$
$$+153.8f_c$$

$$\Sigma V = +C_c + C_s - T - N = 0$$
$$N = 153.8f_c = 150,000$$
$$f_c = 975 > 750 \text{ psi } (0.25f_c') < 1,125 \text{ psi } (0.375f_c')$$

$$f_s \text{ (tension)} = 9 \times \frac{2.94}{10.06} \times 975 = 2{,}570 \text{ psi} < 20{,}000 \text{ psi}$$

$$f_s \text{ (compression)} = 18 \times 975 \times \frac{8.06}{10.06} = 14{,}100 \text{ psi} < 20{,}000 \text{ psi}$$

7.60　Assume a W 12 × 65 (A36 steel)

Area $= A_r = 19.11$ sq in., $d = 12.12$ in., $b = 12$ in.

Diameter of enclosing circle for W $= 17.1$ in. Use spiral outside diameter $= 22$ in. and inside diameter $= 21.25$ in.

$$A_{col} = \frac{\pi}{4} (25)^2 = 490 \text{ sq in.}$$

$$D_s = 25 - 3 - 0.75 - 1.41 = 19.84 \text{ in.} \quad (18.43 \text{ in. clear inside})$$

Use ten #11 bars, $A_s = 15.6$ sq in. and $p = \dfrac{15.6}{490} = 0.0317$

Clear spacing $= (3.14 \times 19.84/10) - 1.41 = 4.82$

$A_{conc} = A_{col} - A_{st} - A_r = 490 - 15.6 - 19.11 = 455$ sq in.

The ultimate axial load that the column may carry is

$$\frac{P_u}{\phi} = 0.85 f_c' A_{conc} + A_r f_{ry} + A_s f_y$$

$$= 0.85 \times 3.0 \times 455 + 19.11 \times 36 + 15.6 \times 40$$

$$= 1{,}160.3 + 687.6 + 624.0 = 2{,}471.9 \text{ kips}$$

The working stress load is $P_u(C/\phi)$, and C/ϕ may be taken to be 0.40.

$$P = 0.40 \times 2{,}471.9 = 988.8 \text{ kips} < 1{,}000 \text{ kips}$$

If W 12 × 65 is changed to W 12 × 72, $P = 1{,}016$ kips.　　OK

Spiral steel:

$$\rho_s = 0.45 \left(\frac{A_g}{A_c} - 1 \right) \frac{f_c'}{f_y'}$$

$$\rho_s = 0.45 \left(\frac{490}{380} - 1 \right) \frac{3{,}000}{60{,}000} = \frac{0.45}{20} (1.296 - 1) = 0.00652$$

Try ⅜-in. spiral, $a_s = 0.11$

$$g = \frac{4a_s}{\rho_s D_c} = \frac{4 \times 0.11}{0.00652 \times 22} = 3.07 \text{ in.}$$

Use $g = 3$ in. (maximum) $< 2\frac{2}{6}$

7.61 The area and moment of inertia of W 12×65 are $A_t = 19.1$ in.^{2} and $I_t = 533.0$ in.^{4}. Assume an 18 by 18 in. concrete encasement.

$$A_{\text{conc}} = 18 \times 18 - 19.1 = 304.9 \text{ in.}^2$$

The ultimate axial load the above section can carry is

$$\frac{P_u}{\phi} = 0.85 f'_c A_{\text{conc}} + A_s f_y$$

Assume a concrete with $f'_c = 3$ ksi and A36 structural steel.

$$\frac{P_u}{\phi} = 0.85 \times 3 \times 304.9 + 19.1 \times 36 = 777.5 + 687.6$$
$$= 1{,}465.1 \text{ kips}$$

The working stress axial load is $P = C P_u / \phi$ where $C/\phi = 0.40$.

$$P = 0.40(1{,}465.1) = 586 \text{ kips}$$

Check to see if there is slenderness effect (ACI 10.11.4 10.15.3).

$$I_g = \frac{18 \times 18^3}{12} - 533 = 8{,}215 \text{ in.}^4$$

$$E_c = 57{,}000 \sqrt{3{,}000} = 3.12 \times 10^3 \text{ ksi}$$

$$r = \sqrt{\frac{3.12 \times 10^3 \times 8{,}215/5 + 29 \times 10^3 \times 533}{3.12 \times 10^3 \times 304.9/5 + 29 \times 10^3 \times 19.1}}$$
$$= 5.26 \quad (10\text{-}10)$$

Assume $k = 1$, $k l_u / r = 1 \times 20 \times 12/5.26 = 45.63 < 100$. No slenderness effect.

7.62 Assume a 12-in. extra strong pipe, with $f_y = 36$ ksi. OD = 12.75 in., ID = 11.75 in., thickness of wall = 0.5 in., $A = 19.2$ in.^{2}, $I = 362$ in.^{4}

ACI requires that thickness of pipe is larger than

$$h \sqrt{\frac{f_y}{8 E_s}} \quad \text{or} \quad 12.75 \sqrt{\frac{36}{8 \times 29{,}000}} = 0.159 < 0.500 \text{ in.}$$

Assume a concrete with $f'_c = 3$ ksi.

$$A_c = \frac{\pi}{4} (11.75)^2 = 108.38 \text{ in.}^2$$

The ultimate axial load is

$$\frac{P_u}{\phi} = 0.85f'_c A_c + f_y A_s = 0.85 \times 3 \times 108.38 + 36 \times 19.2$$
$$= 967.57 \text{ kips}$$

The working stress load is

$$P = 0.40(967.57) = 387.03 \text{ kips} > 350 \text{ kips}$$

Check to see if there is a slenderness effect.

$$E_c = 57,000 \sqrt{3,000} = 3.12 \times 10^3 \text{ ksi}$$
$$I_g = 0.785 \left(\frac{11.75}{2}\right)^4 = 935.19 \text{ in.}^4$$
$$r = \sqrt{\frac{3.12 \times 10^3 \times 935.19/5 + 29 \times 10^3 \times 362}{3.12 \times 10^3 \times 108.38/5 + 29 \times 10^3 \times 19.2}}$$
$$= 4.21 \qquad (10\text{-}10)$$

Assume $k = 1$, $kl_u/r = 1 \times 20 \times 12/4.21 = 57.01 < 100$. No slenderness effect.

7.63 Find the area of the footing required to satisfy the soil conditions. Total dead and live load is

$$P = 350 + 125 = 475 \text{ kips}$$

Assume a footing 2'-0" deep. Then the weight of the soil above footing and the weight of the footing are

$$\text{Soil weight} = 2.5 \times 0.120 = 0.300 \text{ kips per sq ft}$$
$$\text{Footing weight} = 2.00 \times 0.150 = 0.300 \text{ kips per sq ft}$$
$$\text{Total} = 0.600 \text{ kips per sq ft}$$

Net bearing soil capacity $= 2.75 \times 2 - 0.600 = 4.9$ kips per sq ft

Required area of footing $= \dfrac{475}{4.90} = 96.94$ sq ft

Use a footing: 9'-10" by 9'-10" ($A = 96.63$ sq ft). Design the footing.

$$P_u = 1.4 \times 350 + 1.7 \times 125 = 702.5 \text{ kips}$$

Soil pressure is

$$w_u = \frac{702.5}{96.63} = 7.27 \text{ kips per sq ft}$$

Punching shear usually controls, so

$$v_c = 4 \sqrt{f'_c} = 4 \sqrt{4,000} = 253.00 \text{ psi} \qquad \text{ACI 11.10.3}$$

The critical section is at $0.5d$ from the face of the column (see ACI 11.10.2). Perimeter of the critical section is

$$b_p = 4(2\%_{12} + d) = 6.67 + 4d$$
$$V_u = P_u - w_u(2\%_{12} + d)^2 = 702.5 - 7.27(1.67 + d)^2$$
$$v_u = \frac{V_u}{\phi b_p d} = \frac{702.5 - 7.27(1.67 + d)^2}{0.85(6.67 + 4d)d}$$

$$v_c = \frac{253.0 \times 144}{1,000} = 36.36 \text{ kips per sq ft}$$

The above equation yields $d = 1.56$ ft or $d = 18.72$ in. Use $d = 19$ in. Check beam shear. Critical section at d from face of column.

$$v_c = 2 \sqrt{f'_c} = 2 \sqrt{4,000} = 126.50 \text{ psi}$$

The distance of the critical section from the edge of the footing is

$$l = \frac{118 - 20}{2} - 19 = 30 \text{ in. or } 2.5 \text{ ft}$$
$$V_u = 2.5 \times 7.27 = 18.18 \text{ kips per ft of width}$$
$$v_u = \frac{V_u}{\phi b d} = \frac{18,180}{0.85 \times 12 \times 19}$$
$$= 93.81 \text{ psi} < 126.5 \qquad \text{OK} \qquad \text{ACI 11.1.1}(a)$$

Determine flexural steel. The critical section at face of wall [see ACI 15.4.2(a)].

$$\frac{M_u}{\phi} = \frac{1}{2} \times \frac{7.27 \times 4.08^2}{0.90} = 67.23 \text{ ft-kips}$$

$$\text{Minimum steel ratio } \rho = \frac{200}{f_y} = \frac{200}{50,000} = 0.004$$

Use minimum steel ratio $A_s = 12 \times 19 \times 0.004 = 0.912$ sq in. per ft of width.

Total steel $A_s = 0.912 \times 9.83 = 8.96$ sq in. Use nine #9 bars. Development length required is

$$l_d = \frac{0.04 A_b f_y}{\sqrt{4,000}} = \frac{0.04 \times 1.0 \times 50,000}{63.25} = 31.62 \text{ in.} \qquad \text{ACI 12.5}$$

but not less than $0.0004 d_b f_y = 0.004 \times 1.13 \times 50,000 = 22.6$ in.
Development provided $= (9.83 - 4.08 - 0.25)12$

$$= 66.00 \text{ in.} \quad \text{OK}$$

7.64 Use soil pressure from column loads $= 5,500$ psf.

$$\text{Length of footing} = \frac{700,000}{9 \times 5,500}$$

$$L = 14.17 \text{ ft}$$

$$M_{xy} = 9 \times \frac{5.91^2}{2} \times 5,500 = 862,000 \text{ ft-lb}$$

$$M_{PQ} = 14.17 \times \frac{3.33^2}{2} \times 5,500 = 432,000 \text{ ft-lb}$$

$$d_{(xy)}^2 = \frac{M}{Rb} = \frac{862,000}{223 \times 9} = 429, \, d = 20.7 \text{ in.}$$

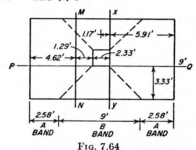

Fig. 7.64

Use $d = 31.5$ in., over-all depth $= 34.5$ in., with weight under
500 psf.

$$V = \left[2.58 \times 9 + \frac{(9 + 2.58 + 2.33)}{2} \times 2.04 \right] \times 5,500$$

$$V = 127,800 + 77,500 = 205,300 \text{ lb}$$

$$v = \frac{V}{bd} = \frac{205,300}{59.5 \times 31.5} = 109.5 \text{ psi} < 110 \text{ psi (no stirrups)}$$

Long steel for 9-ft width:

$$A_s = \frac{M}{f_s jd} = \frac{862,000 \times 0.85 \times 12}{20,000 \times 0.874 \times 31.5} = 15.92 \text{ sq in.}$$

Use sixteen #9 bars, $A_s = 16.0$ sq in.

$$u = \frac{V}{\Sigma_0 jd} = \frac{205,300}{56.8 \times 0.874 \times 31.5} = 131.5 \text{ psi} < 233 \text{ psi}$$

Short steel for 14-ft 2-in. length

$$A_s = \frac{0.85 \times 432,000 \times 12}{20,000 \times 0.874 \times 31.5} = 7.98 \text{ sq in.}$$

$$A_s \text{ for } B \text{ Band} = 7.98 \times \frac{2}{(s+1)}, \text{ where } s = \frac{14.17}{9} = 1.575$$

$$= 7.98 \times \frac{2}{2.575} = 6.20 \text{ sq in.}$$

Use eleven #7 bars on 10-in. centers, $A_s = 6.60$ sq in.

$$u = \frac{77,500}{30.24 \times 0.874 \times 31.5} = 93 \text{ psi} < 301 \text{ psi}$$

$$A_s \text{ for } A \text{ Bands} = \frac{7.98 - 6.20}{2} = \frac{1.78}{2} = 0.89 \text{ sq in.}$$

Use three #6 bars on 10-in. centers in each 31-in. end A Band.
$A_s = 1.32$ sq in.

7.65

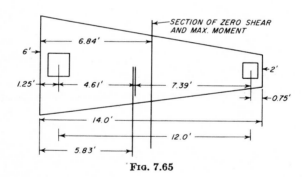

Fig. 7.65

Center-of-gravity distance of column loads from center
line 120 kips load $= \dfrac{12 \times 75}{(120 + 75)} = \dfrac{900}{195} = 4.61$ ft

Use 3,500 psf as working pressure on underside of concrete
slab from column loads.

$$\text{Area of footing} = \frac{195,000}{3,500} = 56 \text{ sq ft}$$

Trapezoidal footing 14 ft long, width 2 to 6 ft, and A of
56 sq ft.

Center of gravity of footing from 6 ft end = **5.83 ft**

$$2 \times 14 \qquad = 28 \times 7 \text{ ft} \quad = 196$$
$$2 \times \frac{14 \times 2}{2} = 28 \times 4.67 \text{ ft} = \underline{130.7}$$
$$\overline{56 \; (5.83 \text{ ft})} \; = \overline{326.7}$$

Place center line of 15 × 15 column 15 in. from 6 ft **end of** footing

Place center line of 12 × 12 column 9 in. from 2 ft end of footing

The center of gravity of column loads and the center **of** gravity of upward pressure coincide within **0.03 ft.** Say OK

Find distance from left end to point of zero shear

$$6 \times 3{,}500 = 21 \text{ kips per sq ft at 6 ft end}$$

Upward pressure varies from 21 kips to 7 kips in **14 ft or** by 1 kip per ft

$$+21x - 1x^2/2 - 120 = 0$$
$$+42x - x^2 - 240 = 0$$
$$x^2 - 42x + (21)^2 = -240 + (21)^2 = +201$$
$$(x - 21) = \pm \sqrt{201} = \pm 14.16$$
$$x = +21 - 14.16 = 6.84 \text{ ft} = \text{point of maximum \textbf{moment}}$$
$$M_{(max)} = (21 - 6.84) \text{ kips} \times \frac{6.84^2}{2} + 6.84 \text{ kips} \times \frac{6.84}{2}$$
$$\times \tfrac{2}{3} \times 6.84 - 120 \times 5.59 \quad \text{ft} = 330.5 + 106.5 - \mathbf{671}$$
$$= 234 \text{ ft-kips}$$

Width 8.5 in. inside face of 15 × 15 column (point *A*)

$$= 6 - \frac{2.58 \times 4}{14} = 6 - 0.74 = 5.26 \text{ ft}$$

Width 8.5 in. inside face of 12 × 12 column (point *B*)

$$= 2 + \frac{1.96 \times 4}{14} = 2 + 0.56 = 2.56 \text{ ft}$$

Width at point of maximum moment $= 6 - \dfrac{6.84 \times 4}{14}$

$$= 6 - 1.95 = 4.05 \text{ ft}$$

Shear at point $A = 21 \times 2.58 - \dfrac{2.58^2 \times 1}{2} - 120$

$\qquad = 54.2 - 3.33 - 120 = 69.13$ kips

Shear at point $B = 7 \times 1.96 + 1 \times \dfrac{1.96^2}{2} - 75$

$\qquad = 13.7 + 1.92 - 75 = 59.38$ kips

$k = \dfrac{1,350}{1,350 + 20,000/9} = 0.377,\ j = 0.874$

$R = \dfrac{1,350}{2} \times 0.874 \times 0.377 = 223$

$v = \dfrac{V}{bd}$ and $d = \dfrac{V}{vb} = \dfrac{59,380}{110 \times 12 \times 2.56} = 17.5$ in. $+ 3$ in.

$\qquad\qquad\qquad\qquad\qquad\qquad\qquad\qquad\quad = 20.5$ in. over all

$\qquad\qquad$ weight $= 248$ psf $< 500\quad \therefore$ OK

Moment concrete could take $= M_c = Rbd^2$

$\qquad = 223 \times 4.05 \times 17.5 \times 17.5$

$\qquad = 275$ ft-kips > 234 ft-kips $\quad \therefore$ OK

$A_s = \dfrac{M}{f_s jd} = \dfrac{234,000 \times 12}{20,000 \times 0.874 \times 17.5} = 9.18$ sq in.

Use six #11 bars, $A = 9.36$ sq in.

$\qquad$ Maximum lateral bending $= 3,500 \times \dfrac{2.38^2}{2} \times 12$

$\qquad\qquad = 119,000$ in-lb

$A_s = \dfrac{M}{f_s jd} = \dfrac{119,000}{20,000 \times 0.874 \times 17.5} = 0.388$ sq in. per ft

Use #4 bars 6 in. on centers, $A_s = \mathbf{0.40}$ sq in. per ft.

For longitudinal bottom steel, and transverse steel top and bottom, use #4 bars at suitable spacings.

7.66   6 to 12 triangle $= 1, 2, \sqrt{5}$

$\cos^2 \theta = \dfrac{(2)^2}{5} = 0.80,\ \sin^2 \theta = \dfrac{(1)^2}{5} = 0.20$

$N_1 = \dfrac{1,200 \times 380}{1,200 \times 0.20 + 380 \times 0.80}$

$\qquad = \dfrac{455,000}{240 + 304} = \dfrac{455,000}{544} = 836$ psi

Fig. 7.66

$$N_2 = \frac{1{,}200 \times 380}{1{,}200 \times 0.80 + 380 \times 0.20} = \frac{455{,}000}{960 + 76}$$

$$= \frac{455{,}000}{1{,}036} = 439 \text{ psi}$$

Dressed size of timbers = 5.5 × 5.5 in.

A 4-in. diameter washer has 11 sq in. bearing area

Washer bearing = 4,500/11 = 409 psi < 439 psi, ∴ OK

$$6 \text{ by } 6 \text{ bearing against } 6 \text{ by } 6 = \frac{9{,}000}{4.9 \times (5.5 - 1)}$$

$$= \frac{9{,}000}{4.9 \times 4.5} = 408 \text{ psi} < 836 \text{ psi, } \therefore \text{ OK}$$

2 × 2.45 in. = 4.90 in. > 4 in. (diameter of washer)

7.67   4 by 6 bearing on plane of notch in 6 by 6 = $\dfrac{5{,}000}{3.5 \times 5.5}$

= 260 psi < 836 psi (see preceding answer), ∴ OK

7.68   Let N be the load per square foot.

For bearing:

$$V = 1.33 \times 14 \times N \times \tfrac{1}{2} = 9.33N$$

$$s_b = \frac{V}{bl} = \frac{9.33N}{2.62 \times 6} = 380$$

$$N = 641 \text{ psf}$$

For shear:

$$s_s = \frac{3}{2}\frac{V}{A} = \frac{3 \times 9.33N}{2 \times 2.62 \times 11.5} = 120$$

$$N = 259 \text{ psf}$$

For bending:

$$f = \frac{M}{S} = \frac{\tfrac{1}{8}(1.33N) \times 14 \times 14 \times 12}{\tfrac{1}{6} \times 2.62 \times 11.5 \times 11.5} = 1{,}600$$

$$N = 236 \text{ psf}$$

For deflection:

$$\frac{\text{Span}}{360} = \frac{5wl^4}{384EI} = \frac{14 \times 12}{360} = \frac{5(1.33N) \times 14^4 \times 12^3}{384 \times 1{,}600{,}000 \times 2.62 \times 11.5^3/12}$$

$$N = 216 \text{ psf}$$

This is the largest loading to satisfy all four conditions.

7.69 From preceding answer, deflection is most critical.

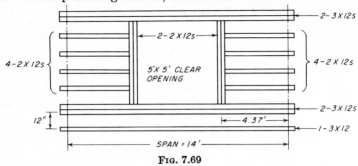

Fig. 7.69

Δ for one 3 by 12 $= \dfrac{5 \times 250 \times 14^4 \times 1{,}728}{384 \times 1{,}600{,}000 \times 333} = \dfrac{5wl^4}{384EI}$

$= 0.406$ in. $<$ allowable $= \dfrac{14 \times 12}{360} = 0.467$ in.

Δ for two concentrated loads on one 3 by 12

$F =$ concentrated load $= 2 \times 250 \times \dfrac{4.37}{2} = 1{,}092$ lb

$a = 4.37$ ft
$l = 14$ ft
$\Delta = \dfrac{Fa(3l^2 - 4a^2)}{24EI}$

$= \dfrac{1{,}092 \times 4.37(3 \times 14 \times 14 - 4 \times 4.37 \times 4.37)1{,}728}{24 \times 1{,}600{,}000 \times 333}$

$= 0.330$ in. $<$ 0.467 in.

The two 3 by 12s will have a $\Delta = 0.368$ in. $<$ 0.467 in.

$V = 7 \times 250 + 1{,}092 = 2{,}842$

$v = \dfrac{3}{2}\dfrac{V}{A} = \dfrac{3 \times 2{,}842}{2 \times 5.25 \times 11.5} = 71$ psi < 120, $\therefore$ OK

$f = \dfrac{(\tfrac{1}{8} \times 250 \times 14 \times 14 \times 12) + (1{,}092 \times 4.37 \times 12)}{\tfrac{1}{6} \times 5.25 \times 11.5 \times 11.5}$

$= \dfrac{73{,}500 + 57{,}500}{115.7} = 1{,}132$ psi $< 1{,}600$, $\therefore$ OK

7.70 If four 2 by 8s are placed on edge side by side

$I = 4 \times \tfrac{1}{12} \times 1.62 \times \overline{7.5}^3 = 4 \times 57 = 228$ in.^{4}

If placed in the form of an I or hollow box of 10.75 in. depth

$$I = \tfrac{1}{12} \times 7.5 \times 10.75^3 - \tfrac{1}{12}(7.5 - 2 \times 1.62) \times 7.5^3$$
$$= 777 - 149 = \mathbf{628} \text{ in.}^4 \text{ (use)}$$

7.71 One 2 by 8, $A = 1.62 \times 7.5 = 12.2$ sq in.

$$Q = \text{statical moment} = 12.2 \times 4.56 + 12.2 \times \frac{3.75}{2}$$
$$= 55.7 + 22.8 = \mathbf{78.5} \text{ in.}^3$$

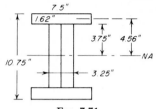

Fig. 7.71

If placed in the form of an I,

$$v = \frac{VQ}{bI} \text{ or } \frac{V}{v} = \frac{bI}{Q} = \frac{3.25 \times 628}{78.5} = \mathbf{260} \text{ sq in.}$$

If placed side by side on edge,

$$v = \frac{3V}{2A} \text{ or } \frac{V}{v} = \tfrac{2}{3}A = \tfrac{2}{3} \times 4 \times 12.2 = \mathbf{32.4} \text{ sq in. (use)}$$

7.72 Dense southern yellow pine has the following properties:

$$E = 1,600,000 \text{ psi}$$

$$\frac{P}{A} \text{ for } \frac{L}{d} < 10 = 1,285 = f_c$$

$$\frac{P}{A} \text{ for } \frac{L}{d} > 10 = 1,285C$$

$$k = \frac{\pi}{2}\sqrt{\frac{E}{4f_c}} = \frac{3.14}{2}\sqrt{\frac{1,600,000}{4 \times 1,285}} = 27.8 \text{ or } 28$$

$$C = \left[1 - \frac{1}{3}\left(\frac{L}{kd}\right)^4\right]$$

$$\frac{L}{d} = \frac{6 \times 12}{5.5} = 13.1$$

$$C = \left[1 - \frac{1}{3}\left(\frac{13.1}{28}\right)^4\right] = 0.98$$

Allowable $\dfrac{P}{A} = 1{,}285 \times 0.98 = 1{,}260$ psi

$$N = 1{,}260 \times (5.5 \times 9.5) = 65{,}800 \text{ lb} > 60{,}000 \text{ lb}$$

Use **6 by 10.**

7.73   Try a 6 by 6.

$$\frac{L}{d} = \frac{12 \times 12}{5.5} = 26.2$$

$$C = \left[1 - \frac{1}{3}\left(\frac{26.2}{28}\right)^4\right] = 0.75$$

$$\frac{P}{A} = 1{,}285 \times 0.75 = 963 \text{ psi}$$

$$N = 963 \times (5.5 \times 5.5) = 29{,}100 \text{ lb} > 27{,}000 \text{ lb}$$

$\therefore$ **6 by 6** is OK.

7.74   Try four 2 by 10s.

Box is $1.62 + 9.5 = 11.12$ by 11.12 in. outside

$$\frac{L}{d} = \frac{12 \times 12}{11.12} = 13$$

$$C = [1 - \tfrac{1}{3}(1\tfrac{3}{28})^4] = 0.98$$

Allowable $\dfrac{P}{A} = 1{,}285 \times 0.98 = 1{,}260$ psi

$$N = 4 \times (1.62 \times 9.5) \times 1{,}260 = 77{,}600 \text{ lb} > 75{,}000 \text{ lb}$$

$\therefore$ **four 2 by 10s** are OK.

7.75 4×8 is 3.62 in. $\times$ 7.5 in. dressed size, $A = 27.2$ sq in.

         2×4 is 1.62 $\times$ 3.62 dressed size, $A = 5.89$ sq in.

$$\underline{5.89}$$
$$38.98 \text{ sq in}$$

Least dimension is $(1.62 + 1.62 + 3.62) = 6.86$ in. < 7.5 in.

$$\frac{L}{d} = \frac{10 \times 12}{6.86} = 17.5$$

$$C = \left[1 - \frac{1}{3}\left(\frac{17.5}{28}\right)^4\right] = 0.949$$

Allowable $\dfrac{P}{A} = 1{,}285 \times 0.949 = 1{,}220$ psi

$$N = 38.98 \times 1{,}220 = \textbf{47{,}500 lb}$$

LAND SURVEYING

8.01

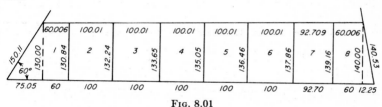

Fig. 8.01

$$150.11 \times \cos 60° = \ \ 75.05 \text{ ft}$$
$$150.11 \times \sin 60° = 130.00 \text{ ft}$$
$$140.53 \times \cos 85° = 12.25 \text{ ft}$$
$$140.53 \times \sin 85° = 140.00 \text{ ft}$$
$$75.05 + 60 + 5 \times 100 + 60 + 12.25 = 707.30 \text{ ft}$$

South dimension of lot 7 = 800 − 707.30 = 92.70 ft
Lots 1 to 8 change in vertical dimension (N-S) from 130.00 ft to
140.00 ft in 800 − (75.05 + 12.25) = 712.70 ft

$$\frac{10}{712.70} \times \ \ 60 \ \ = 0.84 \text{ ft} + 130.00$$

$$= 130.84 \text{ ft east dimension lot 1}$$

$$\frac{10}{712.70} \times 160 \ \ = 2.34 + 130.00$$

$$= 132.24 \text{ ft west dimension lot 3}$$

575

$$\frac{10}{712.70} \times 260 \quad = 3.65 + 130.00$$

$$= 133.65 \text{ ft east dimension lot 3}$$

$$\frac{10}{712.70} \times 560 \quad = 7.86 + 130.00$$

$$= 137.86 \text{ ft west dimension lot 7}$$

$$\frac{10}{712.70} \times 652.70 = 9.16 + 130.00$$

$$= 139.16 \text{ ft east dimension lot 7}$$

North dimension of lots 1 to 8 is longer than the corresponding south dimension by $\frac{10 \times 10}{2 \times 712.70} = 0.07$ ft. Add the following amounts to the south dimensions of lots 1, 3, 7, and 8 to get the north dimensions

$$\text{Lots 1 and 8 add } \frac{0.07}{712.7} \times 60 = 0.006 \text{ ft}$$

$$\text{Lot } \quad 3 \quad \quad \text{add } \frac{0.07}{712.7} \times 100 = 0.010 \text{ ft}$$

$$\text{Lot } \quad 7 \quad \quad \text{add } \frac{0.07}{712.7} \times 92.7 = 0.009 \text{ ft}$$

8.02 Beginning at the northwest corner of the intersection of South Mountain Avenue and Salter Place in the town of Maplewood, Essex County, N.J., and running along the said Salter Place N62°54′30″E a distance of one hundred ninety seven and ninety eight hundredths (197.98′) feet to an iron pipe in the ground which is the southeast corner of the property; thence turning and running N27°05′30″W a distance of one hundred and thirty eight and ninety five hundredths (138.95′) feet along the land now or formerly belonging to James Martin to an iron pipe at land now or formerly belonging to Joseph Reed; thence turning and running N62°54′30″E along the land of said Reed a distance of exactly fifty (50.00′) feet to an iron pipe at land now or formerly belonging to Charles A. Poe; thence turning and running along the land of said Poe S27°05′30″E a distance of one hundred and thirty eight and ninety five hundredths

(138.95′) feet to a drill hole in the concrete drive way at Salter Place; thence turning and running S62°54′30″W along the north-westerly line of Salter Place a distance of exactly fifty (50.00′)

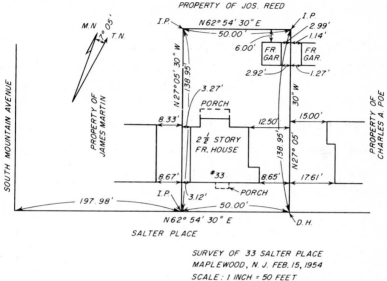

SURVEY OF 33 SALTER PLACE
MAPLEWOOD, N.J. FEB. 15, 1954
SCALE: 1 INCH = 50 FEET
SIGNATURE, ADDRESS, SEAL
(OF LICENSED LAND SURVEYOR)

FIG. 8.02

feet to the iron pipe which is the southeast corner of the property; all bearings being magnetic and the parcel containing a calculated area of 6947.50 square feet more or less.

8.03 The plan should show U.S. pierhead and bulkhead lines, any state pierhead and bulkhead lines, high and low water lines, tributary water courses, drainage ditches, etc., and any structures such as piers, wharves, bulkheads, jetties, groins, etc. The survey must conform to any scale and other special requirements established by whatever state department has jurisdiction over riparian properties.

8.04 Distances and bearings between monuments and/or markers of undisputable permanence take precedence over recorded values. Owner of property is entitled to 144.76 ft.

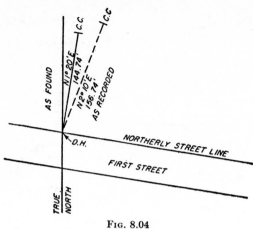

Fig. 8.04

8.05

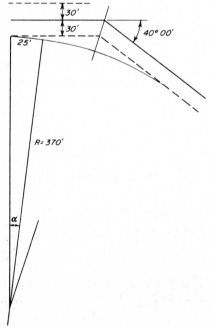

Fig. 8.05

Central angle for an arc of $25' = \alpha = \dfrac{25}{2 \times 370\pi} \times 360°$

$= 3.87° = 3°52'12''$. $(3°52'12'') \times 10 = 38°42'$.   Last central angle increment $= 1°18'$

Chord for $25'$ arc $= 2 \times 370 \times \sin 1°56'06''$
$$= 740 \times 0.03377 = \mathbf{24.99} \text{ ft}$$

$\dfrac{1.30°}{3.87°} \times 25 = \mathbf{8.40}$ ft $=$ last chord

Deflection angles are

1°56.1'	9°40.5'	17°24.9'
3°52.2'	11°36.6'	19°21.0'
5°48.3'	13°32.7'	20°00.0'
7°44.4'	15°28.8'	

8.06

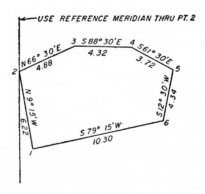

Fig. 8.06

Line	Bearing	Chains × sin bearing =		W	E
2-3	N66°30′E	4.88	0.91706		+ 4.47
3-4	S88°30′E	4.32	0.99966		+ 4.32
4-5	S61°30′E	3.72	0.87882		+ 3.27
5-6	S12°30′W	4.34	0.21644	− 0.94	
6-1	S79°15′W	10.30	0.98245	−10.12	
1-2	N 9°15′W	6.22	0.16074	− 1.00	
				−12.06	+12.06

No correction needed

Line	Bearing	Chains × cos bearing =		N	corr.	S	corr.
2-3	N66°30′E	4.88	0.39875	+1.95	−0.01		
3-4	S88°30′E	4.32	0.02618			−0.11	
4-5	S61°30′E	13.72	0.47716			−1.78	
5-6	S12°30′W	4.34	0.97630			−4.24	+0.01
6-1	S79°15′W	10.30	0.18652			−1.92	
1-2	N 9°15′W	6.22	0.98700	+6.14	−0.02		
				+8.09		−8.05	

Corrected: +8.06　　　　−8.06

Line	Corrected latitudes	Corrected departures	DMD	Double area +	Double area −
2-3	+1.94	+ 4.47	+ 4.47	+ 8.672	
3-4	−0.11	+ 4.32	+13.26		− 1.459
4-5	−1.78	+ 3.27	+20.85		− 37.113
5-6	−4.25	− 0.94	+23.18		− 98.515
6-1	−1.92	−10.12	+12.12		− 23.270
1-2	+6.12	− 1.00	+ 1.00	+ 6.120	

DMD of (2-3) = departure of (2-3)

DMD of (3-4) = DMD of (2-3) + departure of (2-3) + departure of (3-4)

Double area = latitude × DMD

+14.792　　−160.357
+ 14.792

Twice area = 145.565
Area (sq chains) = 72.78
Area (acres) = 7.278

8.07 A parcel of land may be described by:

a. Metes and bounds

b. Stating its location and size in a rectangular system of land subdivision

c. Giving the coordinates of the property corners with reference to a plane coordinate system

d. A deed reference to a description in a previously recorded deed

e. Reference to block and individual property numbers appearing on a recorded map

8.08

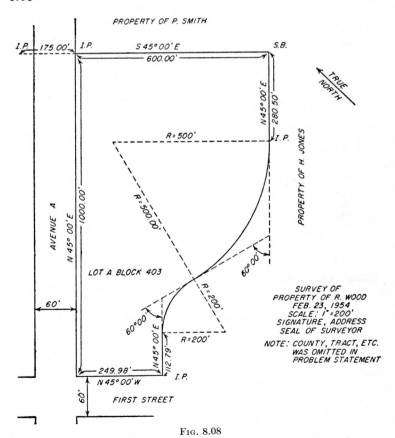

Fɪɢ. 8.08

8.09   Beginning at the northeast corner of the intersection of Avenue A and First Street and running along the said Avenue A N45°00′E, which is also the true bearing of Avenue A as shown on a recorded file map No. 62 by E. R. Sullivan dated May 16, 1922, a distance of one thousand (1000.00′) feet to an iron pipe in the ground which is the northwest corner of the property; thence turning and running S45°00′E a distance of six hundred (600.00′) feet along the land now or formerly belonging to Paul Smith to a stone bound at land now or formerly belonging to Harold Jones; thence turning and running S45°00′W along the land of said Jones a distance of 4.25 chains or two hundred eighty and fifty hundredths (280.50′) feet to an iron pipe in the ground; thence continuing along the land of said Jones along a curve to the right of radius five hundred ($R = 500.00′$) feet a distance along the arc of five hundred twenty three and sixty hundredths (523.60′) feet to a point of reverse curvature, the long chord of said curve being five hundred (500.00′) feet long; thence continuing along the land of said Jones along a curve to the left of radius two hundred ($R = 200.00′$) feet a distance along the arc of two hundred nine and forty four hundredths (209.44′) feet to a point of tangency, the long chord of said curve being two hundred (200.00′) feet long; thence continuing along the afore-mentioned tangent S45°00′W a distance of one hundred twelve and seventy nine hundredths (112.79′) feet to an iron pipe on the northeasterly side of First Street; thence turning and running along First Street N45°00′W, which is also the true bearing of First Street as shown on a recorded File Map No. 62 previously described, a distance of two hundred forty nine and ninety eight hundredths (249.98′) feet to the point of beginning; all bearings being true bearings and the parcel containing a calculated area of 473,313 square feet more or less.

8.10

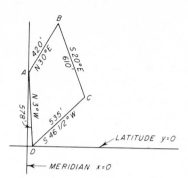

Fig. 8.10

Line	Bearing	Distance	× sin bearing =	W	corr.	E	corr.
AB	N30°E	420	0.50000			210.00	−0.08
BC	S20°E	610	0.34202			208.63	−0.08
CD	S46½°W	535	0.72537	388.07	+0.14		
DE	N3°W	578	0.05234	30.25	+0.01		
				418.32	+0.15	418.63	−0.16

Corrected 418.47   418.47

Line	Bearing	Distance	× cos bearing =	N	corr.	S	corr.
AB	N30°E	420	0.86603	363.73	+0.10		
BC	S20°E	610	0.93969			573.21	−0.16
CD	S46½°W	535	0.68835			368.27	−0.11
DE	N 3°W	578	0.99863	577.21	+0.17		
				940.94	+0.27	941.48	−0.27

Corrected 941.21   941.21

Line	W-E departure	N-S latitude	x coordinate	y coordinate
AB	209.92	363.83	$A = 0$	$A = 577.38$
BC	208.55	573.05	$B = 209.92$	$B = 941.21$
CD	388.21	368.16	$C = 418.47$	$C = 368.16$
DE	30.26	577.38	$D = 30.26$	$D = 0$

$$2 \text{ Area} = \frac{Xa}{Ya} \times \frac{Xb}{Yb} \times \frac{Xc}{Yc} \times \frac{Xd}{Yd} \times \frac{Xa}{Ya} = (2A_1 - 2A_2)$$

$$
\begin{aligned}
Xa \cdot Yb &= 0 \times 941.21 = 0 \\
Xb \cdot Yc &= 209.92 \times 368.16 = 77{,}284 \\
Xc \cdot Yd &= 418.47 \times 0 = 0 \\
Xd \cdot Ya &= 30.26 \times 577.38 = \underline{17{,}472} \\
2A_2 &= 94{,}756 \text{ sq ft} \\[4pt]
Ya \cdot Xb &= 577.38 \times 209.92 = 121{,}204 \\
Yb \cdot Xc &= 941.21 \times 418.47 = 393{,}868 \\
Yc \cdot Xd &= 368.16 \times 30.26 = 11{,}140 \\
Yd \cdot Xa &= 0 \times 0 = \underline{0} \\
2A_1 &= 526{,}212 \text{ sq ft} \\
2A_2 &= \underline{94{,}756} \\
2 \text{ Area} &= \overline{431{,}456} \\
\text{Area} &= \overline{215{,}728} \text{ sq ft}
\end{aligned}
$$

With original distances measured only to the nearest foot and with angles measured to only the nearest degree or one-half degree, an area expressed to the nearest square foot is unrealistic. It is better to state the area as 215,700 sq ft.

8.11 Assume that the sights are long enough to ensure that the vertical cross hair will set upon the point observed without causing angular variation. If the angle is measured once, the accuracy of measurement is probably not closer than ± 5 sec, but if the angle is measured by eight repetitions, the total angle divided by eight should produce a measured angle good to 1 sec of angle. There is little point in going beyond eight repetitions for the average observer since personal errors nullify the gain. If one half the readings are made with the telescope in normal position and the other half with the telescope inverted, error due to lack of very close adjustment of the instrument is eliminated. The act of repetition uses angles in different parts of the graduated circle, thus reducing or eliminating very small errors in graduation. Both verniers should be read on each set of readings, thus eliminating eccentricity of the graduated circle. Care should be used so that systematic errors in the clamps are avoided. The tangent screws should be turned against and not with the springs. The instrument should always be turned either clockwise or counterclockwise to avoid any twist in the tripod. Protection of the instrument against sun or wind should be provided for best results. Rapidity of observation reduces eyestrain and increases the accuracy of sighting. Standing too near or too long at one tripod leg may cause settlement of the tripod and nullify other precautions.

8.12 Simpson's rule:

$$A = \frac{d}{3}(h_{e_L} + 2\Sigma h_{(\text{odd})} + 4\Sigma h_{(\text{even})} + h_{e_R})$$

$h_{(\text{odd})}$			$h_{(\text{even})}$		
h_3	=	8.9	h_2	=	4.6
h_5	=	10.4	h_4	=	15.3
h_7	=	18.5	h_6	=	13.2
h_9	=	17.6	h_8	=	23.4
h_{11}	=	3.2	h_{10}	=	9.3
$2\Sigma_{(\text{odd})} = 58.6 \times 2 = 117.2$			$4\Sigma_{(\text{even})} = 65.8 \times 4 = 263.2$		

$$A = {}^{25}\!/_{3}(0 + 117.2 + 263.2 + 0) = {}^{100}\!/_{12} \times 380.4 = \textbf{3,170 sq ft}$$

1c

Trapezoidal rule

$$A = \frac{d}{2}(h_{e_L} + 2\Sigma h + h_{e_R})$$

$$= {}^{25}\!/_{2}[0 + 2(58.6 + 65.8) + 0] = {}^{100}\!/_{4} \times 124.4 = \textbf{3,110 sq ft}$$

2a

8.13

3.398	= F.S. on B	4.465	= B.S. on B
-3.148	= B.S. on A	-4.035	= F.S. on A
0.250		0.430	

True difference in elevation between A and B
$$= (0.250 + 0.430) \times \tfrac{1}{2} = 0.340 \quad \textbf{1b}$$

$4.465 - 0.340 = 4.125 =$ rod reading on A to give level sight with instrument 4.465 ft above B. **2c**

Move cross hair up. **3a**

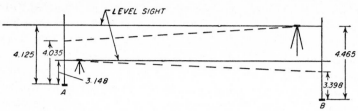

FIG. 8.13

8.14   The survey and plot plan submitted to the client should
show the situation as it should be and also the 0.2-ft encroachment as it exists. Many surveyors would show these lines in two
colors (e.g., black for the deed description and red for the "as-
constructed" condition) on a hard copy, or if a tracing is made
the encroachment might be shown as a dotted line. In any case,
the plan submitted would be a true description of the existing
condition.

The best advice the surveyor can give his client is to consult
a lawyer. It is not within a surveyor's province to presume to
give legal advice. The courts have continuously frowned on any
such attempt on the part of the surveyor.

A condition occurs here which definitely requires legal assist-
ance. A title search may indicate that an easement has been
secured from the city. Legal action may involve securing one if
none exists. Such action is not simple. The laws of adverse
possession do not apply against a municipality. Since the street
involves the public interest, public hearings might be required
before legislation permitting the encroachment could be passed.

8.15   Even with a carefully adjusted instrument, it is unlikely
that the line of sight and the axis of the level tube are truly
parallel. It is more likely that the line of sight is inclined slightly
up or down. The error in reading is proportional to the distance
from the instrument to the rod. If the backsight distance equals
the foresight distance, and since one reading is added and the
other subtracted, the net result due to the error is zero. It is
not necessary that each backsight distance equal each foresight
distance. Since the proportionality is a direct similar triangle
relationship, it is only required that the sum of the backsight
distances equal the sum of the foresight distances.

Curvature of the earth and atmospheric refraction will also enter the problem. Normally, since the correction is about 0.001 ft in 200 ft, it can be ignored. However, on long runs where the aggregate unbalance of backsight and foresight distances is large, this may be a substantial figure. Since the error varies as the square of the distance it is necessary for *exact* work to make *each* backsight distance equal *each* foresight distance. For most work, balancing the aggregate distances gives satisfactory results.

8.16   *a.* The line is construed as running to the definite object since the definite object has a higher call over the distance (see answer to 8.25*c*).

     b. "Northerly" in the absence of other information may mean anything from N45°W to N45°E. With other information it may permit one to tie down the bearing to a specific value.

     c. Where the "metes and bounds" and the area disagree, the metes and bounds will prevail over the area since it is assumed that as long as the boundaries are visible, the intentions of the contracting parties to the sale is quite evident.

8.17   *a.* (See answer to 8.07.)

     b. 1. By "a point of beginning" which should be monumented and referenced by ties or distances along street lines from well-established monuments or easily located and fixed points such as street intersections.

          2. Lengths and bearings (preferably true bearings) of each side such that they give a closed traverse. If magnetic bearings are used, the declination of the needle and the date of the survey should be stated.

          3. The names of abutting property owners.

          4. The area of the property.

          5. All buildings and improvements located by offsets from the property lines.

          6. All (if any) easements or encroachments.

          7. Coordinates of local grid if determined.

          8. Direction of true meridian or magnetic meridian and declination.

9. Any town, county, or other political subdivision lines.
10. Street lines, names, and widths where possible.
11. Boundary monuments and lines and ties to same.
12. Any fences, walls, driveways, etc.
13. Title showing location, scale, and date.
14. Signature of surveyor and, where required by law, his address and seal.
15. Lot, plat, block, or other necessary identification.
16. Occupancy of structures and types of structures.

c. In order of decreasing precedence:

1. Natural monuments and boundaries.
2. Artificial monuments.
3. Calls for adjoiners.
4. Courses and distances.
5. Area. One exception is that this item will take precedence over item 4, courses and distances, in cases where a will states that equal or other specified divisions of land to several heirs are intended.

d. A party wall is a wall built along a property line which usually is (but not necessarily so) in the center of the wall. To the surveyor, the wall becomes a physical monument and the center of it becomes the property line whether this checks or does not check with the recorded distances from other points.

8.18 Present bearing angle $= 180° - (88°30' + 2°00' + 1°30')$
$$= 88°00'$$
Present magnetic bearing $=$ **N88°00′E** (see Fig. 8.18)   **d**

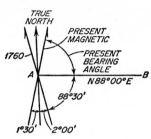

Fig. 8.18

8.19 $Z'' = p'' \sec \phi$

$p = 90° - 89°04'26'' = 0°55'34'' = 3{,}334''$

$\log 3{,}334 \quad\quad = 3.5229656$

$\log \sec 40°40' = 0.1200366$

$\log Z'' \quad\quad\quad = \overline{3.6430022}$

$\quad Z'' \quad\quad\quad = 4{,}395 = \quad 1°13'15''$

Horizontal angle $\quad\quad = \quad 45°15'30''$

Bearing $AB \quad\quad = \overline{\textbf{N46°28'45''W}}$

8.20

Point	B.S.	H.I.	F.S.	Elevation, ft
B.M. 713	10.736	751.610		740.874
T.P.1 H	6.943	754.637	3.916	747.694
T.P.1 L	7.897	754.634	4.873	746.737
T.P.2 H	8.337	760.543	2.431	752.206
T.P.2 L	9.746	760.538	3.842	750.792
T.P.3 H	5.173	760.247	5.469	755.074
T.P.3 L	7.549	760.245	7.842	752.696
T.P.4 H	3.411	755.470	8.188	752.059
T.P.4 L	4.963	755.466	9.742	750.503
B.M. 714			7.238	748.232
				748.228
Sum B.S.	75.491			748.230 mean
Sum F.S.	60.779			740.874
	14.712 ÷ 2 = 7.356 ⟵			⟶ 7.356 check

In the above sums, the B.S. on B.M. 713 and F.S. on B.M. 714 have been taken twice.

8.21

$$a = 3,040 \qquad\qquad b = 926$$
$$c = 2,384 \qquad\qquad d = \underline{1,592}$$
$$e = \underline{68} \qquad\qquad\quad\; 2,518$$
$$5,492$$
$$\underline{2,518}$$
$$2,974 = \frac{AB \times JH}{2}$$

where H is toward G from B and JH is the altitude of the correction angle.

$JH = 2 \times 2,974/375 = 15.86'$
$AB = S77°40'E$ and $EF = N15°30'W$ Angle at $B = 62°10'$
$BH = 15.86/\sin 62°10' = 15.86/0.88431 = \mathbf{17.93'}$
$BJ = 15.86/\tan 62°10' = 15.86/1.89400 = 8.37'$
$AJ = 375.00 - 8.37 = 366.63'$
tan angle (between AB and AH) $= 15.86/366.63 = 0.04326$
Angle $= 2°29'$
$AH = AJ/\cos 2°29' = 366.63/0.99906 = \mathbf{366.97'}$
Bearing $AH = S77°40'E + 2°29' = \mathbf{S80°09'E}$

8.22 .Horizontal distance transit to spire $= \sqrt{(117.5)^2 + (18.95)^2}$
$\qquad\qquad\qquad\qquad\qquad\qquad\qquad\;\; = 119.02 \text{ ft} \qquad \mathbf{1b}$
Altitude of vane above telescope $= 119.02 \tan 40°11'$
$\qquad\qquad\qquad\qquad\qquad\qquad\quad\; = 100.52 \text{ ft} \qquad \mathbf{2d}$

B.M.	246.18
B.S.	6.82
Altitude	100.52
Elev. vane	**353.52** ft **3a**

8.23   *a.* The equal altitude method may be used. Sight the sun several hours before noon or sight a circumpolar star several hours before either upper or lower culmination. Note the altitude and horizontal angle (from a transit line) and the watch time. When the sun or star has passed the meridian and is approaching the same altitude once more, follow it with the telescope and record the horizontal angle when the heavenly body is at the same altitude once more. The meridian will lie midway between the two observations. Greater precision can be obtained by taking a series of observations east and west of the meridian.

b. The line described will locate Polaris. However, if this line was roughly horizontal, then Polaris was near elongation and it would have been impossible to make the two equal altitude observations. It is clear that the observations were made on some other star.

8.24   Bearing $AB = \tan^{-1}\dfrac{1,000}{0} = $ east

Distance $AB = 1,000.00$ ft

$AC = 1,000 \times \dfrac{\sin 50°00'}{\sin 91°00'} = 766.15$ ft   **1c**

$BC = 1,000 \times \dfrac{\sin 39°00'}{\sin 91°00'} = 629.41$ ft

tan distance of curve $= R \tan 44'30' = 196.54$ ft   **3d**

$629.41 - 350.00 = 279.41$ ft
$279.41 - 196.54 = 82.87$ ft
$766.15 - 400.00 = 366.15$ ft
$366.15 - 196.54 = 169.61$ ft
$366.15 - 200.00 = 166.15$ **2b**

Length of curve $= 200 \times 1.5533430 = 310.67$ ft

All dimensions are shown in Fig. 8.24

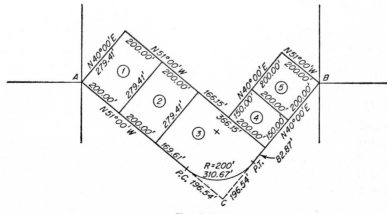

Fɪɢ. 8.24

8.25   Angle $AOB = \cos^{-1} 900/1,000 = 25°50'30''$
From a table of lengths of circular arcs for radius 1

$$25° = 0.4363323$$
$$50' = 0.0145444$$
$$30'' = \underline{0.0001454}$$
$$0.4510221 \times 1,000 = 451.02 \text{ ft} = \text{arc } AB$$

Station of $B = (47 + 50.25) + (4 + 51.02) = \mathbf{52 + 01.27}$
Chord $AB = 2 \times 1,000 \times \sin \text{ angle } AOB/2$
$$= 2,000 \times 0.22360 = 447.20 \text{ ft}$$

Set transit at PC and back sight $0°00'$ on back tangent. Plunge telescope and turn $12°55'15''$ deflection angle to the right and tape $447.20'$ from the PC to set the intersection point B.

$''$	=	inch, in., i.; second, sec of angle
$''^2$	=	square inch, s.i., sq in., in.^{2}
$''^3$	=	cubic inch, c.i., cu in., in.^{3}
$''^4$	=	inches 4th, i.^{4}, in.^{4}
$'$	=	foot, ft., f, ft; minute, min of angle
$'^2$	=	square foot, s.f., sq ft, ft^2
$'^3$	=	cubic foot, c.f., cu ft, ft^3
$'^4$	=	feet 4th, f.^{4}, ft^4
I	=	moment of inertia ($''^4$ or $'^4$)
$\#$	=	pound, lb., lb
^{k}	=	kip (1,000-pound unit)
^{T}	=	ton (2,000-pound unit)
$''\#$	=	inch pound, pound inch, i.p., in-lb
$'\#$	=	foot pound, pound foot, f.p., ft-lb
$'^k$	=	foot kip, kip foot, f.k., ft-kip
p.s.i.	=	pounds per square inch, $\#/''^2$, psi
p.s.i.a.	=	pounds per square inch absolute, $\#a/''^2$, psia
p.s.i.g.	=	pounds per square inch gage, $\#g/''^2$, psig
p.s.f.	=	pounds per square foot, $\#/'^2$, psf
p.c.i.	=	pounds per cubic inch, $\#/''^3$, pci
p.c.f.	=	pounds per cubic foot, $\#/'^3$, pcf
k.s.f.	=	kips per square foot, $k/'^2$, ksf
π	=	ratio of circumference to diameter of a circle or 3.1416
c.c.	=	cubic centimeter, cc
gm	=	gram
kg	=	kilogram or 1,000 grams
l	=	liter, ltr., or 1,000 c.c.
°	=	degree of temperature or of angle
°F	=	degree Fahrenheit
°R	=	degree Rankin [°F (absolute)]
°C	=	degree centigrade
°K	=	degree Kelvin [°C (absolute)]
h	=	hour of time, hr., hr

m = minute of time, min., min; mile(s)

s = second of time, sec

t = time (h, m, s)

m.p.h.   = miles per hour, mph

f.p.s.   = feet per second, fps

a = acceleration ($'/s^2$) or ft per sec per sec, fps per sec;
             absolute (abs.) or gage pressure abs.

g = acceleration due to gravity or 32.2 ($'/s^2$); gage (for
             pressure above atmospheric)

v = velocity (units per sec), f/s or ($'/s$)

v_0 = initial velocity

M = mass or (w/g), weight/gravity

F = force or ma (mass $\times$ acceleration)

f = coefficient of friction

HP   = horsepower, hp (33,000 $'^\#/\text{min}$) or (550 $'^\#/\text{sec}$)

STUDY TEXTS

Basic Fundamentals

CHEMISTRY, MATERIALS SCIENCE, AND NUCLEONICS

Anderson, J. C., and K. D. Leauer, "Materials Science," Van Nostrand-Reinhold, 1969.

Barrow, G., "Physical Chemistry," 3d ed., McGraw-Hill, 1973.

Brady, G., "Materials Handbook,"10th ed., McGraw-Hill, 1971.

Brick, R., R. Gordon, and A. Phillips, "Structure and Property of Alloys," 3d ed., McGraw-Hill, 1965.

Bush, H. D., "Atomic Nuclear Physics," Prentice-Hall, 1962.

DiBennedetto, A., "Structure and Properties of Materials," McGraw-Hill, 1967.

Edelglass, Stephen M., "Engineering Material Science," Ronald, 1966.

Hagelberg, P., "Physics: An Introduction for Students of Science and Engineering," Prentice-Hall, 1973.

Keyser, Carl A., "Materials Science in Engineering," 2d ed., Merrill, 1974.

Quagliano, J. V., and L. M. Vallarino, "Chemistry," 3d ed., Prentice-Hall, 1969.

Sienko, M., and R. Plane, "Chemistry," 3d ed., McGraw-Hill, 1966.

ELECTRICITY

Dawes, C., "Industrial Electricity," vol. II, 3d ed., McGraw-Hill, 1960.

Durney, C., and C. Johnson, "Introduction to Modern Electromagnetics," McGraw-Hill, 1969.

Foecke, H. A., "Introduction to Electrical Engineering Science," Prentice-Hall, 1961.

Grob, B., "Basic Electronics," 3d ed., McGraw-Hill, 1971.

——— and M. Kiver, "Applications of Electronics," McGraw-Hill, 1966.

Leach, D. P., "Basic Electric Circuits," Wiley, 1969.

Zbar, P., "Basic Electricity," 3d ed., McGraw-Hill, 1966.

HYDRAULICS AND FLUID MECHANICS

Allen, T., Jr., and L. Ditsworth, "Fluid Mechanics," McGraw-Hill, 1972.

Dake, J. M. K., "Essentials of Engineering Hydraulics," Wiley, 1972.

King, H., and E. Brater, "Handbook of Hydraulics," 5th ed., McGraw-Hill, 1963.

Shames, I., "Mechanics of Fluids," McGraw-Hill, 1962.

Chemical Engineering

Balzhiser, R. E., M. R. Samuels, and J. D. Eliassen, "Chemical Engineering Thermodynamics," Prentice-Hall, 1972.

Coulson, J. M., and J. F. Richardson, "Chemical Engineering," 2d ed., Pergamon Press, 1964.

Harriott, P., "Process Control," McGraw-Hill, 1964.

Hougen, P. A., K. M. Watson, and R. A. Ragatz, "Chemical Process Principles," 2d ed., Wiley, 1954.

Levenspiel, O., "Chemical Reaction Engineering," Wiley, 1962.

McCabe, W. L., and J. C. Smith, "Unit Operations of Chemical Engineering," 2d ed., McGraw-Hill, 1967.

Meissner, H. P., "Processes and Systems in Industrial Chemistry," Prentice-Hall, 1971.

Newman, J. S., "Electrochemical Systems," Prentice-Hall, 1973.

Perry, R. H., and C. Chilton, "Chemical Engineer's Handbook," 5th ed., McGraw-Hill, 1973.

Peters, M., and K. Timmerhaus, "Plant Design and Economics for Chemical Engineers," 2d ed., McGraw-Hill, 1968.

Prausnitz, J. M., "Molecular Thermodynamics of Fluid Phase-Equilibria," Prentice-Hall, 1969.

Riegel, E. R., "Industrial Chemistry," Reinhold, 1962.

Shreve, R., "Chemical Process Industries," 3d ed., McGraw-Hill, 1967.

Treybal, R. E., "Liquid Extraction," 2d ed., McGraw-Hill, 1963.

Treybal, R. E., "Mass-Transfer Operations," 2d ed., McGraw-Hill, 1968.

Vilbrandt, F. C., "Chemical Engineering Plant Design," 4th ed., McGraw-Hill, 1959.

Williams, D. J., "Polymer Science and Engineering," Prentice-Hall, 1971.

Civil Engineering

Abbett, R. W. (ed.), "American Civil Engineering Practice," 3 vols., Wiley, 1956.

Babbitt, H. E., J. J. Doland, and J. L. Cleasby, "Water Supply Engineering," 6th ed., McGraw-Hill, 1962.

Bowles, J. E., "Foundation Analysis and Design," McGraw-Hill, 1968.

Chellis, R., "Pile Foundations," 2d ed., McGraw-Hill, 1961.

Chow, V. T., "Handbook of Applied Hydrology," McGraw-Hill, 1964.

Clark, J. W., "Water Supply and Pollution Control," 2d ed., International Textbook, 1971.

Drew, D. R., "Traffic Flow Theory and Control," McGraw-Hill, 1968.

Eckenfelder, W. W., Jr., "Industrial Water Pollution Control," McGraw-Hill, 1966.

Foster, N., "Construction Estimates From Take-off to Bid," 2d ed., McGraw-Hill, 1972.

Hall, W. A., and J. A. Dracup, "Water Resources Systems Engineering," McGraw-Hill, 1970.

Hennes, R. G., and M. Ekse, "Fundamentals of Transportation Engineering," 2d ed., McGraw-Hill, 1969.

Horonjeff, R., "Planning and Design of Airports," McGraw-Hill, 1962.

King, H., and E. Brater, "Handbook of Hydraulics," 5th ed., McGraw-Hill, 1963.

Legault, A. R., "Highway and Airport Engineering," Prentice-Hall, 1960.

Leonards, G. A., "Foundation Engineering," McGraw-Hill, 1962.

Linsley, R. K., and J. B. Franzini, "Water Resources Engineering," 2d ed., McGraw-Hill, 1972.

Merritt, F. S., "Standard Handbook for Civil Engineers," McGraw-Hill, 1968.

Metcalf and Eddy, Inc., "Wastewater Engineering: Collection, Treatment, Disposal," McGraw-Hill, 1972.

Nemerow, N. L., "Industrial Waste Treatment," Addison-Wesley, 1963.

Oglesby, C. H., and L. I. Hewes, "Highway Engineering," 2d ed., Wiley, 1963.

Peurifoy, R. L., "Construction Planning, Equipment and Methods," 2d ed., McGraw-Hill, 1970.

Pignataro, L. J., "Traffic Engineering: Theory and Practice," Prentice-Hall, 1973.

Rich, L., "Environmental Systems Engineering," McGraw-Hill, 1973.

Scott, R. F., and J. J. Schoustra, "Soil: Mechanics and Engineering," McGraw-Hill, 1968.

Woods, K., "Highway Engineering Handbook," McGraw-Hill, 1960.

Electrical and Electronic Engineering

Balakrishnan, A. V., "Communication Theory," McGraw-Hill, 1968.

Blake, L. V., "Transmission Lines and Waveguides," Wiley, 1969.

Brenner, E., and M. Javid, "Analysis of Electric Circuits," 2d ed., McGraw-Hill, 1967.

Buban, P., and M. Schmitt, "Understanding Electricity and Electronics," 2d ed., McGraw-Hill, 1969.

Carlson, B., "Communication Systems," McGraw-Hill, 1968.

Chirlian, P. M., "Basic Network Theory," McGraw-Hill, 1969.

Fink, D., and J. Carroll, "Standard Handbook for Electrical Engineers," 10th ed., McGraw-Hill, 1968.

Fitzgerald, A. E., and D. H. Higginbotham, "Electrical and Electronic Engineering Fundamentals," McGraw-Hill, 1964.

Gehmlich, D. K., and S. B. Hammond, "Electromechanical Systems," McGraw-Hill, 1967.

Greenwood, A., "Electrical Transients in Power Systems," Wiley, 1971.

Leach, D. P., "Basic Electric Circuits," Wiley, 1969.

Lloyd, T. C., "Electric Motors and Their Applications," Wiley, 1969.

Matick, R., "Transmission Lines for Digital and Communication Networks," McGraw-Hill, 1969.

Murdock, J. B., "Network Theory," McGraw-Hill, 1970.

Stott, G. S., and G. Birchall, "Electrical Engineering Principles," McGraw-Hill, 1969.

Sunde, E. D., "Communication Systems Engineering Theory," Wiley, 1969.

Taub, H., and D. L. Schilling, "Principles of Communication Systems," McGraw-Hill, 1971.

Watt, J., and F. Stetka, "NFPA Handbook of the National Electrical Code," 3d ed., McGraw-Hill, 1972.

Weedy, B. M., "Electric Power Systems," 2d ed., Wiley, 1972.

Engineering Economics and Business Relations

Barish, N. N., "Economic Analysis for Engineering and Managerial Decision Making," McGraw-Hill, 1962.

Canada, J. R., "Intermediate Economic Analysis for Management and Engineering," Prentice-Hall, 1971.

Grant, E. L., and W. G. Ireson, "Principles of Engineering Economy," 5th ed., Ronald, 1970.

Jelen, F. C., "Cost and Optimization Engineering," McGraw-Hill, 1970.

Riggs, J. L., "Economic Decision Models for Engineers and Managers," McGraw-Hill, 1968.

Thuesen, H. G., W. J. Forbrycky, and G. J. Thuesen, "Engineering Economy," 4th ed., Prentice-Hall, 1971.

Mechanical Engineering

Baumeister, T., and L. Marks, "Standard Handbook for Mechanical Engineers," 7th ed., McGraw-Hill, 1967.

Black, P. H., and O. E. Adams, Jr., "Machine Design," 3d ed., McGraw-Hill, 1968.

Carrier Air Conditioning Co., "Handbook of Air Conditioning System Design," McGraw-Hill, 1965.

Emerick, R., "Heating Handbook," McGraw-Hill, 1964.

Gebhart, B., "Heat Transfer," 2d ed., McGraw-Hill, 1970.

Greenwood, D., "Mechanical Power Transmission," McGraw-Hill, 1962.

Harris, N., "Modern Air Conditioning Practice," McGraw-Hill, 1959.

Holman, J. P., "Thermodynamics," McGraw-Hill, 1969.

Lichty, L. C., "Combustion Engine Processes," 7th ed., McGraw-Hill, 1967.

Phelan, R. M., "Fundamentals of Mechanical Design," 3d ed., McGraw-Hill, 1970.

Power Magazine, "Plant Energy Systems," McGraw-Hill, 1967.

————, "Power Generation Systems," McGraw-Hill, 1967.

Reynolds, W. C., and H. C. Perkins, "Engineering Thermodynamics," McGraw-Hill, 1970.

Rothbart, H., "Mechanical Design and Systems Handbook," McGraw-Hill, 1964.

Stoecker, W. F., "Design of Thermal Systems," McGraw-Hill, 1971.

Woodruff, E., and H. Lammers, "Steam-Plant Operation," 3d ed., McGraw-Hill, 1967.

Structural Engineering

Bowles, J. E., "Foundation Analysis and Design," McGraw-Hill, 1968.

Coull, A., and A. Dyke, "Fundamentals of Structural Theory," McGraw-Hill, 1972.

Gaylord, E. H., and C. N. Gaylord, "Structural Engineering Handbook," McGraw-Hill, 1968.

Johnson, S., and T. Kavanagh, "The Design of Foundations for Buildings," McGraw-Hill, 1968.

Khachaturian, N., and G. Gurfinkle, "Prestressed Concrete," McGraw-Hill, 1969.

LaLonde, W., and M. Janes, "Concrete Engineering Handbook," McGraw-Hill, 1962.

Laursen, H. I., "Structural Analysis," McGraw-Hill, 1969.

Merritt, F., "Structural Steel Designers' Handbook," McGraw-Hill, 1972.

Norris, C. H., and J. B. Wilbur, "Elementary Structural Analysis," 2d ed., McGraw-Hill, 1960.

Shermer, C. L., "Design in Structural Steel," Ronald, 1972.

Tomlinson, M. J , "Foundation Design and Construction," 2d ed., Wiley, 1969.

Winter, G., and A. H. Nilson, "Design of Concrete Structures," 8th ed., McGraw-Hill, 1972.

"Building Code Requirements for Reinforced Concrete," latest ed., American Concrete Institute.

"Code for Arc and Gas Welding in Building Construction," latest ed., American Welding Society.

"Specifications for Railway Bridges," latest ed., American Railway Engineering Association.

"Standard Specifications for Highway Bridges," latest ed., American Association of State Highway Officials.

Land Surveying

Breed, C. B., "Surveying," 3d ed., Wiley, 1971.

Davis, R. E., F. S. Foote, and J. W. Kelly, "Surveying: Theory and Practice," 5th ed., McGraw-Hill, 1966.

Davis, R. E., and J. W. Kelly, "Elementary Plane Surveying," 4th ed., McGraw-Hill, 1967.

Hickerson, T. F., "Route Location and Design," 5th ed., McGraw-Hill, 1967.